WATER QUALITY MANAGEMENT

WATER QUALITY

CHARACTERISTICS ▪ MODELING ▪ MODIFICATION

George Tchobanoglous
Edward D. Schroeder

UNIVERSITY OF CALIFORNIA AT DAVIS

▲▼ ADDISON-WESLEY PUBLISHING COMPANY

Reading, Massachusetts ▪ Menlo Park, California
Don Mills, Ontario ▪ Wokingham, England ▪ Amsterdam ▪ Sydney
Singapore ▪ Tokyo ▪ Mexico City ▪ Bogotá
Santiago ▪ San Juan

THIS BOOK IS VOLUME I IN THE ADDISON-WESLEY SERIES ON WATER QUALITY MANAGEMENT.

Library of Congress Cataloging in Publication Data

Tchobanoglous, George.
  Water quality.

  Includes bibliographies and index.
  1. Water quality.   I. Schroeder, Edward D.   II. Title.
TD365.T38 1985   363.7'394   84-11120
ISBN 0-201-05433-7

Reprinted with corrections February, 1987

Copyright © 1985 by Addison-Wesley Publishing Company, Inc. All rights reserved. No part of this publication may be reproduced, stored in a retrieval system, or transmitted, in any form or by any means, electronic, mechanical, photocopying, recording, or otherwise, without the prior written permission of the publisher. Printed in the United States of America. Published simultaneously in Canada.

CDEFGHIJ-HA-8987

WATER QUALITY

TABLE 3
Derived SI Units Obtained by Combining Base Units and Units with Special Names

QUANTITY	UNITS	QUANTITY	UNITS
Acceleration	m/s^2	Molar entropy	$J/mol \cdot K$
Angular acceleration	rad/s^2	Molar heat capacity	$J/mol \cdot K$
Angular velocity	rad/s	Moment of force	$N \cdot m$
Area	m^2	Permeability	H/m
Concentration	mol/m^3	Permittivity	F/m
Current density	A/m^2	Radiance	$W/m^2 \cdot sr$
Density, mass	kg/m^3	Radiant intensity	W/sr
Electric charge density	C/m^3	Specific heat capacity	$J/kg \cdot K$
Electric field strength	V/m	Specific energy	J/kg
Electric flux density	C/m^2	Specific entropy	$J/kg \cdot K$
Energy density	J/m^3	Specific volume	m^3/kg
Entropy	J/K	Surface tension	N/m
Heat capacity	J/K	Thermal conductivity	$W/m \cdot K$
Heat flux density	W/m^2	Velocity	m/s
Irradiance	W/m^2	Viscosity, dynamic	$Pa \cdot s$
Luminance	cd/m^2	Viscosity, kinematic	m^2/s
Magnetic field strength	A/m	Volume	m^3
Molar energy	J/mol	Wavelength	m

TABLE 4
SI Prefixes

MULTIPLICATION FACTOR	PREFIX*	SYMBOL
$1\ 000\ 000\ 000\ 000 = 10^{12}$	tera	T
$1\ 000\ 000\ 000 = 10^9$	giga	G
$1\ 000\ 000 = 10^6$	mega	M
$1\ 000 = 10^3$	kilo	k
$100 = 10^2$	hecto†	h
$10 = 10^1$	deka†	da
$0.1 = 10^{-1}$	deci†	d
$0.01 = 10^{-2}$	centi†	c
$0.001 = 10^{-3}$	milli	m
$0.000\ 001 = 10^{-6}$	micro	μ
$0.000\ 000\ 001 = 10^{-9}$	nano	n
$0.000\ 000\ 000\ 001 = 10^{-12}$	pico	p
$0.000\ 000\ 000\ 000\ 001 = 10^{-15}$	femto	f
$0.000\ 000\ 000\ 000\ 000\ 001 = 10^{-18}$	atto	a

*The first syllable of every prefix is accented so that the prefix will retain its identity. Thus, the preferred pronunciation of kilometer places the accent on the first syllable, not the second.

†The use of these prefixes should be avoided, except for the measurement of areas and volumes and for the nontechnical use of centimeter, as for body and clothing measurements.

Preface

The concepts and principles used by civil engineers to address water quality management have changed dramatically as the understanding of the impact of the activities of the human race on the environment has increased. At the same time, the application of fundamental scientific principles in engineering education has been stressed increasingly. These changes will continue, possibly at an accelerating rate, into the next century.

The need for introductory textbooks on water quality management that reflect both the changes in understanding of environmental issues and the changes occurring in civil engineering education has become increasingly evident. This book is a response to that need. The focus of the book is on water quality management issues that are central to the civil engineering profession: that is, issues that have been and will continue to be both of importance to society and the primary responsibility of civil engineers. The succeeding two books in the series, *Fundamentals of Water and Wastewater Treatment* and *Hydraulics of Water Quality Management Systems* focus on engineering design for water quality management.

Organization

The book is divided into four parts: Water Quantity and Water Quality, Analytical Methods for Water Quality Management, Modeling of Water Quality, and Modification of Water Quality. A specific set of issues or problems is addressed in each part; and there is a logical progression from the characterization of water quality and its relationship to quantity (Part I), through the development of the analytical methods needed to describe changes in water quality (Part II) and application of the analytical methods in modeling water quality changes in natural and artificial systems (Part III), to the modification of water quality through treatment (Part IV).

The most important issues in water quality management relate to the quantities of water used, the definition of water quality, and the effects of water quality/water

v

quantity interrelationships on human society. These topics are addressed in Part I of the book.

Civil engineers must provide society with quantitative answers to questions of water quality management. No other group is prepared to address the gamut of science and engineering problems, concerning water quality management, that must be identified and solved. Other professions have important but narrower roles in this area, and policy decisions must ultimately be made by the public; but civil engineering is the keystone of the management process because of historical development, the role of civil engineers in public works, data collection, mathematical modeling of systems, and facilities conceptualization, design, construction, and operation. An understanding of analytical approaches to water quality management is necessary to provide quantitative solutions. An introduction to these approaches is presented in Part II.

To consider the movement of contaminants in the aquatic environment and the effects of these contaminants on water quality as a result of this movement (and their reactions or decay) there must be an understanding of the physical, chemical, and biological characteristics of water, and an application of analytical methods to construct mathematical models. The process of constructing water quality mathematical models of streams, estuaries, lakes, and groundwaters is discussed in Part III. The quality of available water is seldom entirely suitable for an intended use; the contaminants in water may eliminate certain uses from consideration unless they are removed, reduced in concentration, or altered in some fashion. Water quality modification is the subject of Part IV.

Objectives

The principal objective in writing this book was to provide an introduction to water quality management that would be suitable for all civil engineering undergraduate students. Recognition was made of the fact that the importance of some subjects varies with geographical location and the structure and development of a given community. For this reason, the second objective was to make the contents of the book broad enough to provide a structure for several types of courses as will be explained below. A third objective was to maintain a quantitative approach to problem identification and solution that is consistent with modern civil engineering education and practice. The fourth objective was to provide students with a practical understanding of the significance of water quality management to society and of the key role of civil engineers in the management process.

The four objectives were used as the basis for the organization of the text. Each of the four parts begins with a brief overview of the material to be covered in their respective chapters. The chapters include sections that describe the significance of the material covered and the necessary theoretical development, detailed example problems that illustrate the concepts presented, and tables and figures that relate concepts to professional practice and methods of analysis. Key ideas, concepts, and issues are presented at the end of each chapter to highlight the material covered and to serve as an aid for self-study. To develop the students' quantitative understanding of the material covered and their analytical skills, numerous discussion topics and problems are given at the end of each chapter. To the extent possible each group of problems dealing with a specific topic or subject area is arranged in order of difficulty. Solutions to all of the numerical problems are presented in a manual available to instructors from the publisher.

Use of this Book

Enough material is provided in this book to support quarter or semester length courses with four different conceptual approaches: 1) water quality management, 2) survey of water quality management, 3) water and wastewater treatment, and 4) water quality modeling. The original objective was to provide a text for the ten-week required senior-level introductory course on water quality management given at the University of California, Davis. In recent years this course has been taught by the authors and four colleagues: Takashi Asano, Ray Krone, Jerry Orlob, and Robert Smith. Each one of the group prefers to emphasize slightly different aspects of water quality management, and the preferences of each have changed from year to year. On the basis of this teaching experience, the use of the manuscript by colleagues at three other universities, and the comments provided by reviewers, the four course outlines discussed below were developed.

Water Quality Management

An introductory water quality management course, outlined below, begins with material related to the students' previous experience in chemistry, physics, fluid mechanics, and hydraulics, and then progresses through characterization of water quality and water quality modeling to an introduction to treatment of water and wastewater.

CHAPTER	TOPICS
1	Sources and Uses of Water
2	Physical, Chemical, and Biological Characteristics of Water
3	Significance of the Characteristics of Water
5	Stoichiometry, Reaction Kinetics, and Materials Balances
6	Mathematical Modeling of Physical Systems
7	Movement of Contaminants in the Environment
8, 9, 10	Modeling of Water Quality in Natural Systems
11	Introduction to Water and Wastewater Treatment

Chapters 2, 5, 6, and 7 are the critical chapters in the above outline. Considerable emphasis must be given to Chapter 2 because few civil engineering students have significant quantitative knowledge of water quality characteristics. Normally, two to three weeks should be devoted to this material. Chapter 5 provides a structural basis for the process models developed in Chapter 6. If the instructor chooses, Chapter 5 can be used as a reference and most of the allotted time can be spent developing the modeling concepts. The emphasis placed on the material in Chapters 8, 9, and 10 is also flexible, and instructors can make choices based on their perception of students' needs. A broad but quantitative introduction to water and wastewater treatment is provided in Chapter 11. Sufficient depth is given to enable students to investigate Chapters 12 through 15 on their own.

Survey of Water Quality Management

A survey course on water quality management, suitable for juniors in civil engineering or a mixed group of engineering and nonengineering majors, can be developed using the five chapters listed below.

CHAPTER	TOPICS
2	Physical, Chemical, and Biological Characteristics of Water
3	Significance of the Characteristics of Water
4	Water Quality: Standards and Global Perspectives
7	Movement of Contaminants in the Environment
11	Introduction to Water and Wastewater Treatment

Restricting the course of these five topics will allow the instructor to place increased emphasis on the qualitative aspects of water quality management; global issues, such as acid rain and distribution of anthropogenic compounds; development of management policy; and interaction of water, air, and land quality. Other ecological topics selected by the instructor can also be included.

Water and Wastewater Treatment

An outline of an introductory course on water and wastewater treatment is presented below.

CHAPTER	TOPICS
1	Sources and Uses of Water
2	Physical, Chemical, and Biological Characteristics of Water
3	Significance of the Characteristics of Water
5	Stoichiometry, Reaction Kinetics, and Materials Balances
6	Mathematical Models of Physical Systems
11	Introduction to Water and Wastewater Treatment
12	Physical Treatment Methods
13	Chemical Treatment Methods
14	Biological Treatment Methods
15	Synthesizing Water and Wastewater Treatment Systems

The material in Chapter 2 will provide the framework for nearly all of the conceptual material presented in this course. Two to three weeks should be allotted to the presentation of the physical, chemical, and biological characteristics of water. A thorough understanding of this material will allow students to gain a much greater understanding of the methods of treatment presented later in the course. It is intended that instructors will expand on the examples of the significance of water quality characteristics, given in Chapter 3, from their own experience. For a ten-week course Chapters 1, 3, and 11 can be given less emphasis in class, although students will benefit from careful reading of this material. Thus there is more than enough material in these ten chapters to offer a full semester course on water and wastewater treatment.

Water Quality Modeling

An increasing number of schools are offering an undergraduate course on either water quality or environmental quality modeling. Such a course can be developed using this book and the outline below.

CHAPTER	TOPICS
2	Physical, Chemical, and Biological Characteristics of Water
5	Stoichiometry, Reaction Kinetics, and Materials Balances
6	Mathematical Modeling of Physical Systems
7	Movement of Contaminants in the Environment
8	Water Quality in Rivers, Estuaries, and Ocean Discharges
9	Water Quality in Lakes and Reservoirs
10	Water Quality in Groundwater Systems

Chapters 3 and 4 provide explanations of the importance of the modeling processes which is excellent background material for students taking a course on water quality modeling. Instructors can cover this material in lecture, with expansion from their own experience, or can assign these two chapters as outside reading if time does not allow discussion of the material in class.

Supporting Material

Appendixes A through J, the tables in the text, the information given in the inside front and back covers, and the name and subject indexes were developed to make the book more useful. The book is intended to be self-contained, so it is hoped that all necessary constants and coefficients are provided in the supporting material. Readers are encouraged to become familiar with the contents of the appendixes as soon as they begin to use the book.

Acknowledgments

This book developed from a series of discussions between the authors, beginning in 1978, on the appropriate content of a required introductory course on water quality. During the book's long period of gestation many other individuals made significant contributions to the text. It would be impossible to acknowledge all of the contributors, but those listed below made particular contributions that the authors deeply appreciate.

Professor Takashi Asano, our colleague at Davis and an Assistant Director of the Office of Water Recycling of the California State Water Resources Control Board, devoted countless hours to the review of the text and problems. His contributions were enormous and the results of his comments are seen throughout the book, particularly in the first eleven chapters. Tom Robbins, our editor at Addison-Wesley Publishing Company, was supportive throughout the project. He joined us in the production process at a time when a very rough first draft had been written and patiently continued to help, with an occasional gentle nudge, to bring the book to fruition. Professor James Hunt of the University of California, Berkeley used a draft manuscript in his class during the fall semester, 1983. His comments and those of his students were very helpful in completing the final manuscript. A number of homework problems were contributed by Professor Hunt, which are noted in the text where they occur. Professor Duncan Mara of Leeds University initiated much of the conceptual development of the material in Chapters 3 and 4 on public health, problems of developing countries, and the societal role of environmental engineers.

An early draft of the manuscript was written during a sabbatical leave spent at Leeds University. The hospitality of the faculty and their contributions were greatly appreciated.

Professors Mark Matsumoto and Scott Weber of the State University of New York, Buffalo and JoAnn Silverstein of the University of Colorado reviewed portions of the manuscript and made valuable suggestions. Our colleagues at Davis, Willie Pfeiffer of the Department of Bacteriology and Norma Lang of the Department of Botany reviewed the material dealing with bacteriology and phycology. Their comments and attention to detail resulted in significant improvements to those sections. Robert MacArthur, Miguel Marino, Ron Crites, Dan Chang, and Max Burchett provided thorough reviews of Chapters 9, 10, 12, 13, and 14 and 15, respectively. Their suggestions and comments were extremely helpful. Our students at Davis used versions of the manuscript for four years, and made numerous suggestions for improving clarity and organization; they also did a substantial amount of proofreading. Their comments were very helpful during each phase of the development of this text.

To our technical editor, Ursala Smith, we owe a debt of gratitude beyond measure. Her concern for clarity and attention to detail is reflected throughout the text. The lucid plates of the microorganisms found in Chapter 2 were prepared by Janet Williams.

Coordination of the manuscript typing was done by Dinah Pfoutz. Through all of the drafts to the last Dinah remained cheerful and helpful. Stephanie Wilson and Rosalind MacArthur typed the final versions of Chapters 5 through 14. Rosemary Tchobanoglous helped a great deal with the production of the manuscript.

Davis, California   *G.T.*
                                                                                              E.D.S.

Water Quality Management, Volume II:

FUNDAMENTALS OF WATER AND WASTEWATER TREATMENT

by Edward Schroeder and George Tchobanoglous

The unit operations and processes introduced in Volume I are examined in detail in Volume II. Application of fundamental principles and rational design concepts is emphasized. Special attention is devoted to the selection of design flow and process loading rates. The relationships between water and wastewater characteristics and treatment plant performance are stressed.

Water Quality Management, Volume III:

HYDRAULICS OF WATER QUALITY MANAGEMENT SYSTEMS

by George Tchobanoglous, Gilles Patry, and Edward Schroeder

Hydraulic principles are applied to the analysis and design of water distribution networks; wastewater collection systems; pumping systems for water and wastewater, including pump characteristics and selection; and water and wastewater treatment plants hydraulics. Computer aided design approaches are presented along with extensive summaries of useful design data.

George Tchobanoglous is Professor of Civil Engineering at the University of California, Davis. He received a B.S. degree in Civil Engineering from the University of the Pacific, an M.S. degree in Sanitary Engineering from the University of California, Berkeley, and a Ph.D. in Sanitary Engineering from Stanford University. His principal research interests are in the areas of primary and secondary effluent filtration, biological wastewater treatment, aquatic wastewater management systems, wastewater management systems for small communities, and solid waste management. He has authored or co-authored over 125 technical publications and five textbooks. Professor Tchobanoglous serves nationally and internationally as a consultant to both private concerns and governmental agencies. He is a registered engineer in the state of California.

Edward D. Schroeder is Professor of Civil Engineering at the University of California, Davis. He received his B.S. and M.S. degrees in Civil Engineering from Oregon State University and his Ph.D. in Chemical Engineering from Rice University. His research interests are in the areas of kinetics and stoichiometry of biological wastewater treatment processes, periodic treatment processes, nitrification and denitrification, control of filamentous bulking, biological treatment of hazardous wastes, and overland flow wastewater treatment. He is the author or co-author of over 70 technical papers and one textbook. Professor Schroeder has served as a consultant to many private and government organizations and served on the California Regional Water Control Board, Central Valley Region for three years.

Contents

2

PHYSICAL, CHEMICAL, AND BIOLOGICAL CHARACTERISTICS OF WATER   43

3

SIGNIFICANCE OF THE CHARACTERISTICS OF WATER 163

4

WATER QUALITY: STANDARDS AND GLOBAL PERSPECTIVES   211

PART **III** MODELING WATER QUALITY IN
          THE ENVIRONMENT

7

MOVEMENT OF CONTAMINANTS IN THE ENVIRONMENT   305

8

WATER QUALITY IN RIVERS AND ESTUARIES AND NEAR OCEAN OUTFALLS   337

9

WATER QUALITY IN LAKES AND RESERVOIRS   383

10

WATER QUALITY IN GROUNDWATER SYSTEMS   405

PART **IV**   MODIFICATION OF WATER QUALITY

11

INTRODUCTION TO WATER AND WASTEWATER TREATMENT   443

12

PHYSICAL TREATMENT METHODS 465

13

CHEMICAL TREATMENT METHODS   559

14

BIOLOGICAL TREATMENT METHODS 595

15

SYNTHESIZING WATER AND WASTEWATER TREATMENT SYSTEMS   677

APPENDIXES

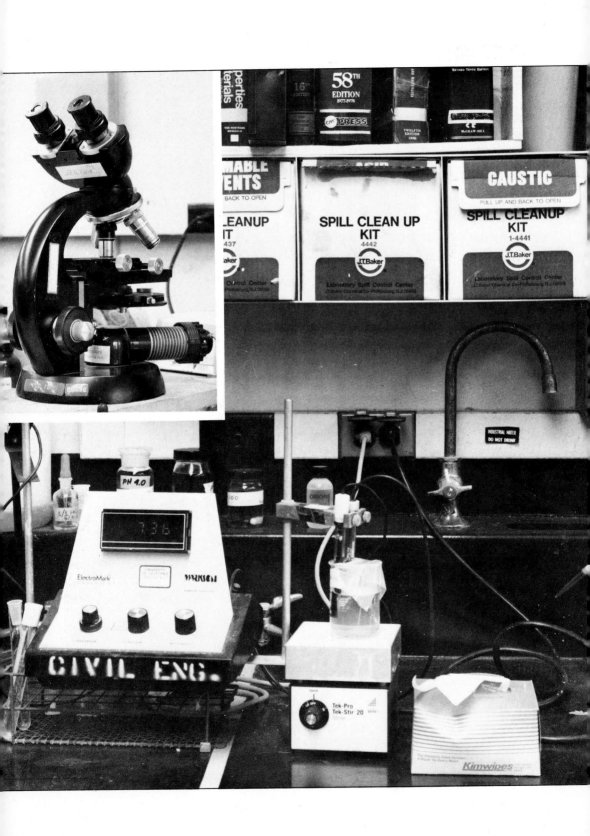

PART **I**

Water Quantity and Quality

Water is an essential element in the maintenance of all forms of life, and most living organisms can survive only for short periods without water. This fact has resulted in the development of direct relationships between abundance of water, population density, and quality of life. As well as being in abundant supply, the available water must have specific characteristics. Water quality is defined in terms of those characteristics. Throughout history the most important of the characteristics has been the concentration of dissolved solids (salts) because of the relationship between salt and land productivity. As population densities increased, health-related characteristics, such as the presence of disease-causing (pathogenic) microorganisms, became more important. With industrial development, characteristics such as temperature (of cooling water) and specific ion content became significant. More recently, the introduction of anthropogenic (manufactured) chemicals that have an impact on health when present in trace amounts has become a problem.

Two of the key concerns in the development of water resources, regardless of the use, are quantity and quality. Within limits, the quantity of water is more important than the quality of water. The quantities of water required in various uses are considered in Chapter 1. Because each use of water has an individual set of constraints, an absolute definition of water quality cannot be made. Rather, as discussed in Chapter 2, the quality of a water is assessed in terms of its physical, chemical, and biological characteristics. The specific characteristics used to assess water quality, based on intended use, are examined in Chapter 3. Finally, global water quality concerns are introduced in Chapter 4.

1

Sources and Uses of Water

The major issues associated with the use of water by humans are quantity and quality. Within limits, water quantity is more important than water quality in determining the extent and type of development possible in a given geographic location. Except for the presence of toxic materials, the constituents of water are not a major deterrent to domestic use until the concentration of dissolved constituents exceeds 1000 g/m^3. Bacteria and sediment can be removed satisfactorily by inexpensive treatment processes. To an extent, the use of a water of poor quality can be offset by increased usage, or where the impact of specific constituents can be managed.

Although the focus of this book is on water quality, it is important to know what quantities of water are needed for the support of modern societies. Such information is necessary to optimize the use of available water for the betterment of human beings. From this perspective, the subjects to be considered in this chapter include (1) the importance of water, (2) water needs versus water availability, (3) sources of water, (4) municipal water use, (5) wastewater sources and flow rates, (6) agricultural water use, and (7) ecological water requirements.

1.1
THE IMPORTANCE OF WATER

A plentiful supply of water is clearly one of the most important factors in the development of modern societies. Availability of water for cleansing is directly related to the control or elimination of disease. The convenience of water available in the home improves the quality of life. Inexpensive water allows individuals and communities to beautify their surroundings and to use water as a carrier for household wastes. The benefits of plentiful water have been utilized to such an extent in the United States that some negative reactions have been voiced concerning a perceived waste. The residents of most communities could, in fact, reduce water consumption by 10 to 25 percent without significantly changing their life-style. Decreases in orders of magnitude would not be possible, however. Forced reductions in water use due to droughts have been experienced in most areas of the world. Use of bathwater by more than one person, saving bath and washer water for lawn and shrub watering, and flushing toilets only once or twice a day result in more than inconvenience. Although people in areas such as the sub-Sahara region in Africa would still consider such conditions luxurious, drought restrictions do greatly affect the quality of life and agricultural, commercial, and industrial activities.

1.2
WATER NEEDS VERSUS WATER AVAILABILITY

The need for water varies with culture, geography, type of community, and season. The quantities required are considered in Section 1.4. The availability of water varies with geography and with the provisions made to supply a constant source of water. Geographic location affects both the quantity and distribution of rainfall. Representative rainfall data for various locations are given in Fig. 1.1 and in Table 1.1. As seen in Fig. 1.1, both the rainfall pattern and mean annual rainfall for Lebanon,

TABLE 1.1

Average Precipitation and Evapotranspiration Data for Davis, Calif., and England

	DAVIS, CALIF.		ENGLAND	
MONTH	Precipitation, mm	Evapo-transpiration, mm	Precipitation mm	Evapo-transpiration, mm
January	93	78	85	5
February	74	86	62	15
March	56	112	54	31
April	32	129	58	56
May	14	161	62	76
June	4	178	53	97
July	0.3	189	78	97
August	0.5	176	77	79
September	6	147	72	48
October	20	123	86	25
November	44	91	90	10
December	84	75	82	3
Total	427.8	1545	859	542

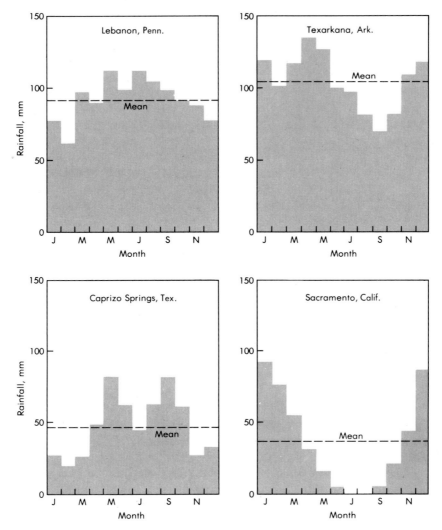

FIGURE 1.1

Rainfall distribution at selected locations in the United States. Note both the variation in mean annual values and the seasonal patterns.

Penn., are quite different from the corresponding values for Sacramento, Calif. A similar difference exists in the rainfall data for England and Davis, Calif., reported in Table 1.1.

In areas blessed with abundant rainfall throughout the year, such as in Lebanon, Penn., there is rarely a worry about the need for water; instead, the concern is with the time intervals between and the intensity of storms. In the United States, agricultural irrigation is relatively minor in the midwest and east, but is a major factor in most western states.

When the growing season is dry (as in California) crops and lawns must be watered intensively. During the hot summer months, water use in arid areas may exceed 2.3 m³/capita · d, compared to a normal demand of 0.4 m³/capita · d, because of the need to water lawns and shrubs. Larger pipes, pumps, wells, treatment facilities, and reservoirs are required in arid areas, thus greatly increasing the cost of supplying water. The need for water and the long dry periods between rainfall events provided the impetus for the development of the large water resource management projects in the western United States.

1.3
SOURCES OF WATER

Most of the water used or affected by humans can be classified as fresh water because the concentration of dissolved constituents is low. The definition of fresh water is not precise, but a value of 1500 g/m³ total dissolved solids (TDS) is an approximate upper limit. Brackish waters may have TDS values of up to 5000 g/m³, waters containing higher TDS concentration being termed "saline." Seawater is usually in the 30,000- to 34,000-g/m³-TDS range. Fresh waters are derived from surface sources and groundwater aquifers. Surface waters include lakes, rivers, and those waters stored as ice or snow. Surface waters tend to be turbid, a property caused by the presence of clays and other light-scattering colloidal particles, and treatment for turbidity removal is usually necessary prior to uses other than in irrigation. Groundwaters usually have higher TDS concentrations than surface waters because of mineral pickup from soil and rocks, and many groundwaters are noted for high concentrations of particular ions or elements such as calcium, magnesium, boron, and fluoride. Because of their high quality with respect to potability and their minimum treatment requirements, groundwaters are often preferred sources of water for individual homes and small communities. Large communities often find surface-water sources in remote mountain areas attractive because of their more desirable chemical characteristics and reliability. Cities such as New York, Denver, Los Angeles, and San Francisco transport water over long distances rather than rely on less desirable and often inadequate local surface and groundwater supplies.

1.4
MUNICIPAL WATER USE

Municipal water use is generally divided into four categories: (1) domestic, (2) commercial and industrial (nondomestic), (3) public service, and (4) unaccounted system losses and leakage. Typical per-capita values for these uses are reported in Table 1.2. Each of these uses is considered separately in the following discussion. In addition, the factors affecting water use and the variations that can be expected in water use are also considered. Estimation of municipal water use is illustrated in Example 1.2, presented at the end of this section.

Domestic Water Use

Domestic water use encompasses the water supplied to housing areas, commercial districts, institutional facilities, and recreational facilities. The uses to which this water is put include drinking, washing, bathing, culinary, waste removal, and yard watering.

Housing Areas

The per-capita water supplied to individual residences and apartments varies from 150 to 480 L/capita · d, averaging about 220 L/capita · d. This quantity includes the water used for the purposes cited above. Referring to Table 1.2, domestic water usage is about 36.7 percent of the total usage in a community. Factors that influence the quantity of water used are discussed at the end of this section. The distribution of water use in an individual residence is reported in Table 1.3.

In general, new developments have more water-using appliances than older ones. Garbage disposals, dishwashers, number of bathrooms, and landscaping can be expected to increase with time in any given community. Many communities are, however, wisely insisting on water conservation measures in new construction; such measures include the use of showers rather than bathtubs, of low-flow shower heads, and of low-water-use toilets (2 to 6 liters per flush versus 16 to 20). In some communities, kitchen garbage-disposal units are banned, and in regions with restricted seasonal precipitation, landscaping with drought-tolerant plants is encouraged.

TABLE 1.2

Typical Municipal Water Use in the United States

| | FLOW, L/capita · d | | |
USE	Range	Average	Percent based on average flow
Domestic	150–480	220	36.7
Commercial and industrial (nondomestic)	40–400	260	43.3
Public service	20–80	30	5.0
Unaccounted system losses and leakage	40–160	90	15.0
	250–1120	600	100

TABLE 1.3
Typical Distribution of Domestic Household Water Use in the United States

USE	FLOW, L/capita·d	PERCENT OF TOTAL
Toilet	88	40.0
Hand and body washing	75	34.1
Kitchen	16	7.3
Drinking*	10	4.5
Clothes washing	16	7.3
House cleaning	3	1.4
Garden watering[†]	10	4.5
Car washing	2	0.9
Total	220	100

*Includes running tap to obtain cold water, spillage, etc.

[†]Areas not requiring extensive irrigation.

TABLE 1.4
Typical Water-Use Values for Commercial Facilities

FACILITY	UNIT	FLOW, L/unit·d Range	FLOW, L/unit·d Typical
Airport	Passenger	8–15	10
Automobile service station	Vehicle served	30–60	40
	Employee	35–60	50
Bar and cocktail lounge	Customer	5–20	8
Boarding house	Resident	80–200	150
Hotel	Guest	150–220	190
	Employee	30–50	40
Industrial building (excluding industry and cafeteria)	Employee	30–65	55
Laundry (self-service)	Machine	1500–2500	2000
	Wash	180–200	190
Motel	Person	90–150	120
Motel with kitchen	Person	190–220	120
Office	Employee	30–65	55
Public lavatory	User	10–25	15
Restaurant (including toilet)			
Conventional	Meal	30–40	35
Short-order	Meal	10–30	15
Tavern	Seat	60–100	80
Rooming house	Resident	80–200	150
Department store	Toilet room	1600–2400	2000
	Employee	30–50	40
Shopping center	Parking space	2–8	4
	Employee	30–50	40
Theater			
Indoor	Seat	8–15	10
Drive-in	Car	10–20	15

Source: Adapted in part from Refs. [1.9] and [1.12].

Commercial Facilities

The per-capita usage for commercial facilities will vary with the type of activity (e.g., stationery store versus restaurant). Representative water usage values are presented in Table 1.4. Based on past records, it has been proposed that a value based on floor area is appropriate. Typical values range from 10 to 15 $L/m^2 \cdot d$, with a typical value being 12 $L/m^2 \cdot d$.

Institutional Facilities

The per-capita water usage for institutional facilities is based on some measure of the size of the facility. Per student, per bed, and per cell are representative units used to define water usage for schools, hospitals, and prisons. Water usage in schools will vary significantly depending on whether the students live in dormitories. Representative water-use values for institutional facilities are reported in Table 1.5.

Recreational Facilities

The per-capita water usage at recreational facilities is quite seasonal. Some typical water-use values are reported in Table 1.6.

TABLE 1.5

Typical Water-Use Values for Institutional Facilities

SOURCE	UNIT	FLOW, $L/unit \cdot d$	
		Range	Typical
Hospital, medical	Bed	500–1000	650
	Employee	20–60	40
Hospital, mental	Bed	300–550	400
	Employee	20–60	40
Prison	Inmate	300–600	450
	Employee	20–60	40
Rest home	Resident	200–450	350
	Employee	20–60	40
School, day:			
With cafeteria, gym, and showers	Student	60–120	80
With cafeteria, but without gym and showers	Student	40–80	60
Without cafeteria, gym, and showers	Student	20–65	40
School, boarding	Student	200–400	280

Source: Adapted in part from Refs. [1.9] and [1.12].

TABLE 1.6

Typical Water-Use Values for Recreational Areas

		FLOW, L/unit · d	
SOURCE	UNIT	Range	Typical
Apartment, resort	Person	200–280	220
Bowling alley	Alley	600–1000	800
Cabin, resort	Person	130–190	160
Cafeteria	Customer	4–10	6
Camp			
Pioneer type	Person	60–120	80
Children's (toilet and bath)	Person	140–200	160
Day (with meals)	Person	40–80	60
Day (without meals)	Person	30–70	50
Trailer	Trailer	400–600	500
Campground (developed)	Person	80–150	120
Cocktail lounge	Seat	50–100	75
Coffee shop	Customer	15–30	20
	Employee	30–50	40
Country club	Member present	250–500	400
	Employee	40–60	50
Dining hall	Meal served	15–40	30
Dormitory, bunkhouse	Person	75–175	150
Fairground	Visitor	2–6	4
Hotel, resort	Person	150–240	200
Laundromat	Machine	1500–2500	2000
Park, picnic (with toilets)	Person	20–40	30
Store, resort	Customer	5–20	10
	Employee	30–50	40
Swimming pool	Customer	20–50	40
	Employee	30–50	40
Theater			
Indoor	Seat	8–15	10
Drive-in	Car	10–20	15
Visitor center	Visitor	15–30	20

Source: Adapted in part from Refs. [1.9] and [1.12].

Commercial and Industrial (Nondomestic) Water Use

The amount of water provided by public water supply agencies to commercial and industrial users is usually limited, although in some communities, industries such as canneries utilize public supplies. The largest industrial water use is for cooling, which accounts for approximately 50 percent of all nonagricultural water use in the United States. In 1975, the steam electric-power industry (Fig. 1.2) used over 1.22×10^{11} m^3/yr, compared to 2.2×10^{11} m^3/yr for agricultural irrigation and 0.3×10^{11} m^3/yr for domestic use [13].

FIGURE 1.2

Water cooling towers used in conjunction with a nuclear power generation facility. Note size of automobiles in foreground.

Nonagricultural industrial use varies widely according to industry, geographic location, age of plant, and cost of water. As noted above, the steam electric-power industry is the largest industrial water user, and cooling requirements of other industries such as steel, oil refining, and chemical production make cooling-water requirements the dominant concern in the industry. The amount of water required is different in each situation. Source and receiving water temperatures are the major factors in determining flow. Two general types of cooling are in common use: once-through and evaporative. Once-through cooling is the simplest and least expensive (see Example 1.1). Evaporative cooling requires construction and operation of cooling towers. Slime growth and scale formation in heat-exchange surfaces are major problems in all cooling systems. Chemical control is the conventional method of keeping surfaces clean, friction losses minimized, and heat transfer rates maximized, but the chemicals used in these systems can result in significant pollution problems. Chromate compounds, used for corrosion control, have been banned by the U.S. Environmental Protection Agency (USEPA). Chlorine, used to control slimes, must be removed before discharge, and control of scale by lowering the pH with acid is difficult if chemical corrosion-control measures cannot be taken.

Food processing is also a major water-using industry. Water is used for washing and carrying products through the plant and is incorporated into some products. Because the water is in contact with food, good sanitation must be maintained and water must be bacterially safe. Wastewater from food processing operations is usually high in soluble organic wastes such as sugars and acids. Meat processing wastewater contains blood, manure, grease, and other difficult-to-handle materials. Poultry processing wastewaters are a particular problem because of grease and feathers.

EXAMPLE 1.1

WATER REQUIREMENTS FOR COOLING

A 1-MW (megawatt) steam power plant is approximately 33 percent efficient; that is, for every 3 MW of energy put into the system about 1 MW of electrical output and 2 MW of heat are generated. Determine the amount of once-through cooling water required to maintain a constant-temperature power system if, to protect the receiving water habitat, the increase in the temperature of the cooling water can be no more than 0.3°C. Assume that all heat generated is absorbed by the cooling water.

SOLUTION:

1. Determine the increase in the temperature of the cooling water expressed in terms of degrees centigrade times cubic meters per second (°C · m³/s).
   a. Convert the waste heat to megajoules (MJ). For every megawatt produced, 2 MW of heat must be dissipated. Because 1 MW = 10^6 J/s, the amount of heat dissipated is

   $$H_d = 2 \text{ MW} \times (10^6 \text{ J/s})/\text{MW}$$
   $$= 2 \times 10^6 \text{ J/s}$$

   b. The heat capacity of water is

   $$C_p = 4.186 \text{ J/g} \cdot °C$$
   $$\approx 4.186 \times 10^6 \text{ J/m}^3 \cdot °C$$

   c. The increase in temperature for each cubic meter per second of cooling water is

   $$\Delta T_c = \frac{2 \times 10^6 \text{ J/s}}{4.186 \times 10^6 \text{ J/m}^3 \cdot C}$$
   $$= 0.48°C \cdot \text{m}^3/\text{s}$$

2. Determine the minimum amount of once-through cooling water required to accommodate the heat generated by the steam power plant. To find the minimum amount of cooling water needed, the allowable increase in the cooling water temperature is equated to the increase per cubic meter per second found in step 1. Thus,

$$(0.3°C)Q_c = 0.48°C \cdot m^3/s$$
$$Q_c = 1.6 \ m^3/s$$

COMMENT

If river water was being used as the source of the once-through cooling water, the minimum flow in the river, if the entire flow was used for cooling, would have to be $1.6 \ m^3/s$. Note that the flow required in this example is relatively large: $1.6 \ m^3/s$ is approximately the same quantity of water as that used for domestic purposes by a community of 628,000 people in the United States.

Public Service Water Use

Public water uses include the water used for public buildings (Fig. 1.3), for flushing and cleaning streets, for watering parks and greenbelts, and

FIGURE 1.3

Public toilet for men, dating from early Roman times. Wastes were removed with water running in channels located in front of and below toilets. (Ruins at Philippi, Greece.)

for fire protection. On a per-capita basis, public uses of water amount to about 30 L/capita · d. Although the amount of water used for fighting fires is only a small fraction of the total supplied on an annual basis, the short-term demand rates can be very high and often dictate the sizing of pipes in small water supply systems. In the past, the following equation, developed by the National Board of Fire Underwriters, has been used to estimate the fire demand required in downtown business districts.

$$Q = 3.86\sqrt{P}\,(1 - 0.01\sqrt{P}\,) \qquad\qquad (1.1)$$

where

Q = flow rate, m^3/min

P = population in thousands

More recently, the following equation has been proposed as a guide for estimating water demands for fire fighting [1.4].

$$F = 320\ C\sqrt{A} \qquad\qquad (1.2)$$

where

F = required fire flow, m^3/d

C = coefficient related to type of construction

A = the total floor area (including all stories, but excluding basements) in the building under consideration, m^2. For fire-resistive buildings the six largest successive floor areas are used if the vertical openings are unprotected; if the vertical openings are protected properly, only the three largest successive floor areas are considered.

Values for the coefficient C are 1.5 for wood-frame construction, 1.0 for ordinary construction, 0.8 for noncombustible construction, and 0.6 for fire-resistive construction. Interpolation between these values is used for construction that does not fall into one of the four categories. The computed value is then adjusted up or down for (1) occupancy, (2) sprinkler protection, and (3) exposure [1.4]. The maximum fire flow determined using Eq. (1.2) must not exceed 43,600 m^3/d for wood-frame and for ordinary and heavy-timber construction and 32,700 for noncombustible construction and for fire-resistive construction for any one location. The required fire flow rate must be available in addition to the coincident maximum daily flow rate. The duration during which the required fire flow should be available varies from 2 to 10 hr, as summarized in Table 1.7.

Because a city will be penalized in its fire insurance rates if the needed flows cannot be met for the specified durations, most cities provide storage reservoirs to meet fire demands. Additional details on fire protection may be found in Refs. [1.4] and [1.7].

TABLE 1.7

Duration of Required Fire Flows Based on Flow

FIRE FLOW		DURATION,
gal/min	m^3/d	hr
2,000 or less	10,900	2
3,000	16,400	3
4,000	21,800	4
5,000	27,300	5
6,000	32,700	6
7,000	38,200	7
8,000	43,600	8
9,000	49,100	9
10,000 and greater	54,500	10

Source: Adapted from Ref. [1.4].

Unaccounted System Losses and Leakage

Approximately 15 percent of the water withdrawn for municipal use cannot be accounted for in a water-balance analysis. In most water supplies, leakage within the distribution system accounts for most of the loss. Leakage is a function of pipe size, joint construction, temperature, and system age. Cast-iron pipe with joints formed on-site should be expected to leak 0.01 to 0.05 m^3/d per kilometer per millimeter of pipe diameter. Average per-capita demand, as measured by water input, increases with system age because of increased leakage at connections and taps as well as increased personal use. In some older systems that have not been maintained, the amount of water lost to leakage will often exceed the domestic use. Repair of old and leaky systems can often eliminate the need for bringing a new source on-line.

Factors Affecting Water Use

The quantity of water used varies widely from location to location in the United States. For this reason, it is difficult to generalize on the factors that affect the observed variations. Factors that are of importance include geographic location, type of community, economic status of the community, water pressure, cost, need for conservation, and water system management.

Geographic Location

The use of water will vary with geographic location, depending on local climatic conditions—principally, temperature and rainfall. Water use for lawns and plants, for bathing, and for air conditioning is greater in dry, as opposed to humid, locations. Water use is lower where it rains throughout the year, as in the eastern and northern United States.

Type of Community
The use of water will vary with the mix of residential, commercial, and industrial areas within a community. Water use will also vary with the time of year, especially where seasonal industries such as fruit and vegetable canneries are supplied with municipal water.

Economic Status of the Community
Water use will vary with the economic status of the community. Wealthier communities generally will have a higher use as compared with communities where many of the residents are on a fixed income or unemployed. In some locations, it has been found that residents give up lawn watering in economic hard times.

Water Pressure
Water use is generally higher in locations with higher pressures. In many older communities, the original water system was installed when the level of development was considerably less than it is today. Because much of the water system is undersized for today's development and corresponding water demand, pressure is reduced.

The 650-home community of Stinson Beach, Calif., located on the coast 32 km north of San Francisco, can be used as an example of what can happen when an older system is modified to improve the water pressure. When a new line was put into service between two storage reservoirs, numerous leaks developed in the water systems in the older portions of the community (Fig. 1.4). Many of the old lines ruptured because they could not withstand the higher water pressure. In fact, many of the lines that were uncovered were found to be old wood stave pipes.

Cost
Water usage is closely related to cost. Communities where the water is metered use considerably less water than those without meters. Over-watering is common where water is inexpensive. In domestic supply systems, the price paid for excessive use includes larger pipes, pumps, collection systems, and wastewater treatment facilities. Because source capacity is often stressed, system reliability also becomes a problem.

Need for Conservation
Under conditions of drought, it has been demonstrated repeatedly that communities can reduce water usage by 20 to 30 percent without any adverse effects. In many communities where the disposal of wastewater is by means of on-site systems, greater water conservation measures may be

FIGURE 1.4

The community of Stinson Beach, Calif. Water-distribution storage reservoirs are located in the hills above the community.

needed to ensure the continued successful operation of the on-site treatment and disposal systems. Several levels of water reduction that are possible with the use of water conservation devices are reported in Table 1.8.

Water System Management

The operation and maintenance of a water supply and distribution system can have a significant effect on the computed amount of water

TABLE 1.8

Typical Water Use Without and With Conservation Appliances

	FLOW, L/person · d		
USE	Without Conservation Appliances	With Conservation Appliances (Level 1)	With Conservation Appliances (Level 2)
Toilet	88	20	10*
Hand and body washing	75	40	10†
Kitchen	16	8	4
Drinking	10	8	6
Clothes washing	16	8	6
House cleaning	3	3	–
Garden watering	10	6	–
Car washing	2	–	–
Total	220	94	36

*2-L flush toilets.   †Air-assisted shower.

served to each customer. If maintenance is neglected, system losses will be high, and the reported average per-capita usage rate will be higher than the true rate. The importance of proper operation and maintenance cannot be overstressed, especially now when many U.S. systems are over 50 years old, and monies for renovation are difficult to obtain.

Fluctuations in Water Use

The data on water use, presented previously in this section, are based on average annual values. Unfortunately, average annual values are of little use in the design of water supply systems. What must be known is how the average values vary so that the hydraulic capacity of the system can be sized to meet the peak demands.

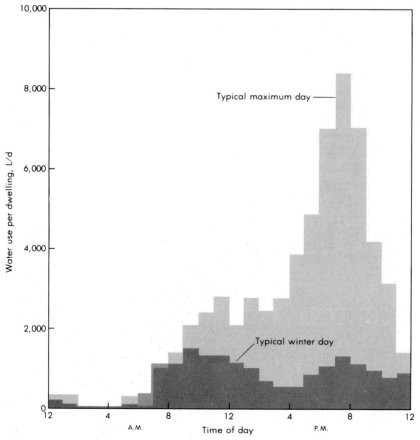

FIGURE 1.5

Daily water-use patterns: maximum day and winter day. Note that the total consumption per dwelling corresponds to the area under the curves.

Source: Adapted from Ref. [1.6].

Daily Variations

Demands for water vary greatly throughout the day and throughout the year. Typical flow variation patterns are shown in Figs. 1.5 and 1.6. As shown in these figures, water usage is very low during nighttime hours (2100 to 0600 hr), increases sharply during the breakfast and morning cleaning period, has a second peak in the evening, and flutters through the TV commercial breaks. Not surprisingly, more water is used on hot than on cold days, and variation is greater in small than in large communities. An excellent example of the impact of TV on a public

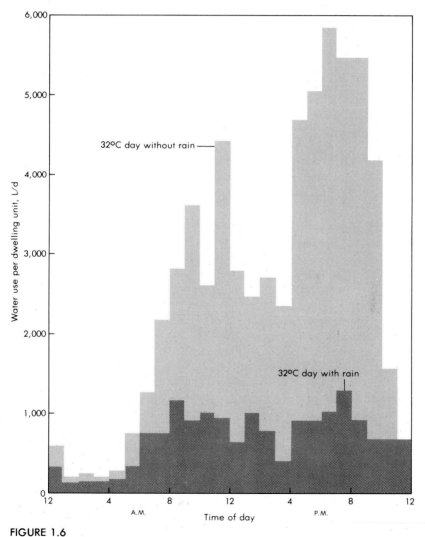

FIGURE 1.6

Daily water-use patterns in high-density residential area: maximum and minimum day. Source: Adapted from Ref. [1.6].

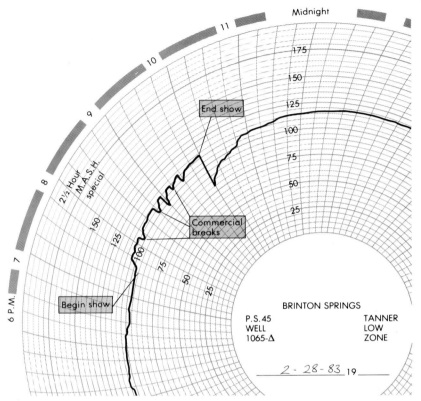

FIGURE 1.7

Effect of TV commercial breaks on pressure in water distribution systems. As demand and flow increase, distribution line pressure decreases.

Source: Courtesy of W. E. Evenson, Superintendent, Salt Lake City Department of Water Utilities.

water supply system is shown in Fig. 1.7, a copy of the pressure record for the Salt Lake City service area during the February 28, 1983, M*A*S*H special.

Design Factors

Important facts to note when examining Figs. 1.5 and 1.6 are the peak-to-minimum ratios and the duration of the peak flow. Facilities must be constructed to supply peak demand, and therefore capital costs are affected by relatively short time periods. Typical design values for the fluctuations that can be expected in water supply systems are presented in Table 1.9. Equation (1.3) has been proposed to estimate peak consumption as a function of the annual average [1.2]:

$$p = 180t^{-0.10} \qquad\qquad (1.3)$$

where

p = percentage of the annual average demand corresponding to time, t

t = time, d $(2/24 \leq t \leq 360)$

The values determined using Eq. (1.3) are most applicable to small communities, since the variation is damped as communities increase in size, and where industries utilize the public water supply.

TABLE 1.9

Typical Fluctuations in Water Use in Municipal Systems

	PERCENT OF AVERAGE FOR YEAR		
FLOW	Range	Typical	Eq. (1.3)*
Daily average in maximum month	110–140	120	128
Daily average in maximum week	120–170	140	148
Maximum day	160–220	180	180
Maximum hour	225–320	270[†]	

*$p = 180t^{-0.10}$ [1.2].

[†]1.5 × maximum-day value.

EXAMPLE 1.2

ESTIMATION OF MUNICIPAL WATER DEMANDS

Estimate the water requirements for a community of 200,000 persons. Use Eq. (1.1) to estimate the required fire flows.

SOLUTION:

1. Determine the average daily demand using the data given in Table 1.2.

Q_{avg} = 0.6 m³/person · d (200,000 persons)

= 1.2×10^5 m³/d

2. Determine the maximum daily demand using the data given in Table 1.9.

$Q_{peak\ day}$ = 1.2×10^5 m³/d × 1.8

= 2.16×10^5 m³/d

3. Determine the maximum hourly demand using the data given in Table 1.9.

$Q_{peak\ hour}$ = 1.2×10^5 m³/d × 2.7

= 3.24×10^5 m³/d

4. Using Eq. (1.1) determine the fire flow demand for the high-value district.

$$Q = 3.86\sqrt{P}\,(1 - 0.01\sqrt{P}\,)\ \text{m}^3/\text{min}$$
$$= 3.86\sqrt{200}\,(1 - 0.01\sqrt{200}\,)\ \text{m}^3/\text{min}$$
$$= 46.87\ \text{m}^3/\text{min}$$
$$= 0.67 \times 10^5\ \text{m}^3/\text{d}$$

5. Check to see if the coincident peak day plus fire flow demand or the peak hourly demand controls the design of the distribution system.
   a. Coincident peak day plus fire flow:

$$Q_{c+f} = 2.16 \times 10^5\ \text{m}^3/\text{d} + 0.67 \times 10^5\ \text{m}^3/\text{d}$$
$$= 2.83 \times 10^5\ \text{m}^3/\text{d}$$

   b. Peak hourly demand:

$$Q_{\text{peak hour}} = 3.24 \times 10^5\ \text{m}^3/\text{d}$$

   c. Because the peak hourly demand is the larger of the two demand values, it will control the design of the distribution system.

COMMENT

Referring to Table 1.7, adequate storage or pump capacity will have to be provided in the water supply system to meet the coincident flow plus fire flow demand for a period of 10 hr.

1.5
WASTEWATER SOURCES AND FLOW RATES

Once the water supply of a community is used for the purposes cited previously, it is discharged as wastewater. Although the term "sewage" is used interchangeably with the term "wastewater," the latter is used exclusively in this text. With the exception of some public uses and leakage, wastewater is collected from all of the locations where water is supplied, including residential areas, commercial and industrial facilities, and institutional and recreational facilities. Ultimately, all wastewaters are returned to the environment following treatment of various degrees—from none to extensive (see Part IV). To predict the impact resulting from such discharges as those discussed in Part III, the quantities of wastewater flow and their variation must be known. Estimation of wastewater flow rates is illustrated in Example 1.3, presented at the end of this section.

Wastewater Flow Rates

In general, wastewater flow rates are quite similar to water demand rates, with two exceptions: (1) water used for irrigation (of lawns, shrubs, and gardens) generally does not enter the sanitary sewer, and (2) when sewer pipes are located below the water table, substantial infiltration may occur. The result is that dry-weather and wet-weather flows are often quite different. In many older communities, the storm drainage system is combined with the wastewater system. In these cases, the wet-weather to dry-weather flow ratio is very pronounced, and treatment of the wet-weather flow becomes an exasperating and very expensive problem.

Residential Districts

Domestic dry-weather flow is composed of the wastes from sinks, baths, showers, toilets, and other sources found in houses and businesses. Approximately 80 percent of the house or per-capita domestic water demand is accounted for as wastewater, but other sources (infiltration, roof drainage, industrial discharges) often result in wastewater flows that are greater than the water demand rates. Typical discharges are reported in Table 1.10. Apartment dwellers use somewhat less water than home dwellers. As noted above, the water demand curve and the wastewater flow-rate curve are generally quite similar.

TABLE 1.10

Typical Wastewater Flows from Residential Areas

SOURCE	UNIT	FLOW, L/unit·d	
		Range	Typical
Apartment			
High-rise	Person	140–300	200
Low-rise	Person	200–320	260
Hotel	Guest	120–220	180
Individual residence			
Typical home	Person	180–360	280
Better home	Person	250–400	320
Luxury home	Person	300–600	380
Older home	Person	120–250	180
Summer cottage	Person	10–200	160
Motel			
With kitchen	Unit	360–720	400
Without kitchen	Unit	300–600	380
Trailer park	Person	120–200	150

Source: Adapted in part from Refs. [1.9] and [1.12].

Commercial and Industrial Facilities

If water-use records are unavailable and water-use projections have not been made, the flow from commercial and industrial facilities or areas is usually estimated in terms of cubic meters per hectare-day $(m^3/ha \cdot d)$. Because the range of values reported in the literature is so great (50 to 1500 $m^3/ha \cdot d$), every effort should be made to obtain records from similar facilities at other locations. The water-use data reported in Table 1.4 can also be used for the purposes of estimation by assuming that about 95 percent of the water reported used eventually becomes wastewater.

Institutional Facilities

Representative flows from institutional facilities can be estimated by assuming from 95 to 100 percent of the water-use values reported in Table 1.5. As shown, the greatest variation occurs with schools, depending on whether they are boarding or day type.

Recreational Facilities

Mirroring water usage almost directly, flows from most recreational facilities and areas are quite seasonal. Representative flows can be estimated by assuming from 95 to 100 percent of the water-use values reported in Table 1.6. Because the flows from such facilities will vary with the type of facility, geographic location, and climate, actual records should be used whenever possible.

Infiltration and Inflow

The two major components of the extraneous flow found in sewers are infiltration and inflow [1.3, 1.8].

Infiltration

Because many sewers are located below the natural or seasonal water table, extraneous groundwater can enter the sewer at service connections, at locations where the pipe is broken, at defective joints, at connections to manholes, and through manhole walls (Fig. 1.8). Collectively, water that enters the sewer in the manner described above composes what is known as *infiltration*.

Depending on the condition of the sewer and its location relative to the location of the groundwater, infiltration can vary from 0.009 to 0.9 $m^3/d \cdot mm \cdot km$ (cubic meters per day per millimeter of pipe diameter per kilometer). Expressed on an area basis, the infiltration may vary from 0.2 to 30 $m^3/ha \cdot d$. When substantial amounts of water accumulate on streets, as occurs in heavy rains, inflow through manholes, in addition to infiltration, may result in flows higher than 450 $m^3/ha \cdot d$ for short periods. Because gravity sewers have limited capacity and surcharging results in wastewater flowing in the streets, prediction of the maximum

FIGURE 1.8

Water infiltrating into a municipal sewer at a displaced pipe joint. In older or poorly maintained systems, increases in flow due to infiltration as large as 10 times the dry-weather flow are not uncommon. Courtesy CUES WEST, Sacramento, CA.

flow is an extremely important design consideration. Average and peak infiltration allowances used for the design of new sewers are given in Fig. 1.9.

Inflow

Water discharged to sewers from sources such as roof leaders; cellar, yard, and area drains; foundation drains; cooling water discharges; drains from springs and other wet areas; manhole covers; cross connections from storm sewers; catch basins; storm water; surface water runoff; and street wash waters and drainage is known as *inflow*. In analyzing wastewater flow rates it is important to distinguish between inflow and infiltration, because control measures are quite different.

Fluctuations in Wastewater Flow Rates

Just as the demand for water fluctuates on an hourly, daily, weekly, monthly, and yearly basis, so do wastewater flow rates. Both short-term and seasonal rate fluctuations are important.

Short-Term Fluctuations

Short-term fluctuations observed in wastewater flow rates tend to follow a diurnal pattern, as shown in Fig. 1.10. Minimum flow occurs in the early morning hours when water use is minimal (Fig. 1.6). The flow rate starts to increase around 6 A.M. and reaches a peak value around 12 noon. The flow rate then drops off in the early afternoon. A second peak

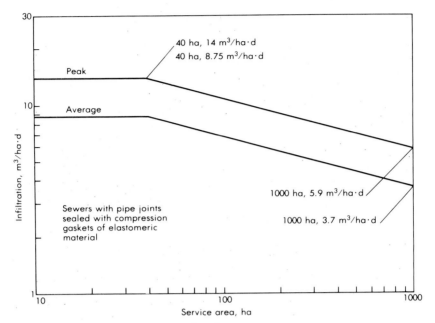

FIGURE 1.9

Average and peak infiltration allowances for new sewers.

Source: Adapted from Ref. [1.9].

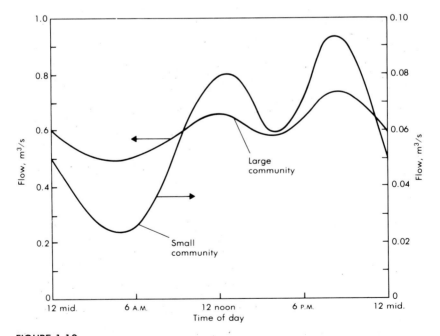

FIGURE 1.10

Typical variation in daily municipal wastewater flow rates.

occurs in the early evening hours between 6 and 9 P.M. In many communities, the magnitude of the second peak will exceed the value of the morning peak. In general, where extraneous flows are excluded from the sewer system, the wastewater flow-rate curves will closely follow water-use curves. In general, the curves will be displaced by a time period corresponding to the travel time in the sewers, as most wastewater flows are measured at the wastewater treatment facility.

Seasonal Fluctuations

Seasonal fluctuations are common in communities with seasonal commercial and industrial activities such as food processing, in resort communities, and in small communities with college or university campuses. Where seasonal flows are anticipated, special attention must be devoted to assess the magnitude of the expected peak flows. Also, as noted previously, infiltration and inflow will vary seasonably.

Peak Flow Rates

Peak flow rates are important in the design of wastewater collection systems. In the absence of more reliable information, the following equation can be used to estimate the peaking factor for wastewater flow rates:

$$PF = 5(P)^{-0.16} \tag{1.4}$$

where

PF = peaking factor

P = population in thousands $(1 < P < 200)$

Beyond a population of about 200,000 Eq. (1.4) should be used with great caution.

EXAMPLE 1.3

ESTIMATION OF WASTEWATER FLOW RATES

Estimate the average and peak wastewater flow rates and the overall peaking factor for the Divine Gardens development shown in Fig. 1.11. Data on the expected saturation population densities and wastewater flows for the various housing areas within the development are given below. The commercial and industrial allowances are 40 and 60 $m^3/ha \cdot d$, respectively. The corresponding peaking factors for these two areas are 1.8 and 2.2. The school to be built in the development will have an enrollment of 1800 students. The average flow rate per student is 40 L/student $\cdot$ d, based on the use of low-flush toilets and other water-conserving devices. The peaking factor for the school is 4.5.

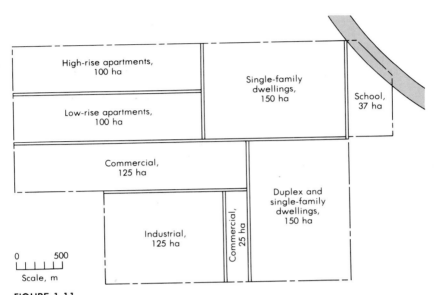

FIGURE 1.11

Land use for the Divine Gardens development for Example 1.3.

TYPE OF HOUSING	SATURATION POPULATION, Persons/ha	WASTEWATER FLOW, L/Person · d
Single-family dwellings	40	300
Duplexes	60	260
Low-rise apartments	110	230
High-rise apartments	160	190

SOLUTION:

1. Set up a computation table for estimating the average and peak wastewater flows. The required computations are summarized in Table 1.11. Although the table is self-explanatory, some additional comments will help to clarify some of the entries.

   a. The peaking factor for the domestic flow is based on the total population rather than on populations for the individual areas. The total population is used because the wastewater flow from the entire area is being estimated. When sewers are designed to serve the individual areas, the population and corresponding peaking factor for each individual area must be used.

   b. As noted above, the infiltration allowance is based on the total area, since the wastewater flow from the entire development is being estimated.

TABLE 1.11
Estimation of Average and Peak Wastewater Flows from the Divine Gardens Development for Example 1.3

LAND USE (1)	TOTAL AREA, ha (2)	POPULATION DENSITY, persons/ha (3)	TOTAL POPULATION (4)	AVG. UNIT FLOW Basis (5)	AVG. UNIT FLOW Value (6)	AVG. FLOW, m³/d (7)	PEAKING FACTOR (8)	PEAK FLOW, m³/d (9)
Single-family dwellings	150	40	6,000	L/person · d	300	1,800	2.8*	5,040
Mixed residential dwellings	150	50†	7,500	L/person · d	280†	2,100	2.8*	5,880
Low-rise apartments	100	110	11,000	L/person · d	230	2,530	2.8*	7,084
High-rise apartments	100	160	16,000	L/person · d	190	3,040	2.8*	8,512
School	37		1,800	L/person · d	40	72	4.5	324
Commercial	150			m³/ha · d	40	6,000	1.8	10,800
Industrial	125			m³/ha · d	60	7,500	2.2	16,500
Subtotal	812					23,042		54,140
Infiltration‡				m³/ha · d	3.85	3,126	1.6	5,002
Total						26,168		59,142

*From Eq. (1.4), based on a total population of 40,500 (6000 + 7500 + 11,000 + 16,000).

†Based on average of data given in problem statement.

‡From Fig. 1.9, based on an area of 812 ha.

2. Determine the overall peaking factor for the development.

$$PF = \frac{59{,}142 \text{ m}^3/\text{d}}{26{,}168 \text{ m}^3/\text{d}}$$

$$= 2.26$$

COMMENT

The overall peaking factor, computed in step 2, will decrease as the total area from which flow is being contributed increases.

1.6

AGRICULTURAL WATER USE

Water supply becomes the primary criterion in agricultural land development and in choice of crops. In Table 1.12, irrigated area and water use are shown for several important agricultural states. The most significant column is the last column, in which the water applied per hectare per year is given. California and Washington, two states with highly seasonal rainfall patterns (see rainfall pattern for Sacramento, Calif., in Fig. 1.1) and specializing in crops requiring extensive irrigation, use by far the greatest amount of water. The importance of irrigation to world society is illustrated in Table 1.13, where countries with over 1 million ha of irrigated land are listed. In many of these countries, nearly all of the land used for producing salable crops is irrigated.

TABLE 1.12

Irrigation Water Utilization in the United States in 1970

STATE	IRRIGATED AREA, ha	WATER USE, 10^6 m^3/yr Groundwater	Surface water	APPLICATION RATE, m^3/ha · yr
Alabama	11,000	7	17	2,181
California	3,521,000	23,330	23,450	13,286
Illinois	15,000	21	9	2,000
Indiana	14,000	25	11	2,571
Nebraska	1,659,000	3,702	2,839	3,943
New York	30,000	20	19	1,300
North Carolina	190,000	69	44	595
Texas	3,359,000	10,861	3,456	4,262
Washington	567,000	481	7,282	13,691
Wisconsin	40,000	46	27	1,825

Source: From Ref. [1.10].

Groundwater Usage

In most cases, groundwater is a major source of irrigation water for historical reasons. People moved onto land hoping that wells would be productive. Where this was the case, development continued. For example, the Sacramento and San Joaquin valleys of California are underlain by huge groundwater basins. These basins, relatively close to the surface, made development of large tracts of rich farmland possible in the nineteenth century. Overuse of the groundwater resources and a desire to increase the amount of land under production (Fig. 1.12) led to the development of the massive reservoir and canal systems of the federally sponsored Central Valleys Project and the state water project. In Washington, agricultural development followed a different pattern. The

TABLE 1.13

Extent of Irrigated Agriculture Throughout the World

COUNTRY	CULTIVATED LAND, 10^3/ha	IRRIGATED LAND, 10^3/ha
Afghanistan	7,980	2,900
Argentina	26,028	1,555
Australia	44,610	1,581
Bulgaria	4,516	1,001
Chile	4,632	1,500
China	111,167	76,500
Egypt	2,852	2,852
India	164,610	38,969
Indonesia	18,000	6,800
Iran	16,727	5,251
Iraq	10,163	4,000
Italy	14,409	3,500
Japan	5,446	2,626
Korea	2,311	1,070
Mexico	23,817	4,200
Pakistan	21,700	12,400
Peru	2,979	1,116
Philippines	11,148	1,090
Spain	20,626	2,435
Sudan	7,000	2,520
Thailand	11,200	3,170
U.S.A.	192,318	21,489
U.S.S.R.	232,609	11,500
Venezuela	5,214	1,000
Vietnam	5,083	3,040

Source: From Ref. [1.5].

FIGURE 1.12

Photograph taken in the Central Valley, Calif., showing extensive agricultural development.

Source: Cartwright Aerial Surveys, Inc., Sacramento, CA.

western third of the state receives substantial precipitation through much of the growing season, while the rich land east of the Cascade Mountains has a precipitation pattern much like the Central Valley of California. The Columbia River flows through the eastern area and was recognized from the beginning as the best source of water available. Development of the Columbia was the reason agriculture evolved in Washington.

Consumptive Use in Agriculture

Consumptive use of water by crops is a function of climate, temperature, length of growing season, and solar radiation, as well as of plant species. A number of empirical relationships exist for estimating consumption. One of the more commonly used is the Blaney–Criddle equation, presented here in a form modified to include SI units [1.1].

$$U = 45.7 K_{BC} \Sigma (T + 17.8) p/100 \tag{1.5}$$

where

U = consumptive use, mm

K_{BC} = Blaney–Criddle consumptive use coefficient (annual, seasonal, monthly, etc.)

T = mean temperature for time increment, °C

p = percent of annual daytime hours occurring during time increment

Consumptive use coefficients for various crops and recorded values in the Central Valley of California are given in Table 1.14. The factors 45.7 and 17.8 in Eq. (1.5) result from the conversion of the original equation to degrees Celsius. Note that the summation term allows use of time increments with relatively constant mean temperatures. Values for p by month for latitudes from 46° south to 60° north are presented in Table 1.15. The use of Eq. (1.5) is illustrated in Example 1.4.

TABLE 1.14

Consumptive Use Coefficients and Typical Values

CROP	BLANEY–CRIDDLE CONSUMPTIVE USE COEFFICIENT, K_{BC}	CONSUMPTIVE USE IN CENTRAL VALLEY OF CALIFORNIA*	
		Total, m/season	Rate, m/d
Alfalfa	0.90	1.02	0.0076
Beans	0.70	0.43	0.0056
Corn	0.85	0.66	0.0089
Cotton	0.70	0.66	0.0076
Grains	0.85	0.46	0.0056
Oranges	0.55	0.71	0.0046
Walnuts	0.70	0.61	0.0056
Grass	0.85	0.92	0.0076
Potatoes	0.75	0.31	0.0041
Rice	1.10		
Sugar beets	0.75	0.84	0.0076
Tomatoes	0.70	0.51	0.0051
Grapes	0.60		

*Based on a 250- to 300-d growing period.   Source: From Ref. [1.5].

EXAMPLE 1.4

ESTIMATION OF CONSUMPTIVE USE

Estimate the consumptive use for corn with a four-month growing season starting on the first of May. Assume that the following mean monthly

TABLE 1.15

Percentage of Total Yearly Daytime Hours Occurring During Each Month

LATITUDE 0° NORTH	JAN	FEB	MAR	APR	MAY	JUN	JUL	AUG	SEP	OCT	NOV	DEC
0	8.50	7.66	8.49	8.21	8.50	8.22	8.50	8.49	8.21	8.50	8.22	8.50
5	8.32	7.57	8.47	8.29	8.65	8.41	8.67	8.60	8.23	8.42	8.07	8.30
10	8.13	7.47	8.45	8.37	8.81	8.60	8.86	8.71	8.25	8.34	7.91	8.10
15	7.94	7.36	8.43	8.44	8.98	8.80	9.05	8.83	8.28	8.26	7.75	7.88
20	7.74	7.25	8.41	8.52	9.15	9.00	9.25	8.96	8.30	8.18	7.58	7.66
25	7.53	7.14	8.39	8.61	9.33	9.23	9.45	9.09	8.32	8.09	7.40	7.42
30	7.30	7.03	8.38	8.72	9.53	9.49	9.67	9.22	8.33	7.99	7.19	7.15
32	7.20	6.97	8.37	8.76	9.62	9.59	9.77	9.27	8.34	7.95	7.11	7.05
34	7.10	6.91	8.36	8.80	9.72	9.70	9.88	9.33	8.36	7.90	7.02	6.92
36	6.99	6.85	8.35	8.85	9.82	9.82	9.99	9.40	8.37	7.85	6.92	6.79
38	6.87	6.79	8.34	8.90	9.92	9.95	10.10	9.47	8.38	7.80	6.82	6.66
40	6.76	6.72	8.33	8.95	10.02	10.08	10.22	9.54	8.39	7.75	6.72	6.52
42	6.63	6.65	8.31	9.00	10.14	10.22	10.35	9.62	8.40	7.69	6.62	6.37
44	6.49	6.58	8.30	9.06	10.26	10.38	10.49	9.70	8.41	7.63	6.49	6.21
46	6.34	6.50	8.29	9.12	10.39	10.54	10.64	9.79	8.42	7.57	6.36	6.04
48	6.17	6.41	8.27	9.18	10.53	10.71	10.80	9.89	8.44	7.51	6.23	5.86
50	5.98	6.30	8.24	9.24	10.68	10.91	10.99	10.00	8.46	7.45	6.10	5.65
52	5.77	6.19	8.21	9.29	10.85	11.13	11.20	10.12	8.49	7.39	5.93	5.43
54	5.55	6.08	8.18	9.36	11.03	11.38	11.43	10.26	8.51	7.30	5.74	5.18
56	5.30	5.95	8.15	9.45	11.22	11.67	11.69	10.40	8.53	7.21	5.54	4.89
58	5.01	5.81	8.12	9.55	11.46	12.00	11.98	10.55	8.55	7.10	4.31	4.56
60	4.67	5.65	8.08	9.65	11.74	12.39	12.31	10.70	8.57	6.98	5.04	4.22

TABLE 1.15 *(Cont.)*

SOUTH

0	8.50	7.66	8.49	8.21	8.22	8.50	8.49	8.21	8.50	8.22	8.50
5	8.68	7.76	8.51	8.15	8.05	8.33	8.38	8.19	8.56	8.37	8.68
10	8.86	7.87	8.53	8.09	7.86	8.14	8.27	8.17	8.62	8.53	8.88
15	9.05	7.98	8.55	8.02	7.65	7.95	8.15	8.15	8.68	8.70	9.10
20	9.24	8.09	8.57	7.94	7.43	7.76	8.03	8.13	8.76	8.87	9.33
25	9.46	8.21	8.60	7.84	7.20	7.54	7.90	8.11	8.86	9.04	9.58
30	9.70	8.33	8.62	7.73	6.96	7.31	7.76	8.07	8.97	9.24	9.85
32	9.81	8.39	8.63	7.69	6.85	7.21	7.70	8.06	9.01	9.33	9.96
34	9.92	8.45	8.64	7.64	6.74	7.10	7.63	8.05	9.06	9.42	10.08
36	10.03	8.51	8.65	7.59	6.62	6.99	7.56	8.04	9.11	9.51	10.21
38	10.15	8.57	8.66	7.54	6.50	6.87	7.49	8.03	9.16	9.61	10.34
40	10.27	8.63	8.67	7.49	6.37	6.76	7.41	8.02	9.21	9.71	10.49
42	10.40	8.70	8.68	7.44	6.23	6.64	7.33	8.01	9.26	9.82	10.64
44	10.54	8.78	8.69	7.38	6.08	6.51	7.25	7.99	9.31	9.94	10.80
46	10.69	8.86	8.70	7.32	5.92	6.37	7.16	7.96	9.37	10.07	10.97

Source: From Ref. [1.1].

TABLE 1.16

Computation of Consumptive Water Use for the Cultivation of Corn in Example 1.4

MONTH	MEAN MONTHLY TEMPERATURE, T, °C	$(T + 17.8)$	DAYTIME HOURS, p,*%	CONSUMPTIVE USE FACTOR, $(T + 17.8)p/100$	BLANEY–CRIDDLE COEFFICIENT, K_{BC}	CONSUMPTIVE USE, mm
May	15.6	33.4	10.02	3.35	0.85	130.1
June	17.2	35.0	10.08	3.53	0.85	137.1
July	20.0	37.8	10.22	3.86	0.85	149.9
August	21.1	38.9	9.54	3.71	0.85	144.1
						561.2

*From Table 1.15.

temperatures apply and that the corn will be grown at latitude 40° north.

MONTH	MEAN MONTHLY TEMPERATURE, °C
May	15.6
June	17.2
July	20.0
August	21.1

SOLUTION:

1. Set up a computation table for estimating the consumptive use. The required computations are summarized in Table 1.16.

2. The total required consumptive use over the four-month period is

   $U = 561.2$ mm

COMMENT

In practice, irrigation systems have to be designed to meet the maximum monthly water demand value rather than the seasonal average.

1.7
ECOLOGICAL WATER REQUIREMENTS

Maintaining ecological quality often requires significant amounts of water. An excellent example is provided by the case of the Sacramento-San Joaquin River delta in California [1.11]. The delta contains over 1600 km of channels that provide habitat for game fish and waterfowl, water for irrigation of some of California's richest land, and excellent recreational facilities for tens of thousands of people each year. Tidal action through the delta results in large differences in salinity over the tidal cycle, particularly at the downstream end. When river flow is low, saline water moves up the rivers and affects the delta ecology. To protect the delta, many people feel that reservoir releases to maintain water quality (i.e., low salinity) and ecological balance are quite reasonable.

Opposed to this viewpoint are those who see all fresh water reaching the ocean (except that necessary for migratory fish) as wasted. In California, the need for fresh water is great because of the long, dry growing season. Thus, two very realistic but strikingly opposed views appear. Normally, groups concerned with maintaining environmental quality would side with the delta inhabitants, but in this case, the decision is more difficult because the delta is no more natural than Disneyland. The diked-off islands are all well below river level and will gradually become uneconomical to operate as dike maintenance and pumping costs rise. At some point in the future, the delta will be of minor agricultural significance, and the environment will be greatly different.

Water quantity has ecological impact in a number of other ways. Flood flows flush out spawning areas, leaving clean new gravels washed out of the hills. Controlling flows by dams prevents both cleaning and renewal. High flow rates sweep debris from river channels and wash down new gravels needed for spawning of many fish. In past years, dilution was considered to be an acceptable "solution to pollution," and the self-purifying capacity of a stream was included in design. This concept is no longer acceptable today. Because all deleterious material is not removed in wastewater treatment, the role of dilution is still significant, and the assimilative capacity must be considered in water quality management.

KEY IDEAS, CONCEPTS, AND ISSUES

- Quantity and quality are the two major issues involved in the use of water.
- Within limits, water quantity is more important than water quality.
- The quantity of water used for municipal purposes is small compared with the quantities used for cooling of power-generation facilities and for agriculture.
- Typically, the design of municipal water distribution systems is based on the peak hourly demand.
- Wastewater, after varying degrees of treatment, is discharged to the environment. To assess the impact of discharging wastewater to the environment, the sources and quantities of flow must be known.
- Massive amounts of water (2.2×10^{11} m^3/yr) are used for irrigation in the United States, especially in the western states. Groundwater is the most common source of irrigation water in the United States.
- The consumptive use by various crops can be estimated with reasonable accuracy using the Blaney–Criddle equation.
- Because ecological water requirements can be significant, special attention should be given to this area in water resources planning.

DISCUSSION TOPICS AND PROBLEMS

1.1. Estimate the required fire flow for a three-story wood-frame apartment building with a total floor area of 300 m^2. Because an automatic sprinkler system is to be installed, assume that the required fire flow can be reduced by 25 percent.

1.2. Estimate the fire flow for a school of fire-resistive construction with a total floor area of 2300 m². Because of unfavorable exposure (e.g., crowding) the required fire flow should be increased by 15 percent.

1.3. Estimate the fire flow for a small shopping center of ordinary, noncombustible, fire-resistive construction. The total floor area is 5000 m². What cost factors must be considered in assessing whether the installation of a sprinkler system is justified?

1.4. Why is Eq. (1.1) no longer favored for the estimation of fire flows in downtown business districts? Present both a qualitative and quantitative analysis.

1.5. In Eq. (1.2) fire flows are computed in units of cubic meters per day. Should this value be considered as a total volume per day or as an instantaneous flow rate? What difference does the interpretation of the rate make with respect to reservoir, pipe, and pump size?

1.6. Determine the origin of the term "fire plug."

1.7. A steel plant is to be constructed at a site where cooling water can be taken from a groundwater aquifer. It is planned to discharge the used cooling water to a small river. Summer flows and temperatures have historically ranged from 30 to 38 m³/s and 17 to 23°C, respectively. The state Department of Environmental Quality has set a maximum stream temperature requirement of 24°C below the proposed discharge. Groundwater temperature is known to be constant at 15°C. Heat dissipation requirements are 2×10^8 J/s. Determine the maximum groundwater pumping rate required.

1.8. A state department of parks is developing a campground in an environmentally fragile area. The objective is to provide access for as wide a range of citizens as possible (that is, the elderly, the physically handicapped, families, and backpackers) without causing irreparable damage to native flora. A decision has been made to limit total occupancy to 200 persons per day, to be distributed between 40 tent or trailer spaces and a 20-unit lodge and adjoining coffee shop and visitor center. Estimate the water demand for the campground.

1.9. Estimate the water demand for a temporary construction camp in a remote area of Saudi Arabia. Because of the temporary nature of the facilities, fire protection will not be provided. Assume the following data are applicable. State all assumptions clearly.

Number of workers = 1500 persons

Number of worker dependents = 650 persons

Support staff (total) = 150 persons

1.10. Estimate the average, peak, and peak-plus-fire-flow demand for a community of 5000 persons with the following characteristics:

ITEM	NO. OF UNITS	TOTAL PERSONS, OTHER	TOTAL FLOOR AREA, m²
Residential			
Apartments	400	800	20,000
Individual homes	1300	4200	1,500,000
Commercial			
Bars and taverns	3		200
Businesses	30		5,500
Fire station	1		100
Motel (40 units)	1		1,000
Offices	30		5,000
Restaurants			
Conventional	2		400
Short-order	2		200
Service stations	4		1,500
Institutional			
Schools	2	1350	5,000
Hospital	1	20 beds	520

1.11. Estimate the percentage reduction that could be achieved in the wastewater flow rates determined in Example 1.3 if (1) low-flush toilets (2 liters per flush), (2) shower-head flow-limiting devices (20 percent reduction), and (3) low-flow controls for washing machines (15 percent reduction) were installed in all of the housing units.

1.12. It has been proposed to develop a new water supply for the Los Angeles area by replacing all of the existing toilets in the area with low-flush toilets (2 liters per flush). Estimate the magnitude of the water supply that would result. If the cost of each new toilet was $150.00 installed, how would the cost of the new water supply compare to the cost of a new water supply in your area. Assume the population of the Los Angeles area is 8 million persons and that there is one toilet for every four persons.

1.13. Residential areas typically have populations of 25 to 50 people per hectare, with 30 being a reasonable figure for communities having parks and commercial areas. Compare community water demand with agricultural usage in an area utilizing irrigation. Note the seasonal nature of agricultural demand.

1.14. Estimate the water lost because of a 1.0-mm-diameter hole in a 150-mm-diameter water main operating under a pressure of 400 kPa. What is the loss at 200 kPa of pressure? For cast-iron pipe, what percentage of the total flow is lost at a pressure of 400 kPa? Assume an orifice coefficient of 0.86, a Manning n of 0.030, and a hydraulic gradient of 0.002.

1.15. Estimate the average, minimum, and maximum water demand rates in your community. Assume that no more than one major fire is occurring.

1.16. Use the curves of Figs. 1.5 and 1.6 to determine a suitable reservoir capacity for a city of 3000 people. Assume that the city is primarily residential with an average of three persons per dwelling unit and that the water treatment facility will operate 8 hr/d. Assume a 20 d emergency storage supply will be required.

1.17. Water demand in most communities fluctuates over a wide range of values throughout a typical day. Discuss the interrelationship of pipe diameter, pipe material (cast iron, concrete, fiber, plastic, etc.), pump size, and storage and treatment plant capacity in meeting variations in demand.

1.18. Application of Eq. (1.3) to a wide range of communities would result in the prediction of peaking factors that were independent of population. Is this a reasonable assumption? Justify your answer.

1.19. In-sink garbage disposals are a common kitchen appliance in the United States. List advantages and disadvantages of their use with respect to environmental quality management. Would use of a sealed, rather than a water-cooled, bearing decrease the water use requirement?

1.20. A small community of 1500 people has a dry-weather wastewater flow of 650 m^3/d and a wet-weather flow ranging up to 6000 m^3/d. What is the probable cause of this wide fluctuation?

1.21. Estimate the wastewater flow from the construction camp described in Problem 1.9.

1.22. Estimate the wastewater flow from the community described in Problem 1.10.

1.23. Obtain weekly dry-weather flow data for the past year from your local wastewater treatment plant. Using these data, estimate the per-capita flow rate for your community or treatment plant service area. How does the computed value compare to the per-capita values given in this text and in two other references? Provide full bibliographic citation information on the references used.

1.24. Using the flow data given in Fig. 1.10, estimate (a) the average flow for the large and small community and (b) the peak-to-average flow ratios. How do the computed peak-to-average values compare with the values that would be obtained using Eq. (1.4) if it is assumed that the average per-capita flow is equal to 300 L/capita · d?

1.25. Estimate the minimum and maximum wastewater flow rates in your community.

1.26. Explain why the peaking factor in Eq. (1.4) decreases with increasing population.

1.27. Briefly discuss the effects of flow-rate peaking on the design of (a) water supply and (b) sewerage systems.

1.28. Mean monthly temperature and precipitation at an orange grove located at N 30° latitude are given below. Estimate the monthly irrigation requirements.

Month	J	F	M	A	M	J	J	A	S	O	N	D
Temperature, °C	13	14	15	17	22	26	28	29	25	20	15	14
Precipitation, mm	112	108	96	82	83	76	71	67	83	91	109	116

1.29. Using Eq. (1.5) and local weather data, estimate the consumptive use for corn in your area. Compare the estimate with local precipitation.

1.30. An arid plateau has been planted intensively in sugar beets (13,000 ha), tomatoes (14,500 ha), and grains (12,000 ha). The only water source is a river with a flow of 400 m^3/s during the summer months. During July, the daytime hours are 11 percent of the total for the year, and the mean temperature is 32°C. Determine the consumptive use and the resulting river flow rate.

1.31. The Colorado River has a mean annual flow rate of 12×10^6 dam^3 (1 dam = 10 m) when it flows into Lake Mead. Lake Mead has a mean surface area of 50,000 ha (1 ha = 10,000 m^2) and an effective annual evaporation of 2.0 m (evaporation minus precipitation). Calculate the mean annual flow leaving Lake Mead. (Courtesy of Jim Hunt.)

1.32. There is considerable concern about the confinement of hazardous wastes in landfills. Such landfills must be designed so that no liquid wastes leave the site. To accomplish a no-flow situation, an on-site reservoir is to be constructed to trap and evaporate all of the runoff. For the following conditions, calculate the reservoir area required to prevent discharge downstream. (Courtesy of Jim Hunt.)

Watershed area = 1000 ha

Annual precipitation = 0.6 m

Fraction of rainfall that runs off land = 0.3

Annual reservoir evaporation = 2.0 m

REFERENCES

1.1. Blaney, H. F., and W. D. Criddle, (1962), *Determining Consumptive Use and Irrigation Water Requirements*, U.S. Department of Agriculture, Technical Bulletin 1275, Washington, D.C.

1.2. Goodrich, R. D., (1942), "Capacity Tests of Ground Water Sources at Laramie, Wyoming," *J. American Water Works Association*, vol. 34, no. 11, p. 1629.

1.3. *Gravity Sanitary Sewer Design and Construction*, (1982), ASCE Manuals and Reports on Engineering Practice No. 60, American Society of Civil Engineers, New York.

1.4. *Guide for Determination of Required Fire Flow*, (1974), 2d ed., Insurance Services Office, New York.

1.5. Hansen, V. E., O. W. Israelsen, and G. E. Stringham, (1979), *Irrigation Principles and Practice*, John Wiley and Sons, New York.

1.6. Linaweaver, F. P., Jr. (1963), "Report on Phase One, Residential Water Use Research Project," Department of Sanitary Engineering, The Johns Hopkins University, Baltimore.

1.7. McKinnon, G. P., and K. Tower, (1976), *Fire Protection Handbook*, 14th ed., National Fire Protection Association, Boston.

1.8. Metcalf & Eddy, Inc., (1981), *Wastewater Engineering: Collection and Pumping of Wastewater*, written and edited by G. Tchobanoglous, McGraw-Hill Book Company, New York.

1.9. Metcalf & Eddy, Inc., (1979), *Wastewater Engineering: Treatment, Disposal, Reuse*, 2d ed., revised by G. Tchobanoglous, McGraw-Hill Book Company, New York.

1.10. Murray, C. R., and E. B. Reeves, (1972), *Estimated Use of Water in the United States in 1970*, U.S. Geological Survey Circular No. 676, Washington, D.C.

1.11. Robie, R. B., "The Delta Decision: Water Rights in a New Era," presented to the Water and Energy Committee of the Los Angeles Chamber of Commerce, February 14, 1972.

1.12. Salvato, J. A., (1982), *Environmental Engineering and Sanitation*, 3d ed., Wiley Interscience Publishers, New York.

1.13. U.S. Water Resources Council, (1978), *The Nation's Water Resources, 1975–2000*, vol. 1, U.S. Government Printing Office, Washington, D.C.

2

Physical, Chemical, and Biological Characteristics of Water

Water quality measures can be classified in a number of ways, but most often are grouped as physical, chemical, and biological. This method of classification will be followed here, but readers should note that two other general categories are obvious: gross and specific. With *gross measures* no distinction is made between the individual physical, chemical, or biological species. Examples of such measures are suspended solids, odor, alkalinity, hardness, and biochemical oxygen demand. In general, the gross parameters are the most easily measured and interpreted, and they are the most commonly used descriptions of water quality. *Specific measures* are necessary when a single characteristic (e.g., toxic organic compound, heavy metal ion, or species of fish) is of concern and, as such, are used to describe water quality as it applies to particular uses.

To provide a perspective for considering the characteristics of water it will be instructive to consider first the hydrologic cycle and its relationship to those characteristics, and then to review the methods used to quantify them.

2.1
THE HYDROLOGIC CYCLE AND THE CHARACTERISTICS OF WATER

The hydrologic cycle as shown in Fig. 2.1 is a schematic representation of the flow of water through the physical environment. Although the cyclic nature of the phenomenon is easily grasped, two factors—time and quality—are often overlooked. The time scale is important because the natural storage of surface water and groundwater can impose significant time delays in the cycle. Lakes have residence times of hundreds of years in many cases, and groundwater basins may be sinks until use provides an outlet. Further, the quality of water at any point in the cycle is a dynamic variable.

43

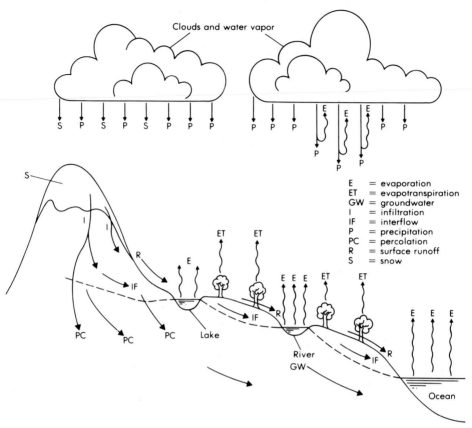

FIGURE 2.1

Schematic representation of the hydrologic cycle.

Water is "pure" only in the vapor state, and impurities begin to accumulate as soon as condensation occurs. Gases dissolve in the droplets forming clouds, and these gases strongly affect water quality. Until recently the most important dissolved gases were oxygen (O_2) and carbon dioxide (CO_2). Increased use of fossil fuels containing sulfur and of internal combustion engines has resulted in the generation of sulfur dioxide (SO_2) and oxides of nitrogen (NO_x). When these gases dissolve in water droplets, mineral acids are produced, and the hydrogen ion content of precipitation may increase by one to three orders of magnitude. Rainwater pH values of 4 have been commonly recorded in Europe, the northeastern United States, and Canada. Because the gases can travel great distances before precipitation occurs, the cause and the effect of acid rain often occur in different countries. Thus control of acid rain is a major political problem (see Chapter 4).

Impurities in Natural Water

Upon reaching the surface, water either percolates into the soil, becoming groundwater, or runs off along the surface in rivulets, streams, and rivers. Minerals dissolve in both groundwater and surface water, but the greater contact with soil and minerals generally results in a higher dissolved-salt concentration in groundwater. Chemical impurities commonly found in water in significant quantities include calcium, magnesium, sodium, potassium, bicarbonate, chloride, sulfate, and nitrate. Trace amounts of other ions such as lead, copper, arsenic, manganese, and a wide spectrum of organic compounds are also common. The organic compounds originate from four principal sources: decaying plant and animal matter, agricultural runoff, wastewater, and the improper management of hazardous wastes discharge. The compounds themselves range from humic materials to the synthetic organics used as detergents, pesticides, herbicides, and solvents. The impurities found commonly in most natural waters are reported in Table 2.1.

Groundwater

In many groundwater basins, the amount of groundwater withdrawn for agricultural irrigation considerably exceeds the rate of recharge. Because water is lost by consumptive use, evapotranspiration, and surface runoff, the level of the groundwater is dropping. At the same time, the concentration of impurities in the groundwater is increasing because of evapotranspiration, the leaching of salts from the soil, and the application of fertilizers. For example, high nitrate concentrations in groundwaters of the western United States are almost universally associated with fertilizer application. Most groundwater basins can be expected to have a useful life of a certain duration, at the end of which salt concentrations will be at a level that severely reduces crop yields. Some methods of groundwater quality management have been developed, but they are in the early stages of application [2.16].

Surface Water

Surface water characteristics also change with time and space. Concentrations of impurities increase because of mineral pickup from surface runoff; silt and debris are carried by surface waters, often resulting in muddy or turbid streams. In areas of slow-moving or stagnant water plants and algae grow, changing the aesthetic characteristics considerably. Surface waters are used for the disposal of most of the world's liquid wastes. Wastes have a major impact on water quality and add greatly to the spectrum of impurities present. Evaporation concentrates these materials, and the residues are deposited in sinks. The major sinks

TABLE 2.1

Summary of the Important Chemical and Biological Impurities Found in Water

ORIGIN	IMPURITY	
	Ionic and Dissolved	
	Positive ions	Negative ions
Contact of water with minerals, soils, and rocks	Calcium (Ca^{+2}) Iron (Fe^{+2}) Magnesium (Mg^{+2}) Manganese (Mn^{+2}) Potassium (K^{+}) Sodium (Na^{+}) Zinc (Zn^{+2})	Bicarbonate (HCO_3^{-}) Carbonate (CO_3^{-2}) Chloride (Cl^{-}) Fluoride (F^{-}) Nitrate (NO_3^{-}) Phosphate (PO_4^{-3}) Hydroxide (OH^{-}) Borates ($H_2BO_3^{-}$) Silicates (H_3SiO_4) Sulfate (SO_4^{-2})
The atmosphere, in rain	Hydrogen (H^{+})	Bicarbonate (HCO_3^{-}) Chloride (Cl^{-}) Sulfate (SO_4^{-2})
Decomposition of organic matter in the environment	Ammonium (NH_4^{+}) Hydrogen (H^{+}) Sodium (Na^{+})	Chloride (Cl^{-}) Bicarbonate (HCO_3^{-}) Hydroxide (OH^{-}) Nitrite (NO_2^{-}) Nitrate (NO_3^{-}) Sulfide (HS^{-}) Organic radicals
Living organisms in the environment		
Municipal, industrial, and agricultural sources and other human activity	Inorganic ions, including a variety of heavy metals	Inorganic ions, organic molecules, color

Source: Adapted in part from Refs. [2.1] and [2.7].

are the ocean, but some inland sinks also exist. Notable among these are the Great Salt Lake in Utah, the Dead Sea in Palestine, and the Caspian Sea on the Iranian-U.S.S.R. border. An interesting sink is the Salton Sea in the Imperial Valley of southern California, formed in 1905 by a major break in the Imperial Canal. Originally viewed as a major disaster, the Salton Sea has become a significant recreational area.

TABLE 2.1   *(Cont.)*

| | IMPURITY | |
Colloidal	Suspended	Gases
Clay Silica (SiO_2) Ferric oxide (Fe_2O_3) Aluminum oxide (Al_2O_3) Magnesium dioxide (MnO_2)	Clay, silt, sand, and other inorganic soils	Carbon dioxide (CO_2)
	Dust, pollen	Carbon dioxide (CO_2) Nitrogen (N_2) Oxygen (O_2) Sulfur dioxide (SO_2)
Vegetable coloring matter, organic wastes	Organic soil (topsoil), organic wastes	Ammonia (NH_3) Carbon dioxide (CO_2) Hydrogen sulfide (H_2S) Hydrogen (H_2) Methane (CH_4) Nitrogen (N_2) Oxygen (O_2)
Bacteria, algae, viruses, etc.	Algae, diatoms, minute animals, fish, etc.	Ammonia (NH_3) Carbon dioxide (CO_2) Methane (CH_4)
Inorganic and organic solids, coloring matter, chlorinated organic compounds, bacteria, worms, viruses	Clay, silt, grit, and other inorganic solids; organic compounds; oil; corrosion products; etc.	Chlorine (Cl_2) Sulfur dioxide (SO_2)

2.2
METHODS OF ANALYSIS

The purpose of this section is to review briefly the methods of analysis used to determine the physical and chemical characteristics of water and wastewater, the units that are used to express the analytical results, and some basic relationships that are useful in the interpretation of analytical results.

Analytical Methods

The methods of analysis used to define the physical and chemical characteristics of water and wastewater samples can be defined as gravimetric, volumetric, and physicochemical.

Gravimetric Methods
As implied by the name, gravimetric analysis is made through a weighing operation. Suspended solids content is the most common parameter determined gravimetrically.

Volumetric Methods
In volumetric methods, the volume of a solution is computed using the following expression, which is based on the principle of conservation of mass:

$$V_1 C_1 = V_2 C_2 \tag{2.1}$$

where

V_1 = volume of solution of known concentration
C_1 = concentration of known solution
V_2 = volume of solution of unknown concentration
C_2 = concentration of unknown solution

Physicochemical Methods
In physicochemical methods, physical properties other than mass and volume are measured. Instrumental methods of analysis such as turbidimetry, colorimetry, potentiometry, polarography, adsorption spectrometry, fluorometry, spectroscopy, and nuclear radiation are representative of the physicochemical methods.

Units of Expression

To develop a clear understanding of the physical and chemical properties of water and wastewater it is necessary to define the units used to express the results of physical and chemical analyses.

Concentration Units
The most commonly used concentration units are as follows:

1. Molality (m):

$$m, \text{mol/kg} = \frac{\text{moles of solute, mol}}{1.0 \text{ kg of solvent}} \tag{2.2}$$

where 1 mol of a compound is equal to the molecular mass expressed in grams.

2. Molarity (M):

$$M, \text{mol/L} = \frac{\text{moles of solute, mol}}{1.0 \text{ L of solution}} \qquad (2.3)$$

3. Mass concentration (*conc*):

$$conc, \text{g/m}^3 = \frac{\text{mass of solute, g}}{1.0 \text{ m}^3 \text{ of solution}} \qquad (2.4)$$

Note that $1.0 \text{ g/m}^3 = 1.0 \text{ mg/L}$ and that both units are in common use.

4. Normality (N):

$$N, \text{eq/L (meq/L)} = \frac{\text{equivalent of solute, eq (meq)}}{1.0 \text{ L of solution}} \qquad (2.5)$$

where

$$\text{equivalent mass, g/eq (mg/meq)} = \frac{\text{atomic (molecular) mass, g(mg)}}{z, \text{ eq (meq)}} \qquad (2.6)$$

(For most compounds z is equal to the number of replaceable hydrogen atoms or their equivalent; for oxidation-reduction reactions z is equal to the change in valence.)

Note that $1.0 \text{ eq/m}^3 = 1.0 \text{ meq/L}$.

5. Parts per million (ppm):

$$\text{ppm} = \frac{\text{mass of solute, g}}{10^6 \text{ g of solution}} \qquad (2.7)$$

Also

$$\text{ppm} = \frac{conc, \text{g/m}^3}{\text{specific gravity of liquid}} \qquad (2.8)$$

Mass Concentration as CaCO$_3$

In much of the older literature dealing with water quality, and even today, the concentrations of the constituents in water, especially of Ca^{+2} and Mg^{+2} (hardness ions), are expressed in terms of calcium carbonate ($CaCO_3$). To convert from milliequivalents per liter of a given ion to milligrams per liter as $CaCO_3$, it is useful to note that

$$\text{milliequivalent mass of } CaCO_3 = \frac{100 \text{ mg/mol}}{2 \text{ meq/mol}}$$

$$= 50 \text{ mg/meq}$$

Thus 2.0 meq/L of calcium, for example, would be expressed as 100 mg/L as $CaCO_3$.

$$Ca \text{ as } CaCO_3 = \frac{2.0 \text{ meq}}{L} \frac{50 \text{ mg } CaCO_3}{1.0 \text{ meq } CaCO_3}$$

$$= 100 \text{ mg/L as } CaCO_3$$

However, because $CaCO_3$ can occasionally have a valance of 1, it is preferable to express the concentrations of the ionic constituents found in water in forms of milligrams per liter and milliequivalents per liter [2.31].

Additional Units

Other units that are often used with liquids include:

1. Density (ρ):

$$\rho, \text{ kg/m}^3 = \frac{\text{mass of solution, kg}}{1.0 \text{ m}^3 \text{ of volume}} \tag{2.9}$$

2. Percent by mass:

$$\% \text{ by mass} = \frac{\text{mass of solute} \times 100}{\text{mass of solute} + \text{solvent}} \tag{2.10}$$

Other Useful Chemical Relationships

Other useful relationships that are commonly used in the chemical analysis of water and in interpretation of the results of such analysis are presented below [2.29, 2.31, 2.34].

Mole Fraction

The ratio of the number of moles of a given solute to the total number of moles of all components in solution is defined as the *mole fraction*. In equation form,

$$x_B = \frac{n_B}{n_A + n_B + n_C + \cdots + n_N} \tag{2.11}$$

where

x_B = mole fraction of solute B

n_B = moles of solute B

n_A = moles of solvent

n_C = moles of solute C

n_N = moles of solute N

The application of Eq. (2.11) is illustrated in Example 2.1.

Electroneutrality

The principle of electroneutrality requires that the sum of the positive ions (cations) must equal the sum of negative ions (anions) in solution; thus

$$\Sigma cations = \Sigma anions \qquad (2.12)$$

where

cations = positively charged species in solution, eq/L
anions = negatively charged species in solution, eq/L

Ionic Strength

In dilute solutions the ions that are present behave independently of each other. However, as the concentration of ions in solution increases, the activity of the ions decreases because of ion interactions. The ionic strength, as defined by G. N. Lewis and M. Randall in 1921, is used as a measure of this interaction. The ionic strength μ of a solution can be determined using Eq. (2.13):

$$\mu = \tfrac{1}{2} \sum_i \left(C_i \times z_i^2 \right) \qquad (2.13)$$

where

C_i = concentration of ionic species i, M
z_i = charge of ionic species i

The Equilibrium Constant

A reversible chemical reaction in which reactants A and B combine to yield products C and D may be written as

$$aA + bB \rightleftharpoons cC + dD \qquad (2.14)$$

When the chemical species come to a state of equilibrium, as governed by the law of mass action, the numerical value of the ratio of the products over the reactants is known as the *equilibrium constant* and is written as

$$\frac{[C]^c [D]^d}{[A]^a [B]^b} = K \qquad (2.15)$$

Brackets are used in Eq. (2.15) to denote molar concentrations. For a given reaction, the value of the equilibrium constant will change with temperature and the ionic strength of the solution.

The Solubility Product

The equilibrium constant for a reaction involving a precipitate and its constituent ions is known as the *solubility product*. For example, the reaction for calcium carbonate ($CaCO_3$) is

$$CaCO_3 \rightleftharpoons Ca^{+2} + CO_3^{-2} \qquad (2.16)$$

Because the activity of the solid phase is usually taken as 1, the solubility product is written as

$$[Ca^{+2}][CO_3^{-2}] = K_{sp} \qquad (2.17)$$

where

K_{sp} = solubility product constant.

Activity Coefficients

When concentrations are above 10^{-4} M, activity coefficients must be considered. The activity of dissolved ions is affected by other ions in solution, and this is accounted for through use of an activity coefficient γ_i. In most field calculations γ_i can be approximated as 1.0, but when more precision is necessary, the Güntelberg equation [2.34] can be used to estimate γ_i:

$$\log \gamma_i = -\frac{0.5(z_i)^2 \mu^{1/2}}{1 + \mu^{1/2}} \quad (\mu < 10^{-1}) \tag{2.18}$$

where

γ_i = activity coefficient for ionic species i

z_i = charge of ionic species i

μ = ionic strength of solution (see Eq. 2.13)

For example, if activity coefficients are used, Eq. (2.16) is written as

$$\left(\gamma_{Ca^{+2}}[Ca^{+2}]\right)\left(\gamma_{CO_3^{-2}}[CO_3^{-2}]\right) = K_{sp} \tag{2.19}$$

However, in dilute aqueous solutions normally encountered in water quality management, activity coefficients are assumed to be equal to 1.

EXAMPLE 2.1

DETERMINATION OF MOLARITY AND MOLE FRACTIONS

Determine the molarity and the mole fraction of a 1-L solution containing 20 g of calcium carbonate ($CaCO_3$). Assume that the density of the $CaCO_3$ is 1.12 g/mL.

SOLUTION:

1. Determine the molarity of the $CaCO_3$ solution.
   a. The moles of $CaCO_3$ are

$$mol_{CaCO_3} = \frac{20 \text{ g}}{100 \text{ g/mol}}$$
$$= 0.20 \text{ mol}$$

   b. The molarity of the $CaCO_3$ solution is

$$M_{CaCO_3} = \frac{0.20 \text{ mol}}{1.0 L}$$
$$= 0.20 M$$

2. Determine the mole fraction of $CaCO_3$.
   a. By definition (see Eq. 2.11)

   $$x_{CaCO_3} = \frac{n_{CaCO_3}}{n_{H_2O} + n_{CaCO_3}}$$

   b. The volume of $CaCO_3$ in the 1-L solution is

   $$V_{CaCO_3} = \frac{20 \text{ g}}{1.12 \text{ g/mL}} = 17.9 \text{ mL}$$

   c. The volume of water is

   $$V_{H_2O} = 1000 \text{ mL} - 17.9 \text{ mL}$$
   $$= 982.1 \text{ mL}$$

   d. The moles of H_2O are

   $$n_{H_2O} = \frac{982.1 \text{ g}}{18 \text{ g/mol}} = 54.56M$$

   e. Substitute for n_{CaCO_3} and n_{H_2O} in Eq. (2.11) as given in Step 2a and solve for x_{CaCO_3}:

   $$n_{CaCO_3} = 0.20M$$
   $$n_{H_2O} = 54.56M$$

   $$x_{CaCO_3} = \frac{0.20}{54.56 + 0.20}$$
   $$= 3.65 \times 10^{-3}$$

COMMENT

Because the mole value for water is so much larger than the combined values of the other constituents found in most waters, the mole fraction of a constituent A is often approximated as $x_A \simeq (n_A/55.56)$.

2.3
SAMPLING

All water quality management decisions are based on the assumption that correct and reasonably accurate and reliable data are available. A management program cannot be developed without knowledge of flow rates, of variation of flow with time, of demand for water, of contaminants contained in the water, of costs of treatment or management operations, and of a myriad other factors and parameters. The most basic, and often most difficult, information to obtain is that related to flow rates and contaminant concentrations. Thus sampling is one of the most basic and important aspects of water quality management. In all cases three questions must be asked: (1) Where should samples be taken? (2) How can representative samples be obtained? (3) What preservation and analytical methods are required to provide reliable and accurate data?

Sample Collection

An understanding of the difficulties and expense of sampling can be developed by considering the river shown in Fig. 2.2. At the point where the picture is taken the width and average depth of the river are approximately 80 and 4 m, respectively. If knowledge of a specific contaminant concentration in the river is needed, the question of where to take the samples must be asked with respect to length, width, and depth. Often contaminants must be traced from their source a long distance downstream. Documentation of contaminant sources requires sampling at a number of points along the length of the river. At each sampling point a "grab" sample might be taken by dropping a container into the river and allowing it to fill, but this would hardly be expected to provide accurate information. Water velocities change considerably with position in the river cross section. Most of the contaminant might be in the high-flow regions or moving down one bank as part of an essentially

FIGURE 2.2

Photograph of the River Seine taken near the Cathedral of Notre Dame, Paris, France.

separate flow resulting from density stratification. Thus a number of samples must be taken at selected depths and positions across the river. Often gathering such samples requires extensive equipment and staff.

Obtaining Representative Samples

Collection of representative samples requires more than taking samples at representative locations. For example, changes of concentration with time are a common occurrence. Wastewater discharge characteristics usually vary widely over the course of a day, and some contaminant concentrations might decrease to unmeasurable values for hours at a time. Determination of characteristics of wastewater requires sampling over time periods when the concentrations are at their peak and at their low values. Often it may be necessary to assign trained staff members to gather several samples per day for a period great enough to provide statistical validity. In the case of wastewater, characteristics change seasonally, and sampling programs must include wet- and dry-weather periods.

Preservation and Analysis of Samples

Analysis of samples raises problems concerning accuracy and reliability. Some measurements can or must be made on-site (Fig. 2.3). Among these are temperature; pH; dissolved-oxygen concentration; and flow rate, or

FIGURE 2.3

Field testing for hydrogen sulfide in sewers.

velocity. However, analysis for many contaminants requires taking samples to a laboratory. Where possible, a mobile laboratory can be used, but in any case care of samples between the time they are obtained and the time they are analyzed is a major concern. Organic materials can be oxidized biologically if left at room temperature and neutral pH. If samples are left unsealed, volatile materials evaporate. Many contaminants break down or react with other materials in the sample. Thus attention must be paid to maintaining stored samples until they can be analyzed. It should be noted that the sampling operation is often much faster than analysis, and samples must be stored for from several hours to several days. Recommended methods of sample preservation and analysis are detailed in Refs. [2.32] and [2.40].

Statistical Considerations

Statistically valid sampling methods should always be used. Numbers of samples and replications, sampling locations, and parameters analyzed can be selected with statistical techniques [2.32]. Variance and range among replicates as well as mean values should be reported so that judgment can be made about accuracy of the results. Many analytical methods are known to have relatively large standard deviations. It is important to know and recognize this fact in making management and design decisions. Ultimately judgment will have to be used in the interpretation of results because it will seldom be possible to obtain all of the data needed or desired.

2.4
PHYSICAL CHARACTERISTICS OF WATER

Most of our impressions of water quality are based on physical rather than on chemical or biological characteristics. We expect water to be clear, colorless, and odorless. Most natural waters are at best cloudy; they are often colored by tannins and other organic materials picked up from decaying plants; and backwaters, sloughs, and swamps are noted for their characteristic odors. Common analyses used to assess the physical impurities in water and wastewater are reported in Table 2.2. Quantitative measurement of these characteristics is necessary for the determination of water quality.

Turbidity

Perhaps the first thing that is noticed about water is its clarity. Very clear natural water allows images to be seen distinctly at considerable depths. The clarity of the water in Lake Tahoe, located on the California-Nevada

TABLE 2.2

Common Analyses Used to Assess the Physical Impurities in Water and Wastewater*

TEST	ABBREVIATION/ DEFINITION	USE
Turbidity	NTU	To assess the clarity of water
Solids		
Total solids	TS	To assess the reuse potential of a
Total volatile solids	TVS	wastewater and to determine the
Suspended solids	SS	most suitable type of process for its
Volatile suspended solids	VSS	treatment; TDS tests assess suitability
Total dissolved solids	TDS	of water sources for public, industrial,
(TS − SS)		and agricultural uses
Settleable solids	mL/L	To determine those solids that will settle by gravity in a specified time period; test data used for design of sedimentation facilities
Color	Various light-yellow hues	To assess the presence of natural and synthetic coloring agents in water
	Light brown, grey, black	To assess the condition of wastewater (fresh or septic)
Odor	MDTOC[†]	To determine if odors will be a problem
Temperature	°C	To design and carry on biological and other treatment processes; to determine the saturation concentrations of various gases

*Detailed procedures for the tests listed in this table are presented in Ref. [2.32].
[†]Minimum detectable threshold odor concentration.

border, is famous. In that lake a Secchi disk, a black and white disk used to assess light penetration, can be distinguished at a depth of 35 m (Fig. 2.4), although this clarity is rapidly changing because of increased turbidity and eutrophication resulting from shoreline development. When colloidal particles accumulate, light is scattered and the water appears turbid. Measurement of turbidity (turbidimetry) is, not surprisingly, accomplished by determining light transmission using standard light sources. The test has little meaning except in relatively clear waters but is very useful in defining drinking-water quality in water treatment.

Solids

All contaminants of water, other than dissolved gases, contribute to the solids load. Solids can be classified by their size and state, by their chemical characteristics, and by their size distribution.

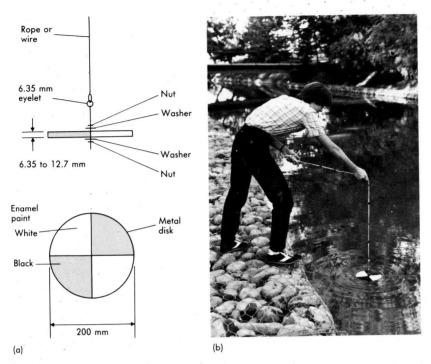

FIGURE 2.4

Secchi disk used to assess light penetration in water: (a) construction details and (b) application.

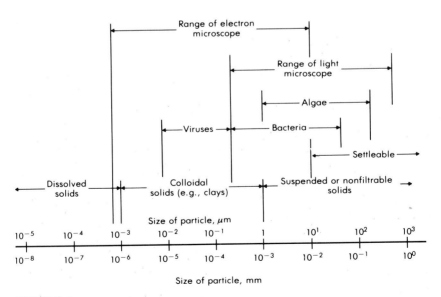

FIGURE 2.5

Particle sizes in water and wastewater.

Sizes of Solids

The solid and semisolid materials in water and wastewater can be classified by their size and state as settleable, suspended, colloidal, or dissolved (Fig. 2.5). Settleable solids are determined by noting the quantity of solids that have settled in an Imhoff cone (Fig. 2.6) after 1 hr. The difference between suspended and settleable solids is a practical one. Suspended solids do not settle out in the Imhoff cone test and would not be expected to be removed by conventional gravity settling. Clays are typical colloidal suspended solids. In practice, the term colloidal solids is rarely used. Instead, solids are divided into two broad groupings: *dissolved* (including colloidal and small suspended particles) and *suspended* (including settleable). The distinction is made using a membrane filter with a pore size of about 1.2 μm. Any particle passing the filter is considered dissolved, and any particle retained is considered suspended (see Example 2.2). In some cases, the more accurate terms, *filtrable* and *nonfiltrable*, are now being used. The sum of dissolved and suspended (filtrable and nonfiltrable) solids is the *total solids* content. The interrelationship of the various types of solids is depicted in Fig. 2.7.

Chemical Characteristics of Solids

Solids are also characterized as being *nonvolatile* or *volatile* (in some texts the terms "fixed" and "volatile" are used). Volatile solids are by definition those solids that volatilize at a temperature of 550°C. In many cases volatile solids are considered to be organic, and thus the test is sometimes used to estimate the organic/inorganic characteristics of the solids.

FIGURE 2.6

Imhoff cone used to determine settleable solids in wastewater. Solids that accumulate in the bottom of the cone are reported as mL/L.

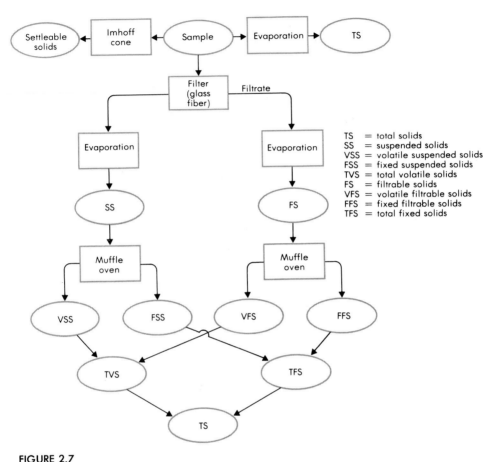

FIGURE 2.7

Interrelationships of solids found in water and wastewater. In much of the water quality literature, the solids passing through the filter are called dissolved solids.

Distribution of Filtrable Solids

Typical data on the particle sizes found in the water and wastewater samples are reported in Table 2.3. The percentage distribution of filtrable solids by size range, reported in the table, is based on the mass of solids retained on a filter with a pore size of 0.1 μm. Percentages are used so that comparisons can be made between samples.

In the river and lake samples both the total mass and distribution of filtrable solids will vary depending on the time of year. In the untreated wastewater samples, these parameters will vary with the time of day. For wastewater, as shown in Table 2.3, the mass distribution of solids is bimodal. It is interesting to note that, in untreated wastewater, from 15 to 30 mg/L of filtrable solids will be found in the size range between 0.1

TABLE 2.3

Typical Data on the Distribution of Filtrable Solids in Various Water and Wastewater Samples

SAMPLE (date, time)	CONC.,* mg/L	PERCENT OF MASS RETAINED IN SIZE RANGE*					
		> 0.1 < 1.0	> 1.0 < 3.0	> 3.0 < 5.0	> 5.0 < 8.0	> 8.0 < 12.0	> 12.0
Surface water[†]							
Sacramento River (8/9/83, 4 P.M.)	56.3	36.8	1.95	0	19.7	0	41.6
American River (8/15/83, 2 P.M.)	8.8	28.4	47.7	0	4.0	11.4	8.5
Lake Berryessa (8/15/83, 11 A.M.)	13.5	51.8	21.5	0	3.7	23.0	0
Folsom Lake (8/15/83, 10 A.M.)	3.5	50.0	0	0	28.6	21.4	0
Untreated wastewater[‡]							
UCD (7/14/82, 11 A.M.)	62.2	12.5	12.9	5.8	3.8	6.1	58.8
UCD (7/21/82, 4 P.M.)	129.9	16.1	25.1	0	0	0	58.8
Las Vegas (8/3/83, 8:30 P.M.)	146.1	14.2	32.4	6.9	0	6.5	40.0
Las Vegas (8/3/83, 2 P.M.)	284.0	1.8	32.6	11.5	11.1	1.8	41.2
Los Banos (8/8/83, 2 P.M.)	268.0	20.5	18.7	6.7	3.0	10.1	41.0

*Mass of solids retained on a filter with a pore size of 0.1 μm.

†Samples collected in the vicinity of Sacramento, Calif.

‡Samples collected at University of California, Davis, Las Vegas, Nev., and Los Banos, Calif.

Source: Adapted from Ref. [2.36].

(a)   (b)

(c)   (d)

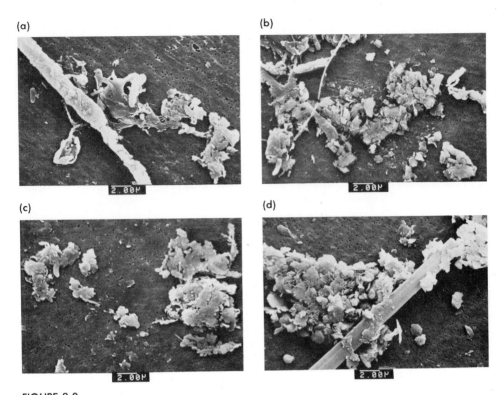

FIGURE 2.8

Micrographs of filtrable solids (in surface waters) retained on a membrane filter with a pore size of 0.1 μm: (a) American River, (b) Folsom Lake, (c) Sacramento River, and (d) Lake Berryessa. The length of the bar shown is 2 μm.

Source: Courtesy of Audrey Levine, University of California, Davis, CA.

and 1.0 μm. In most cases, a large fraction of these solids is not measured in the conventional suspended solids test because the pore size of the Whatman GFC filter used in the test is about 1.2 μm [2.36].

Micrographs of Solids

Micrographs of the particles found in water and wastewater are presented in Figs. 2.8 and 2.9. The size of the individual particles can be estimated using the length of the bar shown in the micrographs which correspond to 2 μm. In the surface water samples shown in Fig. 2.8, the material found is primarily inorganic, with some freshwater microorganisms present. Wastewater samples from which the settleable solids have been removed are shown in Fig 2.9. The solids have been separated into two fractions. It is important to note the amorphous nature of the

(a)

(b)

(c)

(d)

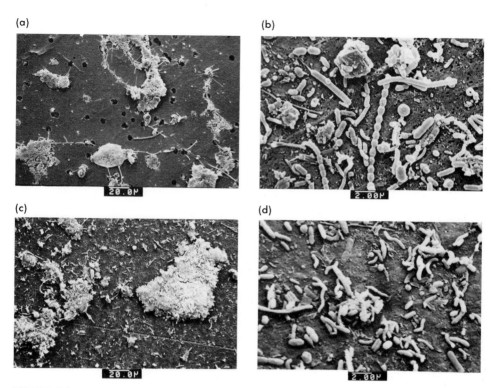

FIGURE 2.9

Micrographs of filtrable solids in settled wastewater taken from (a) and (b) the UC Davis treatment plant and (c) and (d) the Las Vegas, Nev. treatment plant. (a) and (c) Solids retained on a membrane filter with a pore size of 5.0 μm. (b) and (d) Solids retained on a membrane filter with a pore size of 0.1 μm prefiltered through a membrane filter with a pore size of 5.0 μm. The lengths of the bars shown are 20 μm for (a) and (c) and 2 μm for (b) and (d).

Source: Courtesy of Audrey Levine, University of California, Davis, CA.

material retained on the filter with a pore size of 5.0 μm, as compared with the material retained on the filter with a pore size of 0.1 μm.

EXAMPLE 2.2

DETERMINATION OF SUSPENDED SOLIDS

The suspended solids of a river sample are determined gravimetrically using a glass-fiber filter. Determine the suspended solids based on the following laboratory results:

$$Sample\ size = 200\ mL$$
$$Tare\ mass\ of\ filter = 1.3255\ g$$
$$Mass\ of\ filter + solids = 1.3286\ g$$

SOLUTION:

1. Determine the mass of solids retained on the filter.

$$
\begin{array}{r}
\text{Mass filter} + \text{solids} = 1.3286 \text{ g} \\
- \text{Mass filter} = \underline{1.3255 \text{ g}} \\
\text{Difference} = 0.0031 \text{ g}
\end{array}
$$

2. Determine the suspended solids in milligrams per liter.

$$0.0031 \text{ g} = 3.1 \text{ mg}$$

$$SS = \frac{3.1 \text{ mg} \times 1000 \text{ mL/L}}{200 \text{ mL of sample}}$$

$$= 15.5 \text{ mg/L}$$

Odors

Odors associated with water usually result from the presence of decaying organic matter or, in the case of mineral springs, the reduction of sulfates by bacteria to H_2S gas. Decaying organic matter may accumulate in bottom deposits large enough to provide suitable conditions for the anaerobic bacteria that produce noxious gases. Sources of the organics include plant debris washed into streams, dead animals, microorganisms, and wastewater discharges.

In addition to H_2S, which has the classic odor of rotten eggs, a number of other distinct smells are commonly encountered. They are listed in Table 2.4. These odors often occur together, making individual smells difficult to separate and identify. Detection of odors and quantifying of level is very difficult. Individuals have greatly varying sensitivity to odors, and continual exposure to an odor tends to decrease sensitivity.

TABLE 2.4

Categories of Offensive Odors Commonly Encountered in Water

COMPOUND	TYPICAL FORMULA	DESCRIPTIVE QUALITY
Amines	$CH_3(CH_2)_n NH_2$	Fishy
Ammonia	NH_3	Ammoniacal
Diamines	$NH_2(CH_2)_n NH_2$	Decayed flesh
Hydrogen sulfide	H_2S	Rotten egg
Mercaptans	$CH_3SH; CH_3(CH_2)_n SH$	Skunk secretion
Organic sulfides	$(CH_3)_2S; CH_3SSCH_3$	Rotten cabbage
Skatole	$C_8H_5NHCH_3$	Fecal

Source: From Ref. [2.24].

Thus persons working next to an odor source may be less sensitive to it than their occasional guests. On the other hand, knowing that a particular odor may be present increases the probability of its detection. Thus homeowners near an odor source may notice occasional releases at lower levels than nonresidents, a factor making life more difficult.

Odor measurement methods (ASTM Method D1391-51) are designed to determine the minimum detectable threshold odor concentration (MDTOC) on a dilutions basis. Tests are difficult to administer and are liable to considerable subjectivity.

Temperature

Temperature affects a number of important water quality parameters. Chemical and biochemical reaction rates increase markedly with temperature. Gas solubility decreases and mineral solubility increases with temperature. Growth and respiration rates of aquatic organisms increase and decrease with temperature, and most organisms have distinct temperature ranges within which they reproduce and compete.

Lakes vary in temperature from surface to bottom, and fish populations vary accordingly. Cold-water species such as trout stay in the depths, while warm-water species such as bass are found in shallow regions near the edges. Downstream populations can be modified by reservoir releases. Water released from the reservoir surface will be warmer and promote warm-water organisms downstream. Releasing water from the bottom will have the opposite effect. The results go well beyond selection of aquatic species. Where the water is used for irrigation, its temperature must be high enough to induce germination of seeds.

Color

Many of the colors associated with water are not true colors but the result of colloidal suspension; tea is an example of such "coloration." True colors result from dissolved materials, most often organics. Most colors in natural waters result from dissolved tannins extracted from decaying plant material. The result is a slightly brownish tint. Many industrial wastes are colored and if not properly treated can impart color to the receiving stream.

2.5

CHEMICAL CHARACTERISTICS OF WATER: INORGANIC MATTER

Chemical measures of water quality include analysis for the presence of specific ions such as calcium, magnesium, and lead. Gross chemical measures such as alkalinity and hardness are also used to define water quality. Most of the common water quality measures reflect combinations of or interactions between ions. The important chemical constituents of water have been summarized previously in Table 2.1. Common analyses used to assess the chemical characteristics of water and wastewater are reported in Tables 2.5 and 2.6, respectively. It is the purpose of this section to examine in some detail the specific constituents found in water and wastewater and some of the gross measures commonly used to assess the chemical characteristics reported in Tables 2.5 and 2.6.

Major Ionic Species in Water

All natural waters contain dissolved ionic constituents. Based on numerous analyses of surface waters and groundwaters from all over the country, it has been found that the following ionic species represent the principal chemical constituents present in most waters.

CATIONS	ANIONS
Calcium (Ca^{+2})	Bicarbonate (HCO_3^-)
Magnesium (Mg^{+2})	Sulfate (SO_4^{-2})
Sodium (Na^+)	Chloride (Cl^-)
Potassium (K^+)	Nitrate (NO_3^-)

Typically the ionic species that are present will be derived from the contact of the water with various mineral deposits. The most abundant species are the bicarbonates, sulfates, and chlorides of calcium, magnesium, and sodium. The sources of these constituents are summarized briefly in Table 2.7. The distribution of these species will, of course, vary with geographic location and the residence-time history of the water. Potassium, usually present in small amounts, is derived from soil minerals, from decaying organic matter, and from the ashes of burned plants and trees. Nitrate is usually present in small amounts (see discussion of nitrogen compounds later in this section).

In the field of water supply, the above constituents are normally measured when water analyses are performed. In all cases, the results of such analyses should be checked for completeness and accuracy using a cation-anion balance [see Eq. (2.12)]. If significant differences exist, it can be assumed that an error was made in the analysis of the individual constituents or that one or more ionic species were neglected in the analysis. If the error in the charge balance computed using Eq. (2.12) exceeds acceptable limits, the results of the analysis should be reevaluated. Acceptable limits for charge balances for chemical analyses can be

TABLE 2.5
Common Tests Used to Assess the Specific and Gross Chemical Characteristics of Waters for Public Water Supplies*

TEST	ABBREVIATION/DEFINITION	USE
Inorganic constituents		
Dissolved cations[†]		
Calcium	Ca^{+2}	
Magnesium	Mg^{+2}	To determine the ionic chemical composition of water and to assess the suitability of
Potassium	K^+	water for most alternative uses
Sodium	Na^+	
Dissolved anions[†]		
Bicarbonate	HCO_3^-	
Carbonate	CO_3^-	
Chloride	Cl^-	
Hydroxide	OH^-	
Nitrate	NO_3^-	
Sulfate	SO_4^{-2}	
pH	$pH = \log 1/[H^+]$	To measure the acidity or basicity of an aqueous solution
Alkalinity	$\Sigma(HCO_3^- + CO_3^{-2} + OH^-)$	To measure the capacity of the water to neutralize acids
Acidity		To measure the amount of a basic substance required to neutralize the water
Carbon dioxide	CO_2	To assess the corrosiveness of the water and the dosage requirements where chemical treatment is to be used; can be used to estimate pH if the bicarbonate concentration is known
Hardness	Σ(Multivalent cations)	To measure the soap-consuming capacity and scale-forming tendency of the water
Conductivity (at 25°C)	$\mu S/cm$[‡]	To estimate the total dissolved solids or check on the results of a complete water analysis (TDS in $g/m^3 = 0.55$ to $0.7 \times$ conductivity value of sample in $\mu S/cm$)
Radioactivity	Ci	To estimate the presence of radioactive substances
Organic constituents		
Total organic carbon	TOC	To assess the presence of organic matter
Specific organic compounds		To determine the presence of pesticides, solvents, and other organic compounds

*Detailed procedures for the tests listed in this table are presented in Ref. [2.32].

[†] Major cations and anions found in most natural waters.

[‡] Microsiemens per centimeter.

67

TABLE 2.6

Common Tests Used to Assess the Specific and Gross Chemical Characteristics of Wastewater*

TEST	ABBREVIATION/ DEFINITION	USE
		Inorganic constituents
Calcium, magnesium, sodium	Ca^{+2}, Mg^{+2}, Na^+	To determine the sodium adsorption ratio where agricultural reuse is being considered
Chloride	Cl^-	To assess the suitability of the wastewater for agricultural reuse
Nitrogen		
Organic nitrogen	Org N	
Free ammonia	NH_3	
Nitrites	NO_2^-	
Nitrates	NO_3^-	
Total Kjeldahl nitrogen[†]	TKN	To measure the nutrients present and the degree of decomposition in the wastewater; oxidized forms can be taken as a measure of the degree of oxidation
Phosphorus		
Total phosphorus	TP	
Organic phosphorus	Org P	
Inorganic phosphorus, principally PO_4^{-3}	Inorg P	
Sulfate	SO_4^{-2}	To assess the treatability of the waste sludge
Alkalinity	$\Sigma(HCO_3^- + CO_3^{-2} + OH^-)$	To measure the buffering capacity of the wastewater
pH	$pH = \log 1/[H^+]$	To measure the acidity or basicity of an aqueous solution
Heavy metals		To assess the suitability of the wastewater for reuse and for toxicity effects in treatment and receiving waters
Trace elements		May be important in biological treatment and for agricultural use

TABLE 2.6 *(Cont.)*

Organic constituents

Five-day carbonaceous biochemical oxygen demand	BOD_5 ($CBOD_5$)	To measure the amount of oxygen required to stabilize a waste biologically
Ultimate carbonaceous biochemical oxygen demand	BOD_u ($CBOD_u$)	
Nitrogenous oxygen demand	NOD	To measure the amount of oxygen required to oxidize biologically the nitrogen in the wastewater to nitrate
Chemical oxygen demand	COD	To measure the amount of oxygen required to stabilize the waste completely
Total organic carbon	TOC	Often used as a substitute for the BOD_5 test
Specific organic compounds		To determine the presence of pesticides, phenols, surfactants, grease, and other organic materials

Gases

Oxygen	O_2	To assess the performance of aeration facilities
Hydrogen sulfide	H_2S	To assess the production of odors

*Detailed procedures for the tests listed in this table are presented in Ref. [2.32].

†Total Kjeldahl nitrogen = organic nitrogen + ammonia nitrogen.

TABLE 2.7

Sources of Bicarbonates, Sulfates, and Chlorides of Calcium, Magnesium, and Sodium Found in Natural Waters

CONSTITUENT	SOURCE
Calcium bicarbonate [$Ca(HCO_3)_2$]	Dissolution of limestone, marble, chalk, calcite, dolomite, and other minerals containing calcium carbonate
Magnesium bicarbonate [$Mg(HCO_3)_2$]	Dissolution of magnesite, dolomite, dolomitic limestone, and other minerals containing magnesium carbonate
Sodium bicarbonate [$Na(HCO_3)$]	White salt commonly known as baking soda, typically a manufactured product; also present in some natural waters
Calcium sulfate ($CaSO_4$)	Minerals such as gypsum, alabaster, and selenite
Magnesium sulfate ($MgSO_4$)	Heptahydrate form ($MgSO_4 \cdot 7H_2O$) commonly known as Epsom salt or, when found in salt beds or mines, as epsomite; monohydrate form ($MgSO_4 \cdot H_2O$) occurs in a variety of minerals as a double salt with potassium chloride, potassium sulfate, etc.
Sodium sulfate (Na_2SO_4)	Salt lakes, salt beds, caverns, etc.; decahydrate form ($Na_2SO_4 \cdot 10H_2O$) is known as Glauber's salt
Calcium chloride ($CaCl_2$)	Natural brines, salt beds, etc., and a by-product of the chemical industry
Magnesium chloride ($MgCl_2$)	Anhydrous forms found in natural brines, salt beds, etc.
Sodium chloride ($NaCl$)	Salt beds, salt lakes, connate waters, other natural brine

Source: Adapted from Ref. [2.26].

determined using Eq. (2.20), which is derived from a consideration of the standard deviations for routine chemical analyses [2.32].

$$|\Sigma anions - \Sigma cations| \leq (0.1065 + 0.0155\Sigma anions) \qquad (2.20)$$

The preparation of a cation-anion balance and the application of Eq. (2.20) are illustrated in Example 2.3.

EXAMPLE 2.3

CHECKING THE ACCURACY OF WATER ANALYSES

Determine the acceptability of the following water analysis submitted by a commercial analytical laboratory.

CATION	CONC, mg/L
Ca^{+2}	93.8
Mg^{+2}	28.0
Na^+	13.7
K^+	30.2

ANION	CONC, mg/L
HCO_3^-	164.7
SO_4^{-2}	134.0
Cl^-	92.5

SOLUTION:

1. Prepare a cation-anion balance using Eq. (2.12).

	CATION				ANION		
Ion	Conc. mg/L	mg/meq*	meq/L	Ion	Conc, mg/L	mg/meq*	meq/L
Ca^{+2}	93.8	20.0	4.69	HCO_3^-	167.4	61.0	2.74
Mg^{+2}	28.0	12.2	2.30	SO_4^{-2}	134.0	48.0	2.79
Na^+	13.7	23.0	0.60	Cl^-	92.5	35.5	2.61
K^+	30.2	39.1	0.77				
Total			8.36				8.14

*See Eq. (2.6).

2. Check accuracy using Eq. (2.20).

$$|\Sigma anions - \Sigma cations| \leq (0.1065 + 0.0155\Sigma anions)$$
$$|8.14 - 8.36| \leq [0.1065 + 0.0155(8.14)]$$
$$0.22 \leq (0.2327)$$

Because the accuracy is within the allowable limits, the water analysis is acceptable.

Minor Ionic Species in Water

Minor ionic species that will be found in water include:

CATIONS	ANIONS
Aluminum (Al^{+3})	Bisulfate (HSO_4^-)
Ammonium (NH_4^+)	Bisulfite (HSO_3^-)
Arsenic (As^+)	Carbonate (CO_3^{-2})
Barium (Ba^{+2})	Fluoride (F^-)
Boron, as borate (BO_4^{-3})	Hydroxide (OH^-)
Copper (Cu^{+2})	Phosphate, mono- ($H_2PO_4^-$)
Iron, ferrous (Fe^{+2})	Phosphate, di- (HPO_4^{-2})
Iron, ferric (Fe^{+3})	Phosphate, tri- (PO_4^{-3})
Manganese (Mn^{+2})	Sulfide (S^{-2})
	Sulfite (SO_3^{-2})

As with the major ionic species, most of minor ionic species are derived from the contact of water with various mineral deposits. In addition, some of the minor constituents such as ammonium, carbonate, and sulfide may be present because of bacterial and algal activity (see Section 2.7).

Nonionic Species in Water

The principal nonionic mineral found in all natural surface waters and groundwaters is silica, usually expressed as SiO_2. When reference is made to SiO_2 it is assumed that the soluble form is being considered, as opposed to the silica that may be present in the suspended solids. The presence of silica in water is troublesome, especially in industrial applications, where it causes severe scaling problems in boilers and heat exchangers. Reported values of silica range from 1 to 120 mg/L.

Inorganic Species Added by Humans

In addition to the major and minor ionic species found in natural waters, a variety of inorganic species (principally heavy metals) of anthropogenic origin may also be found. The more important of these are shown in the table on the following page.

These constituents are of concern primarily because of their toxicity to microorganisms, plants, and animals. Typically the presence of these constituents is due to the discharge of improperly processed industrial wastes, and high concentrations are often found in wastewater sludges. For example, plating wastes containing chromium and cyanide are common.

CATIONS	ANIONS
Arsenic (As^{+3})	Cyanide (CN^-)
Barium (Ba^{+2})	
Cadmium (Cd^{+2})	
Chromium (Cr^{+3})	
Chromium (Cr^{+6})	
Lead (Pb^{+2})	
Mercury (Hg^{+2})	
Selenium (Se)	
Silver (Ag^{+2})	
Zinc (Zn^{+2})	

Nutrients Added by Humans

Nitrogen and phosphorus are essential for the growth of plants and animals. For this reason these elements are often identified as nutrients, or biostimulants, when discharged in wastewater to surface waters. It should be noted that both organic and inorganic forms of these constituents are of importance. The organic forms are considered further in Section 2.6.

Nitrogen

Nitrogen is a complex element that can exist in seven states of oxidation. From a water quality standpoint, the nitrogen-containing compounds that are of most interest are organic nitrogen; ammonia (NH_3); nitrite (NO_2^-); nitrate (NO_3^-); urea [$CO(NH_2)_2$]; and dinitrogen, or nitrogen gas (N_2).

In nature, nitrogen is cycled between its organic and inorganic forms. Bacteria and plants are responsible for the production of proteins (organic compounds containing nitrogen) from several inorganic forms of nitrogen, including NO_3^-, N_2, and NH_3. Animals, including humans, cannot utilize nitrogen from the atmosphere or from inorganic compounds to produce proteins but must obtain nitrogen in organic form. In turn, proteins are broken down by bacterial activity to urea and NH_3. Urea is also converted enzymatically to NH_3. Ammonia from these two sources is then oxidized bacterially, first to NO_2^- and then to NO_3^-. Nitrite and nitrate can also be converted by bacteria to nitrogen gas. This sequence of events is identified as the *nitrogen cycle*, considered further in Section 3.3. The importance of the various forms of nitrogen will be addressed in numerous locations throughout the remainder of this chapter and in Chapters 3 and 4.

Phosphorus

Phosphorus, like nitrogen, is of great importance in water supply systems and in the aquatic environment. Phosphorous compounds are used for corrosion control in water supply and industrial cooling water systems and in the production of synthetic detergents. As a consequence, the concentration of phosphorus in treated municipal wastewaters has increased from 3 to 4 mg/L in predetergent days to the present values of 10 to 20 mg/L. Because phosphorus is an essential element for the growth of algae and other aquatic organisms, serious problems have resulted when the excess phosphorus in treated effluents is discharged to the aquatic environment. Because of the many problems associated with uncontrolled algal blooms, the discharge of phosphorus to ecologically sensitive waters is now controlled.

The phosphorus-containing compounds of interest with respect to water quality include [2.29]:

Orthophosphates
   Trisodium phosphate (Na_3PO_4)
   Disodium phosphate (Na_2HPO_4)
   Monosodium phosphate (NaH_2PO_4)
   Diammonium phosphate [$(NH_4)_2HPO_4$]

Polyphosphates
   Sodium hexametaphosphate [$Na_3(PO_3)_6$]
   Sodium tripolyphosphate ($Na_5P_3O_{10}$)
   Tetrasodium pyrophosphate ($Na_4P_2O_7$)
   Organic phosphorus

Analytically, orthophosphates can be determined using gravimetric, volumetric, or physicochemical methods. Polyphosphates and organic phosphorus must be converted to orthophosphate before they can be measured [2.32].

Radioactivity in Water

Many natural waters contain low levels of radioactivity, especially waters pumped from great depths. Radioactive substances of anthropogenic origin may also be found in surface water as a result of (1) the worldwide development of the nuclear power industry, (2) the use of radioactive isotopes in medicine and industry, and (3) the past and continued testing of nuclear weapons. With the increasing amounts of radioactive substances found in the environment, it is important to consider the effects of radioactive impurities on water quality, especially in water used for drinking and food preparation.

All of the elements have one or more radioisotopes. Some elements may have several stable forms, others have only one stable form, and a few have only radioactive isotopes. All isotopes, whether of natural or anthropogenic origin, decay by the emission of alpha, beta, or gamma radiation. Alpha rays are charged high-energy particles composed of two protons and two neutrons. Upon acquiring electrons from the environment, alpha rays become helium atoms. Beta rays consist of electrons or positrons (the positive counterpart of electrons). Gamma and X rays are high-energy photons that have no electric charge and travel at the velocity of light. Common radioactive substances, often called *radionuclides*, include iodine 131, strontium 90, cesium 137, and radium 226.

The unit of radioactivity is the *curie* (*Ci*), which corresponds to 3.7×10^{10} atom disintegrations per second (the activity of 1 g of radium). Measurements of radioactivity are made with some type of area monitor, the most common being the Geiger-Müller counter, which is used to count the number of radioactive disintegrations per unit of time. In reporting radioactivity levels a distinction is also made between mass activity (per unit of mass) and volume activity (per unit of volume). In 1975, the U.S. Environmental Protection Agency issued proposed maximum levels for various radioactive substances in water. Exposure limits for synthetic radioisotopes were also set. Representative radioactive concentration levels for water are as follows:

RADIOACTIVE SOURCE	MAXIMUM CONCENTRATION, pCi/L*
Strontium 90	10
Radium 226	5
Gross beta concentration (in the absence of strontium 90 and alpha emitters)	1000

*One picocurie (pCi) $= 3.7 \times 10^{-2}$ atom disintegrations per second.

In general, the levels of radioactivity in most waters are within the limits established by the EPA and other agencies, including the World Health Organization. Where the removal of radioactivity is necessary, it has been found that conventional water and wastewater treatment plants are quite effective. It is important to note, however, that although radioactive substances may be removed, they remain in the sludges and resins used for their removal. As a consequence, waste materials containing radioactive substances require special handling and disposal methods.

pH (Hydrogen Ion Concentration)

When water ionizes, the following simplified relationship applies:

$$H_2O \rightleftharpoons H^+ + OH^- \tag{2.21}$$

The corresponding equilibrium expression for this reaction is

$$\frac{[H^+][OH^-]}{[H_2O]} = K \tag{2.22}$$

where K is the equilibrium constant and the brackets are used to denote molar concentrations. Because the molar concentration of water is essentially constant, it is incorporated into the equilibrium constant. Thus the ion concentration product for water is defined as follows:

$$[H^+][OH^-] = K_w \tag{2.23}$$

where

K_w = equilibrium constant for water.

To satisfy the principle of electroneutrality, the positive ions in solution must be balanced by the negative ions. Thus

$$[H^+] = [OH^-] \tag{2.24}$$

Substituting for OH^- [from Eq. (2.24)] in Eq. (2.23) yields

$$[H^+]^2 = K_w \tag{2.25}$$

Taking the negative logarithm of Eq. (2.25) yields

$$-\log[H^+] = -\tfrac{1}{2}\log K_w \tag{2.26}$$

Using the pK notation from chemistry where

$$pK = -\log K \tag{2.27}$$

the pH of water is defined as

$$pH = -\log[H^+] = \log\frac{1}{[H^+]} = \frac{1}{2}pK_w \tag{2.28}$$

In dilute aqueous solutions the value of K_w at 25 °C is equal to 10^{-14}. Substituting a value of 10^{-14} for K_w in Eq. (2.28), the pH of water is

$$pH = -\log[H^+] = \tfrac{1}{2}pK_w = -\tfrac{1}{2}\log 10^{-14}$$
$$= 7.0$$

Values of K_w as a function of temperature are given in Table 2.8.

When contaminants having ionizable H^+ or OH^- groups dissolve in water, the equilibrium between H_2O, H^+, and OH^- shifts and the pH value increases (becomes more basic) or decreases (becomes more acidic).

The hydrogen ion concentration is important because it affects chemical reactions. As will be shown in the following sections, equilibrium relationships are strongly affected by pH, and many biological systems function only in relatively narrow pH ranges (typically 6.5 to 8.5).

TABLE 2.8

Equilibrium Constants for Water

T, °C	K_w, mol^2/L^2	pH OF WATER AT GIVEN TEMPERATURE
0	1.13×10^{-15}	7.47
5	1.83×10^{-15}	7.37
10	2.89×10^{-15}	7.27
15	4.46×10^{-15}	7.18
20	6.75×10^{-15}	7.09
25	1.00×10^{-14}	7.00
30	1.45×10^{-14}	6.92
35	2.07×10^{-14}	6.84
40	2.91×10^{-14}	6.77

Source: From Ref. [2.6].

pH/Concentration Relationships and Logarithmic-Concentration-versus-pH Diagrams

Many of the important properties of natural waters and wastewaters are due to the presence of weak acids, weak bases, and their salts. The carbonate equilibrium system to be discussed subsequently is preeminent, but ammonia (a weak base) and phosphoric acid are also important. In this discussion, a weak acid will be used as the model. In the case of a weak base, substitution of a cation (e.g., Na^+) for the hydrogen ion or of an anion for the hydroxyl ion is a straightforward process.

Defining a weak acid or a weak base is somewhat arbitrary. In general, acids having equilibrium constant K_a values greater than 1.0 are strongly dissociated in concentrations less than 0.1 M. Acids of real interest in water have K_a values $\ll 1$.

Monoprotic Acids

Consider a monoprotic acid, HA, that is a weak acid with only one exchangeable H^+ ion. The species in solution are HA, A^-, H^+, and OH^-. Four equations can be written to describe a solution of the weak acid in pure water. They are set forth here.

1. Two of the equations are obtained from the equilibrium expressions.
   a. For the weak acid:

$$HA \rightleftharpoons H^+ + A^-$$

$$\frac{[H^+][A^-]}{[HA]} = K_a, \text{mol/L} \qquad (2.29)$$

b. For water:

$$H_2O \rightleftharpoons H^+ + OH^-$$

$$[H^+][OH^-] = K_w, \text{ mol}^2/L^2 \tag{2.30}$$

2. The third equation is obtained from a mass balance for the weak acid.

$$C_T = [HA] + [A^-] \tag{2.31}$$

where

C_T = total concentration of weak acid, mol/L

3. The fourth equation is obtained from a charge balance for pure water containing the weak acid.

$$[H^+] = [A^-] + [OH^-] \tag{2.32}$$

Because there are four equations, the four unknowns [HA], [A⁻], [H⁺], and [OH⁻] can be solved directly (see Example 2.4). The above equations can also be represented on a logarithmic-concentration-versus-pH diagram (hereafter called a log-concentration-versus-pH diagram), as shown in Fig. 2.10. As seen there, the various species are plotted as a function of

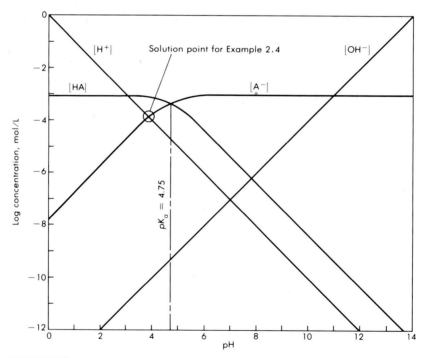

FIGURE 2.10

Log-concentration-versus-pH diagram for a 10^{-3} M solution of acetic acid.

pH, which is considered to be the master variable [2.5, 2.31, 2.34]. The relationships used to plot the curves for [HA] and [A$^-$] are obtained by substituting for [HA] and [A$^-$] from Eq. (2.32) in Eq. (2.29). The desired relationships are

$$[HA] = \frac{C_T[H^+]}{K_a + [H^+]} \tag{2.33}$$

$$[A^-] = \frac{C_T K_a}{K_a + [H^+]} \tag{2.34}$$

The log-concentration-versus-pH diagram shown in Fig. 2.10 was prepared for 10^{-3} M acetic acid which has a $K_a = 1.78 \times 10^{-5}$ (p$K_a = 4.75$), and the general construction of log diagrams is reviewed in Appendix E. The solution of equilibrium problems analytically and with a log-concentration-versus-pH diagram is illustrated in Example 2.4.

EXAMPLE 2.4

SOLUTION OF pH EQUILIBRIUM PROBLEMS

Find the pH of a 10^{-3} M solution of HA using an analytical approach and a log-concentration-versus-pH diagram. Assume $K_a = 1.78 \times 10^{-5}$ and $K_w = 1.00 \times 10^{-14}$.

SOLUTION USING AN ANALYTICAL APPROACH

The analytical solution of equilibrium problems involves the following five steps.

1. Identify the chemical species in solution.

   HA, A^-, H^+, OH^-

2. Define the applicable equilibrium expressions.

   $HA \rightleftharpoons H^+ + A^-$

   $$\frac{[H^+][A^-]}{[HA]} = K_a \tag{2.35}$$

   $H_2O \rightleftharpoons H^+ + OH^-$

   $$[H^+][OH^-] = K_w = 10^{-14} \tag{2.36}$$

3. Define the molar mass balance for the weak acid.

   $$C_T = [HA] + [A^-] = 10^{-3} \ M \tag{2.37}$$

4. Derive the charge balance for the chemical species in solution.

   $$[H^+] = [A^-] + [OH^-] \tag{2.38}$$

5. Solve for the hydrogen ion concentration and the pH.
   a. The simplest method of eliminating the quantities $[A^-]$, $[HA]$, and $[OH^-]$ from these four equations is to substitute for $[A^-]$ and $[HA]$ in Eq. (2.35).
   b. From Eq. (2.38)

$$[A^-] = [H^+] - [OH^-]$$

   c. Substituting for $[A^-]$ in Eq. (2.37) and solving for $[HA]$ yields

$$[HA] = C_T - ([H^+] - [OH^-])$$

   d. Substituting for $[A^-]$ and $[HA]$ in Eq. (2.35) yields

$$\frac{[H^+] \times ([H^+] - [OH^-])}{C_T - ([H^+] - [OH^-])} = K_a$$

   e. Substituting $K_w/[H^+]$ for $[OH^-]$ and simplifying results in a cubic equation for $[H^+]$:

$$\frac{[H^+] \times ([H^+] - K_w/[H^+])}{C_T - ([H^+] - K_w/[H^+])} = K_a$$

$$[H^+]^3 + K_a[H^+]^2 - [H^+](K_w - C_T K_a) - K_a K_w = 0$$

   f. The above cubic equation is an exact expression that can be used to obtain the hydrogen ion concentration for a monoprotic acid. The hydrogen ion concentration can be solved by trial and error. It is, however, not usually necessary to solve the cubic equation to calculate the hydrogen ion concentration of a weak acid in aqueous solution. In most acid solutions, the concentration of $[OH^-]$ is very small; thus

$$[H^+] - [OH^-] \approx [H^+]$$

Because the $[OH^-]$ can be neglected, the equation developed in step 5(d) can be simplified to yield a quadratic expression in $[H^+]$:

$$\frac{[H^+]^2}{C_T - [H^+]} = K_a$$

$$[H^+]^2 + K_a[H^+] - K_a C_T = 0$$

   g. The above quadratic expression can be solved for $[H]$ using the quadratic equation

$$[H^+] = \frac{-K_a \pm \sqrt{K_a^2 + 4K_a C_T}}{2}$$

$$= \frac{1.78 \times 10^{-5} \pm \sqrt{(1.78 \times 10^{-5})^2 + 4(1.78 \times 10^{-5} \times 10^{-3})}}{2}$$

$$= 1.25 \times 10^{-4}$$

$$\text{pH} = 3.9$$

6. If it is necessary to solve the cubic equation to obtain a more precise answer for $[H^+]$, the cubic equation developed in step 5(e) can be solved for $[H^+]$ by trial and error, using the value for $[H^+]$ obtained above as a starting value.

SOLUTION USING A LOG-CONCENTRATION-VERSUS-pH DIAGRAM

In addition to providing visual insight into the importance of the several species in solution as a function of pH, logarithmic-concentration-versus-pH diagrams can be used to solve equilibrium problems. The procedure is illustrated below for the monoprotic acid HA with $K_a = 1.78 \times 10^5$. The procedure for diprotic acids is similar.

1. List all of the chemical species in solution.

HA, A^-, H^+, OH^-

2. Develop a species mass balance for weak acid.

$C_T = [HA] + [A^-]$

3. Prepare a charge balance for species in solution.

$[H^+] = [A^-] + [OH^-]$

4. Using the charge balance developed in step 3, find the pH for which this relationship is satisfied in Fig. 2.10. By inspection, when the value of $[H^+]$ is large, the corresponding value of [OH] is negligible. Thus

$[H^+] \simeq [A^-]$

This relationship is satisfied at a pH value of 3.9. At pH 3.9, the concentration of $[OH^-]$ is three orders of magnitude lower than the corresponding values for $[H^+]$ and $[A^-]$. Thus the approximate relationship used to find the pH is acceptable.

COMMENT

Clearly the use of log-concentration-versus-pH diagrams simplifies the solution of equilibrium problems. In a situation where it is desired to find the pH of a solution containing NaA, using a logarithmic-concentration diagram, it will be necessary to develop another relationship. The needed relationship is known as the *proton condition*. The proton condition is defined in Appendix E.

Polyprotic Acids

Polyprotic acids, particularly carbonic acid (H_2CO_3), are of much more general interest in water quality management than monoprotic acids. The approach to solution is the same as for monoprotic acids, but the problem is a little more complex. The preparation of a log-concentration-versus-pH diagram for a diprotic acid, H_2A, is illustrated in Appendix E.

Carbonate Equilibrium

The most important acid-base system in water is the carbonate system, which controls the pH of most natural waters. The chemical species that compose the carbonate system include gaseous carbon dioxide [$(CO_2)_g$], aqueous carbon dioxide [$(CO_2)_{aq}$], carbonic acid (H_2CO_3), bicarbonate (HCO_3^-), carbonate (CO_3^{-2}), and solids containing carbonate. Because of the importance of the carbonate equilibrium system in the field of water quality management, it is appropriate to consider the subject in greater detail.

In waters exposed to the atmosphere, the equilibrium concentration of dissolved CO_2 is a function of the liquid phase CO_2 mole fraction and the partial pressure of CO_2 in the atmosphere. Henry's law (see Section 2.7) is applicable to the CO_2 equilibrium between air and water; thus

$$x_{CO_2} = K_H P_{CO_2} \qquad (2.39)$$

where

x_{CO_2} = mole fraction of CO_2 at equilibrium in liquid phase

K_H = Henry's law constant, atm^{-1}

P_{CO_2} = partial pressure of CO_2 in atmosphere, atm

The value of K_H is a function of temperature (Table 2.9). Carbon dioxide comprises approximately 0.03 percent of the atmosphere at sea level, where the average atmospheric pressure is 1 atm, or 101.4 kPa. The

TABLE 2.9

Henry's Law Constants for CO_2 and O_2 as a Function of Temperature

T, °C	$K_{H_{CO_2}}$, atm^{-1}	$K_{H_{O_2}}$, atm^{-1}
0	0.001397	0.0000391
5	0.001137	0.0000330
10	0.000967	0.0000303
15	0.000823	0.0000271
20	0.000701	0.0000244
25	0.000611	0.0000222
40	0.000413	0.0000188
60	0.000286	0.0000159

Source: From Refs. [2.5] and [2.28].

concentration of aqueous carbon dioxide is determined using Eq. (2.39) and the definition of mole fraction [Eq. (2.11)]. This computation is illustrated in Example 2.5. Aqueous carbon dioxide $[(CO_2)_{aq}]$ reacts reversibly with water to form carbonic acid.

$$(CO_2)_{aq} + H_2O \rightleftharpoons H_2CO_3 \tag{2.40}$$

The corresponding equilibrium expression is

$$\frac{[H_2CO_3]}{[CO_2]_{aq}} = K_m \tag{2.41}$$

The value of K_m at 25°C is 1.58×10^{-3}. The difficulty of differentiating between $(CO_2)_{aq}$ and H_2CO_3 in solution and the fact that very little H_2CO_3 is ever present in natural waters have led to the use of an effective carbonic-acid value ($H_2CO_3^*$) defined as

$$H_2CO_3^* = (CO_2)_{aq} + H_2CO_3 \tag{2.42}$$

Because carbonic acid is a diprotic acid, it will dissociate in two steps—first to bicarbonate and then to carbonate. The first dissociation of carbonic acid to bicarbonate can now be represented as

$$H_2CO_3^* \rightleftharpoons H^+ + HCO_3^- \tag{2.43}$$

The equilibrium relationship is defined as

$$\frac{[H^+][HCO_3^-]}{[H_2CO_3^*]} = K_1 \tag{2.44}$$

The value of K_1 at 25°C is 4.47×10^{-7} mol/L. Values of K_1 at other temperatures are given in Table 2.10. Note that K_m is unitless and K_1 is expressed in moles per liter. Liters are not a preferred SI unit, but the

TABLE 2.10

Carbonate Equilibrium Constants as a Function of Temperature

T, °C	K_m	K_1, mol/L	K_2, mol/L	K_{sp},* mol²/L²
5		3.02×10^{-7}	2.75×10^{-11}	8.13×10^{-9}
10		3.46×10^{-7}	3.24×10^{-11}	7.08×10^{-9}
15		3.80×10^{-7}	3.72×10^{-11}	6.03×10^{-9}
20		4.17×10^{-7}	4.17×10^{-11}	5.25×10^{-9}
25	1.58×10^{-3}	4.47×10^{-7}	4.68×10^{-11}	4.57×10^{-9}
40		5.07×10^{-7}	6.03×10^{-11}	3.09×10^{-9}
60		5.07×10^{-7}	7.24×10^{-11}	1.82×10^{-9}

*Solubility product constant for $CaCO_3$.
Source: From Refs. [2.6] and [2.20].

practice of defining equilibrium units in terms of moles per liter is long-standing and will be followed in this text.

The second dissociation of carbonic acid is from bicarbonate to carbonate.

$$HCO_3^- \rightleftharpoons H^+ + CO_3^{-2} \tag{2.45}$$

The corresponding equilibrium expression is

$$\frac{[H^+][CO_3^{-2}]}{[HCO_3^-]} = K_2 \tag{2.46}$$

The value of K_2 at 25°C is 4.68×10^{-11} mol/L. Values of K_2 at temperatures other than 25°C are given in Table 2.10.

Application of the aforementioned relationships is illustrated in Example 2.5, in which the pH of rainwater is determined.

EXAMPLE 2.5

CALCULATION OF THE pH OF RAINWATER

Rainwater is normally low in all mineral constituents. Addition of dissolved CO_2 will result in an increase in the H^+ concentration, with the result that the CO_3^{-2} concentration will be negligible. The pH of rainwater is estimated at 25°C as follows:

SOLUTION:

1. Determine the mole fraction of CO_2 in the liquid phase using Eq. (2.39) and the data given in Table 2.9. Assuming the atmosphere contains 0.03 percent CO_2, the mole fraction of CO_2 in the liquid phase is

$$x_{(CO_2)_{aq}} = K_H P_{(CO_2)_{aq}}$$

$$= 6.11 \times 10^{-4} \text{ atm}^{-1} (0.0003 \text{ atm})$$

$$= 1.84 \times 10^{-7}$$

2. Determine the molar concentration of $[CO_2]_{aq}$.

$$x_{CO_2} = [CO_2]_{aq}/([H_2O] + [CO_2]_{aq} + \cdots)$$

$$x_{CO_2} \approx [CO_2]_{aq}/[H_2O]$$

$$x_{CO_2} \approx [CO_2]_{aq}/55.56$$

$$[CO_2]_{aq} = (1.84 \times 10^{-7})(55.56) = 1.02 \times 10^{-5} \text{ mol/L}$$

3. Determine the molar concentration of $[H_2CO_3]$.

$$\frac{[H_2CO_3]}{[CO_2]_{aq}} = K_m$$

$$[H_2CO_3] = (1.58 \times 10^{-3})(1.02 \times 10^{-5})$$
$$= 1.61 \times 10^{-8} \ mol/L$$

4. Determine the molar concentration of $[H_2CO_3^*]$.

$$[H_2CO_3^*] = [CO_2]_{aq} + [H_2CO_3]$$
$$= 1.02 \times 10^{-5} + 1.61 \times 10^{-8}$$
$$= 1.02 \times 10^{-5} \ mol/L$$

5. Determine the pH of rainwater.
   a. The equilibrium expression for $[H_2CO_3^*]$ is

$$\frac{[H^+][HCO_3^-]}{[H_2CO_3^*]} = K_1$$

   b. Consider electroneutrality [see Eq. (2.12)]. In this example the hydrogen ion concentration must be balanced by negative ions. In the case of rainwater, bicarbonate, carbonate, and hydroxide will be assumed to be the only sources of negative ions. Thus

$$[H^+] = [OH^-] + 2[CO_3^{-2}] + [HCO_3^-]$$

   Based on numerous measurements it is known that the pH of rainwater is less than 7.0. But if pH < 7.0, the values of $[OH^-]$ and $[CO_3^{-2}]$ will be negligible and $[H^+] \simeq [HCO_3^-]$. Substituting $[H^+]$ for $[HCO_3^-]$ into the equilibrium expression for $[H_2CO_3^*]$ yields

$$\frac{[H^+]^2}{[H_2CO_3^*]} = K_1$$

   Substitute for K_1 and $[H_2CO_3^*]$ from step 4 and solve for $[H^+]$.

$$[H^+]^2 = (4.47 \times 10^{-7})(1.02 \times 10^{-5})$$
$$[H^+] = 2.13 \times 10^{-6} \ mol/L$$
$$pH = 5.67$$

COMMENT

The alkalinity (see subsequent discussion) of rainwater is quite low, as can be seen from this example. The alkalinity is almost entirely due to the $[HCO_3^-]$ present and is therefore equal to the hydrogen ion concentration. An alkalinity of $2.13 \times 10^{-6} \ M$ provides very little buffering and this is the reason acid rain is produced so easily. Acid rain is considered further in Chapters 3 and 4.

Application of Carbonate Equilibrium

In applying the above carbonate equilibrium relationships to natural systems it is important to determine whether the water will be in contact with a solid source of carbonate such as $CaCO_3$. If, for example, a solid source of calcium carbonate is present, then the solubility product of $CaCO_3$ as given by Eq. (2.48) must also be considered.

$$CaCO_3 \rightleftharpoons Ca^{+2} + CO_3^{-2} \tag{2.47}$$

$$[Ca^{+2}][CO_3^{-2}] = K_{sp} \tag{2.48}$$

Four conditions of importance exist: (1) the open system where gas-liquid equilibrium controls, (2) the closed system with no gas-liquid interface and no solid source of carbonate, (3) the closed system having a solid source of carbonate, and (4) the open system having a solid source of carbonate. The first three systems, which are of most interest in water quality management, are considered further.

The Open System

In the open system, water is only in contact with the atmosphere. This situation exists with rain as it falls toward the earth. The open system is quite simple to work out because the gas-liquid equilibrium condition must be satisfied. This requires that $[H_2CO_3^*] = 1.02 \times 10^{-5}$ mol/L, regardless of pH (see Example 2.5, step 4). Concentrations of HCO_3^- and CO_3^{-2} can then be calculated directly from Eqs. (2.44) and (2.46). At 25°C these relationships are

$$[HCO_3^-] = \frac{(4.47 \times 10^{-7})(1.02 \times 10^{-5})}{[H^+]}$$

$$= 4.56 \times 10^{-12}/[H^+]$$

$$[CO_3^{-2}] = \frac{(4.68 \times 10^{-11})[(4.47 \times 10^{-7})(1.02 \times 10^{-5})]}{[H^+]^2}$$

$$= 2.13 \times 10^{-22}/[H]^2$$

A log-concentration-versus-pH diagram for the open system is shown in Fig. 2.11. Note that as pH increases, the total carbonate in solution becomes larger and larger. Few natural waters have pH values above 9; thus the total concentration would not be expected to exceed 0.005 mol/L or 280 g/m³ as CO_3^{-2}. This is a very large concentration, but one often found in closed systems. Also note that when $[H^+] = [HCO_3^-]$ the pH is 5.65, which is the same value obtained previously for the pH of rainwater in Example 2.5.

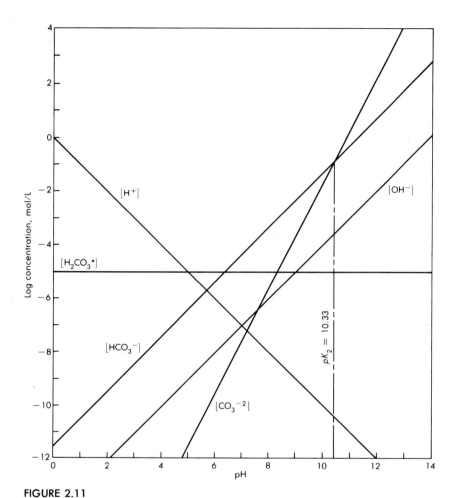

FIGURE 2.11

Log-concentration-versus-pH diagram for an open carbonate system at equilibrium with the carbon dioxide in the atmosphere.

The Closed System Without a Source of Carbonate

This system is characterized by a constant total carbonate concentration value C_T. This constraint also results in a very simple system. The species in solution are $H_2CO_3^*$, HCO_3^-, CO_3^{-2}, H^+, and OH^-. The total carbonate concentration is given by

$$C_T = [H_2CO_3^*] + [HCO_3^-] + [CO_3^{-2}] \tag{2.49}$$

Using Eq. (2.49) and Eqs. (2.44) and (2.46), the following relationships can be derived to predict the concentration of the carbonate species as a

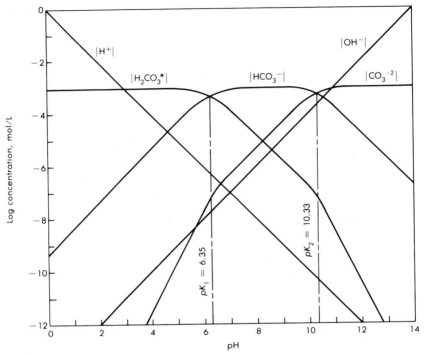

FIGURE 2.12

Log-concentration-versus-pH diagram for a closed carbonate system containing 10^{-3} mol/L total carbonate.

function of pH:

$$[H_2CO_3^*] = \frac{C_T[H^+]^2}{[H^+]([H^+]+K_1)+K_1K_2} \tag{2.50}$$

$$[HCO_3^-] = \frac{K_1C_T[H^+]}{[H^+]([H^+]+K_1)+K_1K_2} \tag{2.51}$$

$$[CO_3^{-2}] = \frac{K_1K_2C_T}{[H^+]([H^+]+K_1)+K_1K_2} \tag{2.52}$$

The distribution of the three carbonate species with $C_T = 10^{-3}$ mol/L is shown in Fig. 2.12 (see also Appendix E).

The Closed System With a Solid Source of Carbonate

This system is controlled by the solid-liquid equilibrium. For a water in contact with solid $CaCO_3$, the value of CO_3^{-2} is constant and can be computed using Eq. (2.53):

$$[CO_3^{-2}] = K_{sp}/[Ca^{+2}] \tag{2.53}$$

Equations (2.46) and (2.44) are then used to determine $[HCO_3^-]$ and $[H_2CO_3^*]$ as a function of pH. The resulting relationships are:

$$[CO_3^{-2}] = 4.57 \times 10^{-9}/[Ca^{+2}] \tag{2.54}$$

$$[HCO_3^-] = \frac{4.57 \times 10^{-9}[H^+]}{[Ca^{+2}]K_2} \tag{2.55}$$

$$[H_2CO_3^*] = \frac{4.57 \times 10^{-9}[H^+]^2}{K_1K_2[Ca^{+2}]} \tag{2.56}$$

In most groundwaters, Ca^{+2} would be associated with other anions in addition to CO_3^{-2} (HCO_3^-, Cl^-, SO_4^{-2}, etc.), and therefore a water analysis would be needed to predict the effects of a pH change. The concepts are the same, however. Adding calcium will result in decreased concentrations of CO_3^{-2}, HCO_3^-, and $H_2CO_3^*$.

Comment on Log-Concentration-versus-pH Diagrams

Log-concentration-versus-pH diagrams are a useful tool in working with many systems other than carbonate equilibrium. Particularly important examples are diagrams used with phosphate and metal ions such as iron or aluminum. Concepts used in developing the diagrams are the same as for the carbonate system, and readers are referred to Refs. [2.2, 2.5, 2.31, and 2.34].

The principal point to be learned from log-concentration-versus-pH diagrams is the relationship that exists between the various carbonate species at various pH values. For example, in the closed system if the measured pH is 6.5, nearly all of the carbonate present is in the form of $H_2CO_3^*$ and HCO_3^-. At pH 8, HCO_3^- accounts for essentially all of the carbonate, and at pH 12 virtually all of the carbonate exists as CO_3^{-2}. This also works in reverse. Given a carbonate analysis of $H_2CO_3^* = 0$, $HCO_3^- = 10^{-4}$ mol/L, and $CO_3^{-2} = 0$, the pH must be in the range of from 7 to 9.

Alkalinity (*A*)

The capacity of water to neutralize acid is termed *alkalinity*. (Correspondingly, *acidity* is the capacity to neutralize bases.) The alkalinity of water is measured by titrating with a strong acid to a pH value near 4.5, corresponding to a solution of pure CO_2. The exact endpoint will depend on the ionic strength of the solution. In nearly all natural waters the presence of carbonate $[CO_3^{-2}]$, bicarbonate $[HCO_3^-]$, and hydroxyl $[OH^-]$ ions accounts for essentially all of the alkalinity. For most practical purposes, the alkalinity can be defined in terms of molar quantities, as

$$A, \text{eq/m}^3 = [HCO_3^-] + 2[CO_3^{-2}] + [OH^-] \tag{2.57}$$

The corresponding expression, in terms of equivalents, is

$$A, \text{eq/m}^3 = (HCO_3^-) + (CO_3^{-2}) + (OH^-) \qquad (2.58)$$

Application of Eq. (2.57) is illustrated in Example 2.6.

A more rigorous definition, in terms of molar concentrations, is as follows:

$$A = [HCO_3^-] + 2[CO_3^{-2}] + [OH^-] - [H^+] \qquad (2.59)$$

Inclusion of the hydrogen ion concentration in the definition for alkalinity can be reasoned as follows. Assume $NaHCO_3$ is added to pure water. The species in solution are Na^+, $H_2CO_3^*$, HCO_3^-, CO_3^{-2}, H^+, and OH^-. The charge balance is

$$[H^+] + [Na^+] = [HCO_3^-] + 2[CO_3^{-2}] + [OH^-]$$

Because $A = [Na^+]$, as there is no other source of carbonate, the correct expression for the alkalinity is given by Eq. 2.59. From this example one very important property of alkalinity can be deduced. That is: The addition or removal of CO_2 does not change the alkalinity, since the concentration of $[Na^+]$ does not change.

EXAMPLE 2.6

DETERMINATION OF ALKALINITY

Determine the alkalinity of a water found to have carbonate, bicarbonate, and hydroxyl ion concentrations of 20, 488, and 0.17 g/m^3, respectively.

SOLUTION:

1. Determine the concentration of each species in equivalents per cubic meter.

CONSTITUENT	ATOMIC MASS, g	EQUIV. MASS, g/eq	eq/m^3
CO_3^{-2}	60	30	0.67
HCO_3^-	61	61	8.00
OH^-	17	17	0.01

2. Determine the alkalinity.

$$
\begin{aligned}
A, \text{eq/m}^3 &= (HCO_3^-) + (CO_3^{-2}) + (OH^-) \\
&= 8.00 + 0.67 + 0.01 \\
&= 8.68 \text{ eq/m}^3
\end{aligned}
$$

Conductivity

Another parameter used to characterize the gross chemical characteristics of water is conductivity. The *conductivity* of a solution is a measure of the ability of that solution to conduct an electrical current. Because the electrical current is transported by the ions in solution, the conductivity increases as the concentration of ions increases. In practice, the *electrical conductance* (conductance of 1.0 cm^3 of solution) is measured by placing two platinum electrodes in a water sample and recording the resistance offered by the solution. Electrical conductance is reported in units of microsiemens per centimeter (newer SI unit) or micromhos per centimeter (older unit). Both units have the same numerical value.

The conductivity of a water sample can be approximated using the following relationship and the data given in Table 2.11 (on page 92):

$$EC \simeq \sum_i (C_i \times f_i) \qquad (2.60)$$

where

EC = electrical conductivity, $\mu S/cm$ (microsiemens/cm)
C_i = concentration of ionic species i in solution, meq/L or mg/L
f_i = conductivity factor for ionic species i (see Table 2.11)

Using the following relationship, the conductivity of a sample can also be employed to estimate the ionic strength μ of a solution [2.32]:

$$\mu \simeq 1.6 \times 10^{-5} \times EC \qquad (2.61)$$

where

μ = ionic strength (see also Eq. 2.13)
EC = electrical conductivity, $\mu S/cm$

Hardness

Multivalent cations, particularly magnesium and calcium, are often present at significant concentrations in natural waters. These ions are easily precipitated and in particular react with soap to form a difficult-to-remove scum. Since 1950, detergents have largely replaced soap in developed countries, and the economic impact of hardness has decreased somewhat. Where these metal ions are in solution, scaling in hot water pipes, water heaters, and appliances continues to be a major problem, however.

For most practical purposes, hardness of water can be represented as the sum of the calcium and magnesium concentrations, given in equiv-

TABLE 2.11

Conductivity Factors for Ions Commonly Found in Water

ION	CONDUCTIVITY FACTOR f_i, μS/cm	
	Per meq/L	Per mg/L
Cations		
Calcium (Ca^{+2})	52.0	2.60
Magnesium (Mg^{+2})	46.6	3.82
Potassium (K^+)	72.0	1.84
Sodium (Na^+)	48.9	2.13
Anions		
Bicarbonate (HCO_3^-)	43.6	0.715
Carbonate (CO_3^{-2})	84.6	2.82
Chloride (Cl^-)	75.9	2.14
Nitrate (NO_3^-)	71.0	1.15
Sulfate (SO_4^{-2})	73.9	1.54

Source: From Ref. [2.32].

TABLE 2.12

Qualitative Classification of Waters According to Level of Hardness*

	HARDNESS	
DESCRIPTION	meq/L	mg/L as $CaCO_3$
Soft	< 1	< 50
Moderately hard	1–3	50–150
Hard	3–6	150–300
Very hard	> 6	> 300

*Typically based on Ca^{+2} and Mg^{+2}.

alents per cubic meter (milliequivalents per liter):

$$\text{Hardness, eq/m}^3 = (Ca^{+2}) + (Mg^{+2}) \tag{2.62}$$

In the older literature, hardness is often expressed as equivalent grams per cubic meter of $CaCO_3$. This method was a matter of convenience that has been superseded by expressing hardness in terms of equivalents. In the literature dealing with water quality, the qualitative classification reported in Table 2.12 is often used to describe the relative hardness of a water.

Two general types of hardness are of interest: carbonate hardness, associated with HCO_3^- and CO_3^{-2}, and noncarbonate hardness, associated with other anions, particularly Cl^- and SO_4^{-2}. The balance between carbonate and noncarbonate hardness is important in water softening (hardness removal) and in scale formation. Because HCO_3^- dissociates at high temperatures (the equilibrium constant increases by a factor of 2.6 between 5 and 60°C), a result of heating hard water is scale formation due to $CaCO_3$ precipitation:

$$Ca^{+2} + 2HCO_3^- \xrightarrow{\Delta H} CaCO_3 + CO_2 + H_2O \tag{2.63}$$

Scale formation plugs pipes, decreases heat-transfer coefficients, and changes frictional resistance to flow.

EXAMPLE 2.7

DETERMINATION OF HARDNESS

Two water samples were analyzed for soluble constituents. Partial results of the analysis are given below. Determine the total hardness (TH), carbonate hardness (CH), and noncarbonate hardness (NCH) of each sample.

CONSTITUENT	CONCENTRATION, eq/m^3	
	Sample 1	Sample 2
Ca^{+2}	5	2
Mg^{+2}	1	4
HCO_3^-	3	7
Cl^-	2	4

SOLUTION:

1. Determine total hardness (TH).
   a. Sample 1:

   $$TH = (Ca^{+2}) + (Mg^{+2}) = 5 + 1 = 6 \text{ eq/m}^3 = 300 \text{ g/m}^3 \text{ as } CaCO_3$$

   where

   $$1 \text{ eq/m}^3 \ CaCO_3 = 50 \text{ g/m}^3 \ CaCO_3$$

   b. Sample 2:

   $$TH = (Ca^{+2}) + (Mg^{+2}) = 2 + 4 = 6 \text{ eq/m}^3 = 300 \text{ g/m}^3 \text{ as } CaCO_3$$

2. Determine carbonate hardness (CH).
   a. Sample 1:

   $$CH = (HCO_3^-) = 3 \text{ eq/m}^3 = 150 \text{ g/m}^3 \text{ as } CaCO_3$$

   b. Sample 2:

   $$CH = (HCO_3^-) = 7 \text{ eq/m}^3$$

   Because the amount of bicarbonate (7 eq/m^3) exceeds the total hardness (6 eq/m^3), the carbonate hardness is taken to be 6 eq/m^3. Thus

   $$CH = 6 \text{ eq/m}^3 = 300 \text{ g/m}^3 \text{ as } CaCO_3$$

3. Determine noncarbonate hardness (NCH).
   a. Sample 1:

   $$NCH = TH - CH$$
   $$= 6 - 3 = 3 \text{ eq/m}^3 = 150 \text{ g/m}^3 \text{ as } CaCO_3$$

   b. Sample 2:

   $$NCH = TH - CH$$
   $$= 6 - 6 = 0 \text{ eq/m}^3$$

2.6

CHEMICAL CHARACTERISTICS OF WATER: ORGANIC MATTER

The amount of organic matter present in most natural waters is low. Typically the source of the organic matter in water is from decaying weeds, leaves, and trees. Humic acid, a high-molecular-mass compound derived from the decomposition of plant matter, is found in most surface waters. At present, most surface waters and many groundwaters also contain organic matter of anthropogenic origin. As noted in Table 2.1, the general category of "organic matter" includes organic matter whose origins could be from both natural sources and from human activities. Natural organic compounds must be distinguished from organic compounds that are solely of anthropogenic origin, such as DDT or some of the more than 100,000 organic compounds synthesized since 1940. The principal classes of organic compounds are reported in Table 2.13.

The presence of organic matter in water is troublesome for many reasons, including (1) color formation, (2) taste and odor problems, (3) oxygen depletion in streams, (4) interference with water treatment processes, and (5) the formation of halogenated compounds when chlorine is added to disinfect water. Because of the significance of the organic matter in water, it is the purpose of this section to introduce the reader to (1) the nature of the characteristics of common organic compounds in water, (2) some of the troublesome organic compounds of anthropogenic origin, and (3) the methods used to measure the concentration of organic matter in water.

Natural Organic Compounds

Most organic compounds are composed of various combinations of carbon, hydrogen, oxygen, nitrogen, phosphorus, and sulfur. The principal organic compounds found in wastewater, and to a much lesser degree in natural waters, include proteins (40 to 60 percent), carbohydrates (25 to 50 percent), and lipids (10 percent). Organic substances of anthropogenic origin are considered separately following this present discussion.

Proteins

Proteins are the principal constituent of animal tissue. They also occur in plants, but to a lesser extent. In addition to carbon, hydrogen, and oxygen, proteins contain nitrogen, which is their distinguishing characteristic, and sulfur. The proportion of nitrogen is typically about 15 percent by mass. Although there are thousands of proteins, they are typically made up of 26 basic building blocks known as *amino acids*, which share a common structure (Table 2.13 on pages 96–97). For example, the formula for the most common amino acid, glycine, is

$C_2H_5NO_2$. The empirical formula most often used to define the composition of bacterial cells is $C_5H_7NO_2$.

Carbohydrates

Containing carbon, hydrogen, and oxygen, carbohydrates are widely distributed in nature. Carbohydrates include sugars, starches, cellulose, and wood fiber. Carbohydrates are the principal constituents of plant tissue and of some animal tissue. Carbohydrates are classified as mono-, di-, tri-, tetra-, and polysaccharides. One of the most common monosaccharides is glucose ($C_6H_{12}O_6$). Low-molecular-mass carbohydrates, such as mono- and disaccharides, are generally soluble in water while polysaccharides are generally insoluble. With the exception of cellulose and wood fiber, most carbohydrates are readily biodegradable. Because they are easily biodegradable, the presence of carbohydrates can lead to the rapid depletion of the oxygen resources of waters (see Biochemical Oxygen Demand later in this section).

Lipids

Those constituents of animal and plant tissue that are insoluble in water, but soluble in ether or other organic solvents, are classified as lipids. Important lipids found in wastewater include fats, oils, greases, and waxes. The fats and oils present in wastewater are typically from butter, lard, margarine, and vegetable fats and oils. Fats are also found in meats, seeds, nuts, and certain fruits. Because of their structure and low solubility, fats are not readily biodegradable.

Synthetic Organic Compounds

As noted above, more than 100,000 organic compounds have been synthesized since 1940. Most of these compounds are not used widely, but some have become ubiquitous in use throughout the world. It is the purpose here to consider briefly those classes of manufactured organic compounds that have found their way into the waters of the world and whose presence is of concern from a health, treatment, and ecological standpoint. Of greatest concern are those organic compounds that may be carcinogenic or that may cause mutations in humans and other living forms at extremely low concentrations.

Surfactants

Glycerides and other greases (hydrocarbon oils) are among the agents responsible for binding dirt to surfaces. Cleaning involves finding a way or an agent that can be used to loosen this material from the surfaces to which it is attached. When soap or synthetic detergents (Fig. 2.13) are added to water, the nonpolar tails of the soap or detergent molecules tend to become embedded (dissolved) in the greaselike material.

TABLE 2.13
The Principal Classes of Organic Compounds

CLASS	DESCRIPTION	FUNCTIONAL GROUP*	REPRESENTATIVE COMPOUNDS
Hydrocarbons	Organic compounds that contain only carbon and hydrogen		
Saturated	Molecules in which carbon atoms share only a single bond	$-\overset{\displaystyle |}{\underset{\displaystyle |}{C}}-\overset{\displaystyle |}{\underset{\displaystyle |}{C}}-$	Ethane (CH_3CH_3) Propane ($CH_3CH_2CH_3$)
Unsaturated	Molecules in which carbon atoms share multiple bonds	$\overset{\diagup}{}C{=}C\overset{\diagdown}{}$ $-C{\equiv}C-$	Ethylene ($CH_2{=}CH_2$) Acetylene ($H-C{\equiv}C-H$)
Aromatic	Molecules in which carbon atoms occur in a hexagonal ring known as a benzene ring	CH_3 (benzene ring)	Benzene (C_6H_6) Toluene ($C_6H_5-CH_3$) Styrene ($C_6H_5-CH{=}CH_2$)
Alcohols and phenols	Organic compounds that contain hydroxyl as their functional group	$R-OH$ $Ar-OH$	Methyl alcohol (CH_3OH) Ethyl alcohol (CH_3CH_2OH) Phenol (C_6H_5OH)
Ethers	Organic compounds in which two functional groups are attached to an oxygen atom	$R-O-R'$ $Ar-O-R$	Methyl ethyl ether ($CH_3OCH_2CH_3$) Ethyl ether ($CH_3CH_2OCH_2CH_3$)
Organic halogen compounds	Organic compounds in which a halogen is the principal atom		Carbon tetrachloride (CCl_4) Chloroform (trichloromethane) ($CHCl_3$)

TABLE 2.13 (*Cont.*)

Aldehydes and ketones	Organic compounds built around the carbonyl group (a carbon linked to oxygen with a double bond)	$$R-\overset{\displaystyle\overset{O}{\|\|}}{C}-R'$$	Formaldehyde (CH_2O) Acetone (CH_3COCH_3)
Carboxylic acids and their derivatives	Organic compounds that contain the carboxyl group	$$R-\overset{\displaystyle\overset{O}{\|\|}}{C}-OH$$	Formic acid (HCOOH) Acetic acid (CH_3COOH) Stearic acid [$CH_3(CH_2)_{16}$ COOH]
Esters	Organic compounds derived from the elimination of water from an acid and an alcohol	$$R-\overset{\displaystyle\overset{O}{\|\|}}{C}-O-R'$$	Ethyl formate (CH_3CH_2COOH) Ethyl acetate ($CH_3CH_2OOCCH_3$)
Amines and amides	Organic compounds that contain nitrogen	$$R-NH_2$$ $$R-\overset{\displaystyle\overset{O}{\|\|}}{C}-NH_2$$	Methylamine (CH_3NH_2) Aniline ($C_6H_{11}NH_2$) Formamide ($HCONH_2$)
Carbohydrates	Organic compounds composed of mono-, di-, tri-, tetra-, and polysaccharides		Glucose ($C_6H_{12}O_6$)
Proteins	Organic compounds that contain nitrogen and are derived from 26 amino acids		Glycine ($C_2H_5NO_2$) Biological cell tissue, approximate ($C_5H_7NO_2$)
Lipids (fats, oils, and waxes) and detergents	Organic compounds of plant and animal origin not soluble in water but soluble in ether		Oleic acid ($C_{17}H_{35}COOH$) Soap [$CH_{17}H_{35}COO^-\ Na^+$]

*R = alkyl group or hydrogen; Ar = aromatic ring (aryl group) when the R-group is primed (e.g., R'- or R"-) the prime symbol means that the two or more R-groups may be identical or different.

Source: Adapted in part from Refs. [2.15] and [2.17].

(a) An ordinary soap—sodium stearate $(C_{17}H_{35}COO^-Na^+)$

$CH_3CH_2CH_2CH_2CH_2CH_2CH_2CH_2CH_2CH_2CH_2CH_2CH_2CH_2CH_2CH_2CH_2C\diagup O \atop \diagdown O^- Na^+$

Nonpolar portion (lyophilic)	Polar portion (hydrophilic)

(b) A typical synthetic detergent—sodium lauryl sulfate $(CH_3(CH_2)_{10}CH_2OSO_2O^-Na^+)$

$$CH_3CH_2CH_2CH_2CH_2CH_2CH_2CH_2CH_2CH_2CH_2CH_2O-\overset{\displaystyle O}{\underset{\displaystyle O}{\overset{|}{\underset{|}{S}}}}-O^-Na^+$$

Nonpolar portion
(lyophilic)

Polar
portion
(hydrophilic)

FIGURE 2.13

Schematic representation of (a) soap and (b) detergent molecules.

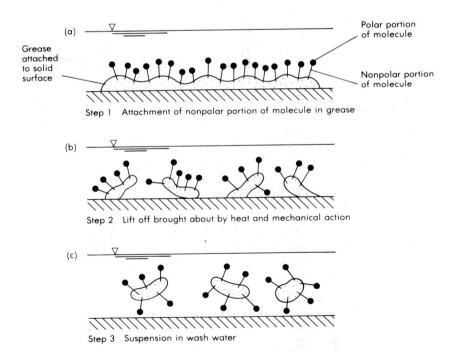

FIGURE 2.14

Schematic representation of the action of soap and detergent molecules in cleansing.

The polar end resists this action, and when assisted by mechanical agitation or boiling, the attached film breaks up into small particles that can be removed easily by washing (Fig. 2.14).

Before 1965, the surfactant in synthetic detergents was primarily of the alkyl-benzene-sulfonate (ABS) type. These detergent compounds were difficult to treat biologically. Thus in 1965 legislation was passed in the United States, and ABS-type detergents were replaced with linear-alkyl-sulfonate (LAS), which is biodegradable (Fig. 2.13). However, in some parts of the world, ABS-type detergents are still in use. The presence of surfactants is measured by noting the color change occurring in a standard solution of methylene-blue dye. The term MBAS (methylene-blue active substances) is commonly used in reporting the presence or absence of detergents in water.

Pesticides and Agricultural Chemicals

The chemicals used in agriculture for the control of diseases and of pest organisms are primarily of anthropogenic origin. The major source of these chemicals in surface waters is surface runoff from agricultural, residential, and park lands. The presence of these substances in water is troublesome because they are toxic to most aquatic organisms, and many are known or suspected carcinogens. In general, agricultural chemicals can be grouped into four major categories based on their molecular structure: chlorinated hydrocarbons, organophosphates, carbamates, and urea derivatives. Representative members of each group are reported in Table 2.14.

Cleaning Solvents

Organic solvents are another group of compounds that are of concern because many are known or suspected carcinogens. Recently, the detection of such compounds in groundwaters near industrial facilities has caused great concern. Many of these solvents have leaked from underground storage tanks into the underlying aquifers. In other locations, solvents have leaked into underlying aquifers from surface impounds where they were stored temporarily after being used. Although there are hundreds of solvents, some of the more common are acetone, benzene, carbon tetrachloride, ethyl alcohol, heptane, methyl alcohol, pine oil, toluol, and trichloroethane 1,1,1. The transport of these constituents in groundwater is considered in Chapter 10.

Trihalomethanes

It has been found that chlorine used for the disinfection of treated water and wastewater can react with some of the organic substances present to form chloroform (a trihalomethane) and other chlorinated hydrocarbons. The organic substances involved in the reaction with chlorine are identified as *precursors*. The formation of trihalomethanes

TABLE 2.14

Classification of Some Common Agricultural Pesticides, Fungicides, and Herbicides

CLASS	TRADE NAME	FORMULA
Pesticides		
Chlorinated	Aldrin	$C_{12}H_8Cl_6$
hydrocarbons	Chlordane	$C_{10}H_6Cl_8$
	DDT	$C_{14}H_9Cl_5$
	Dieldrin	$C_{12}H_8Cl_6O$
	Lindane	$C_6H_6Cl_6$
	Endrin	$C_{12}H_8Cl_6O$
	Heptachlor	$C_{10}H_5Cl_7$
	Methoxychlor	$C_{16}H_{15}Cl_3O_2$
	Toxaphene	$C_{10}H_{10}Cl_8$
Organophosphates	Diazinon	$C_{12}H_{21}N_2O_3PS$
	Malathion	$C_{10}H_{19}O_6PS_2$
	Parathion	$C_{10}H_{14}NO_5PS$
Herbicides		
Carbamate	Carbyl (sevin)	$C_{12}H_{11}NO_2$
Chlorinated	2,4-D	$C_8H_8Cl_2O_2$
hydrocarbons	2,4,5-T	$C_8H_5Cl_3O_3$
	Silvex	$C_9H_7Cl_3O_3$
Ureas	Fenuron	$C_9H_{12}N_2O$
Fungicides	Copper sulfate	$CuSO_4$
	Ferbam	$C_9H_{18}N_3S_6Fe$
	Ziram	$C_6H_{12}N_2S_2Zn$

(THMs) and chlorinated hydrocarbons is of concern because these compounds are suspected carcinogens.

In general, trihalomethane compounds are formed when members of the halogen family—namely, chlorine, bromine, and iodine—react with organic substances. The principal trihalomethanes of concern in water and wastewater treatment are (1) $CHCl_3$ (chloroform), (2) $CHCl_2Br$ (bromodichloromethane), (3) $CHClBr_2$ (chlorodibromomethane), and (4) $CHBr_3$ (bromoform).

The formation of chloroform as a result of the addition of chlorine can be illustrated by the following reactions:

$$CH_3COCH_3 + 3NaOCL \longrightarrow CH_3COCCl_3 + 3NaOH \qquad (2.64)$$

Acetone   Sodium   Trichloroacetone   Sodium
              hypochlorite   hydroxide

$$CH_3COCCl_3 + NAOH \longrightarrow CH_3CO_2Na + CHCl_3 \qquad (2.65)$$

Sodium   Chloroform
acetate

In Reaction 2.64, acetone is the organic precursor in the formation of chloroform. It should be noted that a variety of organic substances of both natural and anthropogenic origin can serve as precursors. Humic acid, produced from decaying leaves, is a natural precursor.

Trihalomethanes are determined by first extracting a sample with an appropriate solvent such as pentane or isooctane. Gas chromotography is then used to determine the THMs in the extract.

Measurement of Organic Matter

Over the years, a number of tests have been developed to measure the organic content of water. In general, the tests may be divided into those used to measure concentrations of organic matter greater than about 1 g/m^3 and those used to measure trace concentrations in the range of 10^{-12} to 10^{-3} g/m^3. The former are used to measure the mixtures of organic compounds found in water. The latter are used to measure the presence of trace quantities of anthropogenic organic compounds.

Laboratory tests commonly run to measure gross amounts of organic matter (greater than 1 g/m^3) include tests for

1. Chemical oxygen demand (COD)
2. Total organic carbon (TOC)
3. Total oxygen demand (TOD)
4. Biochemical oxygen demand (BOD)

The COD test is a chemical test, the TOC and TOD tests are instrumental tests, and the BOD test is a biochemical test involving the use of microorganisms. In the older (prior to 1930) literature dealing with water quality the measures of organic matter most often used were (1) total, albuminoid, organic, and ammonia nitrogen and (2) oxygen consumed. Because these measures are no longer used, they are not considered further in the following discussions.

Trace organics in the range of 10^{-12} to 10^{-3} g/m^3 are determined using instrumental methods including gas chromotography and mass spectroscopy. Within the past 10 years, the sensitivity of the methods used for the detection of trace organic compounds has improved significantly, and detection of concentrations in the range of 10^{-9} g/m^3 is now almost a routine matter.

In addition to the above tests, if the chemical formula of the individual compounds is known, the theoretical oxygen demand (ThOD) can be computed from stoichiometric considerations. The COD, TOC, and TOD tests; the techniques used for the analysis of trace organics; and the determination of the ThOD values are reviewed in the following discussion. Then, using this review as an introduction, the BOD test is considered in detail in the remainder of the section.

Chemical Oxygen Demand (COD)

The COD test is used to measure the organic content of natural waters, municipal wastewaters, and industrial wastes. Functionally, the oxygen equivalent of the organic matter is measured using a strong oxidizing agent (potassium dichromate) in an acidic medium. The test is performed at elevated temperatures, using a catalyst (silver sulfate) to help in the oxidation of certain resistant classes of organic compounds. Using excess dichromate as the oxidizing agent, the reaction involved in the measurement of organic matter may be represented as follows:

$$C_a H_b O_c + Cr_2 O_7^{-2} + H^+ \xrightarrow[\text{catalyst}]{\text{Heat}} Cr^{+3} + CO_2 + H_2O \qquad (2.66)$$

Organic   Dichromate
matter

To determine the amount of organic matter, the amount of dichromate remaining at the end of the test is measured. The difference between the original amount present and the amount remaining is, after appropriate conversion, reported in terms of the equivalent oxygen required to oxidize the organic matter to CO_2, H_2O, NH_4^+, PO_4^{-3}, SO_4^{-2}, etc. Because the reduced nitrogen that is present is converted to ammonia, the COD test is often used to approximate the ultimate carbonaceous biochemical oxygen demand (BOD_u). (See Biochemical Oxygen Demand later in this section.)

The COD test is used extensively because it takes less time (about 3 hr) as compared with other tests such as the BOD test, which takes several days. Unfortunately, the COD test cannot be used to differentiate between biologically oxidizable and inert organic matter. Also, no information is provided about the rate at which the organic material will be oxidized biologically. Further, certain inorganic constituents such as chloride can interfere with the test. Where these constituents are present, the sample must be treated to remove the interference before the COD is determined [2.32].

Total Organic Carbon (TOC)

For smaller amounts of organic matter, the instrumental TOC test has proved to be satisfactory. In one form of this test, a sample is evaporated and oxidized catalytically to carbon dioxide. The amount of carbon dioxide released is measured with an infrared analyzer. To ensure that only organic carbon is measured, the samples to be tested are acidified and aerated to remove the inorganic forms of carbon. In some cases, certain organic compounds may not be oxidized completely, and the measured TOC value will be slightly less than the actual value in the sample.

Total Oxygen Demand (TOD)

In the instrumental TOD test, organic and some inorganic compounds are converted to stable end products, such as CO_2 and H_2O, in a platinum-catalyzed combustion chamber. The TOD is determined by the loss of oxygen in the nitrogen-carrier gas. The results of this test have been correlated to the COD test.

Trace Organics

The general protocol for the detection of trace quantities of organic constituents, as illustrated in Fig. 2.15, involves three steps: isolation,

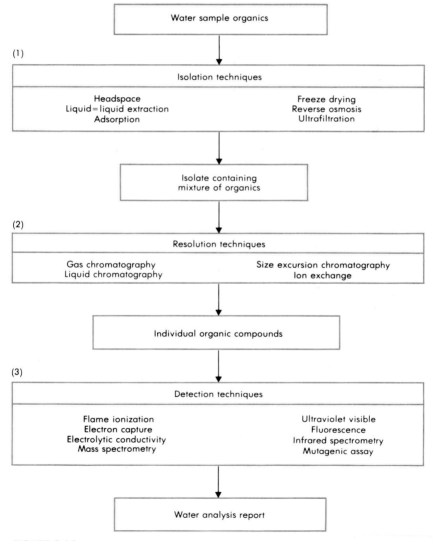

FIGURE 2.15

Protocol for the determination of trace organics using instrumental methods.

Source: From Trussell and Trussell [2.38].

resolution, and detection [2.38]. *Isolation* is used to concentrate the trace organic constituents in the sample. The specific technique used will depend on whether the organic constituents are volatile or nonvolatile. *Resolution* involves the separation of the concentrated organic materials into individual components. Again, depending on the nature of the organic compounds present, several techniques are available (Fig. 2.15). The final step, *detection*, is used to determine quantitatively the concentration of the individual constituents.

Theoretical Oxygen Demand (ThOD)

Where the chemical formula of the organic compound is known, the ThOD can be estimated from stoichiometric considerations. The amount of oxygen required for complete oxidation of an organic compound is equal to the amount required for the oxidation of each element minus the amount of oxygen present initially in the organic compound. Thus it can be shown that

$$C_cH_hO_oN_nP_pS_s + (c + 0.25h - 0.5o + 1.25n + 1.25p + 1.5s)O_2 \rightarrow$$
$$cCO_2 + (0.5h - 0.5n - 1.5p - s)H_2O + nNO_3^- + pPO_4^{-3} +$$
$$sSO_4^{-2} + (n + 3p + 2s)H^+ \qquad (2.67)$$

Because phosphorus and sulfur are present in small amounts in most organic compounds, they can be neglected without serious error. If the nitrogen in the organic compound is not oxidized but is converted to ammonia, the resulting oxygen requirement is essentially equal to the COD and to the BOD_u, or the ultimate carbonaceous biochemical oxygen demand (see later discussion of BOD). The theoretical stoichiometric amount of oxygen required to oxidize the ammonia to nitrate is essentially the same as that required for the biological oxidation of ammonia to nitrate. Determination of the ThOD is illustrated in Example 2.8.

<hr>

EXAMPLE 2.8

DETERMINATION OF THEORETICAL OXYGEN DEMAND

Determine the ThOD of a solution containing 200 g/m^3 of glycine $[CH_2(NH_2)COOH]$. Also estimate the carbonaceous biochemical oxygen demand, the nitrogeneous oxygen demand, and the oxygen equivalent of the compound, assuming that the nitrogen is converted to ammonia.

SOLUTION:

1. Determine the amount of oxygen required to convert (oxidize) glycine to carbon dioxide, ammonia, and water.
   a. Step 1: Balance the carbon in the organic compound with the carbon in carbon dioxide.

$$CH_2(NH_2)COOH \rightarrow 2CO_2$$

b. Step 2: Balance the nitrogen in the organic compound with the nitrogen in ammonia.

$$CH_2(NH_2)COOH \rightarrow 2CO_2 + NH_3$$

c. Step 3: Balance the hydrogen with the hydrogen in water.

$$CH_2(NH_2)COOH \rightarrow 2CO_2 + NH_3 + H_2O$$

d. Step 4: Balance the oxygen with molecular oxygen. *Note:* Molecular masses are shown in parentheses.

$$CH_2(NH_2)COOH + 3/2O_2 \rightarrow 2CO_2 + NH_3 + H_2O$$
$$\quad\quad (75) \quad\quad\quad\quad (32)$$

e. Determine the oxygen required for the first conversion (oxidation) step.

$$O_2 \text{ required} = 1.5 \text{ mol } O_2/\text{mol glycine}$$

2. Determine the amount of oxygen required to oxidize the ammonia to nitrate.
   a. Step 1: Balance the nitrogen in ammonia with the nitrogen in nitrate.

$$NH_3 \rightarrow HNO_3$$

   b. Step 2: Balance the hydrogen with the hydrogen in water.

$$NH_3 \rightarrow HNO_3 + H_2O$$

   c. Step 3: Balance the oxygen with molecular oxygen.

$$NH_3 + 2O_2 \rightarrow HNO_3 + H_2O$$
$$(14 \text{ as } N) \ (32)$$

   d. Determine the oxygen for the second conversion (oxidation) step.

$$O_2 = 2 \text{ mol } O_2/\text{mol } NH_3$$

3. The ThOD is

$$ThOD = 3.5(1.5 + 2.0)\text{mol } O_2/\text{mol glycine}$$

$$= \left(\frac{3.5 \text{ mol } O_2}{\text{mol glycine}}\right)\left(\frac{32 \text{ g}}{\text{mol } O_2}\right)\left(\frac{200 \text{ g/m}^3}{75 \text{ g/mol glycine}}\right)$$

$$= 298.7 \text{ g/m}^3$$

4. Estimate the ultimate carbonaceous biochemical oxygen demand. From steps 1(e) and 1(d) above,

$$CBOD_u = 1.5 \text{ mol } O_2/\text{mol glycine}$$

$$= \left(\frac{1.5 \text{ mol } O_2}{\text{mol glycine}}\right)\left(\frac{32 \text{ g}}{\text{mol } O_2}\right)\left(\frac{200 \text{ g/m}^3}{75 \text{ g/mol glycine}}\right)$$

$$= 128 \text{ g/m}^3$$

5. Estimate the nitrogenous biochemical oxygen demand. From steps 2(d) and 2(c),

NBOD = 2 mol O_2/mol glycine

$$= \left(\frac{2.0 \text{ mol } O_2}{\text{mol glycine}} \right) \left(\frac{32 \text{ g}}{\text{mol } O_2} \right) \left(\frac{200 \text{ g/m}^3}{75 \text{ g/mol glycine}} \right)$$

$$= 170.7 \text{ g/m}^3$$

6. Determine the oxygen equivalent of glycine, assuming nitrogen is converted to ammonia. From step 1(e) above,

$$\text{Oxygen equiv.} = \frac{1.5 \text{ mol } O_2}{1.0 \text{ mol glycine}}$$

$$= \frac{(1.5 \text{ mol } O_2)(32 \text{ g/mol } O_2)}{(1.0 \text{ mol glycine})(75 \text{ g/mol glycine})}$$

$$= 0.64 \text{ g } O_2/\text{g glycine}$$

Dissolved Oxygen and Oxygen Demand

The presence of dissolved oxygen is of fundamental importance in maintaining aquatic life and the aesthetic quality of waters. Because of this importance, oxygen is the most widely used water quality parameter, and determining the impact of contaminants on the oxygen resources of a stream or lake is a major factor in the development of any water quality management plan. The impact is normally measured as *oxygen demand*, a parameter that can be interpreted as a gross measure of the concentration of oxidizable materials present in a water sample and as a potential load on the receiving water.

Mechanisms of Oxidation

Both biochemical and chemical oxidation reactions take place in the aquatic environment, but the principal reactions in natural waters are biochemical. In terms of oxygen demand, the most important type of biochemical reaction involves the oxidation of organic material. In this case the reduced carbon in the organic material is oxidized to its lowest energy state, CO_2, through the metabolic action of microorganisms (principally bacteria). Products other than CO_2 are energy, water, new cell tissue, and minerals.

$$\text{Organic material} + O_2 + \text{nutrients} \xrightarrow{\text{microorganisms}}$$

$$CO_2 + H_2O + \underset{\text{cells}}{\text{new}} + \text{nutrients} + \text{energy} \qquad (2.68)$$

The same basic reactions are carried out at all trophic levels; thus Eq. (2.68) is generally applicable to microorganisms, fish, and zebras. Everyone is aware that they would not survive for a long period of time in a sealed room because the oxygen would gradually be depleted. Most people have undoubtedly lost a goldfish because of oxygen depletion in a small bowl, and this situation provides a good analogy to stream and lake pollution. Goldfish require oxygen for respiration. However, their demands are low, and the rates of oxygen transfer across the surface of small bowls are normally great enough for two or three goldfish to prosper. If the fish are overfed or if the bowl is not regularly cleaned, organic material will accumulate. Bacteria utilize the deposited organics, and because of their high growth rates, the bacterial respiration rate may exceed the oxygen-transfer rate across the water surface in the fish bowl. In this case the dissolved oxygen concentration decreases, often to the point that the fish suffocate.

Biochemical Oxygen Demand (BOD)

The goldfish model can be extended to develop the concept of biochemical oxygen demand (BOD). Proper engineering design of the goldfish bowl and feeding system would include prediction of oxygen transfer potential across the air-water interface (see Chapter 7), the oxygen demand of the excreted and excess food, and the rate at which this oxygen demand would be exerted. The latter two factors, stoichiometry and rate, are the basic components of BOD determinations.

Most organic materials are biodegradable. Certain compounds (e.g., cellulose, lignin, and many synthetic petrochemicals) are very resistant to biological breakdown and can often be considered nondegradable, or recalcitrant. Such materials rarely make up a significant fraction of the total organic mass, however. The amount of oxygen used in the metabolism of biodegradable organics is termed the *biochemical oxygen demand* (*BOD*). In most systems of interest to engineers, the organics are a fairly broad mixture of chemical species and include both soluble and nonsoluble organic materials. Thus, BOD values are usually a measure of the oxygen required for the carbonaceous oxidation of a nonspecific mixture of organic compounds rather than of pure compounds. The principal end products of the carbonaceous oxidation of organic matter are carbon dioxide (CO_2), ammonia (NH_3), and water (H_2O). Unless otherwise noted, reported BOD values are only for the biodegradation of carbonaceous materials. For this reason, and to avoid possible confusion, some governmental agencies are encouraging the use of the term *CBOD* to denote *carbonaceous biochemical oxygen demand*.

Nitrification is the term applied to the biological oxidation of ammonia to nitrate, and the oxygen required is termed the *nitrogenous biochemical oxygen demand* (*NBOD*).

BOD Test Procedures. The BOD of a water sample is determined by placing aliquots with appropriate dilution water in glass-stoppered 300-mL BOD bottles, incubating the bottles at a standard temperature (20°C), and measuring the change in oxygen concentration with time. Concentrations of biodegradable organic material contained in many water samples have an oxygen demand greater than the oxygen saturation value (about 9.2 g/m^3 at 20°C at sea level), and thus dilution of the samples is often required. Some water samples do not have sufficient

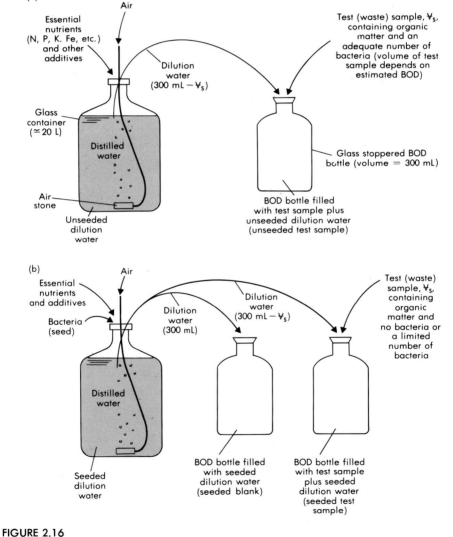

FIGURE 2.16

Procedure for setting up BOD test bottles: (a) with unseeded dilution water and (b) with seeded dilution water.

bacterial populations to carry out the biooxidation of the available organics. A seed culture of suitable organisms must be added, and the effect of the seed on the BOD value must be considered in the analysis. Essential nutrients such as N, P, K, and Fe may be stoichiometrically deficient and must be added prior to measurement. Because BOD measurement is the most widely used water quality test, a standard technique, including the makeup of the dilution water, has been established by regulatory agencies and professional organizations. This technique, as well as other standard procedures, can be found in *Standard Methods* [2.32]. The physical setup of the BOD test bottles is illustrated schematically in Fig. 2.16.

An idealized schematic representation of the BOD concept is presented in Fig. 2.17 for a BOD test performed with seeded dilution water (see Fig. 2.16). As shown in Fig. 2.17, the initial dissolved oxygen concentration of the blank (B_1) and the diluted sample (D_1) are different. This difference is due to the fact that the dissolved oxygen concentration in the undiluted sample is normally less than that of the dilution water. Oxygen uptake resulting from the "seed" is assumed to be the

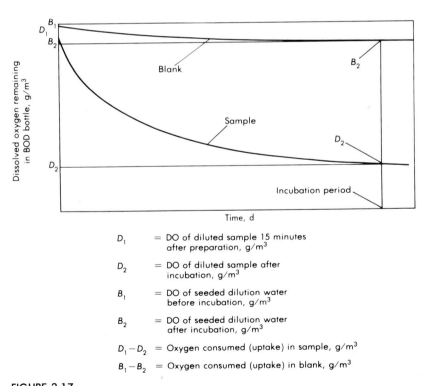

D_1	= DO of diluted sample 15 minutes after preparation, g/m^3
D_2	= DO of diluted sample after incubation, g/m^3
B_1	= DO of seeded dilution water before incubation, g/m^3
B_2	= DO of seeded dilution water after incubation, g/m^3
$D_1 - D_2$	= Oxygen consumed (uptake) in sample, g/m^3
$B_1 - B_2$	= Oxygen consumed (uptake) in blank, g/m^3

FIGURE 2.17

Schematic representation of the oxygen consumed (uptake) in the BOD determination.

same as the uptake in the seeded blank. Thus the difference between the
sample BOD and the blank BOD, corrected for the amount of seed used
in the sample, is the true BOD. With the data reported in Fig. 2.17 the
BOD is computed using the following equation [2.32]:

$$BOD = \frac{(D_1 - D_2) - (B_1 - B_2)f}{P} \tag{2.69}$$

where

BOD = biochemical oxygen demand, g/m^3

D_1 = DO of diluted sample 15 min after preparation, g/m^3

D_2 = DO of diluted sample after incubation at 20°C, g/m^3

B_1 = DO of seeded dilution water before incubation, g/m^3

B_2 = DO of seeded dilution water after incubation at 20°C, g/m^3

f = ratio of seed in sample to seed in control

$= \dfrac{\% \text{ seed in } D_1}{\% \text{ seed in } B_1}$

P = decimal fraction of sample used

$= \dfrac{\text{mL of sample, } Vs}{300 \text{ mL}}$

For an unseeded sample the terms for the blank (B_1 and B_2) are omitted
in the computation of the BOD. The application of Eq. (2.69) is illustrated in Example 2.9.

EXAMPLE 2.9

DETERMINATION OF BIOCHEMICAL OXYGEN DEMAND

A water sample is diluted by a factor of 10 using seeded dilution water
according to the methods described in *Standard Methods* [2.32]. Dissolved oxygen concentration is measured at 1-d intervals, and the results
are listed below. Using these data, determine the sample BOD as a
function of time.

TIME, d	DILUTED SAMPLE DISSOLVED OXYGEN, g/m^3	SEEDED BLANK DISSOLVED OXYGEN, g/m^3
0	8.55	8.75
1	4.35	8.70
2	4.05	8.66
3	3.35	8.61
4	2.75	8.57
5	2.40	8.53
6	2.10	8.49
7	1.85	8.46

SOLUTION:

Determine BOD as a function time using Eq. (2.69).

$$BOD = \frac{(D_1 - D_2) - (B_1 - B_2)f}{P}$$

where

$$f = \frac{90\%}{100\%} = 0.9$$

$$P = \frac{30 \text{ mL}}{300 \text{ mL}}$$
$$= 0.1$$

TIME, d	BOD, g/m³	TIME, d	BOD, g/m³
0	0.00	4	56.4
1	41.5*	5	59.5
2	44.2	6	62.2
3	50.8	7	64.4

$$*BOD = \frac{(8.55 - 4.35) - (8.75 - 8.70)0.9}{0.1}$$
$$= 41.5 \text{ g/m}^3$$

COMMENT

A standard of accuracy for the BOD test is not available. Interlaboratory checks on glucose–glutamic-acid mixtures reported in *Standard Methods* [2.32], had a 15 percent standard deviation corresponding to $\pm 10 \text{ g/m}^3$ at 168 hr. Accuracy of measurement is an important factor, and the difficulty of obtaining precise results must be considered in decision making.

Rate of Oxygen Uptake and BOD Exertion. The stoichiometric amount of oxygen used in the breakdown of organic material is, as has been emphasized, an important parameter in water quality management. Nearly as important is the rate at which the oxygen demand is exerted. The primary cause of oxygen uptake in waters is the metabolism of organic materials by bacteria, and not surprisingly the rate of oxygen uptake clearly parallels the rate of bacterial growth. In a closed system (e.g., the BOD bottle) the three variables—organic concentration, bacterial mass, and dissolved oxygen consumed—are important. The interrelationship of these variables is illustrated graphically in Fig. 2.18. Inactive cultures, the state of most "seeds," do not immediately begin metabolizing organic material but require a "tooling-up" period to change from metabolizing internally stored organics to metabolizing newly available organic materials outside the cell. The time needed to change over from endogenous to

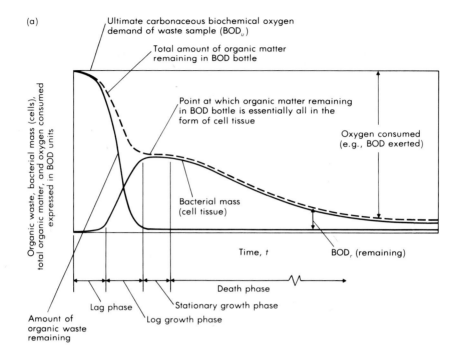

(a)

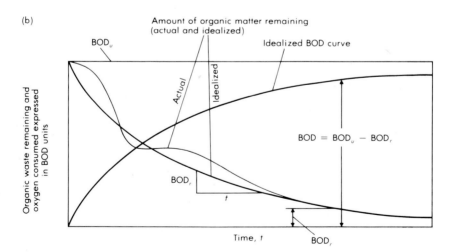

(b)

FIGURE 2.18

Functional analysis of the BOD test: (a) interrelationship of organic waste, bacterial mass (cell tissue), total organic waste, and oxygen consumed in BOD test; (b) idealized representation of the BOD test.

exogenous metabolism may take several hours. Following this lag period, growth and therefore organic substrate and oxygen uptake rates are rapid, virtually exponential, until the organic substrate or some other growth factor becomes limiting. As one or more growth factors become limiting, the rate of growth slows, eventually stops, and the cells return to a resting state. To continue to function, the cells will now begin to consume stored food reserves, including their own cell tissue, utilizing oxygen in the process. When food reserves are exhausted, some of the cells break up or lyse. Examples of lysed cells (empty shells) are shown in the transmission electron micrograph of Fig. 2.19. Contents of the lysed cells are consumed by the remaining bacteria, and oxygen is required for this process also.

The rate of oxygen utilization in this second stage is considerably less than the rate of utilization in the first stage, in which the organic materials present in the sample are metabolized. Most wastewaters are extremely complex in their organic and microbial makeup. Because of this complexity, the measured rate of BOD exertion generally does not correspond well to the distinct phases shown in Fig. 2.18. Nevertheless,

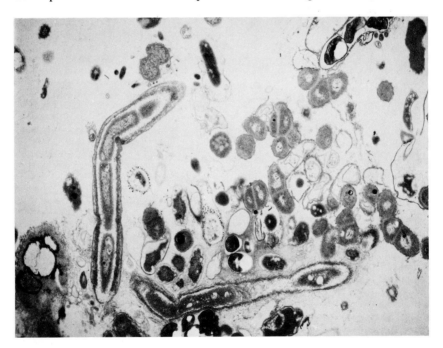

FIGURE 2.19

Transmission electron micrograph of bacterial culture with a large number of lysed and virally infected cells.

Source: Courtesy of JoAnn Silverstein, University of Colorado, Boulder, CO.

The concepts illustrated in Fig. 2.18(a) are extremely important in understanding the BOD process. Busch and his coworkers [2.3, 2.4, 2.13, 2.21, 2.35] have demonstrated that, by properly setting up the BOD test, the distinct phases can be delineated, excellent repeatability of results achieved (with standard deviations of less than 5 percent), and theoretically derived oxygen uptake values obtained.

The First-Order Rate Model for BOD. If a closed system is observed for a long enough period of time, nearly all of the degradable organic matter, including that in cell tissue, will be converted to CO_2, NH_3, and H_2O, and the ultimate BOD (BOD_u) will be approached (Fig. 2.18a). Because of complications induced by the oxidation of ammonia (NH_3) to nitrate (NO_3^-) (see Nitrogenous Oxygen Demand), the BOD_u is very difficult to measure unless the nitrification reaction is suppressed [2.32] or the technique proposed by Busch is applied [2.10, 2.27, 2.35]. In practice, the 20-d BOD value is used as an estimate of BOD_u, or, even more commonly, the BOD_u value is estimated using a first-order rate model.

In the first-order rate model, it is assumed that the oxygen uptake (BOD exertion) is only a function of the BOD remaining. Because the actual curve of the BOD remaining is difficult to model, a simplified first-order curve is assumed, as illustrated in Fig. 2.18(b). For the reaction that occurs in a BOD bottle, it can be shown that

$$\frac{d(BOD_r)}{dt} = -k(BOD_r) \tag{2.70}$$

where

$BOD_r = $ BOD remaining at time t, g/m^3

$k = $ reaction rate constant, t^{-1} (base e)

As in most reaction-rate models, the temperature effects are incorporated into the rate constant. Integrating Eq. (2.70) between the initial limits $BOD_u = BOD_u$ and $BOD_u = BOD_r$ and $t = 0$ and $t = t$ yields

$$\ln \frac{BOD_r}{BOD_u} = -kt \tag{2.71}$$

$$BOD_r = BOD_u e^{-kt} \tag{2.72}$$

where

$BOD_u = $ ultimate carbonaceous BOD, g/m^3

Note that the BOD_r value in Eqs. (2.71) and (2.72) is the BOD remaining to be exerted. In almost all cases, the desired value is the value that has been exerted at time t — BOD_t (Fig. 2.18b). The value of BOD_t is computed using Eq. (2.73):

$$BOD_t = BOD_u - BOD_r$$
$$= BOD_u - BOD_u e^{-kt}$$
$$= BOD_u(1 - e^{-kt}) \tag{2.73}$$

Applying the first-order (or any) model requires determination of the coefficient values. The value of the ultimate BOD_u should be independent of the experimental system, but the rate constant k will depend on the bacterial culture, and possibly other factors, in addition to temperature. Evaluation of these two coefficients is normally accomplished by taking a time series of BOD values (usually 1 d apart) and performing a least-squares fit of the data. (Detailed derivations of the least-squares method are given in most introductory texts on statistical methods.) The result, which gives a fit with minimum derivation from the mean line, is given in Eqs. (2.74) and (2.75):

$$na + b\Sigma y - \Sigma y' = 0 \tag{2.74}$$
$$a\Sigma y + b\Sigma y^2 - \Sigma y'y = 0 \tag{2.75}$$

where

$y = BOD_t$ value, g/m^3
$n =$ number of data points used
$b = -k, d^{-1}$ (base e)
$a = -bBOD_u, g/m^3 \cdot d$
$y' = (y_{n+1} - y_{n-1})/2\Delta t$

The least-squares method is applicable to any first-order expression. It must be remembered that the true BOD curve is only being approximated by a first-order curve and that no insight into the mechanics of the process is provided by the first-order model.

EXAMPLE 2.10

DETERMINATION OF FIRST-ORDER MODEL COEFFICIENTS BY THE LEAST-SQUARES METHOD

For the data below, determine the values of k and BOD_u.

TIME, d	1	2	3	4	5	6
BOD, g/m^3	15.8	26.7	37.4	45.9	50.1	56.1

SOLUTION:

1. Set up a computational table.

Application of Eqs. (2.74) and (2.75) involves setting up a table of data points. Note that the number of data points used is $n - 1$ because of the definition of y'.

t, d	y, g/m^3	y^2, (g/m^3)^2	y', g/m$^3 \cdot$ d	yy', (g/m^3)^2 d^{-1}
0	0	—	—	—
1	15.8	249.6	13.4	211.7
2	26.7	712.9	10.8	288.4
3	37.4	1398.9	9.6	359.0
4	45.9	2106.8	6.4	293.8
5	50.1	2510.0	5.1	255.5
6	56.1	3147.2	—	—
$\sum_{n=5}$	175.9*	6,978.2*	45.3	1,408.4

*Sum based on $t = 1$ to 5 d

2. Determine the values of k and BOD_u. Using Eqs. (2.74) and (2.75),

$$5a + 175.9b - 45.3 = 0$$
$$175.9a + 6978.2b - 1408.4 = 0$$

Solving for a and b yields

$$a = 17.3$$
$$b = -0.23$$

and therefore

$$k = -b = 0.23 \text{ d}^{-1} \text{ (base } e)$$

$$BOD_u = -\frac{a}{b} = 73.8 \text{ g/m}^3$$

Presentation of the data with the fitted curve is shown in Fig. 2.20.

COMMENT

A simple graphical method of determining BOD_u has been developed by Fujimoto [2.11]. In this method an arithmetic plot of BOD_{t+1} versus BOD_t is made. The intersection of this plot with a line of slope $= 1$ gives the ultimate BOD, as shown in Fig. 2.21. When BOD_u has been determined, the rate constant can be calculated using Eq. (2.72) and one of the measured BOD values and the corresponding time.

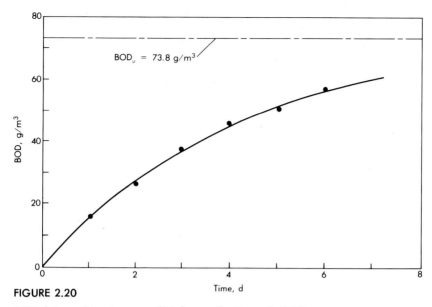

FIGURE 2.20

BOD data and least-squares-fitted curve for Example 2.10.

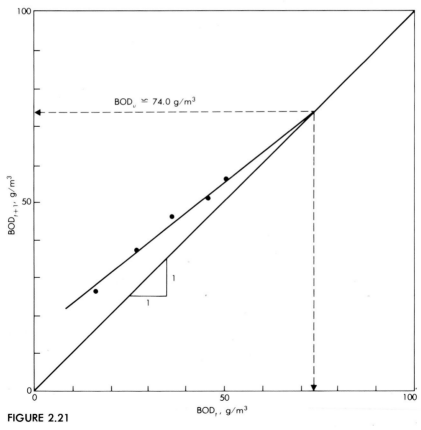

FIGURE 2.21

Graphical solution to first-order BOD model using the method proposed by Fujimoto [2.11].

Temperature Effects on BOD. Reaction rates are strongly affected by temperature changes, but reaction stoichiometry is relatively unaffected. In the BOD-exertion model, temperature effects are accounted for in the rate constant k, using the following modified form of the van't Hoff-Arrhenius relationship:

$$k_T = Ae^{-E/RT} \qquad (2.76)$$

where

k_T = rate constant at temperature T, d^{-1}

A = van't Hoff-Arrhenius coefficient, d^{-1}

E = activation energy, J/mol

R = universal gas constant, 8.314 J/mol·K

T = temperature, K

Because it is convenient to use degrees centigrade and because A is very difficult to evaluate, a modification using a temperature difference is made:

$$\frac{k_{T_1}}{k_{T_2}} = \exp\left(\frac{E}{RT_2} - \frac{E}{RT_1} \right) \qquad (2.77)$$

$$k_{T_1} = k_{T_2} \exp\left[\left(\frac{E}{RT_1 T_2} \right)(T_1 - T_2) \right] \qquad (2.78)$$

In water quality management, temperatures encountered rarely vary from the range 0 to 35°C. In this range the term $\exp(E/RT_1 T_2)$ can be considered approximately constant, and Eq. (2.78) can be rewritten:

$$k_{T_1} = k_{T_2} \theta^{(T_1 - T_2)} \qquad (2.79)$$

where

θ = temperature coefficient

Because 20°C is used as the laboratory reference temperature for the measurement of BOD, the k_{20} value is also used as the reference. Typical k_{20} values (base e) for domestic wastewater are in the range of 0.2 to 0.3 d^{-1} where standard techniques are used. Because the first-order model does not include cell concentration as a parameter, normal population differences can be expected to cause significant rate variation. The most commonly used value of θ is 1.047. Use of $\theta = 1.135$ for $4 < T < 20°C$ and $\theta = 1.056$ for $T > 20°C$ has been suggested [2.10, 2.19, 2.30]. Reported activation energy values are in the range of from 33,500 to 50,250 J/mol for microbial systems. Substituting these values into Eq. (2.76) results in θ values similar to those given above.

EXAMPLE 2.11

EFFECT OF TEMPERATURE ON BIOCHEMICAL OXYGEN DEMAND

A wastewater sample taken at a discharge point of a wastewater outfall has been analyzed. The BOD_u and k_{20} values have been estimated to be 30 g/m^3 and 0.22 d^{-1}, respectively. The summer wastewater discharge rate is 0.5 m^3/s, and the receiving stream has a minimum flow rate of 6 m^3/s. Determine the 5-d BOD value in the stream if summer water temperatures can be expected to reach 25°C. Assume the temperature coefficient θ is equal to 1.056.

SOLUTION:

1. Determine the k_{25} value using Eq. (2.79).

$$k_{25} = 0.22 \text{ d}^{-1} (1.056)^{25-20}$$
$$= 0.29 \text{ d}^{-1}$$

2. Determine the BOD_5 at 25°C using Eq. (2.73).

$$BOD_5 = 30 \text{ g/m}^3 \left[1 - e^{-0.29(5)}\right]$$
$$= 23 \text{ g/m}^3$$

3. Determine the BOD_5 in the stream.

$$\text{Dilution factor} = \frac{Q_d}{Q_d + Q_s} = \frac{0.5 \text{ m}^3 \text{ s}}{(0.5 + 6.0) \text{ m}^3 \text{ s}}$$
$$= \frac{0.5}{6.5} = 0.077$$

$$\text{Stream } BOD_5 = 0.077 \left(23 \text{ g/m}^3\right) = 1.8 \text{ g/m}^3$$

The 5-d BOD. The 5-d, 20°C BOD was used in Example 2.10 because it is the standard reference value used internationally by both design engineers and regulatory agencies. Like many reference values, the BOD_5 (20°C is assumed unless otherwise stated) had a rational beginning and an irrational application. In a report prepared by the Royal Commission on Sewage Disposal in the United Kingdom at the beginning of the century, it was recommended that a 5-d, 18.3°C BOD value be used as a reference in Great Britain. These values were selected because British rivers do not have a flow time to the open sea greater than 5 d and average long-term summer temperatures do not exceed 18.3°C. The temperature has been rounded upward to 20°C, but the 5-d time period has become the universal scientific and legal reference.

That the 5-d BOD has no (and cannot have) theoretical meaning is clear. Equally clear is the fact that the basis for adoption of the BOD_5 is inappropriate. Unfortunately, the desire of engineers for a "number" and of attorneys for a legal reference has resulted in producing a virtually unchallengable, if irrational, product. In almost all cases, the only value reported is the BOD_5. Unless clearly stated otherwise, it must be assumed that any biochemical oxygen demand value reported is the BOD_5 at 20°C.

Nitrogeneous Oxygen Demand

Reduced nitrogen compounds (NH_4^+, NH_3, NO_2^-) can be oxidized aerobically by specific genera of bacteria. These bacteria are chemoautotrophic; that is, they use nitrogen compounds as their energy source and CO_2 as their carbon source. Two distinct steps occur. The first, the oxidation of ammonia nitrogen to the nitrite state ($NH_4^+ \rightarrow NO_2^-$) is carried out by bacteria of the genera *Nitrosomonas*, *Nitrosococcus*, *Nitrosospira*, *Nitrosocystis*, and *Nitrosogluea*. These bacteria can utilize carbon compounds but always require NH_4^+ as an energy source. Nitroso-bacteria are often referred to by the representative genus, *Nitrosomonas*, and this practice will be used here. Oxidation of nitrite to the most oxidized state of nitrogen, nitrate, is carried out by bacteria of the genera *Nitrobacter*, *Nitrospira*, and *Nitrocystis*. The nitrite-oxidizing bacteria are facultative autotrophs; that is, if NO_2^- is absent, they will utilize carbon compounds such as volatile acids. *Nitrobacter* will be used as the representative genus, but in both cases the other groups can be assumed to be present.

Yields of new cells are small for both steps (less than 0.05 g cells/g N oxidized), and the stoichiometry of the reactions can be represented quite accurately using Eqs. (2.80) and (2.81):

$$NH_4^+ + \tfrac{3}{2}O_2 \xrightarrow{\text{\textit{Nitrosomonas}}} NO_2^- + H_2O + 2H^+ \qquad (2.80)$$

$$\dfrac{NO_2^- + \tfrac{1}{2}O_2 \xrightarrow{\text{\textit{Nitrobacter}}} NO_3^-}{NH_4^+ + 2O_2 \rightarrow NO_3^- + 2H^+ + H_2O} \qquad \begin{matrix}(2.81)\\[1.2em](2.82)\end{matrix}$$

Overall, 2 mol of O_2 are required for each mole of ammonia oxidized (4.6 g O_2/g N). Domestic wastewaters typically contain 15 to 50 g/m^3 total nitrogen, which corresponds to a potential oxygen demand of 69 to 230 g/m^3. Exertion of the nitrogenous biochemical oxygen demand (NBOD) is slow and typically is not noted in the BOD test at times less than 5 to 8 d. When nitrification occurs, it complicates measurement of the carbonaceous BOD and makes direct determination of BOD_u values very difficult (Fig. 2.22). When nitrification will cause a problem with the interpretation of the BOD results, it can be inhibited with 2-chloro-6-(trichloromethyl) pyridine and other proprietary chemicals. Inhibition of nitrification is now an accepted procedure [2.32].

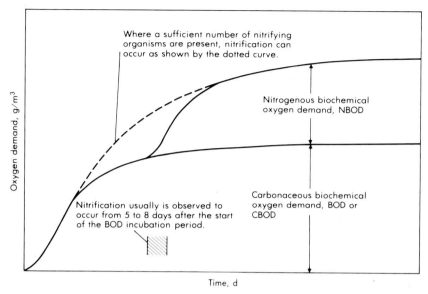

FIGURE 2.22

Definition sketch for the exertion of the carbonaceous and nitrogenous biochemical oxygen demand in a waste sample.

Because nitrification is a separate process from the exertion of carbonaceous BOD, it is always reported separately. Kinetics of nitrification are not well documented, but nitrite is rarely observed in significant amounts, and measured growth rates reported for *Nitrobacter* are higher than those for *Nitrosomonas*. Thus the first step appears to be rate-limiting and affects the overall conversion of ammonia to nitrate.

2.7

CHEMICAL CHARACTERISTICS OF WATER: GASES

Gases found in water include nitrogen (N_2), oxygen (O_2), carbon dioxide (CO_2), hydrogen sulfide (H_2S), ammonia (NH_3), and methane (CH_4). The first three are the common gases of the atmosphere and will be found in all water exposed to the atmosphere. The latter three gases, associated with bacterial metabolism and respiration, are considered further in Section 2.8 and in Chapter 3. Hydrogen sulfide is produced from the bacterial reduction of sulfate under anaerobic conditions, whereas ammonia and methane are derived from the anaerobic biological decomposition of organic matter.

In considering the presence or absence of gases in water the following factors pertain: (1) the solubility of the gas, (2) the partial pressure of the gas in the atmosphere above the liquid, (3) the temperature of the water, and (4) the purity of the water as measured by parameters such as ionic strength, salinity, and suspended solids. The solubility and transfer of gases will be considered further.

Solubility of Gases in Water

In England in 1803, W. Henry reported that there existed a relationship between the solubility of a gas in water and the partial pressure of the gas in the atmosphere above the water. The relationship, known as *Henry's law*, is written as follows:

$$x_g = K_H P_g \tag{2.83}$$

where

x_g = mole fraction of gas at equilibrium in liquid phase

K_H = Henry's law constant, atm^{-1}

P_g = partial pressure of gas in atmosphere, atm

Henry's law constant is a function of the type of gas, the temperature, and the constituents of the liquid. Values for Henry's law constant are reported in Table 2.15. The use of Henry's law has already been illustrated in Example 2.5.

Gas Transfer

The rate at which a gas transfers into solution can be described by Fick's first law, expressed in concentration units (see Chapter 7):

$$\frac{dC}{dt} = K_L a (C_s - C) \tag{2.84}$$

TABLE 2.15

Henry's Law Constants for Several Gases that are Slightly Soluble in Water

	$K_H \times 10^5 \ \text{atm}^{-1}$							
$T°C$	Air	CO_2	CO	H_2	H_2S	CH_4	N_2	O_2
0	2.31	137.4	2.84	1.73	373.1	4.46	1.89	3.92
10	1.82	96.7	2.26	1.57	272.5	3.37	1.50	3.03
20	1.51	70.1	1.87	1.46	207.0	2.66	1.24	2.44
30	1.30	53.8	1.61	1.37	164.2	2.23	1.08	2.11
40	1.15	41.3	1.44	1.33	134.2	1.92	0.962	1.88
50	1.06	35.3	1.31	1.34	113.1	1.73	0.885	1.70
60	0.99	28.6	1.22	1.34	97.1	1.60	0.833	1.59

Source: Adapted from Refs. [2.5] and [2.28].

where

dC/dt = rate of change in concentration of the gas in solution, mg/L · s

$K_L a$ = overall mass transfer coefficient, s^{-1}

C_s = saturation concentration for gas in solution, mg/L

C = actual concentration of gas in solution, mg/L

The value of C_s can be estimated using Henry's law. The term $(C_s - C)$ represents a concentration gradient. As the value of C approaches that of C_s, the transfer rate decreases.

The integrated form of Eq. (2.84) is obtained by integrating between the limits $C = C_0$ and $C = C$ and $t = 0$ and $t = t$, as follows:

$$\int_{C_0}^{C} \frac{dC}{C_s - C} = K_L a \int_{0}^{t} dt$$

The integrated form of Eq. (2.84), which can be used to describe the addition of a gas to a liquid, is

$$\frac{C_s - C_t}{C_s - C_0} = e^{-(K_L a)t} \tag{2.85}$$

where

C_s = saturation concentration of gas, mg/L

C_t = concentration of gas at time t, mg/L

C_0 = initial concentration of gas, mg/L

$K_L a$ = overall mass transfer coefficient, s^{-1}

t = time, s

When gases are removed from solution, Eq. (2.85) is written as follows:

$$\frac{C_0 - C_s}{C_t - C_s} = e^{-(K_L a)t} \tag{2.86}$$

Evaluation of the transfer coefficient $K_L a$ for a given system configuration is usually based on laboratory measurements. The subject of gas transfer is considered further in Chapters 7 and 12.

2.8
BIOLOGICAL CHARACTERISTICS OF WATER

The biological characteristics of water, related primarily to the resident aquatic population of microorganisms, impact directly on water quality. The most important impact is the transmission of disease by pathogenic organisms in water. Other important water quality impacts include the development of tastes and odors in surface waters and groundwaters and the corrosion of and biofouling of heat transfer surfaces in cooling systems and water supply and wastewater management facilities. The impacts of the biological characteristics are considered in Chapter 3. The purpose of this section is to introduce some basic concepts from microbiology, to discuss the characteristics of the principal microorganisms of concern in water and to assess the impact of the water environment on those microorganisms, to outline the methods used in the enumeration of microorganisms, to review the use of indicator organisms and their significance, and to examine the methods used to test the toxicity of water.

Some Basic Concepts from Microbiology

To understand the impact of the microorganisms present in water on humans and on water quality, it will be helpful to review some basic concepts and ideas from microbiology, including (1) the classification used to group microorganisms, (2) the scientific nomenclature used to describe microorganisms, (3) the nature of biological cells, (4) the nutritional requirements of cells, and (5) the environmental effects on microorganisms. References [2.9, 2.25, 2.33, and 2.39] are recommended for further details on these subjects.

Organism Classification

The principal groups of microorganisms found in water may be classified as protists (higher and lower), plants, and animals (Table 2.16). Commonly, the organisms listed in Table 2.16 are characterized as

TABLE 2.16

Simplified Classification of Microorganisms of Interest in Water and Wastewater

KINGDOM	REPRESENTATIVE MEMBERS	CELL CLASSIFICATION
Animal*	Crustaceans Worms Rotifers	Eucaryotic cells (containing a nucleus enclosed within a well-defined nuclear membrane)
Plant*	Rooted aquatic plants Seed plants Ferns Mosses	
Protista Higher	Protozoa Algae Fungi (molds and yeasts)	
Lower†	Blue-green algae Bacteria	Procaryotic cells (nucleus not enclosed in a true nuclear membrane)

*Multicellular with tissue differentiation.

†Uni- or multicellular without tissue differentiation.

procaryotic or *eucaryotic*, depending on whether the nucleus within the cell is enclosed in a well-defined nuclear membrane. Recently it has been proposed to subclassify these microorganisms on the basis of their DNA composition, but this sort of classification is as yet far in the future. For the purpose of this text, the classification given in Table 2.16 will suffice.

Nomenclature for Microorganisms

The nomenclature for microorganisms is based on groups of increasing size:

Kingdom
  Phylum
    Class
      Order
        Family
          Genus
            Species

Following the above classification, the scientific name of any organism includes both the genus and the species names. Given first, the name of the genus is always capitalized and both names are italicized. For example, the scientific name of human beings is *Homo sapiens*. That is, humans belong to the species *sapiens* and to the genus *Homo*. We also belong to the family Hominidae, the order Primate, the class Mammalia, the phylum Chordata, and finally the kingdom Animal.

The Nature of Biological Cells

Because the cell is the basic building block of life, it is appropriate to identify the basic characteristics of cells. In general, most living cells are quite similar. A *membrane* forms the outer boundary, but bacteria, algae, fungi, and plants have a *cell wall* outside the membrane that gives the cell a characteristic shape. If the individual cells are motile, they usually possess *flagella* or some hairlike appendages called *cilia*. The interior of the cell contains various *organelles* and a colloidal suspension of proteins, carbohydrates, and other complex organic matter collectively called *cytoplasm*.

Each cell contains *nucleic acids*, the genetic material vital to reproduction. Ribonuleic acid (RNA), which is important in the synthesis of proteins, is present in the cytoplasmic area. In procaryotic cells there is only a *nuclear area*; in eurcaryotic cells there is a definite *nucleus* enclosed in a membrane. The nucleus (or nuclear area) is rich in deoxyribonucleic acid (DNA), which contains the genetic information necessary for the reproduction of all the cell components. Small strands of DNA called *plasmids* are often found in the cytoplasm of procaryotic cells. Plasmids function like an extra chromosome in that they are reproduced and transmitted to daughter cells. Many plasmids are responsible for specific traits, such as antibiotic resistance, and it is plasmids that are used in recombinant DNA technology to transfer genes from one cell to another. These and other components of cells are illustrated in Fig. 2.23 (on page 128).

Nutritional Requirements of Organisms

To grow, microorganisms, including single-cell bacteria, must extract from the environment those substances needed for the synthesis of new cell material and for the generation of energy for cell maintenance. In general these substances are termed *nutrients*.

To continue to grow properly an organism must have a source of carbon and energy. In addition, elements such as nitrogen, phosphorus, and trace elements including sulfur, potassium, calcium, and magnesium must be available. The two sources of carbon for the synthesis of cell tissue are carbon dioxide and the carbon found in organic matter. If an organism derives its cell carbon from carbon dioxide, it is called *autotrophic*; if it uses organic carbon, it is called *heterotrophic*.

Energy for cell synthesis can be obtained from light and from the chemical oxidation or reduction of inorganic and organic matter. Organisms that use light are known as *phototrophs*, and those that use a chemical energy source are termed *chemotrophs*. A classification of organisms by carbon and energy source is given in Table 2.17.

In addition to carbon and energy, oxygen plays an important role in the growth of cells, as well as being a major component of cell tissue. Many organisms require the presence of molecular oxygen (O_2) for their metabolism. Such organisms are termed *obligate aerobes*. Organisms for which the presence of molecular oxygen is toxic are known as *obligate anaerobes*. These organisms derive the oxygen needed for the synthesis of cells from chemical compounds. There is also a class of organisms that

TABLE 2.17

General Classification of Microorganisms Based on Sources of Energy and Carbon

CLASSIFICATION	ENERGY SOURCE	CARBON SOURCE	REPRESENTATIVE ORGANISMS
Photoautotrophs	Light	CO_2	Higher plants, algae, photosynthetic bacteria
Photoheterotrophs	Light	Organic matter	Photosynthetic bacteria
Chemoautotrophs	Inorganic matter	CO_2	Bacteria
Chemoheterotrophs	Organic matter	Organic matter	Bacteria, fungi, protozoa, animals

can grow in the presence or absence of oxygen. These organisms are known as *facultative anaerobes*. Because of the wide variability of organisms found in nature there are some obligate aerobes that function best at low concentrations of oxygen. These organisms have been termed *microaerophilic*.

Environmental Effects on Organisms

In addition to considering the nutritional requirements of micro-organisms it is also important to consider the effects of environmental factors on microorganisms, especially as those factors affect growth and reproduction. Specifically, it is important to consider the effects of the following factors: (1) chemical composition, (2) pH, (3) oxygen, (4) temperature, and (5) light [2.9, 2.33].

Microorganisms in Water and Wastewater

The principal microorganisms of concern in water and wastewater include bacteria, fungi, algae, protozoa, worms, rotifers, crustaceans, and viruses. Additional details on the microorganisms of concern in water may be found in Refs. [2.22, 2.23, 2.25, 2.33, and 2.41].

Bacteria

Bacteria are single-cell protists. Although there are hundreds of bacteria, most bacteria can be grouped by form, as shown in Fig. 2.24, into four general categories: spheroid, rod, curved rod or spiral, and filamentous. Spherical bacteria, known as *cocci* (singular, coccus), are about 1 to 3 μm in diameter. The rod-shaped bacteria, known as *bacilli* (singular, bacillus) are quite variable in size, ranging from 0.3 to 1.5 μm in width (or diameter) and from 1.0 to 10.0 μm in length. *Escherichia coli*, a common organism found in human feces, is described as being 0.5 μm in width by 2μm in length. Curved rod-shaped bacteria are known as *vibrios* and typically vary in size from 0.6 to 1.0 μm in width (or diameter) and from 2 to 6 μm in length. Spiral bacteria known as *spirilla* (singular, spirillum) may be found in lengths up to 50 μm. Filamentous forms, known under a variety of names, can occur in lengths of 100 μm and longer.

Fungi

Fungi are aerobic, multicellular, nonphotosynthetic, heterotrophic, eucaryotic protists. Most fungi are saprophytes, obtaining their food from dead organic matter. Along with bacteria, fungi are the principal organisms responsible for the decomposition of carbon in the biosphere. Ecologically, fungi have two advantages over bacteria: They can grow in low-moisture areas, and they can grow in low-pH environments. Because of the above characteristics, fungi play an important role in the break-

(a)

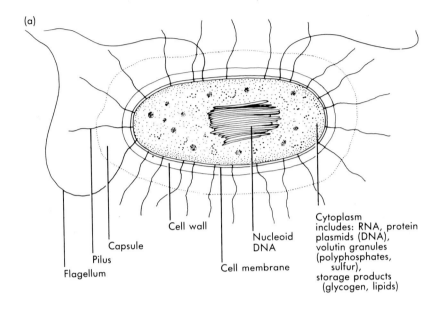

Flagellum
Pilus
Capsule
Cell wall
Nucleoid
DNA
Cell membrane
Cytoplasm
includes: RNA, protein
plasmids (DNA),
volutin granules
(polyphosphates,
sulfur),
storage products
(glycogen, lipids)

(b)

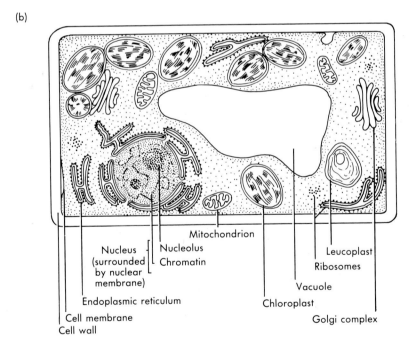

Mitochondrion
Nucleus
(surrounded
by nuclear
membrane)
Nucleolus
Chromatin
Leucoplast
Ribosomes
Vacuole
Chloroplast
Golgi complex
Endoplasmic reticulum
Cell membrane
Cell wall

FIGURE 2.23

Schematic representation of biological cells: (a) bacterial, (b) plant, and (c) animal.

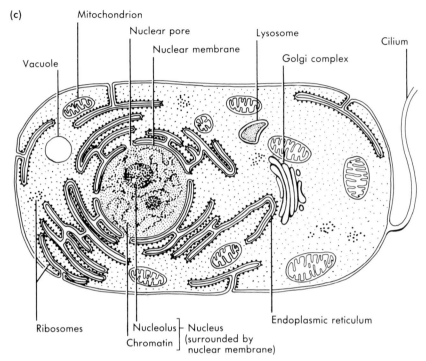

(c)

Vacuole

Mitochondrion

Nuclear pore

Nuclear membrane

Lysosome

Golgi complex

Cilium

Ribosomes

Nucleolus ⎤ Nucleus
Chromatin ⎦ (surrounded by
nuclear membrane)

Endoplasmic reticulum

FIGURE 2.23   (*Cont.*)

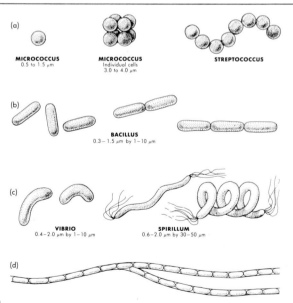

(a)

MICROCOCCUS
0.5 to 1.5 μm

MICROCOCCUS
Individual cells
3.0 to 4.0 μm

STREPTOCOCCUS

(b)

BACILLUS
0.3–1.5 μm by 1–10 μm

(c)

VIBRIO
0.4–2.0 μm by 1–10 μm

SPIRILLUM
0.6–2.0 μm by 30–50 μm

(d)

FIGURE 2.24

**Typical shapes of bacteria: (a) spheroid, (b) rod, (c) curved rod or spiral, and (d)
filamentous (made up of chains of individual cells).**

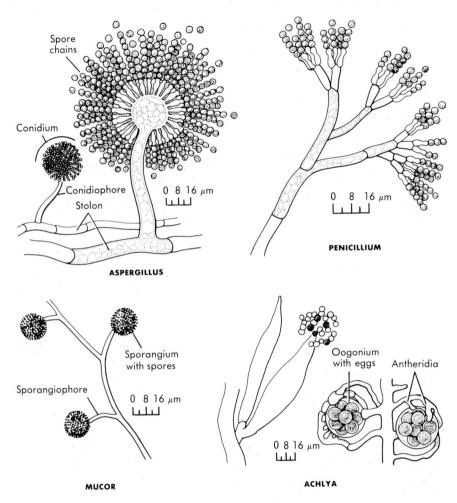

FIGURE 2.25

Representative fungi found in fresh water and wastewater.

TABLE 2.18

Classification of Major Algal Groups

GROUP	DESCRIPTIVE NAME	AQUATIC HABITAT
Chlorophyta	Green algae	Fresh and salt water
Chrysophta	Diatoms, golden-brown algae	Fresh and salt water
Cryptophta	Cryptomonads	Salt water
Euglenophyta	Euglena	Fresh water
Phaeophyta	Brown algae	Salt water
Pyrrhophyta	Dinoflagellates	Fresh and salt water
Rhodophyta	Red algae	Fresh and salt water
Xanthophyta	Yellow-green algae	Fresh and salt water

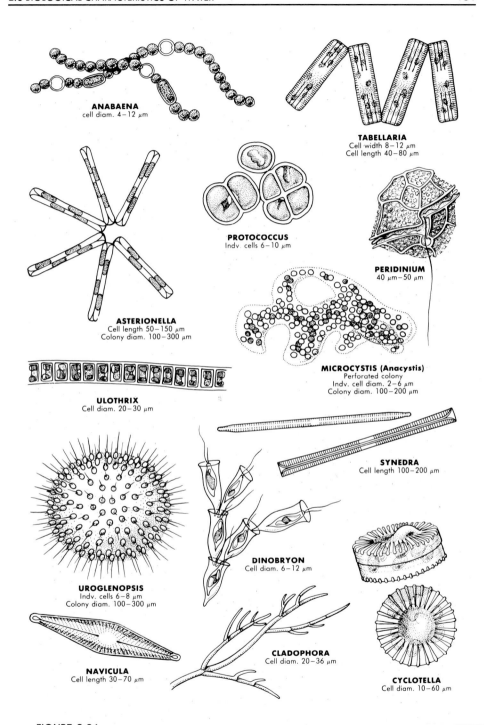

FIGURE 2.26

Representative algal species found in fresh waters.

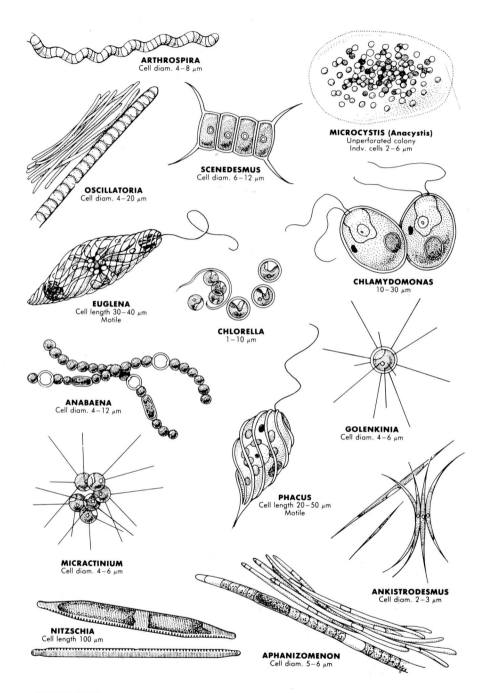

FIGURE 2.27

Representative algal species found in wastewater treatment ponds.

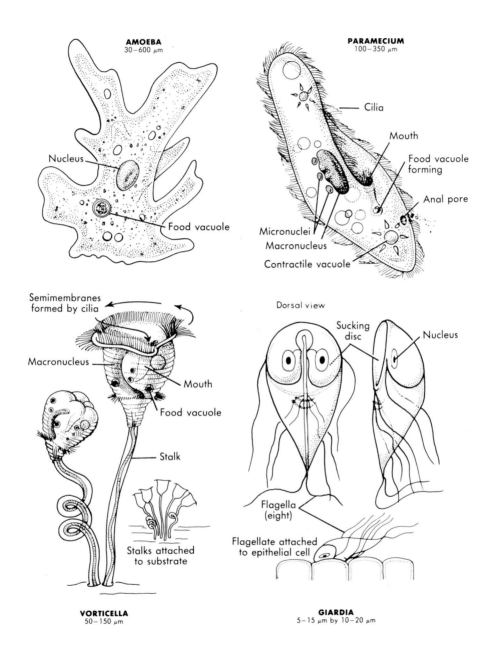

FIGURE 2.28

Representative protozoa found in fresh water and wastewater.

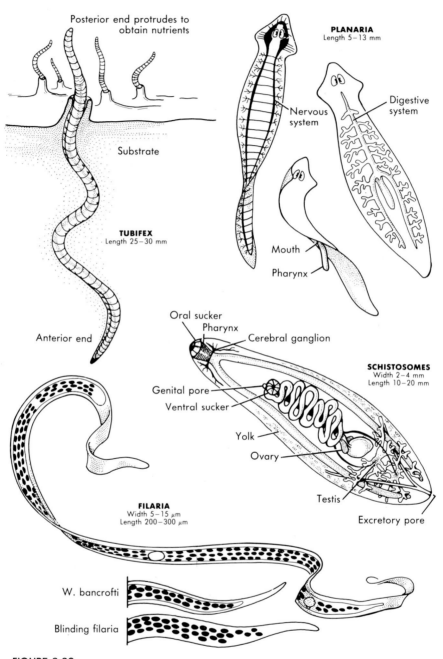

FIGURE 2.29

Representative worms found in fresh water and wastewater.

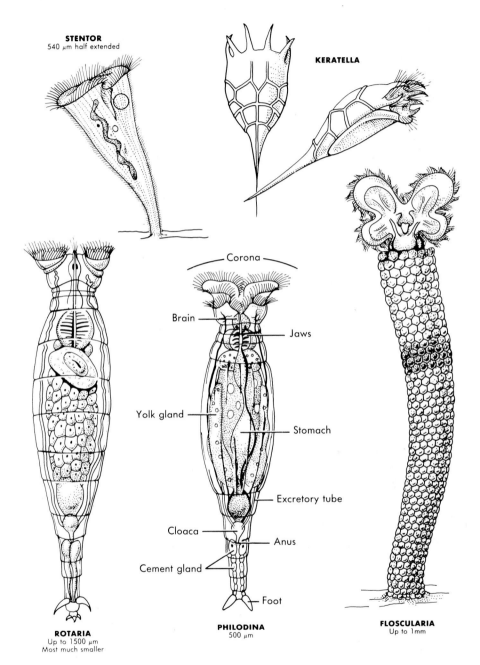

FIGURE 2.30

Representative rotifers found in fresh water and wastewater.

down of organic materials in both terrestrial and aquatic environments. As organic matter is decomposed, fungi release carbon dioxide to the atmosphere and nitrogen to the terrestrial environment. Without the presence of fungi to break down organic material, the carbon cycle would soon cease to exist and organic matter would start to accumulate.

Fungi vary in size from microscopic organisms to mushrooms and are often divided into the following five classes [2.12, 2.22]:

1. Myxomycetes, or slime fungi

2. Phycomycetes, or aquatic fungi

3. Ascomycetes, or sac fungi

4. Basidiomycetes, or rusts, smuts, and mushrooms

5. Fungi imperfecti, or miscellaneous fungi

With respect to water quality, the first two classes are of greatest importance. Some of the important characteristics of the microscopic fungi are illustrated in Fig. 2.25.

Algae

The name *algae* is applied to a diverse group of eucaryotic microorganisms that share some similar characteristics. Typically, algae are autotrophic, photosynthetic, and contain chlorophyll. Although there is much controversy about the classification of individual algal species, the major groupings most often used are summarized in Table 2.18. The principal feature that is used to distinguish algae from fungi is the presence of chlorophyll in the algae. In addition to chlorophyll, other pigments encountered in algae include carotenes (orange), phycocyanin (blue), phycoerythrin (red), fucoxanthin (brown), and xanthophylls (yellow). Combinations of these pigments result in the various colors of algae observed in nature.

Metabolically, algae utilize the CO_2 present in water for the synthesis of cell carbon. Three classes of pigments—chlorophylls, carotenoids, and phycobilins—are used to absorb light energy for photosynthetic cell reproduction and cell maintenance. Oxygen is produced during the photosynthetic process. At night, in the absence of light, algae utilize oxygen. Although respiration also occurs in the presence of sunlight, the amount of oxygen released usually exceeds the amount used during daylight. Simplified reactions to define algal photosynthesis and respiration are as follows:

Photosynthesis:

$$nCO_2 + nH_2O + \text{Nutrients} \xrightarrow{\text{Light}} (CH_2O)_n + nO_2 \qquad (2.87)$$
$$\text{New algal cells}$$

Respiration:

$$(CH_2O)_n + nO_2 \rightarrow nCO_2 + nH_2O \qquad (2.88)$$

Algae are very important microorganisms with respect to water quality. In an aquatic environment, algae will form a symbiotic relationship with bacteria. If allowed to predominate, they can affect the dissolved oxygen balance by causing anaerobic conditions to exist at night. In the absence of bacteria or other sources of CO_2, some algal species can obtain the CO_2 needed for cell growth from the bicarbonate present in the water, as given by the following reaction:

$$2HCO_3^- \xrightarrow{\text{Algae}} CO_2 + CO_3^{-2} + H_2O \qquad (2.89)$$

When the reaction defined by Eq. (2.89) occurs, the pH of the water will generally increase because of the removal of bicarbonate from the water. Depending on local conditions, calcium carbonate ($CaCO_3$) may be precipitated as the pH increases. Many species of algae have been associated with taste and odor problems. Representative algal species found in fresh water and in wastewater treatment lagoons are shown in Figs. 2.26 and 2.27, respectively. The impact of algae on water quality is considered further in Chapter 3.

Protozoa

Protozoa are single-cell eucaryotic mircoorganisms without cell walls. Most protozoa are free-living in nature, although several species are parasitic, living on or in a host organism. Hosts can vary from primitive organisms such as algae to highly complex organisms, including human beings. The majority of protozoa are aerobic or facultatively anaerobic heterotrophs, although some anaerobic types have been reported. The four major groups of protozoa are reported in Table 2.19 and are illustrated graphically in Fig. 2.28.

TABLE 2.19
Classification of Major Groups of Protozoa

| CLASS | MODE OF MOTILITY | TYPICAL MEMBERS | |
		Name	Remarks
Ciliata	Cilia (usually) multiple	*Paramecium*	Free-swimming
Mastigophora	Flagella (one or more)	*Giardia lamblia*	Causative agent for giardiasis
		Trypanosoma gambiense	Causative agent for African sleeping sickness
Sarcodina	Pseudopodia (some with flagella)	*Entamoeba histolytica*	Causative agent for dysentery
Sporozoa	Creeping, often nonmotile flagella at some stages	*Plasmodium vivax*	Causative agent for malaria

One of the protozoa of great interest and concern in drinking water supplies in the United States is *Giardia lamblia*. This flagellated protozoan, which can grow in the upper small intestine, is the cause of the intestinal disease giardiasis. The symptoms of giardiasis are varied depending on the individual but include diarrhea, nausea, indigestion, flatulence, bloating, fatigue, and appetite and weight loss. Unless treated properly, this disease can be chronic. Giardiasis is the most widespread of the protozoan diseases occurring throughout the world. Typically the disease is contracted by drinking surface water contaminated by wild animals or humans [2.8]. Hikers who drink water from mountain streams often acquire the disease. The life cycle of *Giardia lamblia* is illustrated in Fig. 3.9 in Chapter 3.

Worms

A number of worms are of importance with respect to water quality, primarily from the standpoint of human disease [2.18]. Two important worm phyla are the Platyhelminthes and the Aschelminthes (Fig. 2.29). The common name for the phylum Platyhelminthes is flatworms. Free-living flatworms of the class Turbellaria are present in ponds and quiet streams all over the world. The most common form is planarians. Two classes of flatworms are composed entirely of parasitic forms. They are the class Trematoda, commonly known as flukes, and the class Cestoda, commonly known as tapeworms.

The most important members of the phylum Aschelminthes are the nematodes. About 10,000 species of nematodes have been identified, and the list is growing. Most nematodes are free-living. Of greater interest to humans are the parasitic forms. The most serious parasitic forms are *Trichinella*, which causes trichinosis; *Necator*, which causes hookworm; *Ascaris*, which causes common roundworm infestation; and *Filaria*, which causes filariasis.

Rotifers

Rotifers are the simplest of the multicellular animals. The name is derived from the apparent rotating motion of the cilia located on the head of the organism (Fig. 2.30). The cilia are used for motility and for capturing food. Metabolically, rotifers can be classified as aerobic chemoheterotrophs. Bacteria are the principal food source for rotifers.

Crustaceans

Like rotifers, crustaceans are aerobic chemoheterotrophs that feed on bacteria and algae. These hard-shelled, multicellular animals are an important source of food for fish.

Viruses

Viruses are obligate parasitic particles consisting of a strand of genetic material—deoxyribonucleic acid (DNA) or ribonucleic acid

(a)

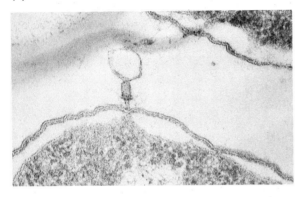

(b)

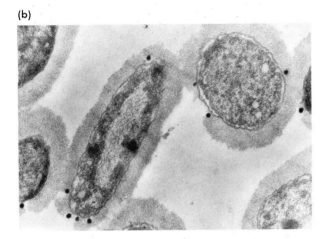

FIGURE 2.31

Electron micrographs of viruses. (a) Ultrathin section of virus (Bacteriophage T2) infecting *E. coli* cell. Bacteriophage DNA has flown from the hexagonal head of the virus through the tail into the cell. The width of the hexagonal head of the virus is approximately 60 nm. (b) Ultrathin section of capsulated *E. coli* cells. The capsule shows up as a gray mass surrounding the cell. Viruses (Bacteriophage K26), that appear as dark dots, have perforated the capsule enzymatically and are ready to infect the cell. The dark dots are the DNA of the individual viruses.

Source: Courtesy of Manfred E. Bayer, Fox Chase Cancer Center, Philadelphia, PA.

(RNA)—within a protein coat (Fig 2.31). The particles do not have the ability to synthesize new compounds. Instead, they invade living cells, where the viral genetic material redirects cell activities toward production of new viral particles at the expense of the host cell growth and maintenance. When the infected cell dies, large numbers of viruses are released to infect other cells.

All viruses are extremely host-specific. Thus a particular viral type can attack only one species of organism. In addition, genetic subspecies will be either particularly susceptible or immune to attack. Examples are the immunity of some individuals to the "common cold" or to herpes simplex I (which causes cold sores). A number of viral diseases are commonly transferred via water. Thus methods of inactivation and enumeration of viruses are of great interest to engineers. Unfortunately, the host specificity of viruses makes their enumeration difficult, and their structural simplicity makes mechanical or chemical methods of inactivation costly. In combination these characteristics make monitoring the effectiveness of treatment both difficult and costly.

Enumeration of Bacteria

Bacteria are normally present in water, and it is their concentration that is significant with respect to potential disease transmission. Enumeration of the bacteria in water is accomplished by growing them, after suitable dilution, in either a solid- or a liquid-culture medium.

Bacterial Counts Using a Solid Medium

One of the most useful techniques for the enumeration of bacteria using a solid-culture medium is known as the *plate count method*. In this method, the first step is to prepare 10-fold dilutions of the sample to be analyzed. The dilutions are prepared as illustrated in Fig. 2.32. The next step is to transfer 1-mL quantities of each dilution, starting with the highest dilution, into separate sterile petri dishes. Replicate samples should always be run. The same pipet may be used throughout, as the dilution effect of the volume carried over from one dilution to the other is insignificant. After the pipeting has been completed, 12 to 20 mL of liquified culture medium is poured into each petri dish. As soon as the medium is added, the sample and the medium are mixed. The mixture is then allowed to solidify. After the mixture has solidified, the petri dishes are inverted and incubated at 35°C ($\pm 0.5°$) for 48 (± 3) hr.

After incubation, the bacterial colonies that develop are counted. Dishes where the number of colonies is between 30 and 300 are considered significant. The bacterial count per milliliter is obtained by multiplying the average count, determined from replicate samples, by the appropriate dilution factor. In applying this test, it is assumed that each colony counted originated from a single bacterial cell. Dishes that have too many colonies or that have colonies that are grown together are not used because of the high probability that the colonies resulted from more than one cell. Thus only selected dilutions are used in estimating bacterial concentrations.

It is important to remember that different bacterial species have different nutrient and environmental requirements. Thus a singe nutrient medium is unlikely to give an accurate estimate of the total population. But this principle of selectivity is useful when it is desired to enumerate one or a very few species of bacteria to the exclusion of others. In this case careful selection of nutrient medium and environmental conditions must be made. The best example in water quality management is the use of the coliform group of bacteria as an indicator of recent pollution and of the probable presence of pathogenic organisms.

Bacterial Counts Using a Liquid Medium

The coliform group of organisms, including *Escherichia coli*, which is found in the intestinal tract of humans as well as of other warm-blooded animals, is capable of fermenting lactose with the production of an abundance of gas. Lactose is a relatively uncommon sugar normally

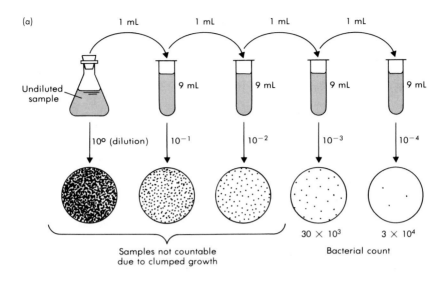

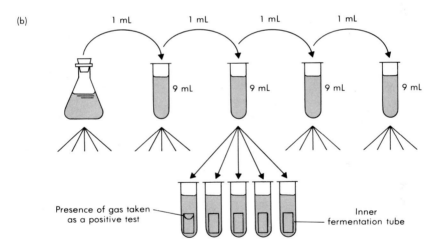

FIGURE 2.32

Illustration of methods used to obtain bacterial counts: (a) use of a solid medium and (b) use of a liquid medium (see also Fig. 2.33).

found only in milk. To enumerate the number of gas-producing organisms in a sample, a test using a liquid lactose medium has been developed. The test involves observing whether gas has formed in an inverted gas-collection tube placed in a large test tube containing the culture medium (Fig. 2.33).

   The principle of the test is based on dilution to extinction. Originally, the test was developed to determine the number of coliform organisms that were present in water samples. For example, if a 10-fold serial

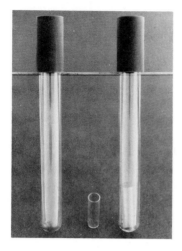

FIGURE 2.33

Multiple-fermentation tubes. Tube on right contains inverted gas-collection tube used in determining bacterial densities (see also Fig. 2.32).

dilution is made and growth, as measured by gas production, is observed in the 10^{-n} dilution but not in the $10^{-(n+1)}$ dilution, then it can be concluded that the sample contains at least 10^n cells per milliliter but less than 10^{n+1} cells per milliliter. Clearly if only one test is made, the results are subject to great statistical variation. The statistical estimate of the number of bacterial cells is improved when multiple tests are conducted at each dilution.

The test in which multiple tests are made at each dilution is known as the multiple-fermentation-tube technique. The test may be described as follows. First a series of dilutions are made, as illustrated in Fig. 2.32. The next step is to transfer a 1-mL sample from each dilution to each of five test tubes containing a suitable lactose culture medium and inverted gas-collection tubes (Fig. 2.33). The inoculated tubes are incubated in a water bath at 44.5°C ($\pm 0.2°$) for 24 (± 2) hr. The accumulation of gas in the inverted gas-collection tubes after 24 hr is considered to be a positive reaction. The results for each dilution are reported as a fraction, with the number of positive tubes over the total number of tubes. For example, the fraction $3/5$ denotes three positive tubes in a five-tube sample. Estimation of bacterial numbers using the test results from the multiple-tube technique is considered subsequently.

Bacterial Counts Using the Membrane-Filter Technique

Where the quality of the water to be tested is high, the membrane-filter technique has been used successfully. The membrane filter apparatus is shown in Fig. 2.34. Operationally, the water sample to be tested is filtered through the membrane. The typical pore size in the membrane is 0.45 μm, so that essentially all of the bacteria of interest will be retained on the filter.

FIGURE 2.34

Membrane filter apparatus used to test for bacteria in relatively clean waters. After centering the membrane filter on the filter support, the funnel top is attached and the water sample to be tested is poured into the funnel. To aid in the filtration process, a vacuum line is attached to the base of the filter apparatus. After the sample has been filtered, the membrane filter is placed in a petri dish containing a culture medium for bacterial analysis.

After a known volume of water has been filtered, the filter is removed and transferred to a small petri dish containing a sterile absorbent pad saturated with a suitable culture medium. The filter membrane is placed face up on the culture medium. After incubation in the inverted position, the bacterial colonies are counted. If the volume of water that was passed though the filter is known, the organism count per unit volume can be determined.

Estimation of Bacterial Densities

Concentrations of coliform bacteria are most often reported as the MPN (most probable number) per 100 mL. Typically the MPN value is determined from the number of positive (lactose-fermenting) tests in a set of five replicates made at three different dilutions (15 samples altogether). For example, a water sample may be diluted to 10, 1, and 0.1 mL/100 mL. After incubating five replicates of each dilution in lactose broth at a standard temperature for a standard time—44.5°C ($\pm 0.2°$) for 24 (± 2) hr—the number of tubes of each dilution in which lactose has been fermented (as noted by the formation of gas) is recorded. The bacterial density is predicted as the value most likely to give the observed distribution of positive and negative values.

Estimation of the MPN is based on the Poisson distribution for extreme values [2.14]. The joint probability y of a given result from a series of three dilutions is given by Eq. (2.90):

$$y = \frac{1}{a}\left[(1 - e^{-n_1\lambda})^{p_1}(e^{-n_1\lambda})^{q_1}\right]\left[(1 - e^{-n_2\lambda})^{p_2}(e^{-n_2\lambda})^{q_2}\right]$$
$$\times \left[(1 - e^{-n_3\lambda})^{p_3}(e^{-n_3\lambda})^{q_3}\right] \tag{2.90}$$

where

y = probability of occurrence of a given result

a = constant for a given set of conditions

n_1, n_2, n_3 = sample size in each dilution, mL

λ = coliform density, number/mL

p_1, p_2, p_3 = number of positive tubes in each sample dilution

q_1, q_2, q_3 = number of negative tubes in each sample dilution

As noted above, three dilutions with five tubes per dilution are normally used. Equation (2.90) can be extended to any number of dilutions by adding additional terms for each dilution. The MPN value is 100 λ and can be found by finding the maximum value of ya for given n_1, p_1, and q_1 values. Thus the value of a need not be known. The evaluation of ya is not difficult using a programmable calculator or microcomputer, and in some cases calculation is the simplest method of estimation. The application of Eq. (2.90) is illustrated in Example 2.12.

EXAMPLE 2.12

DETERMINATION OF MPN USING POISSON DISTRIBUTION

Using the Poisson distribution [Eq. (2.90)], determine the MPN/100 mL for a water sample from which 10, 1, and 0.1 mL/100 mL dilutions resulted in four, three, and five positive-lactose-formation tubes, respectively.

SOLUTION:

1. Determine coefficient values for use in Eq. (2.90).

DILUTION	n_i	p_i	q_i
1	10	4	1
2	1	3	2
3	0.1	5	0

2. Substitute the coefficient values in Eq. (2.90) and determine ya values for selected values of λ.

$$y = \frac{1}{a}\left[(1 - e^{-10\lambda})^4(e^{-10\lambda})^1\right]\left[(1 - e^{-\lambda})^3(e^{-\lambda})^2\right]$$
$$\times \left[(1 - e^{-0.1\lambda})^5(e^{-0.1\lambda})^0\right]$$

λ	ya
0.56	4.490×10^{-11}
0.57	4.521×10^{-11}
0.58	4.541×10^{-11}
0.59	4.550×10^{-11}
0.60	4.548×10^{-11}
0.61	4.535×10^{-11}

3. The maximum value of ya occurs for a λ value of 0.59 organisms per milliliter. Thus the MPN/100 mL is

$$\text{MPN}/100 \text{ mL} = 100(0.59) = 59$$

COMMENT

The most direct method of obtaining a first estimate for λ is to use the Thomas equation given below. Alternatively, where more than three dilutions are involved, the MPN value obtained using the tables given in Appendix G can be used.

More commonly, MPN values are determined using tables such as those found in Appendix G. It should be noted, however, that the tables given in Appendix G are based on a series of three dilutions of five replicates each. In many cases, more than three dilutions will be made. When more than three dilutions have been made, the rules given in Appendix G are used to select the appropriate dilutions to be used in determining the MPN. Values from the tables are reported as MPN/100 mL.

When tables are not available, the Thomas equation can be used to estimate the MPN [2.37]. The Thomas equation is

$$\text{MPN}/100 \text{ mL} = \frac{\text{no. of positive tubes} \times 100}{\left(\begin{array}{c}\text{mL of sample in} \\ \text{negative tubes}\end{array} \times \begin{array}{c}\text{mL of sample in} \\ \text{all tubes}\end{array}\right)^{1/2}} \quad (2.91)$$

In applying the Thomas equation to situations in which some dilutions have all five tubes positive, the count of positive tubes should begin with the highest dilution in which at least one negative result has occurred. The application of the Thomas equation is illustrated along with other methods in Example 2.13.

EXAMPLE 2.13

DETERMINATION OF MPN USING STANDARD TABLES AND THE THOMAS EQUATION

Six water samples were analyzed for coliform bacteria using the five-tube lactose-fermentation technique. The results are given below. Determine the MPN/100 mL for each sample using the MPN tables given in Appendix G and the Thomas equation.

DILUTION, mL/100 mL	NUMBER OF POSITIVE TUBES					
	Water sample					
	1	2	3	4	5	6
10.0	5	4	5	5	5	5
1.00	3	3	5	5	5	5
0.10	1	5	5	5	0	5
0.01	—	—	3	5	3	4
0.001	—	—	2	3	2	3
0.0001	—	—	1	2	1	1

SOLUTION:

1. Determine the MPN values using the tables and data given in Appendix G.

a. The first step is to adjust the reported test results for samples 3, 4, 5, and 6 according to the rules outlined in Appendix G. The modified test results and the dilutions to be used for determining the MPN values are presented below in boldface.

DILUTION, mL/100 mL	WATER SAMPLE					
	1	2	3	4	5	6
10	**5**	**4**	5	5	5	5
1	**3**	**3**	5	5	**5**	5
0.1	**1**	**5**	5	5	**0**	5
0.01	—	—	**3**	5	**5**	**4**
0.001	—	—	**3**	**3**	—	**4**
0.0001	—	—	—	**2**	—	—

b. The MPN values for the five water samples are then determined from Table G.1, using the appropriate dilution factor as a multiplier. The results are as follows:

WATER SAMPLE	MPN/100 mL
1	110
2	59
3	18,000
4	140,000
5	950
6	35,000

2. Determine the MPN value using the Thomas method [Eq. (2.91)] and the modified test data given in step 1(a) above. In applying the Thomas method it is important to remember that the count of positive tubes begins with the highest dilution in which at least one negative result has occurred.

SAMPLE	CALCULATION	MPN/100 mL
1	$400/(2.4 \times 5.5)^{1/2}*$	110
2	$1200/(12 \times 55.5)^{1/2}$	47
3	$600/(0.022 \times 0.055)^{1/2}$	17,248
4	$500/(0.0023 \times 0.0055)^{1/2}$	140,580
5	$500/(0.5 \times 0.55)^{1/2}$	953
6	$800/(0.011 \times 0.055)^{1/2}$	32,525

*Number of positive tubes $(3 + 1) = 4$; milliliters of sample in negative tubes $[(2 \times 1) + (4 \times 0.1)] = 2.4$; milliliters of sample in all tubes $[(5 \times 1) + (5 \times 0.1)] = 5.5$.

COMMENT

As can be seen, the values obtained from the MPN tables and the Thomas equation are quite similar. However, where more than three dilutions are made, use of the Poisson equation is recommended.

The Use of *Escherichia coli* as an Indicator Organism

The coliform group of bacteria includes all aerobic and facultative anaerobic, gram-negative, nonspore-forming, rod-shaped bacteria that ferment lactose with gas formation [2.32]. *Escherichia coli* is the most widely known member of the group because it is a commonly used organism for bacteriological experiments. *Escherichia coli* is a normal inhabitant of the intestine of warm-blooded animals, including humans. Because of the abundance of *E. coli* in the human intestinal tract, it was reasoned in the early 1900s that water could not become polluted with human fecal matter without this organism being present. The presence of this organism was, and is today, taken as an indication of the presence of fecal matter and of the possible presence of pathogenic organisms of human origin.

Although tests are available for specific pathogenic organisms, there is no way of knowing which pathogenic organism is present in a sample. Also the cost of testing for all pathogenic organisms is prohibitive. Hence *E. coli* has come to be used as an indicator organism for human pollution. The key to enumerating the coliform group of organisms is, as noted above, that they ferment (i.e., under anaerobic conditions they metabolize or grow on) lactose. A series of three tests—known as the presumptive, confirmed, and completed tests—are performed to confirm positively the presence of the coliform group [2.32]. The results of the presumptive test (in which multiple-fermentation tubes are inoculated and incubated and gas production is noted) are used to determine the bacterial density, expressed as MPN/100 mL.

Problems associated with the use of *E. coli* as an indicator organism arise from the fact that (1) *E. coli* is not a single species as was thought originally, (2) certain genera of the coliform group such as *Proteus* and *Aerobacter* are normally found outside the human intestinal tract in soil, (3) other organisms found in water that do not represent fecal pollution possess some of the characteristics attributed to *E. coli*, such as the ability to ferment lactose, and (4) *E. coli* identical to that found in humans is also found in the intestinal tract of other warm-blooded animals. Notwithstanding the above-mentioned problems, *E. coli* is still used as an indicator organism, as there is no suitable replacement.

Three other organisms always found in the intestinal tract of humans (and animals) are fecal streptococci, typically *Streptococcus faecalis*; clostridia, typically *Clostridium perfringens*; and certain species of anaerobic lactobacilli. The first of these is now used frequently in conjunction with the coliform test as an indication of fecal contamination. The use of both fecal coliform and fecal streptococci to assess pollution problems is considered further in Section 3.3.

Bioassays

Aquatic bioassays are designed to determine the responses of aquatic organisms to water quality. Bioassays are classified according to (1) the duration of the test—short term is less than 8 d, intermediate term is 8 to 90 d, and long term is greater than 90 d; (2) the method of adding the test solutions (e.g., static, flow-through, or recirculation); and (3) the purpose of the test (e.g., for effluent monitoring or for relative toxicity or growth rate assessment) [2.32]. Short-term static tests are used to monitor effluent discharges. Intermediate-term static and flow-through tests are used to assess the water quality effects on the life stages of organisms with long life cycles. Long-term flow-through tests are used to assess the maximum allowable toxicant concentration (MATC) that does not produce harmful effects with continuous exposure.

In most cases, fish are the organisms studied. The species chosen should be representative of the receiving water being studied and should be the most sensitive to environmental change. For example, choosing suckers for a bioassay study of a trout stream would be inappropriate. Typically, short-term bioassays using fish as the test organism are conducted by adding various percentages of the water and wastewater to be tested to several aquaria containing from 10 to 20 fish each. Survival of the test fish is recorded with time. The most common measure used in reporting bioassay results is the *median lethal concentration* (*LC 50*), which is defined as the concentration of the contaminant that will result in death of 50 percent of the organisms in a specified time period. Usual time periods are 24, 48, and 96 hr.

Procedures for the bioassay test using fish and other organisms are given in *Standard Methods* [2.32]. In addition to the procedures given in *Standard Methods*, biologists often develop modifications suited to a specific site or problem. For example, in-situ tests are often made using cages to contain the test organisms. Cages may be placed above and at several points below a wastewater discharge to determine the effects on the test species under actual field conditions.

KEY IDEAS, CONCEPTS, AND ISSUES

- The quality of a water can be defined in terms of its physical, chemical, and biological characteristics.

- Water quality can be defined in terms of specific parameters such as the concentration of Ca^{+2}, Mg^{+2}, SO_4^{-2}, and Cl^- or of gross parameters such as suspended solids, alkalinity, hardness, and BOD.

- Gross parameters cannot be used to discriminate between the individual constituents responsible for the overall effect.

- Chemical water quality parameters are determined using gravimetric, volumetric, and physicochemical methods of analysis.

- The principal ionic constituents present in most natural water include Ca^{+2}, Mg^{+2}, K^+, Na^+, HCO_3^-, SO_4^{-2}, Cl^-, and NO_3^-.

- The principal classes of organic matter found in water and wastewater include proteins, carbohydrates, and lipids.

- The BOD test is most commonly used to measure the presence of gross amounts of organic matter in water and wastewater.

- The TOC test is used to measure low concentrations of organic matter in water.

- The measurement of trace amounts of organic compounds requires a three-step process involving (1) isolation, (2) resolution, and (3) detection.
- Organic compounds of anthropogenic origin are troublesome when found in water, even in trace amounts, because they may be carcinogenic or mutagenic.
- The bacteria in water are enumerated using a technique known as the plate count method, in which a solid-culture medium is used, or a technique known as the multiple-tube fermentation method, in which a liquid-culture medium is used. The results from the latter technique are reported as MPN (most probable number) per 100 mL.
- Because pathogenic bacteria are difficult to quantify, *Escherichia coli* is used as an indicator organism. If *Escherichia coli* is found in a water sample, it is taken as an indication that the water may be contaminated by fecal matter and that pathogenic bacteria of human origin may be present.

DISCUSSION TOPICS AND PROBLEMS

2.1. Determine the molality of a solution prepared by dissolving 49 g of sulfuric acid (H_2SO_4) in 500 g of water.

2.2. Determine the molarity of a 25% solution of HCl: (a) by mass and (b) by volume. Assume the density of HCl is 1.2 kg/L.

2.3. As sold commercially, sulfuric acid is approximately 95 percent H_2SO_4 by mass. If the density of H_2SO_4 is 1.83 kg/L, what is the molarity and mole fraction of H_2SO_4 in the commercial product?

2.4. Determine the molality, molarity, and mole fraction of 86-proof (43% ethyl alcohol by volume) Wild Turkey whiskey. Assume the density of ethyl alcohol (C_2H_5OH) is 0.79 kg/L.

2.5. What is the normality of a 1-L solution containing
   a. 1.25 g H_2SO_4
   b. 5.0 mg H_3PO_4
   c. 6.5 mg $Ca(OH)_2$
   d. 10 g H_2S
   State all assumptions.

2.6. Given the reaction $A + B \rightleftharpoons C + D$, determine the equilibrium constant if 0.60 mol of C are formed when 1.0 mol of A and B are present initially.

2.7. If acetic acid (CH_3COOH) is ionized to the extent of 1.33 percent at 20°C, determine the ionization constant. Assume the initial concentration of acetic acid is 0.1 M and that the following reaction is applicable:

$$CH_3COOH \rightleftharpoons H^+ + CH_3COO^-$$

2.8. You are given a box containing NH_3, N_2, and H_2 at equilibrium at 1000 K. The contents are analyzed, and the following concentrations are found: $NH_3 = 1.02$ mol/L, $N_2 = 1.03$ mol/L, and $H_2 = 1.62$ mol/L. Calculate the value of K for the reaction:

$$N_{2(g)} + 3H_{2(g)} \rightleftharpoons 2NH_{3(g)}$$

2.9. Determine the ionic strength of the solution of a solution containing 0.015 M NaCl and 0.020 M CaSO$_4$.

2.10. Determine the ionic strength of the solution and the activity coefficients for each ion in a solution containing 0.01 M NaCl, 0.005 M Ca(OH)$_2$, and 0.01 M MgSO$_4$.

2.11. Determine the ionic strength of a water with the chemical characteristics given in Problem 2.28.

2.12. Determine the ionic strength of a water with the chemical characteristics given in Problem 2.29.

2.13. Determine the suspended solids and the percent volatile matter in a solids sample based on the following data:

Sample size = 25 mL

Tare mass of filter = 1.5325 g

Tare mass of filter plus retained solids = 1.5415 g

Tare mass of filter plus retained ash = 1.5378 g

2.14. Data taken on solids contents of a water sample are given below. All samples were 50-mL aliquots.

Residue in evaporating dish after 4 hr at 105°C = 36 mg

Residue in evaporating dish after 4 hr at 550°C = 34 mg

Residue on Whatman GFC filter after drying at 105°C for 1 hr = 12 mg

Net ash content (tare subtracted) after firing filter and residue at 550°C = 11 mg

Determine the concentrations of total solids, total volatile solids, suspended solids, and volatile suspended solids.

2.15. Explain why inorganic compounds such as MgCO$_3$, which are unstable when exposed to heat, can induce an error in the measurement of volatile solids.

2.16. Use an anion-cation balance to determine if the following sample analysis for waters 1 and 3 is acceptable.

CONSTITUENT CONCENTRATION, g/m^3	WATER SAMPLE			
	1	2	3	4
Ca^{+2}	44.4	120.0	76.0	23.1
Mg^{+2}	15.5	75.0	26.8	16.7
Na^+	13.8	1.86	23.0	2.3
K^+	7.8	15.6	19.6	3.9
Cl^-	11.4	42.7	37.2	?
SO_4^{-2}	33.1	?	240.0	63.5
CO_3^{-2}	—	—	10.4	2.6
HCO_3^-	195.2	156.9	126.5	80.3

2.17. Assuming no other constituents are missing, use an anion-cation balance to find the missing constituent concentrations for waters 2 and 4 of Problem 2.16.

2.18. Check the accuracy of the water analysis given in Problem 2.32.

2.19. Check the accuracy of the water analysis given in Problem 2.33.

2.20. Determine the pH of a water in equilibrium with the atmosphere at 5, 25, and 40°C.

2.21. Determine the hydrogen ion concentration in pure water at the following temperature values:
   a. $T = 22°C$
   b. $T = 33°C$
   c. $T = 42°C$

2.22. What is the pH of a solution containing 2.3×10^{-4} g/m^3 of OH^- at 25°C?

2.23. A 0.01 molar solution of a monoprotic acid is maintained at 20°C. The K_a value for the acid is 1×10^{-8}. Determine the pH of the solution.

2.24. Construct logarithmic-concentration-versus-pH diagrams for the following acid-base systems:
   a. 10^{-2} M H_2CO_3 ($pK_1 = 6.35$, $pK_2 = 10.33$)
   b. 10^{-2} M NH_4Cl ($pK_a = 9.26$)
   Using the diagrams constructed, determine the pH and the concentration of all of the species in solution for the following solutions:
   c. 10^{-2} M $NaHCO_3$
   d. 10^{-2} M $NaHCO_3$ and 0.2 M NH_4Cl
   e. 10^{-2} M $(NH_4)_2CO_3$
   f. 10^{-2} M Na_2CO_3

2.25. Using a logarithmic-concentration diagram, determine the pH of a solution containing 10^{-2} M $NaHCO_3$ and 10^{-2} M Na_2CO_3. Assume the system is closed.

2.26. Hydrogen sulfide is a diprotic acid with $K_1 = 1.1 \times 10^{-7}$ and $K_2 = 1.0 \times 10^{-14}$. Determine, for a closed system, the concentration of $[H_2S]$, $[HS^-]$, and $[S^{-2}]$ and the pH for a 10^{-2} M H_2S solution.

2.27. Phosphoric acid (H_3PO_4) undergoes three dissociations and correspondingly has three dissociation constants—K_1, K_2, and K_3. At 25°C the values are

$$K_1 = 7.5 \times 10^{-3}$$
$$K_2 = 6.2 \times 10^{-8}$$
$$K_3 = 4.8 \times 10^{-13}$$

Determine the fraction of each species present as a function of pH for values between 1 and 14. Assume the total amount of phosphate present is 10^{-3} mol/L. Present the results in graphical form.

2.28. A groundwater is being considered for use as a domestic supply. The water has a temperature 15°C and the following chemical analysis:

CONSTITUENT	CONCENTRATION, g/m^3
Ca^{+2}	190
Mg^{+2}	84
Na^+	75
Fe^{+2}	0.1
Cd^{+2}	0.2
HCO_3^-	260
SO_4^{-2}	64
CO_3^{-2}	30
Cl^-	440
NO_3^-	35

a. Use an anion-cation balance to determine if the sample analysis is reasonable.
b. What is the hardness of the water?
c. Determine the alkalinity of the water.
d. Estimate the sample pH.
e. Would the water be an acceptable domestic supply? Why or why not?

2.29. Water from a well has the measured constituent concentrations listed on the table on the following page:
a. Does the analysis appear complete?
b. What is the hardness of the water?
c. Determine the alkalinity of the water.
d. Determine the pH of the water if the temperature is 20°C.

CONSTITUENT	CONCENTRATION, eq/m^3
Ca^{+2}	1.60
Mg^{+2}	4.69
Na^{+}	2.39
K^{+}	0.03
Cr^{+6}	< 0.0001
Fe^{+2}	< 0.0002
Pb^{+2}	< 0.0001
Mn^{+2}	< 0.0001
CO$_3{}^{-2}$	1.87
HCO$_3{}^{-}$	5.49
Cl^{-}	1.04
SO$_4{}^{-2}$	1.00
NO$_3{}^{-}$	0.30

2.30. Given the following water analysis data measured at 25°C, estimate the Cl^{-} and HCO$_3{}^{-}$ concentrations and determine the total hardness and alkalinity.

CONSTITUENT	CONCENTRATION, g/m^3
Ca^{+2}	40.0
Mg^{+2}	24.3
Na^{+}	23.0
K^{+}	39.1
Cl^{-}	?
SO$_4{}^{-2}$	48.0
HCO$_3{}^{-}$	?
CO$_2$	3
pH	7.9

2.31. Given the following water analysis data, determine the unknown values.

$$Ca^{+2} = 40 \text{ g/m}^3 \qquad Cl^{-} = 35.5 \text{ g/m}^3$$
$$Mg^{+2} = ? \qquad SO_4{}^{-2} = 96 \text{ g/m}^3$$
$$Na^{+} = ? \qquad HCO_3{}^{-} = ?$$
$$K^{+} = 39.1 \text{ g/m}^3 \qquad \text{Alkalinity} = 3.0 \text{ eq/m}^3$$
$$\text{Noncarbonate hardness} = 1.0 \text{ eq/m}^3$$

2.32. For a water having the characteristics given below at 25°C, determine (a) the hardness, (b) the carbonate hardness, and (c) the pH if the water is in equilibrium with CO$_2$ in the atmosphere.

CONSTITUENT	CONCENTRATION, g/m^3
Ca^{+2}	2.00
Mg^{+2}	85.05
K^+	39.10
Na^+	92.00
HCO_3^-	61.00
Cl^-	284.00
CO_3^{-2}	3.00
SO_4^{-2}	288.00

2.33. Based on the following analysis of a groundwater sample, determine
a. The alkalinity in g/m^3 as $CaCO_3$
b. The hardness in g/m^3 as $CaCO_3$
c. The total dissolved solids concentration in grams per cubic meter
d. The *EC* value analytically using Eq. (2.60), and compare to the given value

CONSTITUENT	CONCENTRATION, g/m^3
Ca^{+2}	200.0
Mg^{+2}	12.15
K^+	39.1
Na^+	230.0
HCO_3^-	610.0
Cl^-	71.0
SO_4^{-2}	480.0
$T, °C$	20.0
pH	7.0
$EC, \mu S/cm$	2.6×10^3

2.34. A cooling water has an initial temperature of $10°C$ and is heated to $25°C$ in an industrial process. Using the partial $10°C$ data given below estimate the change in the bicarbonate and carbonate concentrations in grams per cubic meter, assuming constant pH.

CONSTITUENT	CONCENTRATION, g/m^3
Ca^{+2}	40
Mg^{+2}	0
HCO_3^-	122
pH	8.0

2.35. Determine the *EC* value for the water given in Problem 2.28.

2.36. Estimate the *EC* value for the water given in Problem 2.29.

2.37. Use Eq. (2.60) to estimate the *EC* value for waters 1, 2, and 4 in Table 3.2. Compare the calculated and reported values.

2.38. Determine the stoichiometric coefficient-multiplier for converting the change in dichromate concentration to oxygen equivalents for the COD test [Eq. (2.66)].

2.39. A river sample is analyzed for BOD content without dilution. The results obtained are given below. Estimate the BOD_5, rate constant, and ultimate carbonaceous BOD.

TIME, d	DISSOLVED OXYGEN, CONCENTRATION, g/m^3
0	8.60
1	7.15
2	5.96
3	4.99
4	3.92

2.40. A solution is known to contain 50 g/m^3 of $CH_3(CH_2)_8COOH$ and 42 g/m^3 of NH_3—N. Determine the theoretical ultimate carbonaceous CBOD, the NBOD, and the theoretical total oxygen demand.

2.41. What conclusion can be drawn from the results given below for a BOD analysis made using three dilutions?

PERCENT WASTE	BOD, g/m^3
1	200
2	120
3	80

2.42. Determine the 5-d 20°C BOD of a wastewater if the 3-d 10°C BOD is 100 g/m^3. Assume a first-order model is valid, k equals 0.20 d^{-1}, and the temperature coefficient θ is 1.135.

2.43. Determine the carbonaceous and nitrogenous oxygen demand of a solution containing 250 g/m^3 of acetic acid (CH_3COOH) and 200 g/m^3 of glycine (CH_2NH_3COOH).

2.44. A water sample is known to contain 300 g/m^3 of casein ($C_8H_{12}O_3N_2$). Estimate the COD, BOD_u, and BOD_5. If k equals 0.21 at 20°C, estimate the BOD_5 at 25°C. Assume the temperature coefficient θ is equal to 1.056.

2.45. A wastewater is known to contain only inorganic ammonia nitrogen and other inorganic nutrients.
   a. If the NH_3 concentration is 30 g/m^3, estimate the oxygen demand.

b. How much will the alkalinity be decreased by the oxidation of the NH_3?

2.46. A meat processing wastewater containing 2100 g/m^3 BOD_5 is to be discharged to a stream. Minimum flow rate (95 m^3/s) occurs in January when the water temperature is 6°C, and maximum temperature occurs in July (26°C) when the flow rate is 175 m^3/s. If the maximum in stream BOD_5 value is to be 0.5 g/m^3 at ambient temperatures, determine the necessary extent of treatment for a wastewater flow of 2 m^3/s. Assume the reaction rate constant is 0.2 d^{-1} at 20°C and the temperature coefficient θ is equal to 1.135 ($< 20°C$) and 1.056 ($> 20°C$).

2.47. A discharge is found to contain 6 g/m^3 of organic nitrogen as N and 9 g/m^3 of ammonia nitrogen. Determine the amount of oxygen required for oxidation of the nitrogen to NO_3^- and the amount of alkalinity destroyed in the process.

2.48. Review "BOD Progression in Soluble Substrates" by A. W. Busch in *Sewage and Industrial Wastes*, vol. 30, no. 11, p. 1336, 1958. Do the findings presented in the paper support the concept of a first-order BOD exertion? Discuss applications of BOD theory based on the paper.

2.49. Bacterial cells are often represented by the empirical chemical formula $C_5H_7O_2N$. Determine the potential carbonaceous BOD of 1 g of cells.

2.50. A polluted river sample is analyzed using standard BOD procedures. Data obtained are listed below. Determine the BOD_5, BOD_u, and k values that best fit the data.

	DISSOLVED OXYGEN, g/m^3			
	Dilution, mL/300 mL			
TIME, d	0	1	5	30
0	8.90	8.87	8.77	8.11
1	8.83	8.49	7.66	2.70
2	8.76	8.41	7.25	0
3	8.69	8.23	6.39	
4	8.62	8.06	5.83	
5	8.55	7.91	5.35	
6	8.48	7.77	4.95	
7	8.41	7.64	4.60	

2.51. A water sample is analyzed for BOD on a daily basis for 7 d using the dilution method. Results of the analysis show $k = 0.23$ d^{-1} and $BOD_u = 15$ g/m^3. Twenty-day samples have also been analyzed, and the BOD value is consistently 23 g/m^3. What might cause the difference between estimated and measured BOD values?

2.52. A water sample has been analyzed and found to have a BOD_5 (20°C) of 7.00 g/m^3 and a reaction rate constant of 0.20 d^{-1}. Determine the 6-d BOD value at 15°C. Assume the temperature coefficient θ is equal to 1.047. If the sample contains 3.0 g/m^3 of ammonia nitrogen (3 g N/m^3), determine the total ultimate oxygen demand including nitrogenous and carbonaceous concentrations.

2.53. Use the method proposed by Fujimoto (see Example 2.10) to estimate the BOD_u and k values for the following BOD data.

t, d	1	2	3	4	5	6	7
BOD, g/m^3	23	41	54	65	73	79	84

2.54. Demonstrate (or prove) why the method of BOD analysis proposed by Fujimoto is valid.

2.55. Compare the saturation (equilibrium) concentrations (in grams per cubic meter) of CO_2, O_2, and N_2 in New York (sea level), Denver (1600 m), and Mexico City (2200 m). Assume isothermal conditions at 25°C.

2.56. Determine the equilibrium concentration of oxygen in the liquid phase of a covered pure-oxygen-activated sludge plant subject to 3 atm of pressure. For the purposes of this problem, neglect any reactions occurring in the system. The composition of the gas above the wastewater is 80 percent oxygen, 15 percent nitrogen, and 5 percent carbon dioxide by volume.

2.57. A bottle of soda water is produced by increasing the gas pressure and carbon dioxide content. For a bottle having a gas pressure of 2 atm and a gas CO_2 content of 100 percent determine the water pH at 25°C.

2.58. Bacteria have equivalent diameters of 2×10^{-6} m and densities of approximately 1 kg/L. Under optimal conditions, bacteria can divide every 30 min. Determine the mass of bacteria that would accumulate in 72 hr under continuing optimal growth conditions. Can this occur? Explain.

2.59. If the bacteria found in feces have an average volume of 2.0 μm^3, determine the concentration of suspended solids that would be represented by a bacterial density equal to 10^8 organisms per milliliter. Assume the density of the bacteria is 1.05 kg/L.

2.60. Three water samples were tested for coliform bacteria using the lactose-fermentation method. For the test results given below, determine the MPN value using (a) the Poisson distribution and (b) the MPN tables given in Appendix G.

SAMPLE SIZE, mL	NUMBER OF POSITIVE TUBES		
	Sample A	Sample B	Sample C
0.1	5	5	4
0.01	5	4	5
0.001	2	4	2
0.0001	1	1	2

2.61. The following MPN test results were obtained for three water samples. Determine the MPN/100 mL using (a) the Poisson distribution [Eq. (2.90)], (b) the MPN tables given in Appendix G, and (c) the Thomas equation [Eq. (2.91)].

SAMPLE SIZE, mL	NUMBER OF POSITIVE TUBES		
	Sample A	Sample B	Sample C
0.1	2	4	5
0.01	1	2	4
0.001	5	1	2
0.0001	0	1	1

2.62. Four water samples have been analyzed for coliform bacteria using the MPN test procedure. From the results given below, calculate the MPN of each sample using (a) the Poisson equation [Eq. (2.90)], (b) the MPN tables given in Appendix G, and (c) the Thomas equation [Eq. (2.91)].

SAMPLE DILUTION	NUMBER OF POSITIVE TUBES			
	Sample A	Sample B	Sample C	Sample D
10.00	3	5	5	5
1.00	2	5	4	4
0.10	4	5	5	5
0.01		4		5
0.001		3		2
0.0001		2		0

REFERENCES

2.1. Applebaum, S. B., (1968), *Demineralization By Ion Exchange*, Academic Press, New York.

2.2. Benefield, L. D., J. F. Judkins, and B. L. Weand, (1982), *Process Chemistry For Water And Wastewater Treatment*, Prentice-Hall, Englewood Cliffs, N.J.

2.3. Busch, A. W., (1958), "BOD Progression in Soluble Substrates," *Sewage and Industrial Wastes*, vol. 30, no. 11, p. 1336.

2.4. _____, and H. N. Myrick, (1961), "BOD Progression in Soluble Substrates III—Short Term BOD and Bio-Oxidation Solids Production," *J. Water Pollution Control Federation*, vol. 33, no. 9, p. 897.

2.5. Butler, J. N., (1982), *Carbon Dioxide Equilibria and Their Applications*, Addison-Wesley Publishing Company, Reading, Mass.

2.6. _____, (1964), *Solubility and pH Calculations*, Addison-Wesley Publishing Company, Reading, Mass.

2.7. Camp, T. R., and R. L. Meserve, (1974), *Water and Its Impurities*, Dowden, Hutchinson, and Ross, Stroudsbourg, Penn.

2.8. Craun, G. F., (1981), "Outbreaks of Waterborne Disease In the United States: 1971–1978," *J. American Water Works Association*, vol. 73, no. 7, p. 360.

2.9. Doetsch, R. N., and T. M. Cook, (1973), *Introduction to Bacteria and Their Ecology*, University Park Press, Baltimore.

2.10. Flegal, T. M., and E. D. Schroeder, (1976), "Temperature Effects on BOD Stoichiometry and Oxygen Uptake Rate," *J. Water Pollution Control Federation*, vol. 49, no. 12, p. 2700.

2.11. Fujimoto, Y., (1961), "Graphical Use of First-Stage BOD Equation," *J. Water Pollution Control Federation*, vol. 36, no. 1, p. 69.

2.12. Gainey, P. L., and T. H. Lord, (1952), *Microbiology of Water and Sewage*, Prentice-Hall, Englewood Cliffs, N.J.

2.13. Grady, C. P. L., Jr., and A. W. Busch, (1963), "BOD Progression in Soluble Substrates VI—Cell Recovery Techniques in the T_bOD Test," *Proc. 18th Industrial Waste Conference*, Purdue University, Lafayette, Ind., p. 194.

2.14. Greenwood, J., and G. U. Yule, (1917), "On The Statistical Interpretation of Some Bacteriological Methods Employed in Water Analysis," *J. Hygiene*, vol. 16, no. 1, p. 36.

2.15. Hart, H., and R. D. Schuetz, (1966), *Organic Chemistry*, 3d ed., Houghton Mifflin Company, Boston.

2.16. Helweg, O. J., (1977), "A Nonstructural Approach to Control Salt Accumulation in Ground Water," *Ground Water*, vol. 15, no. 1, p. 51.

2.17. Holum, J. R., (1968), *Elements of General Biological Chemistry*, 2d ed., John Wiley and Sons, New York.

2.18. Hynes, H. B. N., (1971), *The Biology of Polluted Waters*, University of Toronto Press, Toronto.

2.19. Ingraham, J. L., (1959), "Growth of Psycrophillic Bacteria," *Bacteriology*, vol. 76, no. 1, p. 75.

2.20. Larson, T. E., and A. M. Buswell, (1942), "Calcium Carbonate Saturation Index and Alkalinity Interpretations," *J. American Water Works Association*, vol. 34, no. 11, p. 1667.

2.21. Lewis, J. W., and A. W. Busch, (1964), "BOD Progression in Soluble Substrates VII—The Quantitative Error Due to Nitrate as a Nitrogen Source," *Proc. 19th Industrial Waste Conference*, Purdue University, Lafayette, Ind., p. 846.

2.22. McKinney, R. E., (1962), *Microbiology For Sanitary Engineers*, McGraw-Hill Book Company, New York.

2.23. Mara, D. D., (1974), *Bacteriology For Sanitary Engineers*, Churchill Livingstone, Edinburgh, Scotland.

2.24. Moncrieff, R. W., (1967), *The Chemical Senses*, 3d ed., Leonard Hill, London.

2.25. Nester, E. W., C. E. Roberts, M. E. Lidstrom, N. N. Pearsall, and M. T. Nester, (1983), *Microbiology*, Saunders College Publishing, Philadelphia.

2.26. Nordell, E., (1961), *Water Treatment For Industrial and Other Uses*, 2d ed., Reinhold Publishing Corporation, New York.

2.27. Parisod, J. P., and E. D. Schroeder, (1978), "Biochemical Oxygen Demand Progression in Mixed Substrates," *J. Water Pollution Control Federation*, vol. 50, no. 7, p. 1872.

2.28. Perry, R. H, and D. W. Green (eds.), (1984), *Chemical Engineers' Handbook*, 6th ed., McGraw-Hill Book Company, New York.

2.29. Sawyer, C. N., and P. L. McCarty, (1978), *Chemistry For Environmental Engineering*, 3d ed., McGraw-Hill Book Company, New York.

2.30. Schroepfer, G. J., M. L. Robins, and R. H. Susag, (1964), "The Research Program on the Mississippi River in the Vicinity of Minneapolis and St. Paul," *Advances in Water Pollution Research*, *1*, Pergamon Press, London.

2.31. Snoeyink, V. L., and D. Jenkins, (1980), *Water Chemistry*, John Wiley and Sons, New York.

2.32. *Standard Methods for the Examination of Water and Wastewater*, (1980), 15th ed., American Public Health Association, New York.

2.33. Stanier, R. Y., E. A. Adelberg, J. L. Ingraham, and M. L. Wheelis, (1979), *Introduction to the Microbial World*, Prentice-Hall, Englewood Cliffs, N.J.

2.34. Stumm, W., and J. J. Morgan, (1980), *Aquatic Chemistry*, 2d ed., Wiley Interscience, New York.

2.35. Swilley, E. L., J. O. Bryant, and A. W. Busch, (1964), "Significance of Transportation Phenomena in Wastewater Treatment Processes,"

Proc. 19th Industrial Waste Conference, Purdue University, Lafayette, Ind., p. 821.

2.36. Tchobanoglous, G., A. D. Levine, and J. K. Koltz, (1983), "Filtrable Solids As A Design Parameter For Wastewater Treatment Processes," *Proc. 6th Symposium On Wastewater Treatment*, Montreal.

2.37. Thomas, H. A., Jr., (1942), "Bacterial Densities From Fermentation Tube Tests," *J. American Water Works Association*, vol. 34, no. 4, p. 572.

2.38. Trussell, R. R., and A. R. Trussell, (1980), "Evaluation And Treatment of Synthetic Organics In Drinking Water," *J. American Water Works Association*, vol. 72, no. 8, p. 458.

2.39. Ward, H. B., and G. C. Whipple, (1918), *Fresh-Water Biology*, John Wiley and Sons, New York.

2.40. *Wastewater Sampling for Process and Quality Control*, (1980), Manual of Practice No. OM-1, Water Pollution Control Federation, Washington, D.C.

2.41. Whipple, G. C., (1927), *The Microscopy Of Drinking Water*, 4th ed., John Wiley and Sons, New York.

3

Significance of the Characteristics of Water

Water quality can be defined in two general ways: by describing the physical, chemical, and biological characteristics of a water and by defining the suitability of a water for a specific use. The characteristics of water were considered in Chapter 2. In this chapter the significance of those characteristics is examined. The approach followed in this chapter is to begin with a general discussion of selected physical, chemical, and biological characteristics of water to illustrate the broad range of water quality problems that may be encountered in practice. The impurities in water and their relationship to public health are considered next. Impurities of importance not related to public health are then considered. And in the last section of this chapter the impact of water quality on the aquatic habitat is discussed.

3.1

PHYSICAL CHARACTERISTICS

The physical characteristics of water that are commonly considered important are given in Table 3.1. While it is possible to cite endless textbook examples of the importance of these characteristics, selected practical examples, many based on the experience of the authors, will be used for the purposes of illustration.

Solids

The solids found in water typically include silt and clay from riverbanks or lake bottoms and organic matter and microorganisms from natural or anthropogenic sources. The removal of solids is of great concern in the production of a clear, safe drinking water, in the process industries, and wherever a water of high quality is required.

TABLE 3.1

The Physical, Chemical, and Biological Characteristics of Various Waters

CHARACTERISTICS	Typical surface water	Typical groundwater	Domestic wastewater (U.S.)	Raw water source	Drinking water
	WATER SOURCE			SELECTED DRINKING WATER QUALITY OBJECTIVES*	
Physical					
Turbidity, NTU	—	—	—	—	< 1
Solids, total, g/m³	—	—	700	—	500
Suspended, g/m³	> 50	—	200	—	—
Settleable, mL/L	—	—	10	—	—
Volatile, g/m³	—	—	300	—	—
Filtrable (dissolved), g/m³	< 100	> 100	500	—	—
Color, units	—	—	—	< 150	< 15
Odor, number	—	—	Stale	—	< 3
Temperature, °C	0.5–30	2.7–25	10–25	< 20	—
Chemical: Inorganic Matter					
Alkalinity, eq/m³	< 2	> 2	> 2	—	—
Hardness, eq/m³	< 2	> 2	—	—	—
Chlorides, g/m³	50	200	> 100	250	250
Calcium, g/m³	20	150	—	—	—
Heavy metals, g/m³	—	0.5	—	—	—
Nitrogen, g/m³	< 10	< 10	40	—	—
Organic, g/m³	5	—	15	—	—
Ammonia, g/m³	—	—	25	—	—
Nitrate, g/m³	< 5	5	0	—	10
Phosphorus, total, g/m³	—	—	12	—	—
Sulfate, g/m³	—	—	—	—	250
pH, unitless	—	6.5–8	6.5–8.5	—	6.0–8.5
Chemical: Organic Matter					
Total organic carbon (TOC), g/m³	< 5	—	150	—	—
Fats, oils, greases, g/m³	—	—	100	—	—
Pesticides, g/m³	< 0.1	—	—	—	0.2
Phenols, g/m³	< 0.001	—	—	< 0.005	0.001
Surfactants, g/m³	< 0.5	< 0.5	—	—	0.5
Chemical: Gases					
Oxygen, g/m³	7.5	≈ 7.5	< 1.0	> 4.0	> 4.0
Biological					
Bacteria, MPN/100 mL	< 2000	< 100	$10^8 - 10^9$	< 5000	< 1.0
Viruses, pfu†/100 mL	< 10	< 1	$10^2 - 10^4$	—	—

*From Ref. [3.15].

†pfu = plaque-forming units.

The Effect of Watershed Characteristics

The treatment of surface waters for use as a water supply source usually involves the addition of chemicals to aid in the removal of suspended solids carried out of the watershed (see discussion in Chapter 13). In practice, the characteristics of each watershed vary with respect to the release of suspended solids. The increase in suspended solids (usually measured as turbidity) with time after a rainfall event for two different watersheds is shown in Fig. 3.1. In the first watershed (Fig. 3.1a), the rise in the concentration of suspended solids is gradual and easy to accom-

(a)

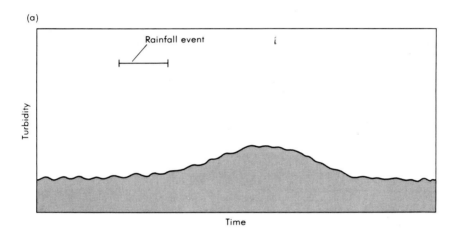

(b)

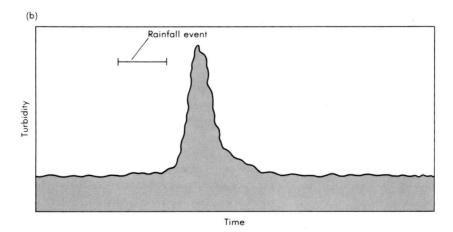

FIGURE 3.1

Increase in turbidity following a rainfall event in streams fed by different watersheds. Peak height and width are functions of watershed size and slope, as well as of storm characteristics.

modate in a water treatment plant. In the other watershed (Fig. 3.1b), the increase in suspended solids is quite pronounced and occurs over a short time period. Waters of this type are more difficult to treat because the response time of the equipment used to feed chemicals is often not fast enough to supply the needed chemicals as the suspended solids move through the plant. In such situations, flow equalization can be used, or the chemical feed equipment can be controlled by upstream turbidity monitors.

The Effect of Solids on Chlorination

Based on numerous field observations, it has been found that it is more difficult to achieve effective bacterial disinfection using chlorine when the concentration of suspended solids is high. It appears that the solids provide the attached bacteria a protective barrier against the action of the chlorine added for disinfection. Thus the removal of suspended solids is of great importance in providing a safe potable water and in treating wastewater that is to be used for irrigation where human contact may occur, such as in watering a golf course.

Odors and Tastes

With the exception of hydrogen sulfide, odors occurring in natural waters are organic in nature. Natural sources of odors include (1) microorganisms of various types, (2) decomposition of natural substances, and (3) reduction of sulfate to hydrogen sulfide. Some of the odors can be so intense that concentrations as low as 1.0 ppb (parts per billion) can be detected readily by human subjects. In general, odors will most commonly be found in surface waters, although the presence of hydrogen sulfide is common in groundwaters drawn from very deep aquifers.

Odors and Tastes in Surface Water Impoundments

One of the serious problems with surface water impoundments is the occasional development of odors and tastes due to the excessive growth of microorganisms. Typically, algae that release taste- and odor-producing substances are the organisms responsible. However, the biological decomposition of leaves and weeds can also lead to the development of odors. This subject is considered further in Section 3.3.

Odors in Sewers

One of the most common problems encontered in existing wastewater collection systems is the development of odors—principally that of hydrogen sulfide. Under reducing conditions, the sulfate present in the wastewater is reduced to hydrogen sulfide according to the following reactions:

$$2CH_3CHOH + SO_4^{-2} \rightarrow 2CH_3COOH + S^{-2} + 2H_2O + CO_2 \qquad (3.1)$$

$\qquad$ Lactate $\qquad$ Sulfate $\qquad$ Acetate $\qquad$ Sulfide
$\qquad\qquad\qquad\qquad\qquad\qquad\qquad\qquad\qquad$ ion

$$4H_2 + SO_4^{-2} \rightarrow S^{-2} + 4H_2O \qquad (3.2)$$

The production of hydrogen sulfide is especially severe in long, flat sewers laid on minimum slope in warm climates. Because of the presence and possible accumulation of toxic concentrations of hydrogen sulfide, sewer workers and others should not be allowed to enter sewers without gas masks. Also, no one should attempt to enter a sewer alone.

Odors from Wastewater Treatment Lagoons

Odors can also develop in wastewater treatment lagoons that are overloaded with organic matter [see Eqs. (3.1) and (3.2)]. Depending on the local meteorological conditions, it has been observed that odors may be measured at undiluted concentrations at great distances from the point of generation. What appears to happen is this: In the early morning hours, under quiescent meteorological conditions, a cloud of odors will develop over the wastewater treatment lagoon. The concentrated cloud of odors can then be transported, without breaking up, over great distances by a weak breeze. In some cases, odors have been detected at distances of up to 25 km from their source. This transport phenomenon has been termed the *puff movement* of odors.

The most common method used to mitigate the effects of the odor puff is to install barriers to induce turbulence, thus breaking up and dispersing the cloud of concentrated odors.

Temperatures

The temperature of a water is important for all of its intended uses. Typical data on the temperature of groundwater is shown in Fig. 3.2. Surface waters are subject to great temperature variations. Over much of the United States, river waters will vary from 0.5 to 3.0°C in the winter to 23 to 27°C in the summer (Fig. 3.3). In some shallow, slow-moving streams, summer temperatures may exceed 30°C. Lakes, reservoirs, ponds, and other impoundments are also subject to temperature changes. Such variations are extremely wide in shallow impoundments.

The temperature of wastewater will generally be warmer than that of the water supply. Depending on the geographic location, the average temperature of wastewater will vary from 10 to 22°C, with a representative value being about 15.6°C.

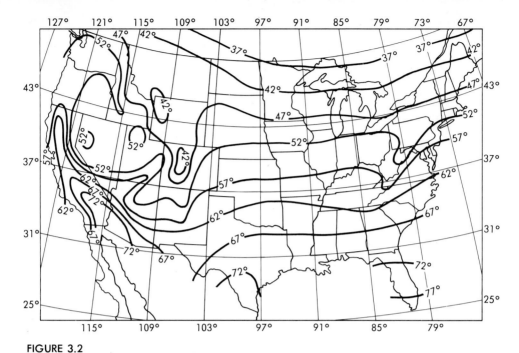

FIGURE 3.2

Approximate temperature of groundwater from nonthermal wells at depths varying from 10 to 20 m. Note: Temperatures are given in degrees Fahrenheit. Source: From W. D. Collins, U.S. Geological Survey, Water Supply Paper 520-F.

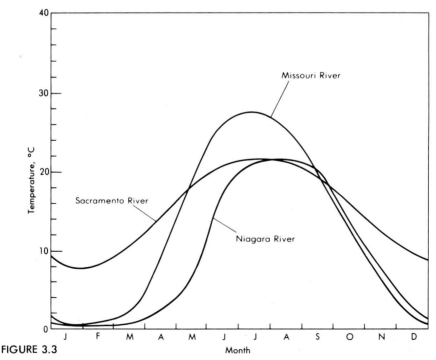

FIGURE 3.3

Generalized monthly variations in temperature in the Missouri River near Blair, Neb.; in the Niagara River at Buffalo, N.Y.; and in the Sacramento River at Sacramento, Calif.

The Effect of Temperature on Reaction Rates

Perhaps the most important effect of temperature is on the rate of chemical and biological reactions. The effect on biological reactions was considered previously in Chapter 2 in connection with the BOD test [see Eq. (2.79)]. The effect of temperature on the rate of chemical reactions is considered in Chapter 5.

Wastewater Treatment Lagoon Design

Consideration of temperature is especially important in the design of treatment facilities, as winter temperatures will usually control the design of the process. The number of treatment plants that have been designed without proper consideration of the effects of winter temperatures is shocking.

A classic example of what can happen when the effect of temperature is not considered properly is told here. A wastewater treatment lagoon was designed to operate as a stratified system during the summer. That is, the warmer water in the lagoon was supposed to stay on the surface while the colder incoming wastewater was supposed to accumulate at the bottom of the lagoon. Unfortunately, temperature measurements were not made initially. Later, when the wastewater treatment lagoon was put into operation, the temperature of the incoming wastewater was found to be 1 to 2°C warmer than the water in the lagoon. As a consequence, the warmer incoming water rose to the surface, and it was not possible to develop a two-phase treatment system.

Color

Color is found mostly in surface waters, although some deep groundwaters may contain color in amounts that are noticeable. As noted previously in Chapter 2, water that has come in contact with decaying organic matter will have a light-yellowish to brownish color. Where industrial wastes are discharged to streams, a wide variety of colors may be encountered, depending on the industrial process.

Color and the Process Industries

The presence of color, especially in large amounts, is troublesome to most process industries. For example, if a soft drink that normally is crystal clear is poured into a glass and is observed to have a yellow color, its consumer acceptability will suffer, although it may be perfectly safe to drink. A similar situation exists with drinking water. Tolerances for color will vary with each use and with the chemical nature of the water.

The Color of Wastewater

Color can also be used to assess the condition of wastewater with respect to how long it has been in a sewer. Fresh domestic wastewater has a light-brown color. As wastewater undergoes anaerobic decomposition, its color will change from a light brown to a light grey. Ultimately, under anaerobic conditions, the color will become black. The black color in wastewater results from the formation of ferrous and other sulfides according to the following equation:

$$H_2S \;+\; Fe^{+2} \;\rightleftharpoons\; FeS \;+\; 2H^+ \tag{3.3}$$

| Hydrogen | Ferrous | Ferrous | Hydrogen |
| sulfide | iron | sulfide | ion |

The hydrogen sulfide and iron in the ferrous form result from reducing anaerobic biological reactions occurring in the sewer.

3.2
CHEMICAL CHARACTERISTICS

The principal chemical constituents found in water are reported in Table 3.2. These chemical characteristics are important because each one affects water use in some manner. Many of the constituents present interact to restrict or enhance specific uses. As noted in Chapter 2, the equilibrium between bicarbonate and carbonate ($HCO_3^- \rightleftharpoons H^+ + CO_3^{-2}$) is a function of pH. As pH rises, the equilibrium increasingly favors the formation of CO_3^{-2}, and the precipitation of carbonate salts is often the result. Precipitated carbonate salts form a hard-to-dissolve scale that reduces pipe-flow and heat-transfer capacity; they also account for the clear blue of marl lakes. The following examples of water quality impacts associated with the constituents of water are presented to illustrate the diversity of problems and engineering challenges encountered in practice.

Water Quality Changes in Rivers

The chemical constituents in a river change along its length as well as with season and time. These changes are important, as they will affect the method of treatment. Usually, both nutrient concentrations (N, P, Fe, etc.) and the level of eutrophication increase with distance from the source. The Truckee and Mississippi rivers serve as examples.

The Truckee River
The Truckee River flows out of Lake Tahoe in California as an oligotrophic (nutrient-deficient) stream. Water quality changes occur, as shown in Table 3.3, in a number of important characteristics because of reservoir releases, wastewater discharges, and natural processes as the river flows to its point of discharge in landlocked Pyramid Lake northeast of Reno, Nevada.

TABLE 3.2
Chemical Characteristics of Selected Community Water Supplies
(Values in g/m³ Unless Otherwise Noted)

CONSTITUENT	AMARILLO, TEX. (1)	BILOXI, MISS. (2)	BLOOMINGTON, ILL. (3)	HONOLULU, HAWAII (4)	OSWEGO, N.Y. (5)	SAN DIEGO, CALIF. (6)	WHEELING, W. VA. (7)
Source	Wells	Wells	Springs	Artesian well	Lake	River	River
Ca^{+2}	40	1.5	36	5.9	42	87	40
Mg^{+2}	34	0.8	5.3	6.6	8.6	29	7
Na^+	25	138	2.4	64	10	99	14
K^+	5	6.2	0.2	2.9	1.3	5.6	2.4
Mn^{+2}	0.00	0	0	0.00	0	0.23	0
Fe^{+2}	0.02	0.9	0.07	0.74	0.33	0.17	0.06
HCO_3^-	302	251	102	81	115	155	29
CO_3^{-2}	0	8	0	0	0	0	0
SO_4^{-2}	33	5.8	27	14	29	282	104
Cl^-	8.2	56	4.2	76	26	97	22
F^-	3.4	0.2	0.1	0	0.1	0.4	0.6
NO_3^-	3.2	1.8	2.3	0.7	1.0	1.0	1.9
SiO_2	63	20	6.1	39	1.2	11	5.2
TDS	364	364	139	251	179	761	224
Hardness, eq/m³	240	7	111	42	140	334	128
Color, units	—	18	0	—	6	—	1
pH, unitless	7.3	8.4	7.8	6.7	7.9	7.9	7.8
Spec. cond., µS/cm, 25°C	545	583	229	388	323	—	345
Turbidity, NTU	—	1	—	—	3.0	—	—
Temperature, °C	17	28	—	—	—	—	—

Source: From Ref. [3.11].

171

TABLE 3.3

**Average Chemical Characteristics of the Truckee River in California and Nevada
July to November 1976**

CHARACTERISTIC OR CONSTITUENT	DISTANCE FROM LAKE TAHOE, km*						
	0	2	10	24	29	52	94[†]
Elevation, m	1900	1900	1860	1770	1740	1615	1342
Total N, g/m^3	0.09	0.34	0.44	0.28	0.25	0.25	1.56
Nitrate (N), g/m^3	0.01	0.01	0.33	0.11	0.08	0.06	0.23
Total phosphate, g/m^3	0.01	0.01	0.02	0.02	0.01	0.02	0.33
Orthophosphate, g/m^3	0.01	0.01	0.01	0.01	0.01	0.01	0.01
Iron, g/m^3	0.10	0.06	0.11	0.12	0.18	0.17	0.45
Sulfate, g/m^3	—	1.8	2.9	3.2	2.7	3.5	13.5
Chloride, g/m^3	2.0	2.1	3.3	3.7	3.3	5.9	8.6
Total dissolved solids, g/m^3	—	78	72	60	77	76	116
Fecal coliform, MPN/100 mL	—	23	6	8	4	4	1531

Source: From Ref. [3.13].

*Major treated wastewater discharges occur at 6, 22, and 84 km, and major tributary flows at 7, 21, 28, 31, 35, and 92 km.

[†] Downstream of the Reno, Nevada, wastewater treatment plant discharge.

The Mississippi River

The Mississippi River can be viewed as a classic example to illustrate the changes that occur in water quality as a river flows from its headwaters to its point of discharge (the Caribbean Sea for the Mississippi River). For example, New Orleans has access to all of the wastewater produced in the entire Mississippi drainage basin. Fortunately for the people of New Orleans, treatment of the wastewater before discharge, combined with the dilution factor provided by the tributary river flows (5000 to 13,000 m^3/s) result in a usable domestic water source. Changes in chemical characteristics of the Mississippi River water shown in Table 3.4 reflect inputs from major tributaries such as the Ohio, Missouri, and Arkansas rivers, as well as from wastewater discharges.

Water Supply Systems

The water supply source for the city of San Francisco is located in the Sierra Nevada mountains of California. Water stored in the Hetch Hetchy reservoir is transported to San Francisco around the lower part of San Francisco Bay in two large reinforced-concrete pipelines. After the pipelines were put into operation, it was noted that their flow capacity was diminishing with time.

Concerned about the continuing reduction in flow capacity, the city engineer had each pipeline removed from service, drained, and inspected

TABLE 3.4

Chemical Characteristics of the Mississippi River

DATE	Q^*, m^3/s	SiO_2, g/m^3	Ca, g/m^3	Mg, g/m^3	Na, g/m^3	K, g/m^3	HCO_3, g/m^3	SO_4, g/m^3	TDS, g/m^2	HARDNESS, eq/m^3	pH
Mississippi River—main stem at Grand Rapids, Minn.											
10/21/65	72	7.5	31	13	4.7	2.3	166	6.8	162	2.7	8.0
12/1/65	53	8.2	34	16	5.4	1.9	184	8.8	177	3.0	8.1
4/1/66	74	7.6	34	13	4.5	2.3	172	6.5	175	2.8	8.1
7/21/66	51	8.2	31	14	—	—	170	5.0	161	2.7	8.1
Mississippi River—main stem at St. Francisville, La.											
10/1–10/65	12,820	8.6	44	11	23	3.7	136	50	242	3.9	7.2
12/1–10/65	6,360	7.8	44	22	28	3.4	168	87	320	5.2	7.6
4/1–10/66	9,160	7.7	54	8	16	3.8	149	56	246	4.3	7.3
7/21–31/66	5,650	7.0	48	16	19	2.8	168	58	248	4.6	7.5

Source: From Ref. [3.17].

*Q = volumetric flow rate.

inside and outside for possible leaks. When the inside of the pipeline was inspected, it was found that the walls of the initially smooth pipe were severely pitted. This pitting accounted for the decrease in the carrying capacity of the pipeline, as pipe friction is a function of the ratio ε/D, where ε is a measure of the roughness of the pipe and D is the diameter.

What had caused the pitting? After investigating a number of possibilities, it was found that the alkalinity and hardness of the water were extremely low. The water equilibrium within the pipeline could be described as a closed carbonate system in equilibrium with solid $CaCO_3$ (i.e., the $CaCO_3$ in the concrete pipe). Because the calcium concentration in the water in the reservoir was less than that needed to satisfy the $CaCO_3$ solubility product, calcium was being leached from the concrete pipes. As the calcium leached from the inner surface of the pipe, pitting resulted. To correct the situation, the pipelines were relined, and lime [$Ca(OH_2)$] is now added at the entrance to the pipelines to adjust the $CaCO_3$ equilibrium so that the water will be noncorrosive.

Mineral Pickup from Municipal Use

There is considerable interest in many of the arid parts of the United States in the reuse of treated domestic wastewater. Where treated wastewater is to be used for crop irrigation, it is important to assess the impact

TABLE 3.5

Typical Mineral Pickup from Domestic Water Usage

CONSTITUENT	INCREMENT ADDED, mg/L	
	Range	Typical
Cations		
Ammonium (NH_4^+)	5–30	15
Calcium (Ca^{+2})	15–40	20
Magnesium (Mg^{+2})	10–30	10
Potassium (K^+)	7–15	10
Sodium (Na^+)	30–80*	50*
Anions		
Bicarbonate (HCO_3^-)	40–180	100
Carbonate (CO_3^{-2})	0–10	—
Chloride (Cl^-)	20–50*	40*
Nitrate ($NO_3^- - N$)	5–20	10
Phosphate (PO_4^{-3})	8–20	12
Sulfate (SO_4^{-2})	10–30	20
Other Data[†]		
Silica (SiO_2)	6–20	10
Hardness as $CaCO_3$	75–225	90
Alkalinity as $CaCO_3$	30–160	80
Total solids	140–480	300

*Excluding the additions from domestic water softeners.

[†] Minor amounts of aluminum, boron, fluoride, iron, and manganese are also added.

of mineral pickup resulting from municipal use. Representative data on the mineral pickup that can be expected in municipal wastewater are presented in Table 3.5. With respect to wastewater reuse for irrigation, two factors must be considered: electrical conductivity and the sodium adsorption ratio (SAR). These factors are considered further in Section 3.5. The increase in conductivity resulting from domestic usage is considered in Example 3.1.

EXAMPLE 3.1

DETERMINATION OF THE EFFECT OF DOMESTIC USAGE ON WATER QUALITY

Given the following analysis of a water supply, estimate the increase in the electrical conductivity that would occur as a result of domestic usage.

| ITEM | CONCENTRATION | |
	mg/L	meq/L
Cations		
Ca^{+2}	85.0	4.24
Mg^{+2}	43.0	3.54
K^+	2.9	0.07
Na^+	92.0	4.00
		11.85
Anions		
HCO_3^-	362.0	5.93
Cl^-	131.0	3.70
SO_4^{-2}	89.0	1.85
NO_3^-	20.0	0.32
		11.80

Conductivity: 1268 μS/cm at 25°C

SOLUTION:

1. Check the approximate correctness of the reported conductivity value using the data given in Table 2.11.

| ION | CONC, mg/L | CONDUCTIVITY | |
		Factor, per mg/L	μS/cm
Ca^{+2}	85.0	2.60	221.0
Mg^{+2}	43.0	3.82	164.3
K^+	2.9	1.84	5.3
Na^+	92.0	2.13	196.0
HCO_3^-	362.0	0.715	258.8
Cl^-	131.0	2.14	280.3
SO_4^{-2}	89.0	1.54	137.1
NO_3^-	20.0	1.15	23.0
		Total	1285.8

Because there is less than 1.5 percent difference between the computed value and the measured value, it is reasonable to assume that the results of the water analysis are accurate.

2. Using the data given in Table 3.5 for the typical mineral pickup resulting from domestic usage, estimate the conductivity of the wastewater.

	CONCENTRATION, mg/L			CONDUCTIVITY	
ION	Initial	Increment added	Final	Factor, per mg/L	$\mu S/cm$
Ca^{+2}	85.0	20	105.0	2.60	273.0
Mg^{+2}	43.0	10	53.0	3.82	202.5
K^+	2.9	10	12.9	1.84	23.7
Na^+	92.0	50	142.0	2.13	302.5
HCO_3^-	362.0	100	462.0	0.715	330.3
Cl^-	131.0	40	171.0	2.14	365.9
SO_4^{-2}	89.0	20	109.0	1.54	167.9
NO_3^-	20.0	10	30.0	1.15	34.5
				Total	1700.3

The increase in conductivity resulting from domestic usage is 414.5 $\mu S/cm$ (1700.3 − 1285.8).

COMMENT

The increase in conductivity is of great interest where it is planned to use treated domestic wastewater for agricultural irrigation. This subject is considered further in Section 3.5.

Corrosion and Scale Formation

Corrosion occurs when a galvanic cell is set up between metal surfaces. This often happens in buried-pipe systems that pass through soils of different characteristics. The results of corrosion in terms of pipe replacement, lost flow capacity, and increased pumping costs (due to increased friction) are enormous. The most common method of corrosion control is to break the electrical circuit involved by coating pipe surfaces with nonconducting materials. A number of coatings are in use, including bituminous materials, cement, and paints.

Another common method is to increase the pH of a water to the point that $CaCO_3$ precipitation occurs to a small extent. The $CaCO_3$ layer deposited on pipe surfaces acts as a protective coating. Because the coatings may be uneven, may interact with other constituents, or may be washed away, the method is not a universal solution, but it has proved to be inexpensive and useful.

Whether a water will be scale-forming (encrustive) (Fig. 3.4) or corrosive can be predicted with the use of the Langelier index (LI) [3.9] and the Ryznar index (RI) [3.18]:

$$LI = pH_{measured} - pH_{sat} \qquad (3.4)$$

$$RI = 2pH_{sat} - pH_{measured} \qquad (3.5)$$

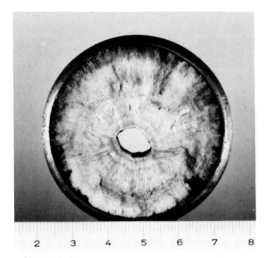

FIGURE 3.4

Section through a water system pipe with excessive buildup of calcium carbonate scale. The reference scale shown in the photograph is in centimeters.

Source: Pipe section provided by R. R. Trusell, J. M. Montgomery Engineers, Pasadena, CA.

where

$$pH_{measured} = \text{actual pH value measured in the water}$$
$$pH_{sat} = \text{pH of the water in equilibrium with solid } CaCO_3$$

The interpretation of these indexes is as follows:
For the Langelier index:

$LI > 0$ Water is scale-forming (that is, supersaturated with respect to $CaCO_3$)

$LI = 0$ Water is neutral

$LI < 0$ Water is corrosive (that is, undersaturated with respect to $CaCO_3$)

For the Ryznar index:

$RI < 5.5$ Heavy scale will form

$5.5 < RI < 6.2$ Scale will form

$6.2 < RI < 6.8$ No difficulties

$6.8 < RI < 8.5$ Water is corrosive

$RI > 8.5$ Water is very corrosive

The difference between these indexes is that Langelier derived his index from an analysis of chemical equilibrium, whereas Ryznar correlated the computed values of the index to field observations.

The value of pH_{sat} can be determined by Eq. (3.6):

$$pH_{sat} = -\log\left(\frac{K_2\gamma_{Ca^{+2}}[Ca^{+2}]\gamma_{HCO_3^-}[HCO_3^-]}{K_{sp}}\right) \qquad (3.6)$$

Calculating the pH_{sat} value, as given in Eq. (3.6), requires knowledge of the activity coefficients for HCO_3^- and $Ca^{+2}(\gamma_{HCO_3^-}, \gamma_{Ca^{+2}})$. In design, these values should be determined accurately. A close approximation of the activity coefficients can be made using the following relationship, proposed by Langelier, in conjunction with Eq. (2.18) [3.9]:

$$\mu = (2.5 \times 10^{-5}) \times TDS, \text{ g/m}^3 \qquad (3.7)$$

The application of these indexes is illustrated in Example 3.2.

<hr>

EXAMPLE 3.2

<hr>

DETERMINATION OF LANGELIER AND RYZNAR INDEXES

Determine the Langelier and Ryznar indexes for the Oswego, N.Y., water supply. The characteristics of the water are summarized in Table 3.2.

SOLUTION:

1. From Table 3.2 the pertinent information for the Oswego, N.Y., water supply is

	CONCENTRATION	
CONSTITUENT	g/m³	mol/L
TDS	179	—
Ca^{+2}	42	1.05×10^{-3}
HCO_3^-	115	1.89×10^{-3}
Conductivity	323 μS/cm	
pH	7.9	

2. Determine the ionic strength.
   a. Based on TDS, using Eq. (3.7):

   $$\mu = 2.5 \times 10^{-5}(179) = 0.00448$$

   b. Based on conductivity, using Eq. (2.6):

   $$\mu = 1.6 \times 10^{-5}(323) = 0.00517$$

3. Determine activity coefficients for HCO_3^- using Eq. (2.18).

   $$\log\gamma_{HCO_3^-} = -\frac{0.5(z_i)^2\mu^{1/2}}{1 + \mu^{1/2}}$$

a. Based on TDS:

$$\log \gamma_{HCO_3^-} = -\frac{0.5(-1)^2(0.00448)^{1/2}}{1+(0.00448)^{1/2}}$$

$$\log \gamma_{HCO_3^-} = -0.0313$$

$$\gamma_{HCO_3^-} = 0.93$$

b. Based on conductivity:

$$\log \gamma_{HCO_3^-} = -0.0335$$

$$\gamma_{HCO_3^-} = 0.93$$

The apparent difference in ionic strength has little effect on the activity coefficients.

4. Determine the activity coefficient for Ca^{+2} based on an ionic strength of 0.00448.

$$\log \gamma_{Ca^{+2}} = -0.125$$

$$\gamma_{Ca^{+2}} = 0.75$$

5. Determine the value of pH_{sat} using Eq. (3.6).

$$pH_{sat} = -\log \frac{(4.17 \times 10^{-11})(0.75)(1.05 \times 10^{-3})(0.93)(1.89 \times 10^{-3})}{5.25 \times 10^{-9}}$$

$$= -\log(1.1 \times 10^{-8}) = 7.96$$

6. Determine the Langelier index using Eq. (3.4).

$$LI = 7.9 - 7.96 = -0.06$$

According to the values proposed by Langelier, this water is corrosive, and scale formation should not be expected.

7. Determine the Ryznar index using Eq. (3.5).

$$RI = 2(7.96) - 7.9 = 8.02$$

According to the values proposed by Ryznar, the water is also corrosive.

Acid Rain

In recent years, emissions from fossil fuel combustion (power plants, automobiles, steel mills, and other sources) and smelters have added large amounts of sulfur dioxide (SO_2) and oxides of nitrogen (NO_x) to the atmosphere. On dissolving in atmospheric moisture, droplets of sulfurous, sulfuric, and nitric acid are formed, and the pH of rainwater often falls considerably below 5.6. The average pH of rainwater in the

TABLE 3.6

**Approximate Minimum pH Values
for Selected Organisms**

ORGANISM	MINIMUM pH
Smallmouth bass	6.0
Lake trout	5.0
Brown trout	5.5
Yellow perch	5.0
Salamander (embryonic)	5.5
Mussel	6.5
Mayfly	6.0
Whirligig	3.5
Water boatman	3.5

Source: From Ref. [3.8].

Adirondack Mountains in the northeastern United States and Canada is about 4.2, and values as low as pH 1.5 have been recorded in the United States.

Effects of acid rain are found in lakes and rivers that contain limited amounts of alkalinity, where the pH may fall below 4.5 with a resultant loss of virtually all aquatic life. Minimum acceptable pH values for selected aquatic organisms are reported in Table 3.6. Toxic metal solubility is greater at low pH values, and leaching of these metal ions from the soil can be a serious additional problem. The corrosive effects of acid rain are having major effects on marble and limestone structures throughout the world. A classic example is the Parthenon in Athens, Greece, which has undergone more erosion since 1960 than in the previous 2000 years.

Water Quality in Aquaculture

Aquaculture may be defined as the growing or husbandry of biological organisms in an aquatic environment. The following examples serve to illustrate the impacts that diverse water quality parameters can have on aquaculture systems.

Aquaculture Using Surface Water

In many parts of the world, surface water is the principal source of water used for aquaculture (Fig. 3.5). Unfortunately, the use of high-quality surface waters has led to serious problems in aquaculture systems, principally those used for the culture of fish. In most fish culture systems in less-developed countries, animal manure is added as a source of organic carbon. Heterotrophic bacteria convert the organic material to CO_2 and cell tissue. Photosynthetic algae utilize the CO_2 in the presence

FIGURE 3.5

Aquaculture facility near Fortaleza, Brazil. Large fish shown in the photo on the right are the Amazonian species known as Tambaqui (*Colossoma macropomun*) and Pirapitinga (*Colossoma bidens*). The small fish is Tilapia.

of sunlight to produce new algal cells. The bacteria and algae become the food source for fish.

Often an imbalance will occur in which the algae population will increase to such an extent that the CO_2 released by the bacteria is insufficient for their needs. When this occurs, many of the algae present will obtain the CO_2 needed for new cell production by splitting the bicarbonate ion [see Eq. (2.89)]. Because the alkalinity of most surface waters is low, the pH will rise, as there is little natural alkalinity (buffer capacity) in the system to resist the increase in pH caused by the release of the carbonate ion. The rise in pH will create an environment within the aquaculture ponds that is toxic to most species of fish. In most cases, it is difficult to define the specific cause of death, as several factors are involved, including pH, ammonia, and heavy metal toxicity and the effects of such toxicity coupled with low-dissolved-oxygen conditions brought about by algal respiration [see Eq. (2.88)] in the evening hours.

Gas Supersaturation

In many of the water-handling operations common to most aquaculture systems, it is possible to supersaturate the water with respect to the gases that are normally present in the atmosphere (e.g., nitrogen, oxygen, and carbon dioxide). Only recently has the importance of this phenomenon been appreciated. It now appears that for sensitive fish, a total gas pressure in the liquid of 102 to 103 percent of the normal saturation level sufficient to cause gas-bubble disease. Supersaturation levels of 115 to 125 percent are lethal for most fish.

The clinical signs of gas-bubble disease include the formation and accumulation of fine bubbles on the fins and body surface of the fish, especially the head, and the presence of gas bubbles in the vascular system. The ill effects of the presence of gas bubbles can be seen in blockage of the capillaries within the vascular system, which may result in anoxia at the tissue level. Also, if too many bubbles are present in the vascular system of the fish, the heart can lose suction, resulting in death. And at the exterior locations where bubbles form and break, the fish are susceptible to secondary bacterial infection.

One way to control the ill effects of gas-bubble disease is to decrease the partial pressure of the gases in solution. Typically, reduction of the total gas pressure is accomplished in a countercurrent-flow, packed-bed gas-stripping tower (see Figs. 12.38 and 12.39). The performance of such a stripping unit can be evaluated using Eq. (2.86).

3.3
BIOLOGICAL CHARACTERISTICS

All natural waters support biological communities, including many organisms that are harmful to humans. Following the format of the previous two sections, it is the purpose here to describe some of the impacts caused by the biological characteristics of water. A detailed discussion of the harmful organisms found in water and of their effects on public health is given in the following section. Topics to be discussed in this section include nutrient cycles, the management of storage reservoirs, and the interpretation of coliform test results.

Nutrient Cycles

All living organisms, in one way or another, are involved in the recycling of elements. For example, animals eat plants, and animal waste products are converted to simpler end products by microorganisms. Similarly, when animals die their remains are decomposed by microorganisms. A wide variety of organisms are capable of carrying out the conversion of complex organic compounds to simple inorganic substances such as ammonia, sulfate, carbon dioxide, and water. The nitrogen, phosphorus, and sulfur cycles, which are of fundamental importance in the recycling of these elements in nature, are considered here. In each case, the role of microorganisms in the process is highlighted.

The Nitrogen Cycle

Nitrogen and its compounds are of great interest in the field of water quality management. Nitrogen, an important constituent of protein and nucleic acids, is the element that is required in greatest quantity, next to carbon and oxygen, for most organisms. Of the hundreds of compounds containing nitrogen, the following are of special significance in water quality management: organic nitrogen, ammonia (NH_3), nitrite (NO_2^-), nitrate (NO_3^-), and nitrogen gas (N_2). The interrelationships that exist between these compounds are best illustrated in a diagram known as the nitrogen cycle (Fig. 3.6).

The atmosphere, as shown in Fig. 3.6, serves as the reservoir for nitrogen. Nitrogen is removed by nitrogen-fixing bacteria and algae and by the action of electrical discharges occurring during electrical storms. Nitrogen is returned to the atmosphere through bacterial denitrification. The addition of nitrogen from the burning organic matter is not considered in Fig. 3.6. Bacterial ammonification of organic matter results in the production of ammonia. The two-step bacterial conversion of ammonia to nitrite and then to nitrate is known as *nitrification*. In turn,

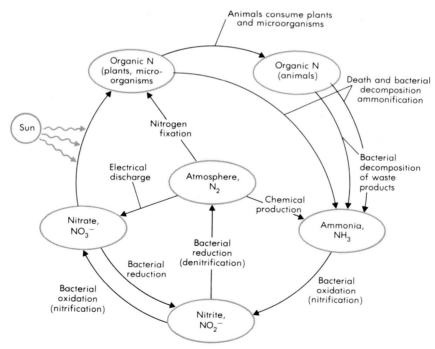

FIGURE 3.6

Simplified representation of the nitrogen cycle in nature.

nitrate is utilized in the production of organic nitrogen in the form of bacteria and plants. Completing the cycle, animals consume the plants and microorganisms.

The Phosphorus Cycle

In a manner similar to nitrogen, phosphorus in the environment cycles between organic and inorganic forms. Organic compounds containing phosphorus are found in all living matter. Orthophosphate (PO_4^{-3}) is the only form of phosphorus that is used readily by most plants and microorganisms. The phosphorus cycle involves two major steps, each of which is bacterially mediated: (1) conversion of organic to inorganic phosphorus and (2) conversion of inorganic to organic phosphorus. Conversion of insoluble forms of phosphorus such as calcium phosphate $[Ca_2(HPO_4)_2]$ into soluble forms, principally PO_4^{-3}, is also carried out by microorganisms. Referring to the nitrogen cycle shown in Fig. 3.6, the organic phosphorus in dead plant and animal tissue and animal waste products is converted bacterially to PO_4^{-3}. The PO_4^{-3} released to the environment is then incorporated into plant and animal tissue containing phosphorus, thus completing the cycle.

Because the concentration of phosphorus in many natural water environments is low, algal growth is usually limited. However, when discharges containing excessive amounts of phosphorus such as untreated and partially treated wastewater and irrigation return waters are added, algal blooms have often resulted. The problem with eutrophication in the Great Lakes is an example of what can happen when large amounts of phosphorus and other nutrients are added to a body of water. (The subject of eutrophication is considered later in this chapter.) At Lake Tahoe, to avoid the problems caused by the discharge of wastes containing phosphorus, all of the wastewater is collected and exported out of the basin for treatment and disposal.

The Sulfur Cycle

Sulfate occurs in most natural waters (e.g., compare the sulfate concentrations in the Mississippi River at the two locations given in Table 3.4). Sulfur is a component of all living matter, principally as a component of the amino acids in proteins. The principal forms of sulfur that are of special significance in water quality management are organic sulfur, hydrogen sulfide (H_2S), elemental sulfur (S^0), and sulfate (SO_4^{-2}). The interrelationships of these forms of sulfur are depicted in Fig. 3.7 in what is known as the sulfur cycle.

In many respects, the sulfur cycle is similar to the nitrogen cycle. Sulfate is taken up by plants and microorganisms for the production of cell tissue. In turn, plants and microorganisms are consumed by animals. The ability to produce hydrogen sulfide from waste products and dead or decaying proteins is common to many organisms. In the presence of free

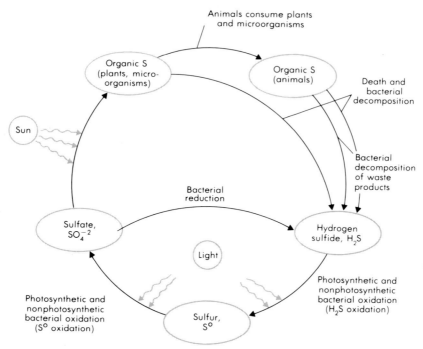

FIGURE 3.7

Simplified representation of the sulfur cycle in nature.

oxygen, hydrogen sulfide is oxidized rapidly to elemental sulfur. Under aerobic conditions, sulfide is oxidized to sulfur and then to sulfate by photosynthetic sulfur bacteria and colorless blue-green algae. An important part of the sulfur cycle is the reduction of sulfate to hydrogen sulfide. The reduction of sulfate to sulfide is carried out by a small group of anaerobic bacteria known collectively as *sulfate-reducing bacteria*. A few bacteria are also capable of reducing elemental sulfur to sulfide.

The sulfur cycle is especially important in water quality management with respect to organism survival, metabolic odor production, and chemical corrosion. Because hydrogen sulfide is toxic to many organisms, the reduction of sulfate to sulfide, which can then accumulate in the environment, is undesirable. Hydrogen sulfide can also combine and precipitate heavy metals such as iron, zinc, and cobalt. Because these elements are required for proper bacterial growth, high levels of H_2S may also inhibit growth. When hydrogen sulfide is produced, the potential for the release of odors always exists.

When hydrogen sulfide is produced in sewers it can lead to their deterioration, especially if they are constructed of reinforced-concrete pipe. Obeying Henry's law, some of the hydrogen sulfide will be released

FIGURE 3.8

Section taken from top of concrete pipe badly corroded due to the release of hydrogen sulfide in the sewer. Note that the reinforcing rods used in the construction of the pipe have corroded away.

to the atmosphere above the wastewater in the sewer. If oxygen is present, sulfuric acid can be produced, according to the following reaction, by bacteria attached to the upper portions of the sewer:

$$H_2S + O_2 \xrightarrow{\text{Bacteria}} H_2SO_4 \tag{3.8}$$

Once formed, the sulfuric acid will react with calcium carbonate in the concrete according to the following reaction:

$$H_2SO_4 + CaCO_3 \rightarrow CaSO_4 + H_2CO_3 \tag{3.9}$$

As the calcium is removed from the concrete, flaking and pitting will result. Ultimately, the structural integrity of the pipe may be damaged. An example of a sewer pipe badly corroded by the formation of hydrogen sulfide is shown in Fig. 3.8.

Management of Storage Reservoirs

Production of a high-quality water is not the only concern in the design of water supply systems. As noted in Chapter 1, water must be available at times of high demand. Peak summer demand rates are often three times the annual average, and potential emergency demands (e.g., fire flows) must be satisfied. In most cases, treated water storage is used to meet most of the demand variation. This solution, however, presents water quality problems, particularly with respect to disinfection, because chlorine residuals last over limited time periods. Because peak demands often coincide with periods of low water availability, untreated (raw) water storage is also often necessary. Changes in water quality during storage or impoundment can be significant. The most common problem is the development of tastes and odors caused by the growth of algae and other microorganisms, known collectively as *plankton*.

Some of the most troublesome planktonic microorganisms that occur frequently in water supplies are listed in Table 3.7. The growth of plankton in reservoirs can be controlled using copper sulfate or more selective algicides. Typical copper sulfate dosages are also reported in Table 3.7. The effective use of these chemicals requires microscopic examination of the water to determine the number and type of organisms involved. Ideally, the control chemicals should be applied just at the time when the number of organisms starts to increase rapidly. Treatment is usually needed when the number of organisms exceeds 500 to 1000 units per milliliter [3.1]. Based on the units given in Table 3.7, a dose of 0.3 mg/L can be used without laboratory control. Because many of the organisms can be controlled with lower dosages, savings are possible with proper control. The required dosage of copper sulfate is influenced by the temperature, the alkalinity, and the carbon dioxide content of the water. The dosage of copper sulfate is usually based on the volume of the upper circulating portion of the reservoir, where the plankton are typically found. If the depth of the circulation pattern is unknown, a rule of thumb is to use an active depth of about 4.5 m [3.1]. The frequency of application should be based on microscopic observation.

Several methods are used to apply copper sulfate to reservoir waters. The most common are (1) the burlap bag method, (2) the spray method, (3) the blower method, and (4) the continuous dosing method. In the burlap bag method, copper sulfate is placed in a bag in a dose equivalent to 3.0 mg/L. The bag is then dragged through the water from the stern of a rowboat or a motorboat. In the spray and blower methods, copper sulfate is either sprayed or blown over the water surface from a boat using tree spraying equipment or air blowers. In many reservoirs, copper sulfate is applied continuously using commercial or locally made dosing equipment.

Interpretation of Coliform Test Results

One of the most difficult problems in the field of water quality management is the proper assessment or interpretation of coliform test results, especially those obtained from stream surveys. To illustrate the problems encountered, tests taken over a period of years in Stinson Beach, California, will be used. However before discussing the situation in Stinson Beach it will be helpful to consider some findings on the ratio of the number of fecal coliforms to fecal streptococci in the feces of warm-blooded animals.

TABLE 3.7

Organisms That May Develop in Surface Water Storage Facilities and the Dosage of Copper Sulfate Needed for Their Control

ORGANISM GROUP	SPECIFIC ORGANISM	PROBLEMS*, ODORS†, AND (TASTES) ASSOCIATED WITH ORGANISM	COPPER SULFATE DOSAGE‡, g/m³
Bacteria			
	Beggiatoa	Decayed, very offensive	5.00
	Cladothrix		0.10–0.15
	Crenothrix	Decayed, very offensive	0.30–0.50
	Leptothrix	Medicinal with chlorine	—
	Sphaerotilis natans	Decayed, very offensive	0.40
	Thiothrix (sulfur bacteria)		Use chlorine
Fungi			
	Achlya		—
	Leptomitus		0.40
	Saprolegnia		0.10–0.20
Algae Chlorophyceae (green)			
	Cladophora	**Septic**	0.50
	Closterium	**Grassy**	0.15–0.20
	Coelastrum		0.05–0.35
	Conferva		0.25
	Desmidium		2.00
	Dictyosphaerium	Grassy, nasturtium, **fishy**	Use chlorine
	Draparnaldia		0.35
	Entomophthora		0.50
	Eudorina	**Fishy**	2.00–10.0
	Gloeocystis	**Septic**	—
	Hydrodictyon	**Septic**	0.10–0.15
	Miscrospora		0.40
	Palmella		2.00
	Pandorina	**Fishy**	2.00–10.0
	Protococcus		Use chlorine
	Raphidium		1.00
	Scenedesmus	**Grassy**	1.00
	Spirogyra	**Grassy**	0.10–0.15
	Staurastrum	**Grassy**	1.50
	Tetrastrum		Use chlorine
	Ulothrix	**Grassy**	0.20
	Volvox	**Fishy**	0.25
	Zygnema		0.50
Cyanophyceae (blue-green)			
	Anabaena	Moldy, grassy, **septic**	0.10–0.15
	Anacystis	Grassy, **septic** (sweet)	0.10–0.25

TABLE 3.7 *(Cont.)*

ORGANISM GROUP	SPECIFIC ORGANISM	PROBLEMS*, ODORS[†], AND (TASTES) ASSOCIATED WITH ORGANISM	COPPER SULFATE DOSAGE[‡], g/m^3
Cyanophyceae (blue-green) *(Cont.)*			
	Aphanizomenon	Moldy, grassy, **septic**	0.10–0.50
	Coelosphaerium	Grassy (sweet)	0.20–0.35
	Cylindrosphermum	Grassy, **septic**	0.10–0.15
	Gloeocapsa	Red, **septic**	0.15–0.25
	Microcystis	Grassy, **septic**	0.20
	Oscillatoria	Grassy, **musty, spicy**	0.20–0.50
	Rivularia	Moldy, grassy, **musty**	—
Diatomaceae (usually brown)			
	Asterionella	Aromatic, geranium, **fishy**	0.10–0.20
	Cyclotella	Geranium, **fishy**	Use chlorine
	Diatoma	**Aromatic**	—
	Fragilaria	Geranium, **musty**	0.25
	Melosira	Geranium, **musty**	0.20–0.35
	Meridion	Aromatic, **spicy**	—
	Navicula		0.05–0.10
	Nitzschia		0.50
	Stephanodiscus	Geranium, **fishy**	0.35
	Synedra	Earthy, **musty**	0.35–0.50
	Tabellaria	Geranium, **fishy**	0.10–0.50
Protozoa			
	Bursaria	Irish moss, salt marsh, fishy	—
	Ceratium	Fishy, septic (bitter), red-brown color	0.25–0.35
	Chlamydomonas	Musty, grassy (sweet)	0.50–1.00
	Cryptomonas	Candied violets (sweet)	0.50
	Dinobryon	Aromatic, violets, fishy	Use chlorine
	E. histolytica (cyst)		0.50
	Glenodinium	Fishy	0.50
	Mallomonas	Aromatic, violets, fishy	0.50
	Peridinium	Fishy, like clamshells (bitter)	0.50–2.00
	Synura	Cucumber, muskmelon, fishy (bitter)	0.25
	Uroglena	Fishy, oily, cod-liver oil	0.05–0.20
Crustacea			
	Cyclops		2.00
	Daphnia		2.00

TABLE 3.7 *(Cont.)*

ORGANISM GROUP	SPECIFIC ORGANISM	PROBLEMS*, ODORS[†], AND (TASTES) ASSOCIATED WITH ORGANISM	COPPER SULFATE DOSAGE[‡], g/m^3
Miscellaneous			
	Bloodworm		Use chlorine
	Chara		0.10–0.50
	Nitella flexilis	Grassy, septic (bitter)	0.10–0.20
	Phaetophyceae (Brown algae)		—
	Potamogeton		0.30–0.80
	Rhodophyceae (Red algae)		—
	Xanthophyceae (Green algae)		—

Source: Adapted from Refs. [3.1], [3.4], and [3.19].

*Without pretreatment, the presence of many of these organisms will result in the clogging of granular-medium filters, thus reducing the length of filter runs.

[†] Odor when algae are present in moderate amounts is shown in standard type. Odor when algae are present in abundant amounts is shown in boldface.

[‡] A chlorine residual of 0.5 to 1.0 g/m^3 will control most of the organisms listed. Exceptions are melosira, crustacea, synura, and cysts of *Entamoeba histolytica*, which require higher dosages.

TABLE 3.8

Estimated Per Capita Contribution of Indicator Microorganisms from Human Beings and Some Animals

ANIMAL	AVERAGE INDICATOR ORGANISM DENSITY PER g OF FECES		AVERAGE CONTRIBUTION PER CAPITA/24 hr		RATIO FC/FS
	Fecal coliform, 10^6	Fecal streptococci, 10^6	Fecal coliform, 10^6	Fecal streptococci, 10^6	
Human	13.0	3.0	2,000	450	4.4
Chicken	1.3	3.4	240	620	0.4
Cow	0.23	1.3	5,400	31,000	0.2
Duck	33.0	54.0	11,000	18,000	0.6
Pig	3.3	84.0	8,900	230,000	0.04
Sheep	16.0	38.0	18,000	43,000	0.4
Turkey	0.29	2.8	130	1,300	0.1

Source: From Ref. [3.14].

The Ratio of Fecal Coliforms to Fecal Streptococci

It has been observed that the quantities of fecal coliforms and fecal streptococci that are discharged by humans are significantly different from the quantities discharged by animals. Therefore, it has been suggested that the ratio of the fecal coliform (FC) count to the fecal streptococci (FS) count in a sample can be used to show whether the suspected contamination derives from human or from animal wastes [3.14]. Typical data on the ratio of FC to FS counts for humans and various animals are reported in Table 3.8. The FC/FS ratio for humans is more than 4.0, whereas the ratio for domestic animals is less than 1.0. If ratios are obtained in the range of 1 to 2, interpretation is uncertain.

The foregoing interpretations are subject to the following constraints:

1. The sample pH should be between 4 and 9 to exclude any adverse effects of pH on either group of microorganisms.

2. At least two counts should be made on each sample.

3. To minimize errors due to differential death rates, samples should not be taken farther downstream than 24 hr of flow time from the suspected source of pollution.

4. Only the fecal coliform count obtained at an incubation temperature of 44°C is to be used to compute the ratio [3.14].

Wastewater Management in Stinson Beach

Stinson Beach, California, is a small coastal community located 32 km north of San Francisco (see Fig. 1.4). Wastewater from all of the 560 homes in the community is discharged to individual on-site disposal systems (septic tanks and leach fields). Before developing an on-site wastewater management district to correct the failed on-site systems, coliform test samples were taken at various locations from the creek that runs through the community. The coliform counts were always high. Based on years of high coliform test results, the county health department reasoned that sewers were needed to correct the high coliform counts, which had to be a consequence of the failed on-site systems located along the creek.

Not convinced that sewers were necessary, the community retained a consultant to study whether they were in fact needed. The consultant took samples from the creek and measured both fecal coliform and fecal streptococci. Based on the ratio of fecal coliform to fecal streptococci and other observations, the consultant came to the conclusion that the high coliform counts were most probably of animal origin. Based on the results of this study, the community elected to develop and implement an on-site wastewater management district rather than constructing a centralized wastewater management system. Though the failed systems along the creek have been corrected, it is interesting to note that the coliform counts are still high in sections of the creek because of the large bird and animal population in the area.

Although controversial, use of the FC/FS ratio is useful as a guide in establishing the source of pollution in rainfall-runoff studies and in water pollution studies conducted in rural areas, especially where septic tanks are used. In many situations where human pollution is suspected on the basis of coliform test results, the actual pollution may, in fact, be caused by animal waste discharges. Establishing the source of pollution can be very important, especially where it is proposed or implied that the construction of conventional wastewater management facilities will eliminate the measured coliform values.

3.4
WATER IMPURITIES AND PUBLIC HEALTH

Water is of fundamental importance to the survival of human beings. From the standpoint of usage, the quantity of water ingested by humans is very small compared with the quantity used for agriculture (see Chapter 1). Yet, if the water ingested by humans is not safe to consume, serious health problems and possibly death can result. It is, therefore, the purpose of this section to examine the health impacts of the chemical and biological impurities in water. Current contaminant limits set by the U.S. Environmental Protection Agency for drinking water are considered at the end of this section.

Health-Related Chemical Impurities

A large number of chemical elements and compounds that directly affect public health are found in natural waters. Because of the introductory nature of this text, only a few examples will be discussed. More detailed information can be found in Refs. [3.3], [3.10], and [3.18].

Inorganic Constituents
Health-related impurities occur both naturally and through human activities. Monitoring of these materials is a comparatively recent development, and new horror stories are turning up almost daily in the news media. Magnesium, nitrate, and sodium are the only constituents commonly found in natural waters that have significant health-related effects. Magnesium has a mild laxative effect. Nitrate in concentrations exceeding 10 g N/m^3 interferes with oxygen utilization in newborn babies. The resulting disease, methanoglobanemia, is serious but easily prevented. In the past, sodium has been implicated in hypertension and high blood pressure. However, based on more recent evidence (1984), it appears that high blood pressure most probably results from an imbalance of the cations in the system.

Trace constituents not commonly found in water can be major health concerns. Arsenic, a heavy metal, is occasionally found in groundwaters. The dental-caries preventative, fluoride (which in excessive amounts causes tooth discolorization and brittleness), and heavy metals such as lead and cadmium discharged in wastewaters are often present in small concentrations.

Organic Constituents

The enormous variety of organic chemicals synthesized in recent years are also a major concern. Many of these compounds that are known or suspected carcinogens (such as trihalomethanes—THMs) are now found in groundwaters, rivers, and lakes. For example, trichloroethylene (TCE) and dibromochloropentane (DBCP) have recently been found in a number of California groundwaters. A soil fumigant, DBCP was used widely until it was found to cause sterility in production process workers. In many cases these compounds accumulate in tissue. Long-term effects of these compounds on fertility, their carcinogenicity, and other health risks are generally unknown, and the compounds are exceedingly difficult to monitor.

Health-Related Biological Impurities

People remove water from streams, lakes, and groundwater aquifers for domestic, commercial, industrial, and agricultural uses. Often they return the used water or wastewater to the same water source. Because microorganisms found in wastewater can survive for varying periods of time in natural waters, water use often becomes a health hazard as well as a necessity. Waterborne and water-related diseases are among the most serious health problems in the world today. Up to 35 percent of the potential productivity of many developing nations is lost because of these diseases. Thus the cost to the world economy is staggering. Many of the classic waterborne infections (typhoid fever, dysentery, and cholera) have been reduced to minor significance in developed nations. Other diseases such as shistosomiasis predominate in tropical climates and may go far to explain why development is generally associated with temperate climates.

Historical Perspective

Water quality management has its roots in disease prevention. Only recently have a few nations had the luxury of discussing such water quality aspects as aesthetics and the preservation of endangered species and of "natural" or "pristine" conditions. These fortunate nations include approximately 25 percent of the world's population. The remaining 75 percent are, to varying degrees, an endangered species themselves. In the worst situations, 50 percent of all children die before age five, and approximately 70 percent of these deaths are related to environmental sanitation [3.2]. Water is in extremely short supply for much of the world's population, either because of its lack or the lack of distribution facilities. In such areas, ponds and streams are often used for washing clothes, carrying away wastes, and drinking. Because access is uncontrolled, drinking water is often taken from a contaminated source. Villages in developing nations often make use of a central common well or spigot. Maintenance of the site is often poor; wells are shallow and unprotected from drainage. Because drainage includes human and animal wastes, these waters are often heavily polluted.

Water-Related Diseases

Areas with central village wells or spigots use very little water per capita [3.24]. People carrying heavy water cans long distances to their

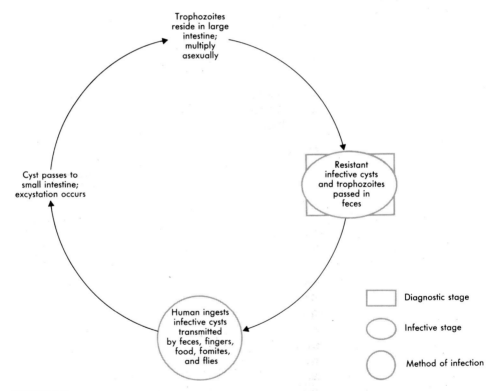

FIGURE 3.9

Life cycle of the infectious agent *Giardia lamblia*, the protozoan that is the cause of the giardiasis. The life cycle of the protozoan *Entamoeba histolytica* is essentially the same. Because cysts are resistant to desiccation and wide temperature changes, these protozoan diseases are very difficult to eliminate from the environment.

Source: Adapted from Ref. [3.10].

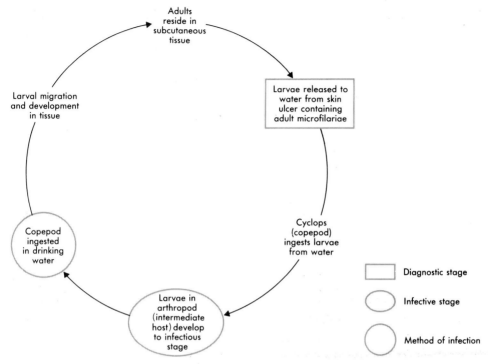

FIGURE 3.10

Life cycle of a filaria, the trematode worm that is the cause of filariasis.

Source: Adapted from Ref. [3.10].

homes are unlikely to wash their hands frequently or to keep the dishes sparkling clean at all times. The result is a high incidence of water-related diseases, particularly the various forms of diarrhea. Because diarrhea weakens the body through dehydration, effects of other health conditions such as malnutrition are enhanced. The result is that large segments of the world's population have serious chronic health problems. Several of the most prevalent diseases and their causes are given in Table 3.9. Examples of the life cycle of the infectious agents of some important water-based diseases and results of infection are shown in Figs. 3.9 through 3.13. Because many of the diseases listed in Table 3.9 can coexist in the human body and because the diet of many of the affected people is deficient, the effects are often greater than might be expected from the clinical characteristics of the individual diseases.

Over 100 virus types are known to occur in human feces, and an infected person may excrete as many as 10^6 infectious particles in 1 g of feces. Thus the potential for disease transmission is very great. Of those

TABLE 3.9

Typical Diseases Associated with Water

CATEGORY AND METHOD OF CONTRACTION	DISEASE	CAUSATIVE AGENT	SYMPTOMS
Waterborne: ingesting contaminated water	Amebiasis, (amoebic dysentery)	Protozoan (*Entamoeba histolytica*)	Prolonged diarrhea with bleeding, abscesses of the liver and small intestine
	Shigellosis (dysentery)	Bacteria (*Shigella*, 4 spp.)	Severe diarrhea
	Cholera	Bacteria (*Vibrio cholerae*)	Extremely heavy diarrhea, dehydration, high death rate
	Gastroenteritis	Virus (enteroviruses, parvovirus, rotovirus)	Mild to severe diarrhea
	Giardiasis	Protozoan (*Giardia lamblia*)	Mild to severe diarrhea, nausea, indigestion, flatulence
	Infective hepatitis	Virus (hepatitis A virus)	Jaundice, fever
	Leptospirosis (Weil's disease)	Bacteria (*Leptospira*)	Jaundice, fever
	Salmonellosis	Bacteria (*Salmonella*, ~ 1700 spp.)	Fever, nausea, diarrhea
	Typhoid fever	Bacteria (*Salmonella typhosa*)	High fever, diarrhea, ulceration of small intestine
Water-washed: washing with water contaminated	Shigellosis (dysentery)	Bacteria (*Shigella*)	Mild to severe diarrhea
	Scabies	Mite	Skin ulcers
	Trachoma	Virus	Eye inflammation, partial or complete blindness
Water-based: worm infections involving water as one stage in cycle	Filariasis	Worm	Blocking of lymph nodes, permanent damage to tissue
	Guinea worm	Worm	Arthritis of joints
	Schistosomiasis	Worm (schistosomes)	Tissue damage and blood loss in bladder and intestinal venous drainage

Source: Adapted from Refs. [3.6], [3.10], and [3.19].

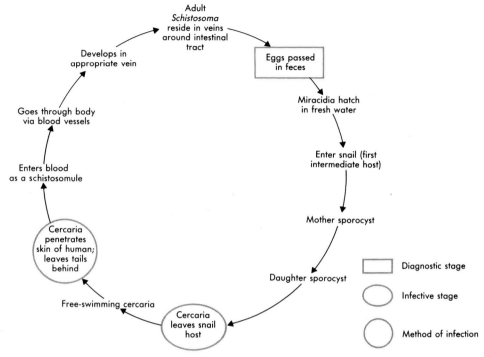

FIGURE 3.11

Life cycle of a schistosome, the worm that is the cause of schistosomiasis. The older name for this disease is bilharziasis. Schistosomas, a genus of trematodes, are also commonly called blood flukes.

Source: Adapted from Ref. [3.10].

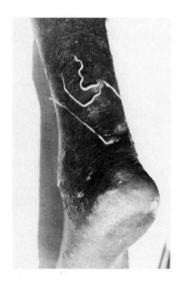

FIGURE 3.12

Adult Guinea-worm emerging from ankle of infected person. There is some evidence that the snake wound on a staff—the symbol of the medical profession—is actually based on a still-used method of winding Guinea-worms out of infected areas on a stick.

Source: *Radiology of Tropical Diseases* by M. M. Reeder and P. E. S. Palmer, ©1981, the Williams & Wilkins Co., Baltimore.

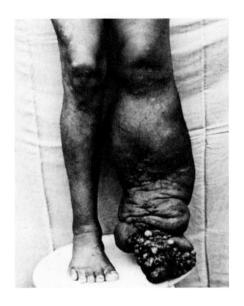

FIGURE 3.13

An advanced case of filariasis. Tissue damage is permanent.

Source: *Radiology of Tropical Diseases* by M. M. Reeder and P. E. S. Palmer, ©1981, the Williams & Wilkins Co., Baltimore.

contacting viruses only a small percentage are infected, and of those infected only about 2 percent may become recognizably ill. Assuming a 1 percent infection rate and a 2 percent illness rate, this means that 1 in every 5000 persons coming in contact with a virus becomes ill, a very high rate if water is contaminated with fecal matter.

Disease Prevention

Clearly, prevention of waterborne diseases is an important aspect of water quality management, and the role of engineers in public health is well established agencies in such as the World Health Organization of the United Nations. Because this text is not directed toward developed nations, large amounts of space will not be devoted toward engineering solutions for developing nations. Much of the material herein will also be applicable to developing nations, but a cautionary note should be made. Technology applicable in the developed countries is not necessarily suitable for developing nations. The reasons are climatic and socioeconomic as well as technological. Equipment maintenance problems vary with climate, and often factors responsible for corrosion, scaling, rotting, or plugging are not foreseen. Sociological impediments to the use of facilities also vary from culture to culture. Quite often a facility is designed that is not used because of local customs, tastes, or beliefs. Every culture has such customs, and there is no reason to think of some as better than others. Thus the installation of even such relatively simple facilities as privies and water spigots must be made with care. Excellent information on problems of developing nations can be found in Refs. [3.3], [3.14], and [3.19].

Protection Against Disease
Protecting society from waterborne diseases must be done in a conservative manner. Epidemic diseases result in serious emotional and economic losses to affected communities. Thus the threat of, rather than actual, infection is used as a basis for control. Threat of waterborne disease is usually identified by the presence of coliform bacteria in water. When significant numbers of coliforms are found in a water supply, further testing is done to determine if other bacterial species definitely associated with the human intestinal tract are present [3.20].

Contaminant Limits

The current guidelines for assessing the suitability of a surface water or groundwater for use as a public water supply are the regulations mandated by the U.S. Environmental Protection Agency. The regulations are delineated in Title 40, Parts 141 and 143 of the Safe Drinking Water Act (SDWA). The primary regulations, as reported in Table 3.10, include maximum permissible levels for inorganic and organic chemicals, turbidity, coliform bacteria, and radiological constituents. Further details on these regulations may be found in Appendix D. The World Health Organization (WHO) standards for drinking water are similar. Further information on the health effects of specific constituents found in water may be found in Refs. [3.11], [3.16], and [3.21].

As reported in Table 3.10, maximum regulations have been set for only six anthropogenic organic compounds and total trihalomethanes. It is anticipated that this list will be expanded as the health risks of other synthetic chemicals are documented. Recently a number of volatile organic chemicals (VOC) have been detected in groundwater supplies. The presence of these chemicals is serious because many of them have been found to pose a carcinogenic or mutagenic risk to humans. The seriousness of the problem is compounded by the long residence time in the groundwater aquifers. The U.S. Environmental Protection Agency is currently considering several alternatives for controlling the presence of these compounds in water. In the future, it is anticipated that regulations will be issued to cover these chemicals. The movement and treatment of chemicals found in the groundwater is considered further in Chapters 10 and 11, respectively.

TABLE 3.10

**Contaminant Limits Set in Interim Primary
Regulations for Drinking Water***

CONTAMINANT	MAXIMUM PERMISSIBLE LEVEL
Inorganic Chemicals	
Arsenic	0.05 mg/L
Barium	1.0
Cadmium	0.010
Chromium	0.05
Lead	0.05
Mercury	0.002
Nitrate (as N)	10
Selenium	0.01
Silver	0.05
Fluoride[†]	1.4–2.4
Organic Chemicals	
Endrin	0.0002 mg/L
Lindane	0.004
Methoxychlor	0.1
Toxaphene	0.005
2,4-D	0.1
2,4,5-TP Silvex	0.01
Total trihalomethanes	0.1
Turbidity	1–5 NTU
Coliform bacteria	1/100 mL (mean)
Radiological	
Gross alpha	15 pCi/L
Radium 226 and 228	5
Gross beta	50
Tritium	20,000
Strontium 90	8

*Drinking water regulations are contained in Parts 141 and 143 of the Safe Drinking Water Act.

[†] Level varies with temperature (see Appendix D).

3.5

IMPURITIES OF IMPORTANCE NOT RELATED TO HEALTH

Other impurities in water are important because their presence affects the use potential of a particular source of water [3.12, 3.16, 3.21]. The purpose of this section is to introduce some important impurities in relation to specific water use categories, that is, agricultural, industrial, and ecological.

Agriculturally Related Impurities

Impurities in water that affect plant growth and soil characteristics, such as permeability, may be termed *agriculturally related*. Most often these impurities have a negative effect on agricultural production (Fig. 3.14). Examples are salinity, which affects the availability of crop water; sodium, which, because of its large radius of hydration, causes clay soils to disperse; and boron, which is toxic to many plants at concentrations below 1 g/m^3.

When clay soils are dispersed the soil pore size is then reduced and permeability to water is greatly decreased. Because the dispersive effect results from the positive sodium ion being adsorbed on negatively charged clay surfaces, replacement by other ions is possible. The interaction in ion exchange of the most abundant minerals in natural waters—calcium (Ca^{+2}), magnesium (Mg^{+2}), and sodium (Na^+)—provides an example. Calcium and magnesium are more tightly bonded to clay surfaces than sodium and so are preferentially adsorbed. Because the adsorption reactions are reversible, an equilibrium between the adsorbed and dissolved ions is approached, and the relative amounts of the adsorbed species will be a function of the amounts in solution. In most

(a)   (b)   (c)

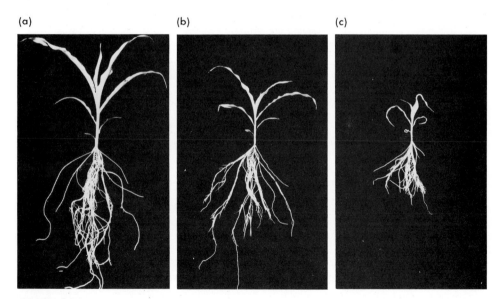

FIGURE 3.14

Effect of salinity on plant growth. The three corn plants shown in this figure were grown from seeds hydroponically in a 1 mM (millimolar) NaCl solution for 14 days. The plants were then transferred to solutions containing: (a) 1 mM NaCl, (b) 75 mM NaCl, and (c) 100 mM NaCl, and grown for an additional seven days. The plant in the 1 mM NaCl solution served as the control.

Source: Courtesy Malcolm Drew Agricultural Research Council, Letcombe Laboratory, United Kingdom.

agricultural irrigation waters, the great bulk of the cations are Ca^{+2}, Mg^{+2}, and Na^+, and other ions can be neglected in determining the risk of soil dispersion. A widely used relationship for estimating water permeability problems is the sodium adsorption ratio (SAR) [3.5, 3.23]:

$$SAR = \frac{(Na^+)}{\sqrt{\frac{(Ca^{+2}) + (Mg^{+2})}{2}}} \qquad (3.10)$$

where

Na^+ = concentration of sodium, eq/m^3
Ca^{+2} = concentration of calcium, eq/m^3
Mg^{+2} = concentration of magnesium, eq/m^3

Acceptable SAR values for irrigation water are a function of particular soil characteristics and are often considered in conjunction with the electrical conductivity. Representative values are reported in Table 3.11. In general, waters having SAR values less than 3 are low risk, while use of waters having SAR values greater than 6 requires considerable care. The use of Eq. (3.10) is illustrated in Example 3.3.

TABLE 3.11
Selected Guidelines for Assessing the Suitability of Water for Irrigation Purposes

POTENTIAL PROBLEM	PARAMETERS*	UNITS†	DEGREE OF RESTRICTION OF USE		
			None	Slight to moderate	Severe
Salinity	EC_w	dS/m	< 0.7	0.7–3.0	> 3.0
(affects availability of crop water)	TDS	g/m^3	< 450	450–2000	> 2000
Permeability	SAR and	SAR and	0–3	0–3	0–3
(affects infiltration	EC_w	dS/m	> 0.7	0.7–0.2	< 0.2
rate of water	SAR and	SAR and	3–6	3–6	3–6
into soil, evaluated	EC_w	dS/m	> 1.2	1.2–0.3	< 0.3
using SAR and	SAR and	SAR and	6–12	6–12	6–12
EC_w together)	EC_w	dS/m	> 1.9	1.9–0.5	< 0.5
	SAR and	SAR and	12–20	12–20	12–20
	EC_w	dS/m	> 2.9	2.9–1.3	< 1.3
Specific ion	Sodium	SAR	< 3	3–9	> 9
toxicity (affects	Chloride	meq/L	< 4	4–10	> 10
sensitive crops)	Boron	mg/L	< 0.7	0.7–2.0	> 2.0

$*EC_w$ = electrical conductivity of irrigation water; SAR = sodium adsorption ratio.
†dS/m = decisiemens per meter.
Source: Adapted from Ref. [3.23].

EXAMPLE 3.3

DETERMINATION OF SODIUM ADSORPTION RATIO

Determine the SAR for the water supplies of Biloxi, Miss., and San Diego, Calif., using the information given in Table 3.2.

SOLUTION:

1. Determine the Ca^{+2}, Mg^{+2}, and Na^+ concentrations in equivalents per cubic meter.

		CONCENTRATION			
		Biloxi, Miss.		San Diego, Calif.	
CONSTITUENT	EQ. MASS, g/eq	g/m^3	eq/m^3	g/m^3	eq/m^3
Ca^{+2}	20.0	1.5	0.075	87	4.35
Mg^{+2}	12.15	0.8	0.066	29	2.39
Na^+	23.0	138	6.00	99	4.30

2. Compute the SAR of the two waters.
   a. For Biloxi, Miss.:

$$SAR = \frac{6.00}{\sqrt{\left(\frac{0.075 + 0.066}{2}\right)}} = 22.6$$

   This is a very high SAR value, and the water would be generally unacceptable for irrigation. Note that the TDS value is moderate and that Biloxi water is generally excellent for domestic use.
   b. For San Diego, Calif.:

$$SAR = \frac{4.30}{\sqrt{\left(\frac{4.35 + 2.39}{2}\right)}} = 2.34$$

   The water supply of the city of San Diego has a very low SAR value and would be an excellent source of irrigation water. The TDS value is very high (761 g/m^3), and San Diego has a much inferior source of domestic water as compared with Biloxi.

Industrially Related Impurities

Impurities that affect industrial water use may be categorized as *industrially related*. These impurities would include materials that cause taste, odor, and color; toxic substances (such as pesticides, herbicides, solvents, and heavy metals); biological growth stimulants; reactive materials (particularly those that cause scaling, corrosion, or reduction in dissolved oxygen concentration); and nonsoluble materials such as colloids that cause turbidity and sediment that results in shoaling and filling of pipes and channels.

Scaling and corrosion result in expenditures of over several billion dollars per year in the United States because of required maintenance and replacement of pipes, boilers, and system appurtenances. Corrosion is often the result of bacterial slime growth that either exposes surfaces or actually utilizes the iron in pipes. The bacterial slimes responsible for the corrosion also reduce heat transfer in boilers and increase friction losses in pipes.

Ecologically Related Impurities

Ecologically related impurities are closely related to the phenomena of eutrophication and acid rain. Impurities that affect the ecosystem through toxicity or growth stimulation are termed *ecologically related*. Microorganisms affecting aesthetic characteristics, such as floating algae, or producing toxic materials such as the blue-green alga, microcystis, are included. Microcystis often occurs in pools in arid and semiarid areas in the southwestern parts of the United States. The toxin produced by the organism causes illness in cattle and humans and can be fatal. Thus, the organism can also be classified as a health-related impurity.

Most impurities in water have some ecological impact, and many of the impacts change markedly with concentration [3.22]. Many heavy metals (e.g., Zn^{+2}) are required in trace amounts for plant and animal growth but are toxic when concentrations approach the 0.1 to 1.0 g/m^3 range. Some ions (e.g., Cl^-) have minimal ecological effect over concentration ranges of several orders of magnitude, although other characteristics of the water, such as taste, may be affected.

Three dissolved gases are particularly important—NH_3, CO_2, and O_2. Ammonia, which occurs in equilibrium with NH_4^+ in water, is very toxic to fish, often being lethal at the 0.1 g/m^3 level. Carbon dioxide occurs in equilibrium with carbonic acid (H_2CO_3), bicarbonate ions (HCO_3^-), and carbonate ions (CO_3^{-2}). The equilibrium relationships of these impurities are important factors in determining the pH and buffering characteristics of water. Thus the carbonate equilibrium condition is significant aside from its ecological effect, and for this reason it was discussed in considerable depth in Chapter 2. Oxygen is required by most living organisms, and the presence or absence of oxygen is probably the most important single factor in the aesthetic quality of a water.

Biostimulants (chemical elements that stimulate biological growth) are the most easily recognized ecologically important impurities because their addition to natural waters results in greatly increased eutrophication rates. Nitrogen as NH_4^+ and NO_3^- and phosphate (PO_4^{-3}) are the most common biostimulants, but any growth-limiting nutrient can act in this manner.

3.6
THE AQUATIC HABITAT

A marvelously wide range of organisms inhabit the waters of the world. In most cases the casual observer sees little other than an occasional fish and possibly some crustaceans, algae, and aquatic plants. These organisms are only a tiny fraction of the diverse groups making up the aquatic ecosystem (see Fig. 3.15). Many organisms, such as the stone fly and mayfly shown in Fig. 3.15, spend their larval stages attached to rocks in swift-running streams. Tubifex worms bury themselves in benthic (bottom) deposits and leave only an antennalike tail waving in the water to absorb oxygen. Single-celled organisms such as the protozoa (see Fig. 2.28), algae (see Fig. 2.26), and bacteria (see Fig. 2.24) are typical representatives of the kingdom Protista.

Organisms found in water usually have relatively tight restrictions in their environments. For example, Pacific salmon require temperatures below 16°C and dissolved-oxygen concentrations above 5 g/m^3 to spawn. These fish will not enter a river on their spawning run until conditions are satisfactory. Kelp, a species of ocean algae, require both a rocky bottom for attachment and clear water to allow light penetration for photosynthesis. In many cases seemingly small changes in water temperature or dissolved constituents will make one species more competitive

FIGURE 3.15

A portion of an aquatic ecosystem illustrating the presence of (a) adult stone fly and stone fly larvae, (b) adult mayfly and mayfly larvae, and (c) tubifex worms.

than others in the vicinity. Changing of riverbanks from shaded to sunlit by logging may result in a temperature increase that allows suckers to replace trout as the dominant fish species. Thus environmental changes, natural or anthropogenic, can be expected to change the type of population present in any natural water.

The Food Chain

Organisms found in water can be classified as *saprophytic* (those that use dead and decaying organic materials as their found source), *chemoautotrophic* (those that obtain energy from the oxidation of inorganic chemicals and carbon from CO_2), *chemophototrophic* (those that use light as an energy source), and *predatory* (those that use other living organisms as a source of food). Some species fall into more than one group. For example, blue-green algae (an inappropriately named group of bacteria) are both chemoautotrophs and chemophototrophs, and many fish are both predators and saprophytic. Under normal conditions members of each group are present in numbers that provide a balance of the various trophic levels. Saprophytic organisms consume dead organic materials; predators consume saprophytes, autotrophs, and smaller predators. Members of each group die and are consumed by saprophytes. The system is an almost endless chain or circle fueled by solar energy trapped in photosynthesis. Environmental changes such as the passing seasons or discharge of wastes may directly stimulate or restrict one segment of the chain more than others, but because the entire chain is connected, all segments will be affected. Thus, natural waters must be considered as complex ecosystems rather than habitats for specific species such as game fish.

Eutrophication

Natural waters may be classified, according to their ability to support life, as oligotrophic, mesotrophic, and eutrophic. *Oligotrophic waters* contain low concentrations of essential nutrients such as nitrogen, phosphorus, and iron, and life forms are present in small numbers. Lake Superior in the Great Lakes, Lake Tahoe in California and Nevada, and Crater Lake in Oregon are examples of oligotrophic waters. Natural processes of growth, death, and decay as well as input of nutrients from runoff result in a gradual increase in nutrient concentrations, and the waters become increasingly productive. The terms mesotrophic and eutrophic are used to describe these progressive states.

Mesotrophic waters are characterized by the abundance and diversity of life forms at all trophic (food chain) levels. *Eutrophic waters* characteristically have fewer species present, but the concentration of algae is particularly high. Dissolved-oxygen levels often fluctuate widely between day and night. The process of moving from oligotrophic through mesotrophic to eutrophic conditions is called *eutrophication*. This process is a natural one, and depending on conditions may occur over relatively short periods (decades) or relatively long periods (geologic time). Human activities nearly always accelerate the rate of eutrophication, and one of the major objectives of water quality management is to reduce or mitigate the effects of human activities on eutrophication.

Water quality management requires an understanding of the physical, chemical, and biological characteristics of water; of the requirements associated with various uses of water; of the methods of predicting changes in water quality due to environmental changes; and of the methods of modification of water quality.

KEY IDEAS, CONCEPTS, AND ISSUES

- The range of water quality problems encountered in the field is extremely diverse, involving one or more of the physical, chemical, and biological characteristics of water.

- The solution of most water quality problems involves a fundamental understanding of the basic phenomenon involved and the application of the basic principles and concepts discussed in Chapter 2.

- The principal inorganic constituents that occur naturally in water that may have serious health impacts are magnesium, nitrate, and sodium.

- The most troublesome organic constituents in water are of anthropogenic origin, including compounds used as pesticides, fungicides, soil fumigants, and solvents. At present, the long-term effects of these compounds on fertility, carcinogenicity, and other health risks are generally unknown.

- Up to 35 percent of the potential productivity of many developing nations is lost because of waterborne and water-related diseases.

- The suitability of a water for irrigation purposes depends on its sodium adsorption ratio (SAR) and electrical conductivity.

- The suitability of a water for industrial purposes varies with the type of industrial application and is, therefore, site-specific.

- The aging of bodies of water because of the accumulation of mineral and organic nutrients is termed eutrophication.

- Human activities nearly always accelerate the rate of eutrophication.

- One of the major objectives of water quality management is to reduce or mitigate the effects of human activities on eutrophication.

DISCUSSION TOPICS AND PROBLEMS

3.1. Briefly explain the differences in water quality problems, and hence the differences in definition of water quality, that would be identified by engineers working for (a) the California Department of Water Resources and (b) the national government of Egypt.

3.2. Dissolved-oxygen concentrations are environmentally important for two primary reasons. Give a brief explanation of the two.

3.3. Determine the Langelier and Ryznar indexes for waters (1), (2), and (6) in Table 3.2. Use 20°C for those temperatures not listed in the table.

3.4. Determine the Langelier and Ryznar indexes for the following three waters:

	WATER		
CONSTITUENTS	A	B	C
Ca^{+2}, g/m^3	100	100	100
HCO_3^-, g/m^3	107	196	152
TDS, g/m^3	850	850	850
pH	6.5	8.3	9.1
T, °C	15	15	15

3.5. Determine the Ryznar index for Mississippi River water in July 1966 at Grand Rapids, Minn., and St. Francisville, La. (see Table 3.4). Assume the water temperature is 15°C.

3.6. Acid rain decreases the pH of soil moisture. What effect does this have on the long-term availability of phosphate, iron, and calcium for agriculture? Is there a significant increase in the required fertilizer application rate?

3.7. Why is the effectiveness of copper sulfate, used to control plankton in water storage reservoirs, reduced if the alkalinity of the water is high and the carbon dioxide content is low?

3.8. How much copper sulfate is needed to provide a free residual level of 2 mg/L in a 10,000 m^3 storage reservoir if the concentration of hydrogen sulfide in the reservoir water is 5.0 mg/L?

3.9. Determine the relative amounts of magnesium ingested per day from drinking Amarillo, Tex., water and from using the recommended dosage of a magnesium-based laxative.

3.10. Why have schistosomiasis and guinea worm remained water-based diseases with a major economic impact despite control efforts?

3.11. Explain why diarrhea is such a catastrophic disease in developing nations.

3.12. Discuss the possible sources of transmission of salmonella to a customer of a fast-foods restaurant. Note which sources are most likely.

3.13. Use the library to determine the principal methods of transmission of hookworms, pinworms, beef tapeworms, and trichinosis.

3.14. Develop a diagram showing the transmission cycle of filariasis.

3.15. Discuss application of the MPN test in setting standards for (a) municipal water supplies, (b) treated municipal water, (c) swimming pools, and (d) treated wastewater discharges.

3.16. Use the library to develop an example of heavy-metal poisoning due to cadmium, mercury, or lead.

3.17. Estimate the electrical conductivity value for the water supplies of Biloxi, Miss., and Wheeling, W. Va., using the conductivity factors given in Table 2.11. The chemical characteristics of the water supplies are given in Table 3.2.

3.18. Irrigation water quality is judged on the basis of SAR, salinity (as TDS or electrical conductivity), and specific ions. Would waters (1) and (5) listed in Table 3.2 be a satisfactory irrigation water?

REFERENCES

3.1. Cox, C. R., (1964), *Operation and Control of Water Treatment Processes*, World Health Organization, Geneva.

3.2. Feachem, R. G., (1977), "Water Supplies for Low-Income Communities: Resource Allocation, Planning and Design for a Crisis Situation," in *Water, Wastes and Health in Hot Climates*, edited by R. G. Feachem, M. McGarry, and D. D. Mara, John Wiley and Sons, Chichester, England.

3.3. _____, M. McGarry, and D. D. Mara, (1977), *Water, Wastes and Health in Hot Climates*, John Wiley and Sons, Chichester, England.

3.4. Hale, F., (1942), "The Use of Copper Sulfate in Control of Microscopic Organisms," Phelps Dodge Refining Corporation, New York.

3.5. Hansen, V. E., O. W. Israelsen, and G. E. Stringham, (1979), *Irrigation Principles and Practice*, John Wiley and Sons, New York.

3.6. Hawkes, H. A., (1971), *Microbial Aspects of Pollution*, Academic Press, London.

3.7. Hesse, L. W., G. L. Hergenrader, H. S. Lewis, S. D. Reetz, and A. B. Schlesinger, (1982), *The Middle Missouri River*, Missouri River Study Group, Norfolk, Neb.

3.8. LaBastille, A., (1981), "How Menacing is Acid Rain?" *National Geographic*, vol. 160, no. 5, p. 652.

3.9. Langelier, W. F., (1936), "The Analytical Control of Anticorrosion Water Treatment," *J. American Water Works Association*, vol. 28, no. 10, p. 1500.

3.10. Leventhal, R., and R. F. Cheadle, (1979), *Medical Parasitology*, F. A. Davis Company, Philadelphia, Penn.

3.11. Lohn, E. H., and S. K. Love, (1952), "The Industrial Utility of Public Water Supplies in the United States," U.S. Geological Water Supply Paper 1299, U.S. Government Printing Office, Washington, D.C.

3.12. McKee, J. E., and H. W. Wolf, (1973), *Water Quality Criteria*, 3d ed., Bulletin 3, California State Water Resources Control Board, Sacramento, Calif.

3.13. McLaren, F. R., (1976), "Water Quality Studies of the Truckee River," Frederick R. McLaren Environmental Engineering, Sacramento, Calif.

3.14. Mara, D. D., (1976), *Sewage Treatment in Hot Climates*, John Wiley and Sons, Chichester, England.

3.15. National Interim Primary Drinking Water Regulations, (1975), *Federal Register Part IV*, pp. 59566–89, Washington, D.C.

3.16. *Quality Criteria for Water*, (1976), U.S. Environmental Protection Agency, Washington, D.C.

3.17. "Quality of Surface Waters of the United States," (1966), U.S. Geological Survey, Parts 5–8, Water Supply Papers, nos. 1993 and 1994.

3.18. Ryznar, J. W., (1944), "A New Index for Determining Amount of Calcium Carbonate Scale Formed by a Water," *J. American Water Works Association*, vol. 36, no. 4, p. 472.

3.19. Salvatto, J. A., (1982), *Environmental Engineering and Sanitation*, 3d ed., Wiley Interscience, New York.

3.20. *Standard Methods for the Examination of Water and Wastewater*, (1980), 15th ed., American Public Health Association, New York.

3.21. *Water Quality Criteria 1972*, (1973), National Academy of Science, Ecological Research Series, U.S. Environmental Protection Agency Report R3-73-033, Washington, D.C.

3.22. Welch, E. B., (1980), *Ecological Effects of Waste Water*, Cambridge University Press, Cambridge.

3.23. Wescot, D. W., and R. S. Ayers, (1984), "Water Quality Criteria," in *Irrigation With Reclaimed Municipal Wastewater: A Guidance Manual*, edited by G. S. Pettygrove and T. Asano, State Water Resources Control Board, Sacramento, Calif.

3.24. White, A. V., (1977), "Patterns of Domestic Water Use in Low-Income Communities," in *Water, Wastes and Health in Hot Climates*, edited by, R. G. Feachem, M. McGarry, and D. D. Mara, John Wiley and Sons, Chichester, England.

Water Quality: Standards and Global Perspectives

The term *standard* may be applied to any definite rule, principle, or measure established by authority. Setting standards for water quality is clearly a difficult task. As has been noted in the preceding chapters, the definition of water quality is closely related to the intended use of the water. Unfortunately, setting standards based on a given use potential is inappropriate. For example, the effect of thermal discharges may be insignificant, or even beneficial, to domestic and agricultural users, but catastrophic with respect to the aquatic community. Hard water may be perfectly satisfactory for drinking, but very unsatisfactory for use in boilers. Thus some other approach must be used in setting water quality standards. It is the purpose of this chapter to examine how standards are set, the issues associated with the setting of standards, and the responsibility of the engineering community in helping to set the standards. Global perspectives in water quality management are considered in the final section of this chapter.

4.1
STANDARDS SET BY GOVERNMENTAL STIPULATION

Two general methods are in use for setting standards: (1) governmental stipulation and (2) a policy of minimum degradation. The latter approach is considered in Section 4.2. Most national governments are deeply involved in the economic management of their countries. Even in nominally "free enterprise" nations such as the United States, prices of many products and commodities are directly controlled or subsidized. Governments often include environmental quality in their economic management programs in a number of ways. For example, a common method of improving conditions in poor or less-developed areas or countries is to subsidize industrial development, and one of the most common subsidies is the minimization of environmental quality standards. In many cases, this means allowing discharge of wastewater (or other waste products) with little if any treatment. Often such discharges have resulted in grossly polluted lakes, rivers, and near-shore ocean waters. In systems that are flushed continually, such as rivers and estuaries, such actions are, in most cases, reversible. The cleanup of the Ohio River, resulting from the work of the Ohio River Sanitation Commission (ORSANCO), is an excellent example of what can be done when a concerted effort is made [4.2]. Although the task is generally difficult and expensive, the cleanup of our streams and rivers must be dealt with eventually. It is hard to conceive of a situation in which it would not be the better choice to provide a subsidy directly for the treatment of wastes prior to discharge.

When lakes are the receiving body of water the result is much worse. Returning Lake Superior or Lake Michigan to a pristine state is an impossible task. The flushing time is too great, and such an effort would be counter to the natural process of eutrophication. Allowing lake pollution as a stimulant to industrial growth is nothing more than trading a nation's heritage for trinkets of passing value. Examples such as allowing the discharge of ore processing tailings into Silver Bay on the northwest corner of Lake Superior provide excellent support for maintaining a very conservative policy [4.16]. In this case, an ore processing plant located on Silver Bay was allowed to discharge tailings to the lake. Movement of the fine particles was much greater than predicted, and deposition over a large lake area and increased turbidity of a portion of the lake were two results. A third result was the introduction of a major source of asbestos fibers into the lake and hence into many community water supplies (most notably that of Duluth, Minn.). Asbestos fibers are carcinogenic if inhaled, but the effects of their presence in drinking water are not known.

An example of a more indirect effect of governmental policy on water quality is provided by noting some results of the opening of the Great Lakes to ocean shipping. Two species of fish, lamprey eels and alewives, migrated into the Great Lakes and were very destructive. The lampreys fed on the eggs of other fish, decimated the natural game fish population, and destroyed the commercial fishing industry. Control was achieved through use of a selective poison. Alewives, small sardinelike fish, had no natural predators in the Great Lakes and rapidly increased in numbers while outcompeting many native species. The unexplained deaths of large numbers of alewives just prior to their spawning period resulted in offensive conditions on beaches (Fig. 4.1). Obviously, the beach environment was not enhanced by mounds of rotting fish. Control by predators such as coho salmon, introduced in 1964, and by heavy

FIGURE 4.1

Dead fish on the shore of Lake Ontario.

Source: NFB Photothèque, Ottawa, Canada.

commercial fishing has been reasonably successful in restoring the balance of fish populations.

Another method of government stipulation is to classify streams, or reaches of streams, for various uses. Classifications might range from "pristine," meaning unaffected by human activities, to "grossly polluted," meaning that the stream reach is an open sewer. Most countries that classify rivers do so in a more innocuous manner, using a number or letter classification (e.g., 1 for pristine; 4 for grossly polluted), but the effect is exactly the same. In many cases, classification came after a stream reach was heavily polluted. But by classifying segments of a stream or river, a type of zoning is accomplished, and upper reaches can be saved. An example is the Aire River in Yorkshire, England, which begins in the Pennine Mountains and flows into the Humber about 30 km east of Leeds. As the river passes through Leeds and Bradford, a heavily industrialized area, it becomes a class-4 stream. Making major improvements in the water quality of the Aire River as it flows through the final 30 km lined with factories, coal mines, and power plants would be very difficult, as well as very expensive. Improvements over a long period of time may be possible, and this is an important consideration, since the Aire is a waste stream discharging into the Humber and therefore affects the quality of the Humber. In the meantime, the upper reaches of the Aire are protected through the imposition of a higher classification.

4.2

STANDARDS BASED ON MINIMUM DEGRADATION

Setting standards by the concept of minimum degradation can be thought of as an approximation of the no-pollutant discharge objective put forth in the Federal Water Pollution Control Act Amendments of 1972 (Public Law 92-500). Wastewater discharges will always have some effect on the receiving stream. Removal of all contaminants (including heat) is not a realistic goal in terms of technology or economics, but minimizing impact or degradation is a reasonable objective. Setting a maximum allowable depletion of dissolved oxygen in a receiving water because of a wastewater discharge would be a statement of minimal degradation. Placing a thermal discharge downstream of fish-spawning areas or setting a maximum increase in temperature over the ambient upstream river temperature would be another example.

A policy of minimum degradation has the advantage of forcing water users to adapt to "natural" conditions. The river or lake is left in a near-natural condition and, therefore, can be used for all beneficial uses. Although simple in concept, minimum degradation policies are difficult to apply. The water quality of lakes and rivers changes because of nonpoint discharges such as drainage from farms, roads, and riparian lands. Reservoirs and other control structures strongly affect water quality. In arid regions, movement of water for irrigation projects can severely degrade water quality, both through loss of water in the stream by diversion and through the return of salt-laden drainage. Cumulative effects of municipal and industrial discharges can be major, and there is great difficulty in parceling out portions of the "allowable degradation" in an equitable manner.

The Federal Republic of Germany (FRG) has established a carrot and stick approach based on a minimum degradation philosophy. Construction grants for wastewater treatment facilities of up to 40 percent are given to dischargers by the FRG. Fees are charged for each kilogram of selected pollutants (e.g., COD, BOD, SS, and THMs) discharged and the fee structure is such that a strong economic advantage exists for effective wastewater treatment. Compliance is established through a series of unannounced samplings. A major advantage of this approach is the virtual elimination of litigation.

4.3

THE U.S. PHILOSOPHY ON STANDARDS

The United States has gravitated toward a modified minimal degradation policy that incorporates a best-available-treatment (BAT) criterion. Under this policy the objective is to minimize effects of discharges on receiving waters, but it is recognized that the extent of wastewater treatment is technically and economically constrained. Best-available-treatment technology is defined by the type of waste under consideration. For municipal wastewaters BAT is usually considered to be secondary treatment (i.e., encompassing the removal of 90 percent of BOD_5 and of suspended solids). However, the definition of BAT is situation-dependent. If phosphorus is a limiting nutrient in the receiving water, phosphorus removal is added to the BAT requirement. The current BAT standard for phosphorus is based on precipitation and granular-medium filtration.

It is important to note that the U.S. Environmental Protection Agency has included economic feasibility as a component in determining BAT. Distillation is always a technical possibility that would produce a noncontaminating or polluting discharge, but economic considerations eliminate it from consideration in most cases.

Stream water quality standards are set, and wastewater discharges are not allowed to violate these standards. Requirements for wastewater discharges are set on a minimal degradation basis, and, in addition, all dischargers are required to use BAT. Defining BAT is not always straightforward, and it must also be recognized that BAT changes with time. Application of the minimum-degradation approach must be flexible, and care must be taken to ensure that protecting water quality remains as the priority rather than simply forcing all discharges to have a particular level of treatment. Overall, the system has worked well, however.

The principal problem with the U.S. approach has been that enforcement is based on a system of fines for non-compliance. Enforcement actions often result in suits and counter-suits that take years to settle and cost millions of dollars. The approach used in the Federal Republic of Germany appears to be considerably more effective.

4.4
THE IMPORTANCE OF SETTING STANDARDS

Water is a resource held in common. Legislated water quality standards are required to prevent misuse of the resource for individual gain. There is no actual benefit to a corporation or a city in maintaining downstream water quality, at least not in the short term. Despite this fact, many outstanding examples exist of sound water quality management based on ethical ideals, civic pride, and philanthropic generosity. Rivers and lakes are often symbols of a community or of a region, and residents, including corporations, often have an emotional stake in maintaining their beauty and quality. However, city councils are nearly always dealing with difficult budget constraints, and corporate leaders must face stockholders' demands for high return on investment. Thus, if maintaining water quality is an important priority, legal constraints must be placed on degradation, or a "tragedy of the commons" [4.6] will surely occur.

In the late nineteenth and early twentieth centuries, a general debate occurred on the need of treating wastewater prior to discharge. The debate was primarily economic, and the focus was on the least expensive mode of operation, treating wastewater prior to discharge or treating lower quality water for specific uses [4.15]. It was recognized that in some cases wastewater treatment was necessary to maintain public health, but aesthetic concerns were not prominent issues. As the population increased, urbanization and heavy industry grew, and an increasing number of rivers became open sewers, especially in the eastern United States. By 1950 the Willamette in Oregon, the Ohio, the Illinois, and many other rivers were examples of the fact that dilution was no longer the solution to pollution. The Willamette River case is particularly noteworthy; an outstanding salmon run was destroyed early in the twentieth century because of pollution resulting from wastewater discharges by two paper mills near the juncture with the Columbia River. Communities and industries along the entire length of the river contributed to the pollution, but the two mills were singularly efficient in eliminating the salmon.

As the public became aware of the impact of water quality on the overall quality of life, pressure was brought to bear on state and federal legislatures to change the system. In many cases, the early leaders of the groups were dismissed as "ecofreaks and Sierra Clubbers." Today there is little question about the correctness of their position in the late 1960s and early 1970s or about the benefits resulting from the changes these individuals and groups brought about. Considering the seriousness of the water quality problems society now faces, such as the presence of toxic materials in water supplies, nitrate buildup, acid rain, and eutrophication, it is fortunate that public concern was and remains aroused.

4.5
BASES FOR ESTABLISHING WATER QUALITY STANDARDS

In general, the role of public policy is to establish goals. Implementation of those goals is achieved by the adoption of standards and the development of criteria, regulations, and guidelines [4.8]. For example, the drinking water standards (see Appendix D) adopted by the U.S. Environmental Protection Agency are designed to protect public health, a goal established by public policy. Criteria represent the quantification of the physical, chemical, biological, and aesthetic characteristics of water quality, usually for a specific application—for example, the number of turbidity units that can be tolerated in a water to be used in the production of paper. The distinction between standards and criteria is that standards are derived from criteria and a variety of other considerations. Also, standards are legally enforceable, whereas criteria are not. Regulations and guidelines are developed to define how the standards shall be met—for example, how many samples must be collected and at what intervals.

Water quality standards typically are established on the basis of one or more of the following factors:

1. Established or ongoing practice (experience, criteria, etc.)
2. Technical attainability
3. Economic attainability
4. Results of biological experiments (e.g., bioassays)
5. Ability to measure parameters reliably
6. Evidence derived from accidental human exposure
7. Educated guess based on available information and judgment
8. Application of mathematical models (e.g., simulation of health risks)
9. Legal enforceability

Considering the factors involved in establishing them, it is clear that the standards must be dynamic in nature and not "set in concrete." However, once established and accepted, standards are considered to be legal requirements and are extremely difficult to change. The continued use of the BOD test is an important example (see Chapter 2). Ideally, the BOD test should be used to measure the amount of oxygen required to convert all of the organic matter to end products and cell tissue. What is actually measured depends on the nature of the waste and the rate at which the waste is converted biologically. Thus the correct interpolation of the measured value and its usefulness will be different for each waste.

4.6
THE RESPONSIBILITY OF THE ENGINEERING PROFESSION

The record of engineers in water quality management is generally very good. Engineers have provided innovative solutions to a great number of problems. In most cases, however, engineers have contributed little to the development of a philosophy of water quality management. There is perhaps a sound case for engineers serving as technical counsel and letting others make the decisions. Certainly it is inappropriate for the engineering profession to assume its members know what is best for society. There has been an increasing trend for engineers to limit their roles to that of "advocate for the client" and to minimize the significance of water quality to society at large. Such attitudes are beneficial to a client in the short run, but not over an extended period of time. More importantly, treating water quality matters in adversary fashion can lead to deterioration of a public trust—the maintenance of water quality. If engineers abdicate their responsibility to the society in the management of water quality, the only long-term solution to the problem may be increased regulation.

4.7

GLOBAL PERSPECTIVES IN WATER QUALITY MANAGEMENT

Water quality management has large-scale, and, in fact, global impacts. The issues associated with acid rain, household detergents, and pesticides serve as excellent examples to illustrate the geographical and global impacts of pollution.

Acid Rain

The acid rain problem is an example of how the impact of industrial development in one region or country can affect a much larger area [4.7, 4.12]. Utility and industrial plants in North America produce approxi-

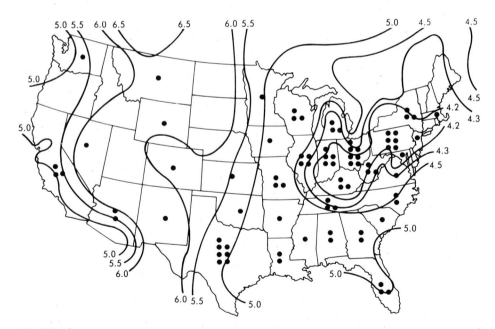

FIGURE 4.2

Distribution of acid rain in the United States and Canada. Each dot represents 500 kilotonnes of sulfur and nitrogen oxide emissions (map prepared for the U.S.-Canadian Memorandum of Intent, U.S. Department Of Energy).

mately 36 tonnes of sulfur oxides per year. Air pollution controls instituted to protect the local environment have forced the use of taller stacks, many of them 400 m high. Introduction of the pollutants into the atmosphere at such altitudes assures that they will be distributed over a large area. In this situation (using tall stacks), dilution with a larger air mass was assumed to be the solution to pollution, but as is often the case, a new problem was created. Thus a concentration of sources of sulfur oxides in industrial and urban areas such as the Ohio River valley in the United States has a major effect over a wide geographical region (Fig. 4.2).

Detergent Use

The use of detergents and the subsequent efforts to control eutrophication in lakes by limiting the input of phosphorus provide another interesting, although in this case more satisfying, example of the large-scale interactions associated with water quality management. When synthetic detergents were becoming widely available in 1948, the primary surface active agent (surfactant) was alkyl-benzene-sulfonate, ABS, a highly branched molecule that is difficult for most bacteria to metabolize. Because of the slow biodegradation rate and the fact that detergents can cause significant foaming at very low concentrations, aesthetic problems began to occur in receiving waters during the 1950s. Stories of beerlike heads on groundwater, foam piling up at dams, and sewage treatment plants lost in piles of suds became widespread. As a result, ABS was replaced in the early 1960s by linear-alkyl-sulfonate, LAS, which has little branching, is somewhat more expensive, and is much more biodegradable (see Chapter 2). This change in the makeup of detergents resulted in a disappearance of the foam and a sharp increase in the available phosphorus in domestic wastewater. Many lakes are phosphorus-limited—that is, phosphorus is the growth-limiting nutrient in the aquatic environment—and introduction of new phosphorus sources greatly increases their eutrophication rate. Several of the Great Lakes are phosphorus-limited, making the control of phosphorus a serious problem for Canada and the United States.

In the late 1960s, when the problem was first recognized, it was not known how common the phosphorus limitation was (nitrogen limitation now appears to be more widespread), but three courses of action were followed: (1) the phosphorus in synthetic detergents was reduced, (2) phosphorus was removed from wastewater, and (3) partial control was applied to nonpoint sources of water containing phosphorus. None of these actions was entirely satisfactory. Removing phosphorus from wastewater requires the addition of large amounts of lime, is expensive, and results in the production of large amounts of a sludge difficult to handle. Satisfactory substitutes for phosphate-based detergents have not been found. The most promising compound, nitrilotriacetic acid (NTA) was found to cause brain damage in infants. At the present time, lower phosphate formulas are used in detergents, and when the concentration of phosphorus becomes a problem, it is removed during wastewater treatment, or the discharge is diverted to a less sensitive point. The change in the phosphorus content in Lake Washington near Seattle provides an excellent example of the beneficial effect of moving a

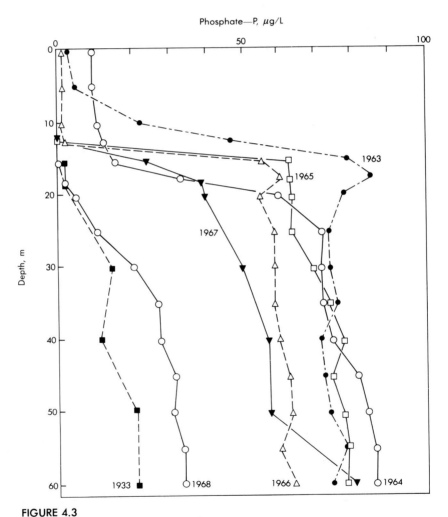

FIGURE 4.3

Depth distribution of phosphate (P) in Lake Washington (Seattle) before, during, and after sewage diversion.

Source: A.D. Hasler, "Man Induced Eutrophication of Lakes" in the Changing Global Environment, edited by S.F. Singer, D. Reidel Publishing Co., Dordrecht, Holland.

wastewater discharge to a less sensitive point. Phosphorus concentrations increased in the lake from the time wastewater was first introduced in 1933 until the discharge was stopped in 1965. By 1968, phosphorus levels were near the 1933 values as shown in Fig. 4.3.

DDT: A Case Study in the Global Movement of Pollutants

Movement of pollutants on a global scale is an increasingly threatening phenomenon. Since 1950, there has been a steady documentation of the accumulation of toxic or potentially toxic materials throughout the world environment. Although DDT has been banned in many countries, it is still used widely in developing nations. Although the use of DDT is decreasing, the reservoir of this very persistent pesticide that has accumulated is great, and the potential for further damage to the worldwide ecosystem is serious. Traces of DDT can be found throughout the environment in plants, animals, soil, and air. It has been estimated that most vertebrates worldwide contain about 1 ppm of DDT on a dry-mass basis and that the total DDT stored in a living organism is on the order of 10^{10} g [4.17].

Production of DDT through the 1960s was about 10^{11} g/yr (mostly in the United States), and its half-life in the environment may be as great as 20 years. Thus a great deal of DDT will be present for the forseeable future. The mechanism of movement of DDT through the environment is not clear, but certainly much of it is airborne. Although the vapor pressure is low (1.5×10^{-7} mmHg at 20°C) [4.14], the resulting equilibrium concentration in the atmosphere of 3×10^{-6} g/m^3 would place the saturation capacity of the atmosphere near 10^{12} g, a value that is of the same order of magnitude as cumulative DDT production through 1974 [4.17].

Where application of DDT has stopped, recovery of the ecosystem has generally occurred. Some organisms, including oceanic birds and fish that receive their DDT as part of the worldwide circulation, will not be greatly helped by local bans. On the other hand, many animal populations will respond rapidly to decreases in DDT use. The California brown pelican population has increased considerably since DDT use was stopped in the United States. Fish in Sebago Lake, Maine, increased in both size and number when DDT application decreased [4.1].

Changes in the Chemical Composition of the Atmosphere

On a larger scale, there is increasing concern about potential changes in the chemical makeup of the atmosphere and worldwide climate because of pollution from modern industrial societies. Estimates of fossil fuel

TABLE 4.1

World Production of Fossil Fuels in 1967

FUEL	PRODUCTION, 10^{15} g/yr
Coal	1.75
Lignite	1.04
Fuel oils	1.63
Natural gas	0.66

Source: From Ref. [4.5].

TABLE 4.2

Major Solid Wastes Discharged to the Ocean from the United States in 1968

WASTE TYPES	AMOUNT, 10^9 kg/yr
Dredge spoil	34.9
Industrial wastes	4.3
Sewage sludge	4.1
Construction debris	0.5

Source: Adapted from Ref. [4.5].

combustion and selected pollutant discharges given in Tables 4.1 through 4.4 can be used to develop an understanding of why engineers and scientists from many disciplines are concerned about long-term prospects for maintaining a livable environment. Combustion of fossil fuels is increasing the CO_2 content of the atmosphere at a rate of 0.25 to 0.3 percent per year [4.10] (see Example 4.1). Increases in CO_2 are believed to result in a warming of the lower atmosphere [4.9]; a 10 percent increase in CO_2 concentration should produce a temperature increase of 0.2 to 0.3°C.

TABLE 4.3

Major Particulate Emissions in the United States from Stationary Operations

PARTICULATE SOURCE	EMISSIONS, 10^{12} g/yr
Fuel combustion	5.9
Crushed stone, sand, and gravel	4.6
Agriculture	1.8
Iron and steel	1.4
Cement	0.9
Forest products	0.7
Lime	0.6
Clay	0.5
Primary nonferrous metals	0.5
Fertilizer and phosphate rock	0.3
Asphalt	0.2

Source: Adapted from Ref. [4.11].

TABLE 4.4

Annual World Lead Budget

	AMOUNT, 10^{12} g/yr
World production	3.5
Northern hemisphere production	3.1
Lead burned as alkyls	0.3
River input of soluble lead to the marine environment	0.2
River input of particulate lead to the marine environment	0.5

Source: From Ref. [4.5].

EXAMPLE 4.1

DETERMINATION OF GLOBAL CO_2 RELEASED TO THE ATMOSPHERE

If it is assumed that the CO_2 content of the atmosphere is increasing at the rate of 0.25 percent per year, estimate the amount of excess CO_2 that is being released to the atmosphere in tonnes.

SOLUTION:

1. Determine the moles of CO_2 in the atmosphere.
   a. The surface area of the earth is

   $$A_s \text{ (earth)} \simeq 5.1 \times 10^{18} \text{ cm}^2$$

   b. The mass of the air above 1 cm^2 of the earth's surface is 1,033 g. This value is obtained by noting that one standard atmosphere is equal to 10.33 m of water (see Physical Constants given inside rear cover).

   c. The total mass of the atmosphere above the earth is

   $$\text{Mass}_{air} = 1,033 \text{ g/cm}^2 \times 5.1 \times 10^{18} \text{ cm}^2$$
   $$= 52.7 \times 10^{20} \text{ g}$$

   d. The average molecular mass of air is 29 g; thus the total moles of air are

   $$M_{air} = 52.7 \times 10^{20} \text{ g}/(29 \text{ g}/M)$$
   $$= 1.8 \times 10^{20} M$$

   e. Assuming the fraction of CO_2 in the atmosphere is 0.0003, the moles of CO_2 in the atmosphere are

   $$M_{CO_2} = 0.0003 \times 1.8 \times 10^{20} M$$
   $$= 5.4 \times 10^{16} M$$

2. Determine excess CO_2 assuming the annual increase in CO_2 is 0.25 percent per year.
   a. The annual increase in CO_2 moles is about

   $$CO_{2_{increase}} = 0.0025/\text{yr} \times 5.4 \times 10^{16} M$$
   $$= 135 \times 10^{12} M/\text{yr}$$

   b. The annual increase in CO_2 in tonnes is

   $$CO_{2_{increase}} = 135 \times 10^{12} M/\text{yr} \times 44 \text{ g}/M$$
   $$= 59.4 \times 10^{14} \text{ g/yr}$$
   $$= 59.4 \times 10^{8} \text{ tonnes/yr}$$

COMMENT

The magnitude of the number is difficult to conceptualize. It can be put in some perspective if it is assumed that there are over 100×10^6 automobiles in the world, and an estimate is made of the amount of gas utilized per year.

Climatic changes brought about by human activities are also of concern. For example, the accumulation of atmospheric particulates results in increased reflection of solar energy and thus tends to cool the atmosphere. Approximately 30 percent of the annual particulate input to the atmosphere can be considered unnatural and, therefore, controllable to some extent.

Global Assessment

In summary, what is happening to the global environment is not understood fully, but it is known that the scale of human activities that impact the global environment is significant. The important question is: How stable is the global environment? If it is unstable or metastable in terms of a livable environment, it is imperative that immediate corrective actions be taken.

KEY IDEAS, CONCEPTS, AND ISSUES

- Two methods for setting water quality standards are in general use: governmental stipulation and a policy of minimum degradation.
- Water quality standards are needed to protect the nation's waters, a resource held in common.
- Standards by their very nature must be dynamic, changing with experience and reflecting the ever-expanding scientific data base.
- The engineering profession must take a leadership role in protecting the nation's waters rather than merely relying on governmental regulations and controls.
- The effects of acid rain are a grim reminder of what can happen when a simplistic approach, such as "the solution to pollution is dilution," is applied to a complex problem involving our national economy.
- Because of the widespread use of chemicals of anthropogenic origin, many water quality problems are no longer limited to national boundaries but are of global concern.

DISCUSSION TOPICS AND PROBLEMS

4.1. Trace the historical development of the U.S. Environmental Protection Agency's coliform standard for drinking water.

4.2. Estimate the mass of phosphate (PO_4^{-3}) discharged per day by your community as the result of detergent usage and determine the mass of algae this quantity will support. Phosphate content can be found on detergent containers. Typical values range from 0 to 18 percent by mass.

4.3. Garrett Hardin's paper, "The Tragedy of the Commons" (see Ref. [4.6]), was written as a rationale for population control, but it offers a number of thoughtful insights applicable to water quality management. Briefly discuss the concept of rivers and lakes as "commons," and explain the concept of the "tragedy."

4.4. Aesthetic values played a small role in water quality management prior to 1950. Is there an economic justification for maintaining aesthetic characteristics of natural waters?

4.5. Using the U.S. Geological Survey data presented in *Water Supply Papers* and population data from a recent census, estimate the fraction of flow that is used in the Ohio River. When values exceed 1 for a river basin, all of the flow is used (on the average) at least once. Does the fraction of usage support the concept of direct reuse?

4.6. Most communities using rivers as water supply sources are distributing mixtures of surface runoff and treated wastewater from upstream sources. Discuss this issue in terms of health hazards and the appropriate maximum treated wastewater fraction.

4.7. Use the library to investigate transmission of viral diseases by water. Choose a particular disease (hepatitis, viral dysentery, etc.) and determine the effect of modern water treatment on control of the disease.

4.8. Explain why acid rain presents unusual problems with respect to control, regulation, and mitigation of damage.

4.9. Use the library to determine the meaning of "the greenhouse effect." Why are environmental scientists concerned about the greenhouse effect? Are there any potential benefits of the greenhouse effect?

4.10. One view of lake management is that high productivity is beneficial. Adding nutrients to lakes will increase fish populations and the protein supply to nearby communities. Evaluate this approach to lake management.

4.11. Determine the characteristics and effects of cadmium and lead poisoning.

4.12. A water resource has a limited capacity for use. For example, a river of a given size has a limited economic value as a source of cooling water and as a receiving water for treated wastes. How should these resources be allocated? Is "first come, first served" a completely satisfactory approach? Is there an inherent right of local community members to have a voice in resource allocation? What level of government should control allocation?

4.13. Protection of endangered species has become a major environmental issue. What role should economics play in the protection of endangered species?

4.14. Estimate the cost per person per day of (a) water, (b) wastewater disposal, (c) purchased beverages (beer, soda, milk, etc.), (d) entertainment, and (e) transportation.

4.15. Organic pesticides tend to be nonsoluble and difficult to degrade biologically. Assuming these two characteristics are intended, explain their value and their effect on water quality management.

4.16. Develop an alternative approach to resource management that would eliminate "the tragedy of the commons" (see Ref. [4.6]) and minimize regulatory requirements. Explain.

4.17. Should rural communities be required to meet the same wastewater discharge standards as metropolitan areas?

4.18. Perform a survey of the active ingredients in pesticides and herbicides available in your supermarket. Identify known carcinogens and USEPA priority pollutants.

REFERENCES

4.1. Anderson, R. B., (1966), "Sebago's Bright Future," Maine Dept. of Inland Fisheries and Game, Augusta.

4.2. Cleary, E. J., (1967), *The Orsanco Story*, Johns Hopkins Press, Baltimore.

4.3. Edmondson, W. T., (1969), "Eutrophication in North America," *NAS Eutrophication Symposium*, p. 124.

4.4. _____, (1970), "Phosphorus, Nitrogen and Algae in Lake Washington After Diversion of Sewage," *Science*, vol. 169, no. 3946, p. 690.

4.5. Goldberg, E. D., (1975), "Man's Role in the Sedimentary Cycle," in *The Changing Global Environment*, edited by S. F. Singer, D. Reidel Publishing Co., Dordrecht, Holland.

4.6. Hardin, G., (1968), "The Tragedy of the Commons," *Science*, vol. 162, no. 3859, p. 1243.

4.7. LaBastile, A., (1981), "How Menacing is Acid Rain?" *National Geographic*, vol. 160, no. 5, p. 652.

4.8. McGauhey, P. H., (1968), *Engineering Management of Water Quality*, McGraw-Hill Book Company, New York.

4.9. Manabe, S., and R. F. Strickler, (1964), "Thermal Equilibrium of the Atmosphere with a Convective Adjustment," *J. Atmospheric Science*, vol. 21, no. 3, p. 241.

4.10. Mitchell, J. M., Jr., (1975), "A Reassessment of Atmospheric Pollution as a Cause of Long-Term Changes of Global Temperature," in *The Changing Global Environment*, edited by S. F. Singer, D. Reidel Publishing Co., Dordrecht, Holland.

4.11. Shannon, L. R., A. E. Vandegrift, and P. G. Gorman, (1970), "Assessment of Small Particle Emissions," Midwest Research Institute Report, Contract CPA-22-69-104, Air Pollution Control Office, Environmental Protection Agency, Washington, D.C.

4.12. Singer, S. F., (1975), "Environmental Effects of Energy Production," in *The Changing Global Environment*, edited by S. F. Singer, D. Reidel Publishing Co., Dordrecht, Holland.

4.13. Sisler, F. D., (1975), "Impact of Land, Air and Sea Pollution on Chemical Stability in the Atmosphere," in *The Changing Global Environment*, edited by S. F. Singer, D. Reidel Publishing Co., Dordrecht, Holland.

4.14. Standen, A., (ed.), (1966), *Encyclopedia of Chemical Technology*, 2d ed., vol. 11, p. 691, Interscience, New York.

4.15. Tarr, J. A., and F. C. McMichael, (1977), "Historic Turning Points in Municipal Water Supply and Wastewater Disposal, 1850–1932," *Civil Engineering*, vol. 47, no. 10, p. 83.

4.16. *Water Quality Source Book*, (1979), Inland Wastes Directorate, Water Quality Branch, Ottawa, Canada.

4.17. Woodwell, G. M., P. P. Craig, and H. A. Johnson, (1974), "DDT in the Biosphere: Where Does It Go?" *Science*, vol. 174, no. 4014, p. 1101.

Analytical Methods for Water Quality Management

In Part I, the focus was on water quality, its definition and quantification. The objective in Part III is to develop models that can be used to study changes in water quality that occur in the natural environment. The physical, chemical, and biological methods and means used to modify water quality are introduced in Part IV. To understand and deal effectively with the material presented in Parts III and IV, it is necessary to review some fundamental concepts and methods that can be used to analyze the physical, chemical, and biological phenomena that bring about changes in water quality. These subjects are considered here in Part II in Chapters 5 and 6.

In Chapter 5, three fundamental concepts are introduced: stoichiometry, reaction kinetics, and materials balances. The rates of the complex chemical and biochemical reactions that occur in water are described by kinetic models or relationships. These expressions are largely empirically derived, although many are partially conceptual as well. Based on the principle of conservation of mass, materials balances are used to define changes that take place in the environment. Most engineering students are familiar with a simple materials balance on a single chemical species, such as water, that is used to derive the equation of continuity in fluid mechanics.

Mathematical descriptions of rivers, estuaries, lakes, and reservoirs require the development of coupled reaction and hydraulic flow models. Typically, flow models are based on an analysis of reactors (vessels) used to carry out chemical and biological reactions under controlled conditions. The most commonly used hydraulic flow models are introduced and discussed in Chapter 6. Various applications of these models are examined in preparation for the discussion of water quality modeling in Part III.

5

Stoichiometry, Reaction Kinetics, and Materials Balances

Development of water quality models involves the coupling of concepts from biological and chemical reaction kinetics and stoichiometry with materials (mass) balances and known physical constraints in a manner that provides a useful predictive model. Emphasis should be placed on practical usefulness because of the complexity of most natural systems and the difficulty of determining the variables accurately. Overly complex models fail because data are unavailable, and extremely simple models fail because variable interaction is ignored. This leaves the modeler in the difficult position of trying to determine the correct level of sophistication for the modeling effort. The decision is most often based on the availability of data, the goal being to make the model as complex as possible. It is the purpose of this chapter to introduce the reader to the basic concepts of stoichiometry, reaction kinetics, and materials balances on which all environmental models are based.

5.1
STOICHIOMETRY

From the standpoint of process selection and design, the controlling stoichiometry and the rate of the reaction are of principal concern. The number of moles of a substance entering into a reaction and the number of moles of the substances produced are defined by the *stoichiometry of a reaction*. Concepts of stoichiometry are generally one of the most fundamental parts of chemistry. Considerable use of stoichiometry was made in Chapter 2 in dealing with equilibrium concepts. Stoichiometry involves the application of the principle of conservation of mass, as illustrated in Example 5.1.

EXAMPLE 5.1

OXIDATION OF A HEXOSE

Hexose sugars have the general empirical formula $C_6H_{12}O_6$. If these compounds can be oxidized to CO_2 and H_2O, demonstrate that mass is conserved.

SOLUTION:

1. Write a balanced equation for the conversion of $C_6H_{12}O_6$. (*Note:* In balancing such chemical equations it is assumed that the carbon is converted to carbon dioxide, the hydrogen in the original compound is balanced with the hydrogen in water, and the oxygen is balanced with molecular oxygen.)

   $$C_6H_{12}O_6 + 6O_2 \rightarrow 6CO_2 + 6H_2O$$

2. Demonstrate that mass is conserved.
   Conservation of mass requires that the total reacting mass ($C_6H_{12}O_6$ + $6O_2$) must equal the total product mass ($6CO_2 + 6H_2O$). Thus

COMPOUND	MOLECULAR MASS, g	STOICHIOMETRIC COEFFICIENT	REACTING MASS, g
$C_6H_{12}O_6$	180	1	180
O_2	32	6	192
			372
CO_2	44	6	264
H_2O	18	6	108
			372

Clearly conservation of mass is met.

The stoichiometric relationship applied in Example 5.1 can be generalized using Eq. (5.1):

$$aA + bB + cC + \cdots \rightarrow pP + qQ + rR \cdots \qquad (5.1)$$

where

$$A, B, C = \text{reactant species}$$
$$P, Q, R = \text{product species}$$
$$a, b, c, p, q, r = \text{stoichiometric coefficients}$$

Assigning the quantitative mass per mole to each reactant and product species, a negative sign to all reactant stoichiometric coefficients, and a positive sign to all product stoichiometric coefficients allows a useful rearrangement of Eq. (5.1):

$$aA + bB + cC + \cdots + pP + qQ + rR \cdots = 0 \qquad (5.2)$$

Applying Eq. (5.2) to the oxidation of hexose in Example 5.1 gives

$$(-1 \text{ mol})(180 \text{ g/mol}) + (-6 \text{ mol})(32 \text{ g/mol}) +$$
$$(6 \text{ mol})(44 \text{ g/mol}) + (6 \text{ mol})(18 \text{ g/mol}) = 0$$

The usefulness of Eq. (5.2) in relating the rates at which materials are formed or disappear is discussed below.

5.2
REACTION KINETICS

The rate at which a reactant disappears or a product is formed in any given stoichiometric reaction is defined as the *rate of reaction*. The rate at which a reaction proceeds is an important consideration in all phases of water quality management. For example, treatment processes may be designed on the basis of the rate at which the reaction proceeds rather than the equilibrium position of the reaction, because the reaction usually takes too long to go to completion. In this case, quantities of chemicals in excess of the stoichiometric, or exact reacting, amount may be used to accomplish the treatment step in a reasonable period of time. What follows is intended to serve as an introduction to the subject of reaction kinetics. Additional details may be found in Refs. [5.1], [5.3], and [5.4].

Classes of Reactions

The two principal classes of reactions that occur in nature are termed homogeneous and heterogeneous (nonhomogeneous).

Homogeneous Reactions
Reactions that occur within a single phase (i.e., liquid, solid, or gas) are defined as *homogeneous*. In homogeneous reactions, the reactants are distributed continuously, but not necessarily uniformly, throughout the fluid, so that the potential for reaction exists at all points in the fluid. Homogeneous reactions may be either irreversible or reversible.
Examples of single irreversible reactions are

$$A \rightarrow P$$
$$A + A \rightarrow P$$
$$aA + bB \rightarrow P$$

Examples of multiple irreversible reactions are

$$A \diagup^{B}_{\diagdown C} \quad \text{(parallel)}$$

$A \rightarrow B \rightarrow C$ (consecutive or series)

Examples of reversible reactions are

$$A \rightleftharpoons B$$
$$A + B \rightleftharpoons C + D$$

Heterogeneous Reactions

Reactions that occur at surfaces between phases are defined as *heterogeneous*. Typically heterogeneous reactions occur between one or more constituents that can be identified with specific sites, such as those on the surface of an ion-exchange resin. Reactions that require the presence of a solid-phase catalyst are also classified as heterogeneous. These reactions are more difficult to study because a number of interrelated steps may be involved [5.6].

Rate and Order of Reaction

The *rate of reaction* r_i is the term used to describe the disappearance or formation of a particular substance or chemical species *i*. For homogeneous reactions, the units of r_i would be moles (or mass) per unit volume and unit time $(\text{mol}/L \cdot t)$, and for heterogeneous reactions the rate of formation would be moles (or mass) per unit area and unit time $(\text{mol}/m^2 \cdot t)$. Reactants have negative rates of reaction, and products have positive rates of reaction.

At constant temperature, it has been observed that the rate of reaction (e.g., disappearance or formation of a reactant or product) typically is some function of the concentration of the reactants. For example, for the following reaction

$$a A + b B \rightarrow c C + d D$$

the overall rate of reaction is defined as

$$\mathbf{r} = k [A]^\alpha [B]^\beta \tag{5.3}$$

where

$\mathbf{r}$ = overall rate of reaction, $\text{mol}/L \cdot t$

k = reaction rate constant

$[\ ]$ = molar concentration of reactants, mol/L

α, β = empirical exponents

The constants α and β are used to define the order of the reaction with respect to the individual reactants A and B. The overall order of the reaction is defined as $(\alpha + \beta)$. The exponents α and β are usually 0, 1, or 2, but fractional powers have also been observed. For example, if the overall rate of reaction of a particular reaction was found to be $\mathbf{r} = k[A]^2[B]$, the reaction is said to be second order with respect to reactant A and first order with respect to reactant B; overall it is a reaction of the third order.

Note that the units of the reaction-rate constant k are a function of reaction order. If the reactions are homogeneous, the zero-order coefficient has units of moles per unit volume and unit time $(mol/L \cdot t)$, while the first- and second-order rate constants have units of time (t^{-1}) and of unit volume per mole and unit time $(L/mol \cdot t)$, respectively.

Stoichiometric Relationships Between Rates of Reaction

In defining the rate of reaction, it is important to identify clearly whether the rate is the overall rate or the rate based on individual reactant or product. For example, if the stoichiometric coefficients for two reactants are different, the rate of reaction expressed in terms of one of the reactants will be different as compared with the rate expressed in terms of the other reactant. Therefore to avoid confusion, the overall rate of reaction is defined on the basis of the stoichiometric coefficients and the individual reaction rates. For the reaction

$$a\text{A} + b\text{B} \rightarrow c\text{C} + d\text{D} \tag{5.4}$$

the overall rate of reaction and the rates of reaction for the individual reactants and products are defined as

$$\mathbf{r} = \frac{r_A}{a} = \frac{r_B}{b} = \frac{r_C}{c} = \frac{r_D}{d} \tag{5.5}$$

where

$$\mathbf{r} = \text{overall rate of reaction}$$
$$a, b, c, d = \text{stoichiometric coefficients}$$
$$r_A, r_B, r_C, r_D = \text{rates of reaction for individual reactants and products}$$

The correctness of the above expression can be reasoned by considering the reaction $a\text{A} \rightarrow b\text{B}$. For this reaction, the following relationship must always be true:

$$\frac{r_A}{r_B} = \frac{a}{b} \tag{5.6}$$

That is, the ratio of the rates of reaction must be equal to the ratio of the stoichiometric coefficients. The application of Eq. (5.5) is illustrated in Example 5.2.

EXAMPLE 5.2

ANALYSIS OF RATES OF REACTION

The overall rate of reaction r for the reaction $3A \rightarrow 2B + C$ is known to be first order with respect to A. Determine the rates of reaction for the individual reactants and products as a function of the reaction-rate constant k and the molar concentration of A.

SOLUTION:

1. The overall rate of reaction is

$$\mathbf{r} = k[A]$$

2. Applying Eq. (5.5) yields

$$\mathbf{r} = \frac{r_A}{a} = \frac{r_B}{b} = \frac{r_C}{c}$$

$$= \frac{r_A}{-3} = \frac{r_B}{2} = \frac{r_C}{1}$$

In effect, the rate at which product C is formed is one-third the rate at which reactant A disappears.

3. Equating the rates defined in steps 1 and 2, the rate of disappearance of A, B, and C, expressed in terms of A, is

$$r_A = -3k[A]$$
$$r_B = 2k[A]$$
$$r_C = k[A]$$

Types of Reactions and Reaction Rates

The most common types of reactions encountered in the field of environmental engineering are (1) irreversible, (2) reversible, (3) saturation, and (4) autocatalytic. Each type of reaction is considered briefly in the following discussion.

Irreversible Reactions

Irreversible reactions are typically defined as single or multiple reactions. In single irreversible reactions there is one reaction step. For example,

$$A \rightarrow P$$
$$A + A \rightarrow P$$
$$aA + bB \rightarrow P$$

The rates of reaction for simple irreversible reactions are obtained as illustrated in Example 5.2.

In multiple irreversible reactions, more than one reaction step must be considered. For example, consider the following series reaction:

$$a A \xrightarrow{1} b B \xrightarrow{2} c C \tag{5.7}$$

For the purpose of analysis the above reaction should be treated as being made up of two steps:

$$a A \xrightarrow{1} b B \tag{5.8a}$$

$$b B \xrightarrow{2} c C \tag{5.8b}$$

If the overall rates of reaction $\mathbf{r}_1, \mathbf{r}_2$ are known, then the rate of reaction of any reactant or product can be defined. Thus for the above reactions,

$$\mathbf{r}_1 = \frac{r_A}{a} = \frac{r_{B_1}}{b} \tag{5.9a}$$

$$\mathbf{r}_2 = \frac{r_{B_2}}{b} = \frac{r_C}{c} \tag{5.9b}$$

The rates of reaction for A, B, and C are

$$r_A = a\mathbf{r}_1 \tag{5.10a}$$
$$r_B = r_{B_1} + r_{B_2} = b\mathbf{r}_1 + b\mathbf{r}_2 \tag{5.10b}$$
$$r_C = c\mathbf{r}_2 \tag{5.10c}$$

If both of the rates of reaction are first order, then

$$r_A = ak_1[A] \tag{5.11a}$$
$$r_B = bk_1[A] + bk_2[B] \tag{5.11b}$$
$$r_C = ck_2[B] \tag{5.11c}$$

The above procedure can always be used to find the rates of reaction in multiple reactions, taking care to observe the sign convention.

Reversible Reactions

For reversible reactions of the form

$$a A \underset{2}{\overset{1}{\rightleftarrows}} b B \tag{5.12}$$

the rates of formation of each species are the sums of two individual rates of formation. For the purpose of analysis the above reaction should be treated as being made up of two steps:

$$a A \xrightarrow{1} b B$$

$$b B \xrightarrow{2} a A$$

Following the procedure presented above in the discussion of irreversible multiple reactions, the rate expression can be written as

$$\mathbf{r}_1 = \frac{r_{A_1}}{a} = \frac{r_{B_1}}{b} \tag{5.13a}$$

$$\mathbf{r}_2 = \frac{r_{B_2}}{b} = \frac{r_{A_2}}{a} \tag{5.13b}$$

Thus the rates of reaction for A and B are

$$r_A = r_{A_1} + r_{A_2} = a\mathbf{r}_1 + a\mathbf{r}_2 \tag{5.14a}$$

$$r_B = r_{B_1} + r_{B_2} = b\mathbf{r}_1 + b\mathbf{r}_2 \tag{5.14b}$$

If both of the rates of reaction are first order, then

$$r_A = ak_1[A] + ak_2[B] \tag{5.15a}$$

$$r_B = bk_1[A] + bk_2[B] \tag{5.15b}$$

Note that for reversible reactions the signs of the stoichiometric coefficients are always positive and the sign is related to the overall direction of the reaction.

Saturation-Type Reactions

Saturation-type reactions have a maximum rate, that is, a point at which the rate becomes independent of concentration A. A typical saturation-rate function is given in Eq. (5.16) for the reaction $aA \rightarrow bB$.

$$r = \frac{k[A]}{K + [A]} \tag{5.16}$$

where

r = rate of reaction, mol/L · t

k = reaction rate constant, mol/L · t

$[A]$ = concentration of reactant A, mol/L

K = half-saturation constant, mol/L

It is important to note that the half-saturation constant K has units of concentration and that the rate coefficient k has units of moles per liter and unit time (mol/L · t).

When $K \ll [A]$ the saturation-rate function may appear to be zero order ($r \rightarrow k$), and when $[A] \ll k$ the reaction may appear to be first order ($r \rightarrow k[A]$). An overall representation of Eq. (5.16) is given in Fig. 5.1.

Often saturation-rate functions are more complex than Eq. (5.16). Two other typical forms often encountered in the modeling of environ-

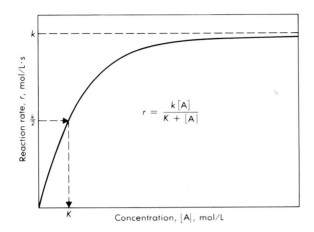

FIGURE 5.1

Graphical representation of saturation-type reaction rate as a function of concentration.

mental engineering systems are given in Eqs. (5.17) and (5.18):

$$r = \frac{k[A][B]}{K+[A]} \qquad (5.17)$$

$$= k\left(\frac{[A]}{K_1+[A]}\right)\left(\frac{[B]}{K_2+[B]}\right) \qquad (5.18)$$

Autocatalytic Reactions

Many reaction rates are functions of the product concentration. An example is bacterial growth in which the rate of increase in bacterial numbers is proportional to the number present. Autocatalytic reactions can be first order, second order, or saturation type, or they can be partially autocatalytic: a function of a reactant and a product. Examples are given below for the reaction $a\mathrm{A} \rightarrow b\mathrm{B}$.

First-order autocatalytic:

$$r = k[B] \qquad (5.19a)$$
$$r_A = ak[B] \qquad (5.19b)$$
$$r_B = bk[B] \qquad (5.19c)$$

Second-order partially autocatalytic:

$$r = k[A][B] \qquad (5.20a)$$
$$r_A = ak[A][B] \qquad (5.20b)$$
$$r_B = bk[A][B] \qquad (5.20c)$$

Effect of Temperature on Reaction-Rate Constants

The temperature dependence of the rate coefficients was discussed in the section on biochemical oxygen demand in Chapter 2. The pertinent equations are as follows:

$$k_T = A\,e^{-E/RT} \tag{2.76}$$

where

k_T = reaction rate constant, variable units

A = van't Hoff-Arrhenius coefficient, variable units

E = activation energy, J/mol

R = universal gas constant, 8.314 J/mol · K

T = temperature, K

To compare the reaction rate constant at two different temperatures, the above equation is usually written as

$$\frac{k_{T_1}}{k_{T_2}} = \exp\left[\left(\frac{E}{RT_1T_2}\right)(T_1 - T_2)\right] \tag{2.78}$$

It is again noted that a customary method of denoting temperature dependence is to assume E/RT_1T_2 is approximately constant so that Eq. (2.78) can be written as

$$k_{T_1} = k_{T_2}\theta^{(T_1 - T_2)} \tag{2.79}$$

where

θ = temperature coefficient

5.3

ANALYSIS OF EXPERIMENTAL DATA

Laboratory and pilot-scale experiments are conducted to obtain data that can be used to develop the rate expressions needed to model chemical and biological reactions. At the present time, four methods are commonly used to determine the order of a reaction from experimental data. They are (1) the method of integration, (2) the differential method, (3) the time-reaction method, and (4) the isolation method. Because the first two methods are used most commonly in the analysis of environmental data, they are considered in detail in the following discussion. Additional details on these and the other two methods may be found in Refs. [5.2] and [5.6].

Method of Integration

The easiest method of determining the order of a reaction is to substitute experimental data on the amount of material remaining at various times in the integrated forms of the various rate expressions and then to solve for the reaction-rate constant. The equation that results in the most consistent k values is assumed to represent the order of the reaction. Alternatively, the order of a reaction can be determined by plotting the experimental data functionally, based on the integrated form of the rate expression. If a straight-line plot is obtained, it is assumed that the order of the reaction corresponds to the reaction plotted. The functional plotting of data is favored by most researchers, as it also allows a visual assessment of the data. The development of the integrated form of several rate equations and the method of plotting the data to determine the specific reaction rate constant are illustrated below.

Typically, rate experiments are conducted in vessels in which there is no inflow or outflow. In chemical engineering terminology a reaction vessel whose contents are well stirred and in which there is no inflow or outflow (Fig. 5.2) is known as a *batch reactor*, and this term will be used here also. For example, the BOD test (see Fig. 2.16) is carried out in a batch reactor. For a batch reactor, as will be shown later in this chapter, the following relationship holds:

$$r_A = \frac{d[A]}{dt} \tag{5.21}$$

It should be noted clearly that the above definition is true only for a

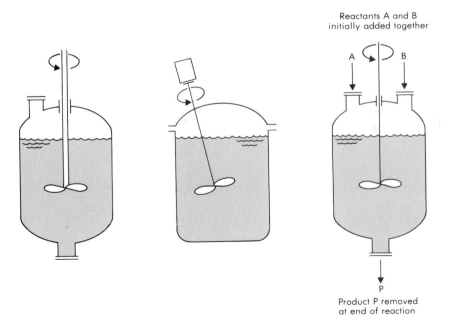

Reactants A and B
initially added together

A B

P

Product P removed
at end of reaction

FIGURE 5.2

Typical examples of batch reactors used to carry out chemical and biological reactions in the laboratory and in industrial operations.

batch reactor; it will be used to analyze the rate equations to be discussed below.

Irreversible Zero-Order Reactions

Consider the following reaction carried out in a batch reactor:

$$A \rightarrow P$$

If the reaction is zero order, then

$$r_A = \frac{d[A]}{dt} = -k \qquad (5.22)$$

Integrating the above expression between the limits from $[A] = [A_0]$ to $[A] = [A]$ and from $t = 0$ to $t = t$ results in

$$[A] = [A_0] - kt \qquad (5.23)$$

The reaction-rate constant can be obtained graphically by plotting $[A]$ versus t (Fig. 5.3a).

Irreversible First-Order Reactions

Consider the following reaction carried out in a batch reactor:

$$A \rightarrow P$$

If the reaction is first order, then

$$r_A = \frac{d[A]}{dt} = -k[A] \qquad (5.24)$$

Integrating the above expression between the limits from $[A] = [A_0]$ to $[A] = [A]$ and from $t = 0$ to $t = t$ results in

$$\ln\left[\frac{A}{A_0}\right] = -kt \qquad \text{or} \qquad \log\left[\frac{A}{A_0}\right] = -\frac{kt}{2.3} \qquad (5.25)$$

The reaction-rate constant is obtained graphically by plotting $-\log[A/A_0]$ versus t (Fig. 5.3b).

Irreversible Second-Order Reactions

Consider the following reaction carried out in a batch reactor:

$$A + A \rightarrow P$$

If the reaction is second order, then

$$r_A = \frac{d[A]}{dt} = -k[A]^2 \qquad (5.26)$$

Integrating the above expression between the limits from $[A] = [A_0]$ to $[A] = [A]$ and from $t = 0$ to $t = t$ results in

$$\frac{1}{[A]} - \frac{1}{[A_0]} = kt \qquad (5.27)$$

The reaction-rate constant is obtained graphically by plotting $1/[A]$ versus t, as shown in Fig. 5.3c.

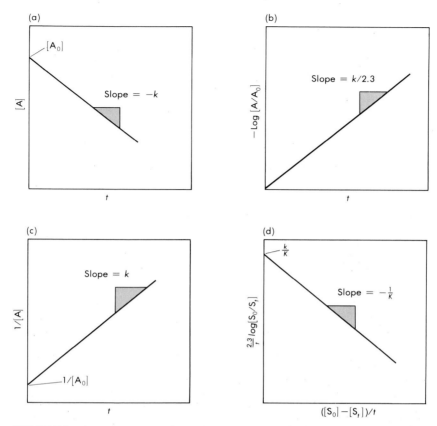

FIGURE 5.3

Graphical analysis for the determination of reaction order and reaction-rate constants: (a) zero-order reaction, (b) first-order reaction, (c) second-order reaction, and (d) saturation-type reaction.

Irreversible Parallel Reactions

Consider the following reaction carried out in a batch reactor:

$$A \underset{2}{\overset{1}{\longrightarrow}} \begin{array}{c} B \\ C \end{array}$$ (5.28)

If the reaction is first order, then

$$r_A = \frac{d[A]}{dt} = -(k_1 + k_2)[A]$$ (5.29a)

$$r_B = \frac{d[B]}{dt} = k_1[A]$$ (5.29b)

$$r_C = \frac{d[C]}{dt} = k_2[A]$$ (5.29c)

Integrating Eq. (5.29a) between the limits from $[A] = [A_0]$ to $[A] = [A]$ and from $t = 0$ to $t = t$ results in

$$\ln\left[\frac{A}{A_0}\right] = -(k_1 + k_2)t \tag{5.30}$$

Dividing Eq. (5.29b) by Eq. (5.29c) yields

$$\frac{r_B}{r_C} = \frac{d[B]}{d[C]} = \frac{k_1}{k_2} \tag{5.31}$$

If the above expression is integrated between the limits $[B] = [B_0]$ and $[B] = [B]$ and $[C] = [C_0]$ and $[C] = [C]$, the following equation is obtained:

$$\frac{[B] - [B_0]}{[C] - [C_0]} = \frac{k_1}{k_2} \tag{5.32}$$

The values of k_1 and k_2 can be determined using Eqs. (5.30) and (5.32).

Other Irreversible and Reversible Reactions

For more complex second-order irreversible reactions (e.g., $r_A = k[A][B]$), irreversible multiple-order reactions (e.g., $r_A = k[A]^{n_A}[B]^{n_B}$), and reversible reactions, the integrated forms may be found in Refs. [5.1], [5.4], [5.5], and [5.6].

Saturation Reactions

Consider a saturation reaction carried out in a batch reactor that can be defined by the following rate expression:

$$r_S = \frac{d[S]}{dt} = -\frac{k[S]}{K + [S]} \tag{5.33}$$

Integrating between the limits from $[S] = [S_0]$ to $[S] = [S_t]$ and from $t = 0$ to $t = t$ results in the following expression:

$$kt = 2.3K \log\frac{[S_0]}{[S_t]} + ([S_0] - [S_t]) \tag{5.34}$$

To obtain the constants graphically, the above equation must be re-arranged:

$$\frac{2.3}{t}\log\frac{[S_0]}{[S_t]} = -\frac{1}{K}\frac{([S_0] - [S_t])}{t} + \frac{k}{K} \tag{5.35}$$

The constants are then obtained by plotting $(2.3/t \log[S_0]/[S_t])$ versus $([S_0] - [S_t])/t$, as shown in Fig. 5.3d.

The Differential Method

The differential method of analysis used to determine the order of a reaction is based on the assumption that the rate of reaction is proportional to the nth power of the concentration; thus

$$r_A = \frac{d[A]}{dt} = -k[A]^n \qquad (5.36)$$

For two different concentrations at two different times, the following expressions hold:

$$\frac{d[A_1]}{dt} = -k[A_1]^n \quad \text{and} \quad \frac{d[A_2]}{dt} = -k[A_2]^n \qquad (5.37)$$

If the log of each of the above equations is taken and the k values are equated, then the value of n can be determined using the following expression;

$$n = \frac{\log(-d[A_1]/dt) - \log(-d[A_2]/dt)}{\log[A_1] - \log[A_2]} \qquad (5.38)$$

It is important to note that the above equation is independent of the units used to express the concentrations.

EXAMPLE 5.3

DETERMINATION OF REACTION ORDER

Determine the order of the reaction and the reaction-rate constant for the following data derived from an experiment carried out in a batch reactor. Solve the problem using both the integral and differential methods of analysis.

TIME, min	[A], mol/L
0	100.0
1	50.0
2	37.0
3	28.6
4	23.3
5	19.6
6	16.9
7	15.2
8	13.3
9	12.2
10	11.1

SOLUTION:

1. Develop the data needed to plot the experimental data functionally, assuming the reaction is either first or second order.

TIME, min	[A], mol/L	$-\log\left[\dfrac{A_t}{A_0}\right]$	$\dfrac{1}{[A]}$
0	100.0	0.00	0.010
1	50.0	0.30	0.020
2	37.0	0.43	0.027
3	28.6	0.54	0.035
4	23.3	0.63	0.043
5	19.6	0.71	0.051
6	16.9	0.77	0.059
7	15.2	0.82	0.066
8	13.3	0.87	0.075
9	12.2	0.91	0.082
10	11.1	0.95	0.090

2. Plot $-\log[A_t/A_0]$ versus t to determine if the reaction is first order (Fig. 5.4a). Because the data plot is not a straight line, the experimental data do not follow a first-order reaction.

3. Plot $1/[A]$ versus t to determine if the reaction is second order (Fig. 5.4b). Because the data plot is a straight line, the reaction is second order. The reaction-rate constant is given by the slope of the line:

$$k = \frac{0.017 \text{ L/mol}}{2.1 \text{ min}}$$
$$= 0.0081 \text{ L/mol} \cdot \text{min}$$

The reaction-rate expression is

$$r_A = -(0.0081 \text{ L/mol} \cdot \text{min})[A]^2$$

4. Determine the reaction order and the reaction-rate constant using the differential method of analysis [Eq. (5.38)].

$$n = \frac{\log(-d[A_1]/dt) - \log(-d[A_2]/dt)}{\log[A_1] - \log[A_2]}$$

a. Use the experimental data obtained at 2 and 6 min.

TIME, min	[A], mol/L	$\left(\dfrac{[A_{t+1}] - [A_{t-1}]}{2}\right) \approx \dfrac{d[A_t]}{dt}$		
2	37.0	$\dfrac{28.6 - 50.0}{2}$	$\approx$	-10.7
6	16.9	$\dfrac{15.2 - 19.6}{2}$	$\approx$	-2.20

b. Substitute and solve for n.

$$n = \frac{\log(10.7) - \log(2.2)}{\log[37.0] - \log[16.9]}$$

$$= \frac{1.029 - 0.342}{1.568 - 1.228}$$

$$= \frac{0.687}{0.34} = 2.02$$

c. The reaction is second order.
d. The reaction-rate constant computed using the two times is

$$\frac{d[A]}{dt} = -k[A]^2$$

(1) At 2 min:

$$-10.7 = -k(37.0)^2$$

$$k = 0.0078 \text{ L/mol} \cdot \text{min}$$

(2) At 6 min:

$$-2.20 = -k(16.9)^2$$

$$k = 0.0077 \text{ L/mol} \cdot \text{min}$$

COMMENT

The reaction-rate constants computed with both methods are essentially the same. The accuracy of the differential method can be improved by plotting the data and determining the slope more accurately.

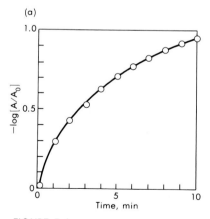

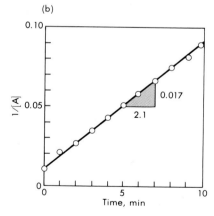

FIGURE 5.4

Graphical analysis to determine the reaction order and the reaction-rate constant for 5.3: (a) first-order example reaction and (b) second-order reaction.

5.4
MASS RATES OF REACTION

In most water quality management problems determination of molar concentrations is not possible, and mass concentrations are used instead. All of the molar rate concepts discussed previously remain valid, except the use of stoichiometric coefficients. Rates of reaction cannot be related unless new, mass-based stoichiometric relations are defined.

Mass-based rates will be written in the same manner as molar-based rates, except mass concentration C_i will be used, as in Eq. (5.39), and, in general, the stoichiometric coefficients will be incorporated into the rate coefficient, as in the BOD relationships [and in Eq. (5.39)]:

$$a\text{A} \rightarrow b\text{B}$$
$$r_\text{A} = -kC_\text{A} \tag{5.39}$$

where

r_A = mass rate of reaction of A, $\text{g/m}^3 \cdot \text{t}$

k = mass-based reaction-rate constant, t^{-1}

C_A = mass concentration of A, g/m^3

5.5
THE MATERIALS BALANCE

The materials balance (or mass balance) is a quantitative description of all materials that enter, leave, and accumulate in a system with defined boundaries. A materials balance is based on the law of conservation of mass (i.e., mass is neither created nor destroyed). The basic materials balance expression is developed on a chosen control volume and has terms for material entering, leaving, being generated, and being accumulated or stored within the volume. If a balance on the species A over the small differential volume ΔV is considered (Fig. 5.5), then the amount of material being accumulated must be equal to the amount of material entering minus the amount leaving plus the amount generated. The general word statement for the materials balance is

$$\begin{array}{cccc} \text{Rate of} & \text{Rate of flow} & \text{Rate of flow} & \text{Rate of mass} \\ \text{accumulation of} & \text{of mass} & \text{of mass} & \text{generation} \\ \text{mass within the} = & \text{into the} - & \text{out of the} + & \text{within the} \\ \text{system boundary} & \text{system boundary} & \text{system boundary} & \text{system boundary} \end{array}$$

$$\tag{5.40}$$

The corresponding simplified word statement for a materials balance is

$$\text{Accumulation} = \text{Inflow} - \text{Outflow} + \text{Generation} \tag{5.41}$$

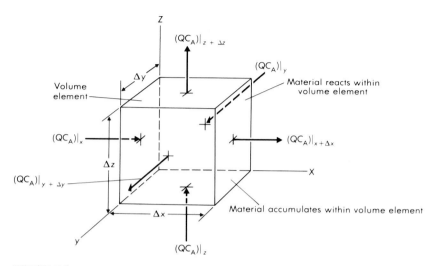

FIGURE 5.5

Definition sketch for materials balance analysis.

Using r_A to indicate the mass rate of generation of A within the volume and assuming that dispersion and diffusion are negligible, Eq. (5.40) can be written in quantitative symbolic form:

$$\underbrace{\frac{\partial C_A}{\partial t} \Delta V}_{\text{Accumulation} =} = \underbrace{(QC_A)|_x + (QC_A)|_y + (QC_A)|_z}_{\text{Inflow}}$$

$$\underbrace{- (QC_A)|_{x+\Delta x} - (QC_A)|_{y+\Delta y} - (QC_A)|_{z+\Delta z}}_{\text{Outflow}} + \underbrace{r_A \Delta V}_{\text{Generation}}$$

$$\text{(5.42)}$$

where

V = volume, m^3

C_A = mass concentration of A, g/m^3

Q = volumetric flow rate in a single direction, m^3/s

r_A = mass rate of generation, g/m$^3 \cdot$ s

Remembering that $Q_x = v_x \Delta y \Delta z$, $Q_y = v_y \Delta x \Delta z$, and $Q_z = v_z \Delta x \Delta y$, where v_x, v_y, and v_z are the velocities in the respective directions, and taking the limit as Δx, Δy, and Δz approach zero, Eq. (5.42) becomes

$$\frac{\partial C_A}{\partial t} = -\left[\frac{\partial v_x C_A}{\partial x} + \frac{\partial v_y C_A}{\partial y} + \frac{\partial v_z C_A}{\partial z} \right] + r_A \qquad \text{(5.43)}$$

the mass rate of generation can be positive or negative. Most of the materials of interest (e.g., oxygen) disappear, and therefore r_A will be negative in most cases. Nitrification results in the production of a material of interest, NO_3^- and, in that case, $r_{NO_3^-}$ would be positive. In writing the mass balance equation, the rate term should always be written as a positive term. The correct sign for the term will be added when the appropriate rate expression is substituted for r_A.

Steady and Transient States

In applying materials balances, two operational states must be considered: steady state and transient (unsteady) state. The primary requirement for steady state is that there be no accumulation within the system (i.e., $\partial C_A/\partial t = 0$). For example, a pump which is discharging a constant volume of water with time is said to be operating at steady state. Stated another way, all rates and concentrations do not vary with time at steady state. In the transient state, the rate of accumulation is changing with time ($\partial C_A/\partial t \neq 0$). The filling of a reservoir or the purging of the contents of a tank are two examples of systems in the transient (unsteady) state.

The Preparation of Materials Balances

In preparing materials balances it is helpful if the following steps are followed, especially as the techniques involved are being mastered [5.2]:

1. Prepare a simplified schematic or flow sheet of the system or process for which the materials balance is to be prepared.
2. Draw a system boundary to define the limits over which the materials balance is to be applied.
3. List all of the pertinent data that will be used in the preparation of the materials balance on the schematic or flow sheet.
4. List all of the equations for the biological or chemical reactions that occur in the process.
5. Select a convenient basis on which the numerical calculations will be based.

If the above steps are followed rigorously, it will be found that fewer omissions and careless errors are made in materials balance analyses. For that reason, engineers with years of experience still follow the above steps.

5.6

MATERIALS BALANCE APPLICATIONS

Mass balances are of fundamental importance in the field of civil engineering. Among the most common applications is the use of the continuity equation in pipe flow and flood routing. In the field of chemical engineering, materials balances are used to design equipment and to evaluate process performance. In water and wastewater treatment, materials balances are used to evaluate process performance. In the field of water quality management, materials balances are used to study the response of the aquatic environment to selected inputs, such as the discharge of wastewater (see Part III). In this case, a chemical species or biological response is monitored rather than a nonreactive fluid.

In this section, the materials balance principle will be illustrated by applying it to some familiar engineering subjects, including fluid flow, flood routing, and the purging of a storage tank. The application of the materials balance principle to a process involving a chemical reaction is also considered.

Fluid Flow

Consider a portion of a fluid-flow system such as shown in Fig. 5.5. If a system boundary is drawn around the pump, the materials balance for the system (remembering that the mass concentration of water is its density) is

$$\frac{d(\bar{\rho}V)}{dt} = Q_1\rho_1 - Q_2\rho_2 + r_w V \qquad (5.44)$$

Accumulation = Inflow − Outflow + Generation

where

$\bar{\rho}$ = mean density of the fluid in the control volume, kg/m^3

V = liquid volume within fixed control volume, m^3

Q_1, Q_2 = volumetric flow rates in and out of control volume, m^3/s

ρ_1, ρ_2 = density of fluid flowing into and out of control volume, kg/m^3

r_w = mass rate of generation, $g/m^3 \cdot s$

If there is no accumulation of fluid in the system ($d(\bar{\rho}V)/dt = 0$) and fluid is neither produced nor lost in the system ($r_w = 0$), Eq. (5.44) reduces to

$$Q_1\rho_1 = Q_2\rho_2 \qquad (5.45)$$

Thus Eq. (5.45) corresponds to the conservation of mass statement for the system shown in Fig. 5.6. Because water, for practical purposes, is considered to be incompressible ($\rho_1 = \rho_2$), Eq. (5.45) reduces to Eq. (5.46), which is the simplest form of the equation of continuity:

$$Q_1 = Q_2 \qquad (5.46)$$

Flood Routing

Flood routing involves a special type of mass balance. Because many engineers are more familiar with flood-routing formulas than with mass

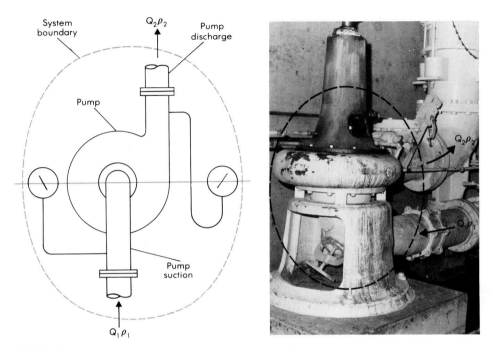

FIGURE 5.6

Definition sketch for materials balance in a fluid-flow system.

balances for chemical species, it is useful to develop the routing expression and compare it with Eqs. (5.41), (5.42), and (5.43).

In flood routing, the objective is to predict changes in river elevation due to flood flows. A stream is divided into short sections (reaches) that can be assumed to have constant elevation (Fig. 5.7). The sections need not be uniform and normally would be expected to vary considerably in length and volume. Flow into section n can be defined as Q_{n-1}, and flow out of section n is defined as Q_n. Assuming unidirectional flow, the routing equation for reach n can be written as

$$\frac{\Delta V}{\Delta t}\rho_w = Q_{n-1}\rho_w - Q_n\rho_w + r_w V_n \tag{5.47}$$

Accumulation = Inflow $\quad$ − Outflow + Generation

where

V = volume, m^3

ρ_w = density of water, kg/m^3

r_w = rate of generation, $kg/m^3 \cdot s$

and other terms are as defined previously. Equation (5.46) must be written in incremental terms because the flow rates are calculated from the depths in each section, and thus Q_{n-1} and Q_n are independent. The

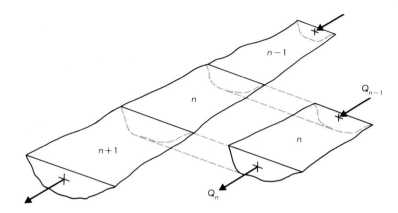

FIGURE 5.7

River segments used in flood routing analysis.

generation rate r_w is normally omitted, but could be used to account for losses due to percolation, input from springs, rainfall, etc. Losses and gains are usually insignificant during floods, however. Equation (5.47) is now seen as a statement of conservation of mass. The storage is calculated over the time increment Δt, as illustrated in Example 5.4.

<hr>

EXAMPLE 5.4

APPLICATION OF MATERIALS BALANCES IN FLOOD ROUTING

A channel reach, rectangular in cross section, has an average width of 150 m. The 100-year flood predicted from historical records and stage-discharge relationships would result in the measured flows shown below

t_i, hr	Q_1, m³/s	Q_2, m³/s
0	1000	1000
1	1100	1000
2	1300	1000
3	1700	1000
4	2500	1100
5	1700	1300
6	1300	1700
7	1100	2500
8	1000	1700
9	1000	1300
10	1000	1100
11	1000	1000

for two successive gauges 10 km apart. Determine the storage/accumulation and the increase in water depth with time. What is the increase in depth after $7\frac{1}{2}$ hr?

SOLUTION:

1. Derive the appropriate materials balance. The constant-density incremental form of Eq. (5.47) is

$$\frac{\Delta V}{\Delta t} = \overline{Q}_1 - \overline{Q}_2$$

Accumulation = Inflow − Outflow

where $\overline{Q}_1$ and $\overline{Q}_2$ are average flows over the time period Δt. The change in volume ΔV can be estimated as $(\Delta V/\Delta t)\,\Delta t$ or $\Delta t(\overline{Q}_1 - \overline{Q}_2)$. The sum of these values is the total quantity stored.

2. Prepare a computation table to determine the increase in water depth. The necessary computations are given below; the results are plotted in Fig. 5.8.

t, hr	$\overline{Q}_1 - \overline{Q}_2$, m^3/s	ΔV^*, $m^3 \times 10^{-5}$	$V - V_0$, $m^3 \times 10^{-5}$	$h - h_0{}^\dagger$, m
0–1	50	1.8	1.80	0.12
1–2	200	7.2	9.00	0.60
2–3	500	18.0	27.0	1.80
3–4	1050	37.8	64.8	4.32
4–5	900	32.4	97.2	6.48
5–6	0	0	97.2	6.48
6–7	− 900	− 32.4	64.8	4.32
7–8	− 1050	− 37.8	27.0	1.80
8–9	− 500	− 18.0	9.0	0.60
9–10	− 200	− 7.2	1.8	0.12
10–11	− 50	− 1.80	0	0

$^*\Delta V = \Delta t(\overline{Q}_1 - \overline{Q}_2)$
$^\dagger h - h_0 = [(V - V_0)/150 \text{ m} \times 10{,}000 \text{ m})]$

3. Determine the increase in depth after $7\frac{1}{2}$ hr. From Fig. 5.8 the depth is 2.0 m.

Purging of a Storage Tank

Another problem often encountered in practice is the purging or filling of a storage reservoir or tank. The transient state of a purging tank (see Example 5.5) is also interesting because it serves to illustrate one of the common features of many materials balance problems—that is, the need

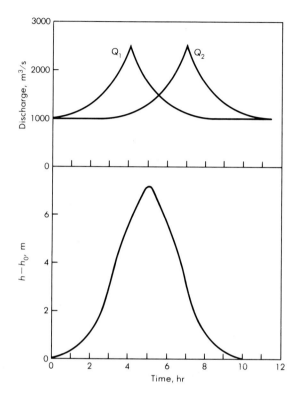

FIGURE 5.8

Input-output hydrographs and change in depth for Example 5.4.

to find an appropriate relationship between the variables of the problem so that a solution can be derived. The filling of a reservoir with a spillway by flood flows is another example. The relationship between the spillway discharge and the height of water over the crest of the spillway must be known to predict the increase in water level in the reservoir.

EXAMPLE 5.5

DETERMINATION OF TIME REQUIRED TO PURGE (EMPTY) A STORAGE TANK

A storage tank with a diameter of 1 m is filled to a depth of 2 m. If a 0.005-m-diameter hole were to develop in the bottom of the tank, how long would it take for the liquid level to drop 1.0 m?

SOLUTION:

1. Write the mass balance for the liquid in the tank.

$$\frac{d(\rho V)}{dt} = 0 \quad - \quad A_h v_h \rho \quad + \quad 0$$

Accumulation = Inflow − Outflow + Generation

where

ρ = fluid density, kg/m³

V = volume, m³

A_h = area of hole in tank bottom, m²

v_h = fluid velocity through hole, m/s

2. Assuming the density of the liquid remains constant, the mass balance is

$$\frac{dV}{dt} = -A_h v_h$$

Substituting $A_t\, dh$ for dV, where A_t is the area of the tank, yields

$$A_t \frac{dh}{dt} = -A_h v_h$$

3. Find a relationship between the liquid velocity through the hole and the geometry of the problem. From basic hydraulic considerations it is known that the velocity through an orifice is related to head of the liquid above it, so that

$$v_h = C\sqrt{2gh}$$

where

C = flow coefficient

g = acceleration due to gravity, 9.81 m/s²

h = depth of the liquid, m

4. Develop an equation that can be used to solve for t.
   a. Substitute for v_h from step 3 and separate the variables:

$$dt = -\frac{A_t}{CA_h(2g)^{1/2}} \frac{dh}{(h)^{1/2}}$$

   b. Integrate between the limits of 0 to t and 0 to -1:

$$\int_0^t dt = -\frac{A_t}{A_h(2g)^{1/2}} \int_0^{-1} \frac{dh}{(h)^{1/2}}$$

$$t = -\frac{2A_t}{CA_h}\left(\frac{h}{2g}\right)^{1/2}\Bigg|_0^{-1} = \frac{2A_t}{CA_h}\left(\frac{1}{2g}\right)^{1/2}$$

5. Substitute given values and solve for t. Assume the value of the flow

coefficient is 0.61.

$$t = \frac{2\pi d_t^2/4}{C\pi d_h^2/4}\left(\frac{1}{2(9.81)}\right)^{1/2}$$

$$= \frac{2(1.0)^2}{0.61(0.005)^2}\left(\frac{1}{2(9.81)}\right)^{1/2}$$

$$= 29{,}604 \text{ s}$$

$$= 8.2 \text{ hr}$$

Processes Involving Chemical Reactions

In the two previous applications of the materials balance principle, the rate processes were zero. For applications occurring in nature, the rate of generation term is very important. This is also true for most water and wastewater treatment processes. The importance of the rate of generation term is illustrated in Example 5.6.

EXAMPLE 5.6

DETERMINATION OF REACTION RATE ORDER

Two experiments are run in a laboratory to determine r_A for the reaction $A + B \rightarrow C$. The experiments are carried out in a laboratory batch reactor (see Fig. 5.2), the contents of which are well stirred and maintained at constant temperature and pH. In the first experiment, the initial concentrations are $[A_0] = 1$ mol/L and $[B_0] = 1000$ mol/L. In the second experiment, $[A_0] = 1$ mol/L and $[B_0] = 1$ mol/L. The results of the two experiments are given below. Using these data, determine the appropriate rate expression and the value of the reaction-rate constant.

EXPERIMENT 1		EXPERIMENT 2	
Time, d	[A], mol/L	Time, d	[A], mol/L
0.01	0.82	5	0.91
0.05	0.37	15	0.77
0.10	0.14	25	0.67
0.15	0.05	45	0.53
		75	0.40

SOLUTION:

1. Write a mass balance for component [A] for a reaction carried out in a batch reactor. A mass balance for component [A] in the liquid

contained in the batch reactor results in

$$\frac{d[A]}{dt} V = Q[A_i] - Q[A] + r_A V$$

Accumulation = Inflow − Outflow + Generation

However, because there is no inflow or outflow the resulting equation is

$$\frac{d[A]}{dt} = r_A$$

2. Derive an appropriate rate equation using a trial-and-error procedure. To begin the solution, an assumption of a simple functional form for r_A is appropriate: $r_A = -k[A]$ or $r_A = -k[A][B]$ would both be good choices, considering the stoichiometric reaction expression.

a. Trial 1:

(1) Assume $r_A = -k[A]$. The integrated form of this expression is (see Section 5.3)

$$[A] = [A_0]e^{-kt}$$

(2) Solve for k at each of the given experimental data points:

EXPERIMENT 1		EXPERIMENT 2	
t, d	k, d^{-1}	t, d	k, d^{-1}
0.01	19.85	5	0.019
0.05	19.89	15	0.017
0.10	19.66	25	0.016
0.15	19.97	45	0.014
		75	0.012

Clearly, the first-order model is incorrect. Values of k for both experiments should be the same, and, although the value of k is constant for experiment 1, the value varies for experiment 2.

b. Trial 2:

(1) Assume $r_A = -k[A][B]$. From the stoichiometry of the reaction $\Delta[A] = \Delta[B]$. Thus the concentration of $[B]$ at any time is $[B] = [B_0] - [A_0] + [A]$.

(2) For experiment 1, the following expression for $[A]$ is obtained when the assumed rate expression and the value for $[B]$ are substituted in the mass balance derived in step 1 and the resulting expression is integrated between the limits of $[A] =$

$[A_0]$ and $[A] = [A]$ and $t = 0$ and $t = t$:

$$[A] = \frac{[A_0]([B_0] - [A_0])e^{-([B_0] - [A_0])kt}}{[B_0] - [A_0]e^{-([B_0] - [A_0])kt}}$$

(3) For experiment 2, because $[B_0] = [A_0]$, $[A] = [B]$ and $r_A = -k[A]^2$, the resulting expression for $[A]$ is:

$$[A] = \frac{[A_0]}{1 + [A_0]kt}$$

(4) Solve for k at each of the given experimental data points:

EXPERIMENT 1		EXPERIMENT 2	
Time, d	k, L/mol · d	Time, d	k, L/mol · d
0.01	0.020	5	0.020
0.05	0.020	15	0.020
0.10	0.020	25	0.020
0.15	0.020	45	0.020
		75	0.020

Thus, the second-order model is valid, at least for the regions used in the experiment. In the first experiment, the value of [B] was nearly constant and contributed very little to the variation in rate over time. Thus the contribution of [B] was not detectable in experiment 1.

COMMENT

This example was structured to illustrate the fact that rate expressions often relate strongly to experimental conditions. The first-order model would be satisfactory as long as [B] changed very little. This is important to remember for two reasons: First, complete trust should never be placed in a rate expression, particularly outside the region for which data are available. Second, in many cases, simple models are satisfactory if care is taken in their application.

5.7
OTHER TYPES OF BALANCES

The materials balance, based on the law of conservation of mass, was introduced in Section 5.5. Other important thermodynamic balances employed in the study of water quality are based on the laws of conservation of energy and momentum. Energy balances (based on internal, mechanical, and electromagnetic energy) are important in the study of the movement of water in deep lakes and reservoirs. Momentum balances are used to model complex wind- and wave-induced water movements. While the application of energy and momentum balances is beyond the scope of this text, it is important to remember that Eq. (5.40) is the starting point for all of these balances. Thus it is of fundamental importance to study and understand the use of Eq. (5.40).

KEY IDEAS, CONCEPTS, AND ISSUES

- The number of moles of a substance(s) entering into a reaction and the number of moles of the substances produced are defined by the stoichiometry of a reaction.
- Stoichiometry involves the application of the principle of conservation of mass to chemical reactions.
- The rate at which a substance disappears or is formed in a stoichiometric reaction is defined as the rate of reaction.
- The two principal types of reaction that occur in nature are classified as homogeneous and heterogeneous.
- Reaction rates are usually functions of the physical and chemical environment. The molar concentration of the reactants and products, the stoichiometric coefficients, temperature, and pH are important variables affecting the rate of reaction.
- All rates of reaction are empirical and are derived from experimental studies. Zero-order reactions are concentration-independent ($r_A = k$); similarly, first-order reactions are functions of one reactant concentration ($r_A = k[A]$).
- A materials balance analysis, based on the principle of conservation of mass, is a quantitative description of all materials that enter, leave, and accumulate in a system with defined boundaries.
- The simplified word statement for a materials balance is

  Accumulation = inflow − outflow + generation

DISCUSSION TOPICS AND PROBLEMS

5.1. A 1-L beaker (batch reactor) is used for laboratory experiments to measure the rate coefficient for the reaction $A \rightarrow 2B$. The system is started by adding 1.00 mol of A to the beaker, and the concentration of A is monitored with time. Use the data given below to determine the reaction rate constant.

TIME, min	[A], mol/L
0	1.00
2	0.67
4	0.45
8	0.20
16	0.04
32	Trace

5.2. A reactant A forms two products, B and C, in competitive reaction, both of which are first order with respect to A:

$$A \begin{array}{c} \nearrow^{1} B \\ \searrow_{2} C \end{array}$$

Plot $[B]/[A_0]$ versus $[C]/[A_0]$ where $[A_0]$ is the initial value of [A]. Determine the region within which the reactions can occur. Assume a batch reaction system is used.

5.3. A 1-L batch reactor is used to determine the rate coefficients for the following parallel reaction:

$$A \begin{array}{c} \nearrow^{1} B \\ \searrow_{2} C \end{array}$$

If both reactions are known to be first order, use the data given below to determine k_1, k_2, and [C].

t, min	[A], mol/L	[B], mol/L
0	1.00	0
2	0.55	0.30
4	0.30	0.47
8	0.09	0.61
16	0.01	0.66

5.4. A consecutive reaction $A \xrightarrow{1} B \xrightarrow{2} C$ is known to be first order for both reactions ($r_1 = k_1[A]$, $r_2 = k_2[B]$).
   (a) Determine [A], [B], and [C] as functions of time for a batch reaction system.
   (b) For the data below evaluate [B] and [C] as a function of time.

$$[A_0] = 10 \text{ mol/L}$$
$$[B_0] = [C_0] = 0$$
$$k_1 = 0.05 \text{ s}^{-1}$$
$$k_2 = 0.01 \text{ s}^{-1}$$

5.5. If bacteria are allowed to grow unrestricted at their optimal temperature, a typical rate of division is a half-hour. If bacteria average 2×10^{-6} m in diameter and have a density of 1100 kg/m^3, determine the mass after 72 hr of a culture starting with one cell. Note that the mass of the earth is 6×10^{24} kg.

5.6. A reaction is termed *autocatalytic* if the rate is a function of the product concentration. For the reaction $A \rightarrow B$ with a reaction rate $r = k[B]^{\alpha}$, plot $[B]/[A_0]$ versus time in a batch reaction system. Use $\alpha = 0.5$, 1.0, 2.0 and $[B_0] = 0.01$ mol/m^3, $[A_0] = 10$ mol/m^3, and $k = 2(mol/m^3)^{1/2}$ d^{-1}, 2 d^{-1}, and 2 $m^3/mol \cdot d$, respectively. Note the 1:1 stoichiometry.

5.7. A reaction is partially autocatalytic in that the product C affects the reaction rate as shown below:

$$2A \rightarrow C + \text{other products}$$
$$r_A = -k[A][C]$$

Develop an expression for r_A in terms of [A] and $[C_0]$, and plot [A] and [C] versus t for $[A_0] = 10$ mol/m^3, $[C_0] = 1$ mol/m^3, and $k = 1$ $m^3/mol \cdot d$ in a batch reaction system.

5.8. A batch reaction system is used to carry out the reaction

$$2NO_2^- + O_2 \rightarrow 2NO_3^-$$

If the rate of reaction of $NO_2^- - N$ can be described by a saturation-type expression

$$r_{NO_2^-} = -\frac{k_{NO_2^-} C_{NO_2^-}}{K_{NO_2^-} + C_{NO_2^-}}$$

where

$$k_{NO_2^-} = 7 \text{ g/m}^3 \cdot d$$
$$K_{NO_2^-} = 0.1 \text{ g/m}^3$$

plot $C_{NO_2^-}$ and $C_{NO_3^-}$ in grams per cubic meter as a function of time if $C_{NO_{2_0}} = 3.0$ g/m^3 and $C_{NO_3^-} = 0$. All concentrations are in terms of N.

5.9. Given the reaction $A \rightarrow B$, plot the function $\eta = (C_{A_0} - C_A)/C_{A_0}$ versus time for a batch-reaction system for the rate relationships given below:
(a) $r_A = -k$
(b) $r_A = -k(C_A)^{0.5}$
(c) $r_A = -k(C_A)$
(d) $r_A = -k(C_A)^2$

5.10. A reaction $3A \rightarrow B$ is carried out in a batch reactor. Molar concentrations of A and B are monitored with time. Determine the appropriate reaction model based on the following data:

t, d	[A], mol/m^3	[B], mol/m^3
0.0	15.0	1.0
0.5	13.2	1.6
1.0	10.8	2.4
1.5	8.2	3.3
2.0	5.6	4.1
2.5	3.6	4.8
3.0	2.2	5.3
3.5	1.3	5.6
4.0	0.7	5.8
4.5	0.4	5.9
5.0	0.2	5.9

5.11. The data given below were obtained for the reaction A → B using a laboratory-scale reaction system. Determine the reaction order and the value of the reaction rate constant.

TIME, hr	C_A, g/m^3
0	50.0
1	35.6
2	25.8
3	18.5
4	12.8
6	7.3

5.12. The reaction-rate data given below were obtained using a batch reaction system for the reaction A → products. Determine an appropriate rate expression and the rate coefficient.

TIME, min	C_A, g/m^3
0	30.00
0.5	12.00
1	7.50
2	4.29
4	2.31
8	1.20
16	0.61
32	0.31

5.13. A chemical reaction occurring in a batch reactor can be described by the following rate equation:

$$r_A = -k_2 \left(C_{A_0} - 0.5 C_A \right)^2$$

where C_{A_0} and C_A are the mass concentrations (in grams per cubic meter) of A present initially and at time t, respectively. Determine the value of k_2 if $C_{A_0} = 100$ g/m^3 and C_A ($t = 10$ d) $= 10$ g/m^3.

5.14. The complex set of reactions given below is to be carried out in a batch reactor. Determine the concentrations of each material as a ratio with the initial concentration of A ($[A]_i$). All reactions are first order.

$$2A \xrightarrow{1} B \qquad k_1 = 0.10 \text{ d}^{-1}$$

$$2A \xrightarrow{2} 3C \qquad k_2 = 0.15 \text{ d}^{-1}$$

$$3C \xrightarrow{3} D \qquad k_3 = 0.20 \text{ d}^{-1}$$

$$C \xrightarrow{4} E \qquad k_4 = 0.10 \text{ d}^{-1}$$

5.15. The following data were obtained when glucose ($C_6H_{12}O_6$) was added to a batch culture of microorganisms. Determine the reaction order for the disappearance of glucose.

TIME, min	CONCENTRATION, g/m^3	
	Glucose	Cells
0	100	1500
10	67	1516
20	50	1525
30	40	1530
40	33	1534
50	29	1535

5.16. A reaction sequence A $\xrightarrow{1}$ B $\xrightarrow{2}$ C is carried out in a batch reactor. Both reactions are first order with respect to the reactant concentration, and the rate coefficients are equal to 0.1 and 0.2 d^{-1}, respectively. Plot $[A]/[A_0]$, $[B]/[A_0]$, and $[C]/[A_0]$ as functions of time.

5.17. Verify that the equation derived for [A] in step 2, trial 2(2) in Example 5.6 is correct.

5.18. A town requiring 1.0 m^3/s of drinking water has two sources, a local well with 60 g/m^3 nitrate and a distant reservoir with 10 g/m^3 nitrate. What flow rates of well and reservoir water are needed to meet the 45 g/m^3 drinking water standard and minimize the use of more expensive reservoir water? (Courtesy of Jim Hunt).

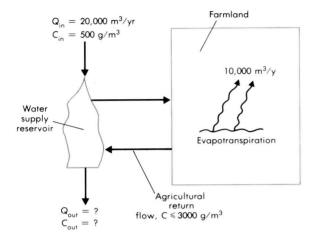

$Q_{in} = 20,000$ m³/yr
$C_{in} = 500$ g/m³

Farmland

10,000 m³/y

Evapotranspiration

Water supply reservoir

$Q_{out} = ?$
$C_{out} = ?$

Agricultural return flow, $C \leqslant 3000$ g/m³

FIGURE 5.9

Definition sketch of the irrigation system for Problem 5.19.

5.19. A reservoir has been constructed to supply irrigation water to adjacent farmland. The water requirement for the farmland is composed of the water lost by evapotranspiration (10,000 m³/yr) and the water needed to flush out salt carried in by the irrigation water. The salty return water is called *agricultural return flow*, and the salt concentration cannot exceed 3,000 g/m³; otherwise plants are killed. If the agricultural return flow enters the reservoir as indicated in Fig. 5.9, determine (a) the water flow rate out of the reservoir, (b) the salt concentration in the reservoir (which is equal to the salt concentration leaving the reservoir), and (c) the pumping rate needed to irrigate the farmland and flush salts back into the reservoir. Assume steady-state conditions, no rainfall, no reservoir evaporation, and no salt lost by evapotranspiration. (Courtesy of Jim Hunt).

5.20. A water storage tank receives a constant feed rate of 0.2 m²/s, and the demand varies according to the relationship $0.2(1 - \cos \pi t/43,200)$ m³/s. The tank is cylindrical with a cross-sectional area of 1000 m². If the depth at $t = 0$ is 5 m, plot the water depth as a function of time.

5.21. A large tank having a floor area of 1000 m² and a sidewall depth of 10 m is used as an equalization reservoir. Flow out of the basin is 0.3 m³/s, while flow into the basin is $0.3(1 + \cos \pi t/43,200)$ m³/s. Plot the hourly values of water depth versus time, assuming $h = h_0 = 5$ m at $t = 0$.

5.22. Water is being pumped into a 3-m-diameter tank at the rate of 0.65 m³/min. At the same time, water leaves the tank at a rate that is dependent on the height of the liquid in the tank. The relationship governing the flow from the tank is $q = (2$ m²/min$)h$(m). If the tank was initially empty, develop a relationship that can be used to define the height of the liquid in the tank as a function of time. What is the steady-state height of the liquid in the tank?

REFERENCES

5.1. Churchill, S. W., (1974), *The Interpretation and Use of Rate Data: The Rate Concept*, McGraw-Hill Book Company, New York.

5.2. Glasstone, S., (1946), *Textbook of Physical Chemistry*, 2d ed., D. Van Nostrand Company, New York.

5.3. Kirkbride, D. G., (1947), *Chemical Engineering Fundamentals*, 1st ed., McGraw-Hill Book Company, New York.

5.4. Levenspiel, O., (1972), *Chemical Reaction Engineering*, 2d ed., John Wiley and Sons, New York.

5.5. Moore, W. J., (1972), *Physical Chemistry*, 4th ed., Prentice-Hall, Englewood Cliffs, N.J.

5.6. Smith, J. M., (1981), *Chemical Engineering Kinetics*, 3d ed., McGraw-Hill Book Company, New York.

6

Mathematical Models
of Physical Systems

Most of the systems of concern in water quality management have significant flowthrough. The actual systems vary from lakes, rivers, and estuaries to treatment process units, but, with few exceptions, all have continuous flow inputs and outputs, and, therefore, Eq. (5.40) applies. A number of approaches to the analyses of these systems are in use. Most of the approaches begin with an assumption about the hydraulic characteristics of the system. Two basic hydraulic models are used to simulate natural systems. They are the complete-mix model and the plug-flow model. These two ideal models are used to define an envelope within which other models fall. Often these models are combined in various ways to simulate (model) large, complex systems with spatially varying characteristics. It is the purpose of this chapter to develop a mathematical basis for the analysis of these two hydraulic models and various combinations thereof. The material developed in this chapter will be applied in Part III to analyze water quality changes in natural systems including rivers, lakes, and groundwater aquifers.

6.1

HYDRAULIC MODELS OF NATURAL SYSTEMS

The objective of water quality modeling is to study the behavior of natural systems in response to external and internal inputs. As a first step in the modeling process, it is important to be able to predict the hydraulic performance of the system. In predicting hydraulic performance it has become common practice to use models that have been developed to describe the hydraulic characteristics of reactors used to carry out chemical and biological reactions.

The five principal reactor models that are of interest with respect to water quality modeling are (1) the batch reactor, (2) the complete-mix

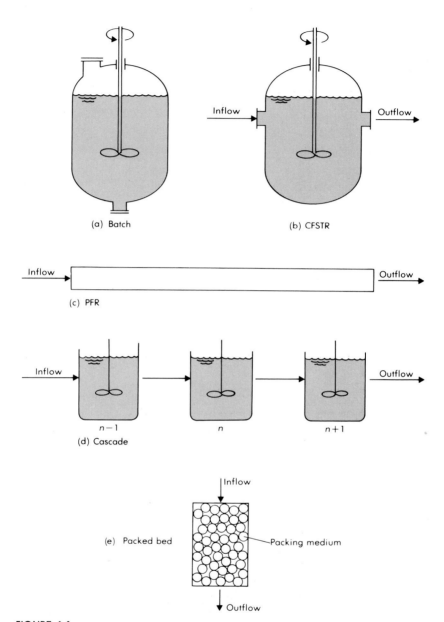

FIGURE 6.1

Definition sketch for various types of reactors used to carry out chemical and biological reactions: (a) batch reactor, (b) continuous-flow stirred-tank reactor (CFSTR), (c) plug-flow reactor (PFR), (d) cascade reactor, and (e) packed-bed reactor.

reactor, (3) the plug-flow reactor, (4) the cascade of complete-mix reactors, and (5) the packed-bed reactor. Each of these models is illustrated schematically in Fig. 6.1 and described briefly below. The complete-mix, plug-flow, and cascade reactors are considered in detail in the following sections of this chapter.

Batch Reactors

A reactor in which flow is neither entering nor leaving is defined as a batch reactor. The BOD test discussed in Chapter 2 is carried out in a batch reactor. Batch reactors are sometimes used to model shallow lakes that are mixed completely.

Complete-Mix Reactors

In the complete-mix reactor, fluid particles that enter the reactor are instantaneously dispersed throughout the reactor volume. The fluid particles leave the reactor in proportion to their statistical population. In the field of chemical engineering, the complete-mix reactor is known as the *continuous-flow stirred-tank reactor* (CFSTR). The complete-mix model is used to study lakes and reservoirs with continuous inputs and outputs.

Plug-Flow Reactors

In a plug-flow reactor (PFR), fluid particles pass through the reactor and are discharged in the same sequence in which they entered the reactor. Each fluid particle remains in the reactor for a time period equal to the theoretical detention time. Plug-flow reactors are often identified as *tubular reactors*. Plug-flow models are used to study river and estuary systems.

Cascade of Complete-Mix Reactors

The cascade of complete-mix reactors is used to model the flow regime that exists between the hydraulic flow patterns corresponding to the complete-mix and plug-flow reactors. If the cascade is composed of one reactor, the complete-mix flow regime prevails. If the cascade consists of an infinite number of reactors in series, the plug-flow regime results.

Packed-Bed Reactors

Reactors filled with some type of packing medium are known as packed-bed reactors. In the field of water quality management, packed-bed-reactor models are used to study the movement of water and contaminants in groundwater systems. When the pore volume of the medium is filled with a liquid, the flow is said to be *saturated*. When the pore volume is partially filled, the flow is said to be *unsaturated*. Flow in a porous medium is considered further in Chapter 10.

6.2

CONTINUOUS-FLOW STIRRED-TANK REACTOR MODELS

Continuous-flow stirred-tank reactors (CFSTRs), illustrated in Fig. 6.2, have no concentration gradients within the system. Material entering is uniformly dispersed instantaneously throughout the reactor. The result is that the concentration of any material leaving the reactor is exactly the same as the concentration at any point in the reactor. A materials balance for material A for the reactor shown in Fig. 6.2 would result in Eq. (6.1):

$$\frac{dC_A}{dt} V = QC_{A_i} - QC_A + r_A V \qquad (6.1)$$

Accumulation = Inflow − Outflow + Generation

where

C_A = concentration of material A in reactor, g/m^3

V = reactor volume, m^3

Q = volumetric flow rate, m^3/s

C_{A_i} = input concentration of material A, g/m^3

r_A = rate of reaction of material A, $g/m^3 \cdot s$

Because the characteristics of CFSTRs are understood most easily through

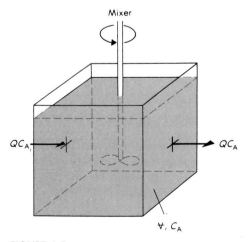

FIGURE 6.2

Continuous-flow stirred-tank reactor.

the consideration of examples, Examples 6.1, 6.2, and 6.3 are presented to introduce the concepts involved in the use of the CFSTR reactor model developed in Eq. (6.1).

EXAMPLE 6.1

RESPONSE OF CFSTRs TO TRACER INPUT

A conservative (nonreactive) material is injected into the input flow of a CFSTR (Fig. 6.2) on a continuous basis, beginning at time $t = 0$ and resulting in a constant input tracer concentration of C_{T_i}. Determine C_T, the reactor output concentration, as a function of time, and plot the tracer-output response curve (i.e., reactor output versus time).

SOLUTION:

1. Write the materials balance equation for the CFSTR.

$$\frac{dC_T}{dt} V = QC_{T_i} - QC_T + r_T V$$

Accumulation = Inflow − Outflow + Generation

where

C_T = concentration of tracer in reactor, g/m^3

V = reactor volume, m^3

Q = volumetric flow rate, m^3/s

C_{T_i} = initial concentration of tracer, g/m^3

r_T = rate of tracer generation, $g/m^3 \cdot s$

2. Because the tracer is conservative (i.e., nonreactive), the generation term is zero since $r_T = 0$. The materials balance derived in step 1 can be rearranged and integrated as shown below. The limits of integration are from $C = 0$ to $C = C_T$ and from $t = 0$ to $t = t$.

$$\frac{dC_T}{dt} = \frac{Q}{V}(C_{T_i} - C_T)$$

$$\int_0^{C_T} \frac{dC_T}{C_{T_i} - C_T} = \frac{Q}{V} \int_0^t dt$$

$$\ln\left(\frac{C_{T_i} - C_T}{C_{T_i}}\right) = -t\left(\frac{Q}{V}\right)$$

$$C_T = C_{T_i}(1 - e^{-t(Q/V)})$$

3. Plot the tracer response curve. Because numerical data are not given, assume $C_{T_i} = 1$, and plot C_T/C_{T_i} versus the term $t(Q/V)$. The tracer response curve for the reactor output is plotted in Fig. 6.3.

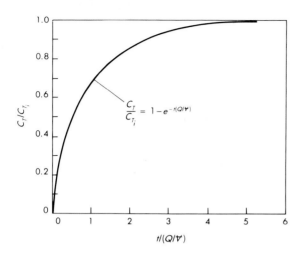

FIGURE 6.3

Tracer-output response curve for CFSTR subjected to tracer step input for Example 6.1.

Hydraulic Detention Time

In Example 6.1, the term V/Q appeared, which has units of time, and will be referred to as the *mean hydraulic detention time* θ_H. It is a useful reference parameter and will appear often in subsequent discussions. In examining Fig. 6.3, note that the value of C_T/C_{T_i} is 0.95 after three hydraulic detention times ($t = 3\theta_H$). In practice, the conditions existing after three hydraulic detention times are often considered to be a satisfactory approximation of the final, or steady-state ($dC_T/dt = 0$), conditions. Thus if a CFSTR is subjected to a step change in input characteristics, the new steady-state value is assumed to have been achieved when three detention times have passed.

EXAMPLE 6.2

RESPONSE OF CFSTRs TO STEP INCREASE IN REACTANT CONCENTRATION

A reaction A $\rightarrow$ B, known to be first order ($r_A = -kC_A$), is to be carried out in a CFSTR. Water is run through the reactor at a flow rate Q m^3/s, and at $t = 0$ the reactant A is added to the input stream on a continuous basis. Determine the output concentration of A as a function of time, and plot reactor-output response curves for reactant A.

SOLUTION:

1. Write the materials balance equation for the system

$$\frac{dC_A}{dt}V = QC_{A_i} - QC_A + r_A V$$

Accumulation = Inflow − Outflow + Generation

where the terms are as defined previously in the development of Eq. (6.1).

2. Assume the rate of generation is

$$r_A = -kC_A$$

where k = reaction-rate constant

3. Rearrange the materials balance given in step 1 and integrate the resulting expression to find C_A. The limits of integration are from $C = 0$ to $C = C_A$ and from $t = 0$ to $t = t$.

$$\int_0^{C_A} \frac{dC_A}{C_{A_i} - C_A(1 + k\theta_H)} = \frac{1}{\theta_H} \int_0^t dt$$

$$-\frac{1}{k\theta_H + 1} \ln\left[\frac{C_{A_i} - (1 + k\theta_H)C_A}{C_{A_i}} \right] = \frac{t}{\theta_H}$$

$$C_A = \frac{C_{A_i}\left(1 - e^{-(1 + k\theta_H)t/\theta_H}\right)}{1 + k\theta_H}$$

4. As t approaches infinity, the steady-state solution is approached:

$$C_A \approx \frac{C_{A_i}}{1 + k\theta_H}$$

5. Plot output response curves.

Output response curves for reactant A (i.e., C_A/C_{A_i} versus t/θ_H) are shown in Fig 6.4 for a range of $k\theta_H$ values. Note that as the reaction parameter $k\theta_H$ increases, both the time to approximate steady state and the steady-state value of C_A decrease.

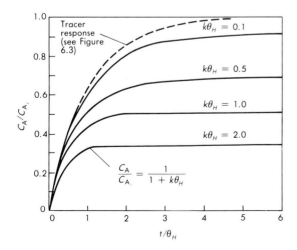

FIGURE 6.4

Output response curves for input of a first-order reactant in CFSTR in Example 6.2.

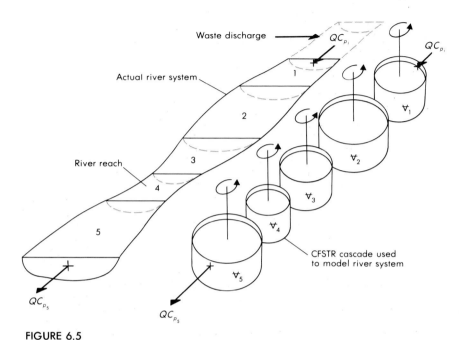

FIGURE 6.5

River reach segmented for analysis as a cascade series of connected reactors for Example 6.3.

Application of the CFSTR Model

Application of the CFSTR model in water quality management is made in two ways. Some treatment facilities are designed as stirred tanks (contents of the tank are mixed completely), and the model applies directly. Another use is made in describing natural systems. In the latter case, a physical system is divided conceptually into a series or group of stirred tanks. For example, a reach of a river (Fig. 6.5) might be considered to behave as a series of CFSTRs. The analysis of a river reach as a series of CFSTRs is examined in Example 6.3. Two-dimensional systems can be used also, but flows are difficult to predict in open bodies of water, such as lakes. Many rivers have deltas made up of a large number of channels. In this situation, two-dimensional grids are more feasible because flows can be measured or predicted.

EXAMPLE 6.3

ANALYSIS OF RIVER REACH AS SERIES OF CFSTRs

The river reach shown in Fig. 6.5 has been divided into five segments based on measured velocities and depths. An industrial facility is planned just upstream of the first segment, and it is necessary to estimate the effect of the wastewater discharge. A series of dye experiments have been run, and each of the segments was found to behave as an approximate CFSTR. The pollutant is expected to disappear according to a first-order reaction. For the data given below, determine the steady-state pollutant concentration in each segment.

$$p = \text{pollutant} \qquad V_1 = 8.64 \times 10^5 \text{ m}^3$$
$$C_{p_i} = 30 \text{ g/m}^3 \qquad V_2 = 25.92 \times 10^5 \text{ m}^3$$
$$r_p = -kC_p \qquad V_3 = 17.28 \times 10^5 \text{ m}^3$$
$$k = 0.2 \text{ d}^{-1} \qquad V_4 = 8.64 \times 10^5 \text{ m}^3$$
$$Q = 5 \text{ m}^3/\text{s} \qquad V_5 = 25.92 \times 10^5 \text{ m}^3$$

SOLUTION:

1. Assuming steady-state conditions ($dC_{p_i}/dt = 0$), write mass-balance equations for each segment.

1. $\quad 0 \ = QC_{p_i} - QC_{p_1} + (-kC_{p_1})V_1$
2. $\quad 0 \ = QC_{p_1} - QC_{p_2} + (-kC_{p_2})V_2$
3. $\quad 0 \ = QC_{p_2} - QC_{p_3} + (-kC_{p_3})V_3$
4. $\quad 0 \ = QC_{p_3} - QC_{p_4} + (-kC_{p_4})V_4$
5. $\quad 0 \ = QC_{p_4} - QC_{p_5} + (-kC_{p_5})V_5$

$$\text{Accum} = \text{Inflow} - \text{Outflow} + \text{Generation}$$

where

$$Q = \text{volumetric flow rate, g/m}^3$$
$$C_{p_i} = \text{influent concentration of pollutant, g/m}^3$$
$$C_{p_1}, C_{p_2}, C_{p_3}, C_{p_4}, C_{p_5} = \text{concentration of pollutant leaving each reactor, g/m}^3$$
$$V_1, V_2, V_3, V_4, V_5 = \text{reactor volume, m}^3$$

2. Rearrange the equations developed in step 1 and solve for C_{p_n}:

$$C_{p_1} = \frac{C_{p_i}}{1 + k\theta_{H_1}}$$

$$C_{p_2} = \frac{C_{p_i}}{(1 + k\theta_{H_1})(1 + k\theta_{H_2})}$$

$$C_{p_3} = \frac{C_{p_i}}{(1 + k\theta_{H_1})(1 + k\theta_{H_2})(1 + k\theta_{H_3})}$$

$$C_{p_4} = \frac{C_{p_i}}{(1 + k\theta_{H_1})(1 + k\theta_{H_2})(1 + k\theta_{H_3})(1 + k\theta_{H_4})}$$

$$C_{p_5} = \frac{C_{p_i}}{(1 + k\theta_{H_1})(1 + k\theta_{H_2})(1 + k\theta_{H_3})(1 + k\theta_{H_4})(1 + k\theta_{H_5})}$$

3. Solve for the concentrations. Substituting numerical values and solving for the individual pollutant concentrations yields

$C_{p_1} = 21.4 \text{ g/m}^3$

$C_{p_2} = 9.7 \text{ g/m}^3$

$C_{p_3} = 5.4 \text{ g/m}^3$

$C_{p_4} = 3.4 \text{ g/m}^3$

$C_{p_5} = 1.76 \text{ g/m}^3$

6.3

PLUG-FLOW REACTOR MODELS

Plug-flow reactors (PFRs), illustrated in Fig. 6.6, are ideally mixed in the lateral direction and unmixed longitudinally. Although this is an unrealistic assumption for most real-world systems, the PFR is used to define a limit and, as will be shown, can be approximated closely, if that is desirable. Because longitudinal mixing does not occur in a PFR, the mean hydraulic residence time θ_H is the true hydraulic residence time. An effluent (or output) tracer signal is exactly the same as the input, except that it is transposed in time by θ_H, as shown in Fig. 6.7.

FIGURE 6.6

Plug-flow reactor used in activated sludge wastewater treatment system in Stuttgart, W. Germany. (See Chapter 14.)

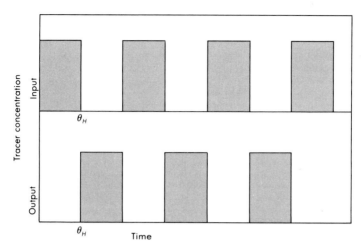

FIGURE 6.7

Tracer input and output response curves for an ideal PFR.

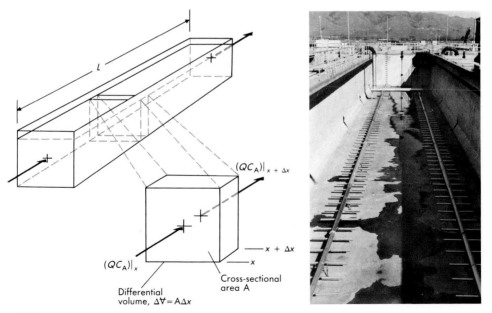

FIGURE 6.8

Definition sketch for the analysis of a PFR.

A materials balance on a reactive material must be made on an incremental volume, $\Delta V = A \Delta x$, as shown in Fig. 6.8, because a longitudinal concentration gradient exists. For the reaction $A \rightarrow B$, the mass balance over the increment is given by Eq. (6.2):

$$\frac{\partial C_A}{\partial t} \Delta V = (QC_A)|_x - (QC_A)|_{x+\Delta x} + r_A \Delta V \tag{6.2}$$

Accumulation $=$ Inflow $-$ Outflow $+$ Generation

where

C_A = concentration of material A in volume element, g/m^3

V = volume, m^3

Q = volumetric flow rate, m^3/s

r_A = rate of reaction of reactant A, $g/m^3 \cdot s$

Because $\Delta V = A \Delta x$, Eq. 6.2 can be written as Eq. (6.3) by taking the limit as Δx approaches zero. A more detailed derivation of Eq. 6.3 is given in Appendix H.

$$\frac{\partial C_A}{\partial t} = -\frac{Q}{A}\frac{\partial C_A}{\partial x} + r_A \tag{6.3}$$

Generally, the flow rate Q is nearly constant, and the following transformation can be made:

$$\frac{1}{A}\frac{\partial QC_A}{\partial x} = \frac{Q}{A}\frac{\partial C_A}{\partial x} = \frac{\partial C_A}{\partial \theta_H}$$

where

$$\frac{A\,\partial x}{Q} = \partial \theta_H$$

After substitution of the above transformation into Eq. (6.3), the steady-state ($\partial C_A/\partial t = 0$) result is

$$\frac{dC_A}{d\theta_H} = r_A \tag{6.4}$$

Equation (6.4) is the same form as that derived for the batch reaction system in Example 5.6. Care should be taken to avoid confusion between the non-steady-state accumulation rate, dC_A/dt; the longitudinal concentration gradient, $dC_A/d\theta_H$; and the generation rate r_A. Only in special cases can these parameters be equated.

EXAMPLE 6.4

PERFORMANCE OF PLUG-FLOW REACTORS

A plug-flow reactor (Fig. 6.8) is to be used to carry out the reaction $A \rightarrow B$. The reaction is first order, and the rate is characterized as

$r_A = -kC_A$. Determine the steady-state effluent concentration as a function θ_H.

SOLUTION:

1. Write the steady-state mass balance.
   The steady-state mass balance is given by Eq. (6.4). Substituting $-kC_A$ for r_A yields:

$$\frac{dC_A}{d\theta_H} = -kC_A$$

2. Determine the steady-state effluent concentration as a function of θ_H. Integration of the above equation between the limits $C = C_{A_i}$ and $C = C_A$ and $\theta_H = 0$ and $\theta_H = \theta_H$ leads to the steady-state solution:

$$\int_{C_{A_i}}^{C_A} \frac{dC_A}{C_A} = -k \int_0^{\theta_H} d\theta_H$$

$$C_A = C_{A_i} e^{-k\theta_H}$$

Application of a series of PFRs to a problem such as that given in Example 6.3 results in a very simple expression for a first-order reactant:

$$C_n = C_i e^{-k \sum_{j=1}^{n} \theta_{H_j}} \tag{6.5}$$

Clearly, if a river reach behaves as a PFR, segmenting is not necessary. Using Eq. (6.5) and the parameter values of Example 6.3 results in a final concentration value of $C_A = 0.55$ g/m³. This value is 70 percent less than the value obtained using a series of CFSTRs (1.76 g/m³). Thus comparison of the performance of the two configurations is of interest.

6.4
COMPARISON OF REACTOR MODEL PERFORMANCE

Comparison of CFSTR and PFR reactors is most easily made for steady-state conditions. For convenience, the volume or residence time necessary to achieve a selected effluent or output reactant value will be used as the basis for comparison. Thus the expressions derived above for CFSTR and PFR reactors must be arranged to solve for θ_H.

1. For first-order reactions ($r_A = -kC_A$):
   CFSTR

$$\theta_{H_{CFSTR}} = \frac{1}{k}\left(\frac{C_{A_i}}{C_A} - 1\right) \tag{6.6}$$

PFR

$$\theta_{H_{\text{PFR}}} = -\frac{1}{k} \ln \frac{C_A}{C_{A_i}} \tag{6.7}$$

2. For second-order reactions ($r_A = -kC_A^2$):
   CFSTR

$$\theta_{H_{\text{CFSTR}}} = \frac{C_{A_i} - C_A}{kC_A^2} \tag{6.8}$$

   PFR

$$\theta_{H_{\text{PFR}}} = \frac{C_{A_i} - C_A}{kC_{A_i}C_A} \tag{6.9}$$

Note: In higher-order reactions, stoichiometry and inclusion of other reactants increases complexity. For simplicity, the example is limited to simple second-order reactions.

3. For saturation-type reaction ($r_A = -kC_A/K + C_A$):
   CFSTR

$$\theta_{H_{\text{CFSTR}}} = \frac{(C_{A_i} - C_A)(K + C_A)}{kC_A} \tag{6.10}$$

   PFR

$$\theta_{H_{\text{PFR}}} = \frac{1}{k}\left(K \ln \frac{C_{A_i}}{C_A} + C_{A_i} - C_A \right) \tag{6.11}$$

The first-order and simple second-order systems can be compared directly, and the results are shown in Fig. 6.9. Comparison of the configurations for the saturation-type reaction requires values for C_{A_i} and K. The ratio is quite insensitive to the choice of values, however, and the example shown in Fig. 6.10 is typical.

Performance Comparisons

Performance comparisons are of the most interest in treatment process selection and design, where the engineer can influence results. In water and wastewater treatment systems, the output C_A/C_{A_i} value is usually fairly low. For example, wastewater treatment plants are usually required to remove 80 to 90 percent of the entering BOD_5, corresponding to $C_A/C_{A_i} = 0.1$ to 0.2. In most systems, the removal is accomplished in two steps—a primary stage, in which about 30 percent of the BOD_5 is removed, and a secondary stage, which removes about 60 percent. Considering the secondary stage as a separate unit, $C_A/C_{A_i} = 0.1/0.7 = 0.14$, and for any of the three reaction models used in Figs. 6.9 and 6.10, the PFR is much more efficient in this region.

The reason PFRs are more efficient than CFSTRs for most reaction systems is that the average reaction rate is higher. If the reaction rate was

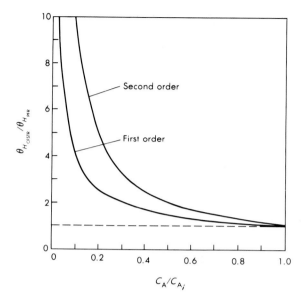

FIGURE 6.9

Comparison of the ratio of CFSTR and PFR hydraulic detention times (volumes) for first and second-order reactions.

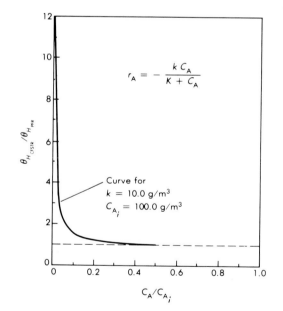

FIGURE 6.10

Comparison of the ratio of CFSTR and PFR hydraulic detention times (volumes) for a saturation-type reaction.

calculated at various points along the length of a PFR, the values would gradually decrease. The lowest value would be at the outlet where the reactant concentration was the lowest. Comparing this situation with that of the CFSTR, where the reaction rate must be the same everywhere, and therefore must be the lowest value, explains why PFRs are more efficient than CFSTRs.

Use of Reactors in Series to Improve Performance

Placing CFSTR systems in a series or cascade arrangement can make them more efficient (see Fig. 6.1d). The first-order, steady-state value of C_A/C_{A_i} for the nth CFSTR in an identical series is given by

$$\frac{C_A}{C_{A_i}} = \frac{1}{(1 + k\theta_H)^n} \qquad (6.12)$$

where θ_H is the residence time in a single unit. The total residence time required, $\theta_{H_T} = n\theta_H$, is given by Eq. (6.13), again for a first-order reaction:

$$\theta_{H_T} = \frac{n}{k}\left[\left(\frac{C_{A_i}}{C_{A_n}}\right)^{1/n} - 1\right] \qquad (6.13)$$

If $n = 3$, the total CFSTR volume to achieve a 90 percent removal value is only 1.5 times the PFR volume, as compared with 3.9 times the PFR volume for the single unit. In practice, it is rarely economical to construct more than three or four reactors in series, but other advantages of CFSTRs often make the cascade configuration useful.

6.5
NONIDEAL FLOW MODELS

Most flow systems deviate considerably from either the ideal CFSTR or PFR, although these two configurations define the effective performance envelope. Common causes of deviation include dispersion (the longitudinal transport of material due to turbulence and molecular diffusion), dead volume, short circuiting, and density stratification. The latter phenomenon almost always occurs when two flows meet—for example, a tributary entering a larger stream, tidal flow in an estuary, or a river discharging into a lake. Very slight differences in density, due to sediment load or temperature, can, depending on the time of year, result in the maintenance of distinctly separate flows for long distances. The effect of density differences due to sediment transport is illustrated in the photos shown in Fig. 6.11 of the juncture of the American and the Sacramento rivers at Sacramento, California.

Dead volume often develops in corners, below inlet and outlet weirs, and on the inside edge of bends. The effect is similar to short circuiting, in that the effective residence time is decreased, but short circuiting usually involves only a portion of the flow and is the result of density stratification rather than a physical characteristic of the system.

(a)   (b)

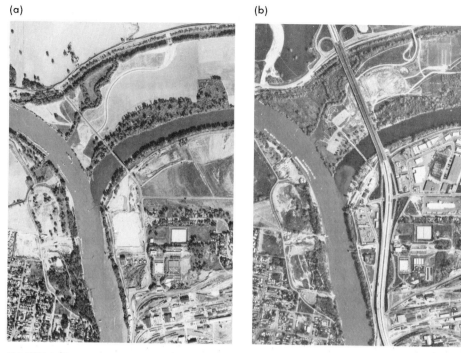

FIGURE 6.11

Two photos of the American River (flowing from east to west) joining the Sacramento River (flowing from north to south) at Sacramento, CA: (a) taken on October 22, 1963; (b) taken on April 9, 1981. Note the separation in flows caused by density differences and time of year. The American River is silt free, whereas the Sacramento River is silt laden. Note also the development that has occurred in the intervening years.

Source: Cartwright Aerial Surveys Inc., Sacramento, CA.

Dispersion

Dispersion is most easily represented as a function of the concentration gradient:

$$F_{A_x} = -D_x \frac{\partial C_A}{\partial x} \tag{6.14}$$

where

F_{A_x} = mass flux of material A in x direction, g/m$^2 \cdot$ s

D_x = coefficient of dispersion in x direction, m^2/s

C_A = concentration of material A, g/m^3

Only one-dimensional problems are considered here, but in reality, all dispersion problems are three-dimensional. The coefficient of dispersion varies with direction and degree of turbulence [6.1].

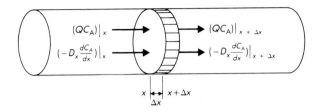

FIGURE 6.12

Definition sketch for the analysis of a PFR with dispersion.

A one-dimensional mass balance including dispersion (Fig. 6.12) results in Eqs. (6.15) and (6.16):

$$\frac{\partial C_A}{\partial t} A \Delta x = \left(uA_x C_A - A_x D_x \frac{\partial C_A}{\partial x} \right)\bigg|_x$$

Accumulation =   Inflow

$$- \left(uA_x C_x - A_x D_x \frac{\partial C_A}{\partial x} \right)\bigg|_{x+\Delta x} + r_A A_x \Delta x \qquad (6.15)$$

$$-\qquad\qquad \text{Outflow} \qquad\qquad + \text{Generation}$$

where

A_x = cross-sectional area in x direction, m^2

u = average velocity, m/s

r_A = rate of reaction of material A, g/m$^3 \cdot$ s

and other terms are as defined previously. Taking the limit of Eq. (6.15) as Δx approaches zero yields the following two expressions:

$$\frac{\partial C_A}{\partial t} = D_x \frac{\partial^2 C_A}{\partial x^2} - u \frac{\partial C_A}{\partial x} + r_A \qquad (6.16a)$$

$$\frac{\partial C_A}{\partial t} = D_x \frac{\partial^2 C_A}{\partial x^2} - \frac{\partial C_A}{\partial \theta_H} + r_A \qquad (6.16b)$$

In Eq. (6.16b) the hydraulic detention $\partial \theta_H$ time has been substituted for the term $\partial x / u$. For convenience, the subscript x will be dropped from the dispersion coefficient ($D_x = D$), but the directional nature of the term should be kept in mind.

Dispersion of Tracers

For the case of a conservative material (e.g., a tracer input), the reaction-rate term $r_A = 0$, and Eq. (6.16b) reduces to Eq. (6.17):

$$\frac{\partial C_A}{\partial t} = D \frac{\partial^2 C_A}{\partial x^2} - \frac{\partial C_A}{\partial \theta_H} \qquad (6.17)$$

Equation (6.17) can be solved for small amounts of dispersion. The solution of Eq. (6.17) for a unit pulse input results in a symmetrical output curve defined by Eq. (6.18) [6.5]:

$$C_T = \frac{1}{2\sqrt{\pi(D/uL)}} \exp\left[-\frac{(1 - t/\theta_H)^2}{4(D/uL)}\right]$$

(6.18)

where

C_T = dimensionless tracer concentration based on a unit pulse input

t = time, s

θ_H = hydraulic detention time, s

L = characteristic length (usually defined as the distance between the input and the point where measurements are being made), m

and other terms are as defined previously. In a plot of concentration versus time, the unit pulse-input is defined as having a width of zero and an area of 1.0. Thus, the initial concentration value of the tracer C_T is infinite and concentration values greater than 1.0 are possible at times greater than zero.

When dispersion is large, the output curve becomes increasingly nonsymmetrical, and the problem becomes sensitive to the boundary conditions. In environmental problems, a wide variety of entrance and exit conditions are encountered, but most can be considered approximately open; that is, the flow characteristics do not change greatly as the boundaries are crossed.

For the open system, the pulse-input solution to Eq. (6.17) is

$$C_T = \frac{1}{2\sqrt{\pi\left(\dfrac{t}{\theta_H}\right)(D/uL)}} \exp\left[-\frac{(1 - t/\theta_H)^2}{4\left(\dfrac{t}{\theta_H}\right)(D/uL)}\right]$$

(6.19)

Defining the extent of dispersion is difficult. For practical purposes, dispersion is small when $D/uL < 0.025$ and large when $D/uL > 0.2$. Evaluation of the coefficient of dispersion D is often done experimentally. Because the systems encountered in water quality management are large, experimental work is difficult and expensive. Thus, values of D encountered in the literature are often very rough estimates. An appropriate value of D for water for large Reynolds numbers is [6.2]

$$D = 10^{-6} N_R^{0.875}$$

(6.20)

where

D = coefficient of dispersion, m^2/s

N_R = Reynolds number

Typical values for N_R found in open channel flow are in the range of from 10^4 to 10^5, and the corresponding range for D is from 0.003 to 0.02 m^2/s.

Dispersion with Reaction

The steady-state solution of Eq. (6.16b) for a first-order reaction ($r_A = -kC_A$) is independent of inlet and outlet conditions. The following solution has been developed for this situation [6.8]:

$$\frac{C_A}{C_{A_i}} = \frac{4a \exp\left(\dfrac{uL}{2D}\right)}{(1+a)^2 \exp\left(\dfrac{auL}{2D}\right) - (1-a)^2 \exp\left(-\dfrac{auL}{2D}\right)} \qquad (6.21)$$

where

$$a = [1 + 4k\theta_H(D/uL)]^{1/2}$$

k = first-order reaction rate constant, s^{-1}

and other terms are as defined previously.

The earlier statement that ideal CFSTRs and PFRs can be used to define an envelope for nonideal reactors is illustrated in Fig. 6.13, which represents a graphical solution of Eq. (6.21). Values of the rate coefficient can be derived from laboratory experiments, and satisfactory values of u and L can be calculated or measured. Accurate estimation of the

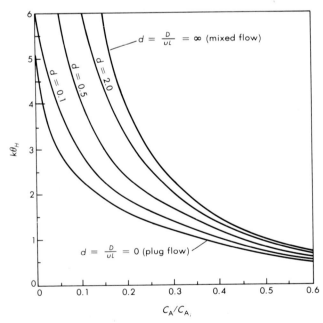

FIGURE 6.13

Graphical solution of the Wehner and Wilhelm equation [Eq. (6.21)].

dispersion coefficient is difficult, and more than one method should be used, if possible.

As shown in Fig. 6.13, the value of $k\theta_H$ must be of the order of 1 or more to obtain significant removals. Biological systems are characterized by slow reaction rates and, for most natural systems, the $k\theta_H$ term is very small. The result is that the dispersion model does not fit river data very well. Tidal systems behave differently, however, and dispersion models work reasonably well in their evaluation.

Reactors in Series

Reactors-in-series models are often used to describe nonideal flow (see Fig. 6.1d). The steady-state response of a cascade of CFSTRs to the introduction of a first-order reactant has been discussed in the previous section. Consideration of the response to a pulse input of tracer is of interest as well. If a "slug" of tracer is added at $t = 0$, the response of the first reactor is given by Eq. (6.22):

$$C_1 = C_0 e^{-t/\theta_H} \tag{6.22}$$

where C_0 is the initial concentration of the tracer in reactor 1, and θ_H is the residence time in each of the n equal volume units. A mass balance on the second reactor results in Eq. (6.23):

$$\frac{dC_2}{dt}\theta_H = C_1 \quad - \quad C_2$$

Accumulation = Inflow − Outflow

$$\frac{dC_2}{dt}\theta_H = C_0 e^{-t/\theta_H} - C_2 \tag{6.23}$$

Because $C_2 = 0$ when $t = 0$, integration of Eq. (6.23) results in Eq. (6.24). The solution procedure for Eq. (6.23) is given in detail in Appendix I.

$$C_2 = C_0 \frac{t}{\theta_H} e^{-t/\theta_H} \tag{6.24}$$

The solution for the ith tank for n tanks in series is a straightforward extension of Eqs. (6.23) and (6.24):

$$C_i = \frac{C_0}{(i-1)!}\left(\frac{t}{\theta_H}\right)^{i-1} e^{-t/\theta_H} \tag{6.25}$$

Tracer-response curves for three CFSTRs in series subjected to a slug input of tracer added to the first reactor are shown in Fig. 6.14. As expected, the response curve for the first CFSTR is the inverse of the curve given in Fig. 6.3.

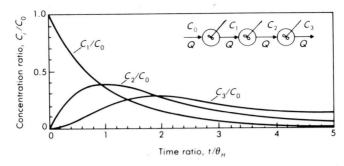

FIGURE 6.14

Tracer response curves for a cascade of three CSFTRs subjected to a slug input of tracer in the first reactor.

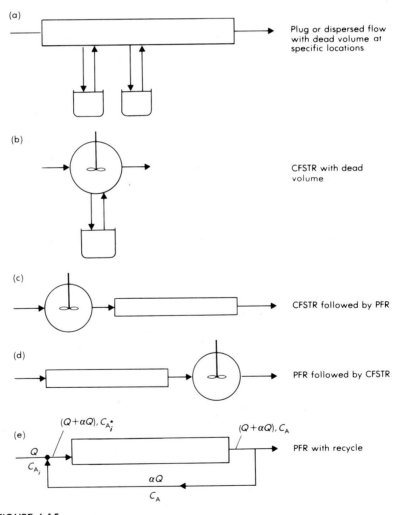

(a) Plug or dispersed flow with dead volume at specific locations

(b) CFSTR with dead volume

(c) CFSTR followed by PFR

(d) PFR followed by CFSTR

(e) PFR with recycle

FIGURE 6.15

Examples of multiparameter systems used to model streams, lakes, estuaries, and treatment processes.

Multiparameter Models

Multiparameter models are also very useful. Most multiparameter models utilize combinations of CFSTRs and PFRs to fit response curves to field or laboratory data. Some examples are shown in Fig. 6.15. Obviously, many other configurations are possible. The dead volumes shown in Figs. 6.15(a) and 6.15(b) are not strictly dead because some interchange is taking place. Interchange may be of any magnitude and quite often is insignificant. In some cases, the interchange is not convective and has properties similar to interphase transfer. An example is the transport of oxygen (or other materials) across density gradients formed by temperature or salinity differences.

If dead volume exists, the primary effect is to decrease the reaction time, and the response curve does not change shape. Thus the CFSTR with dead volume would behave as a CFSTR with a smaller than intended volume.

EXAMPLE 6.5

ANALYSIS OF CFSTR FOR DEAD VOLUME

Results of tracer studies run on a 15-m^3 tank described as a CFSTR by its manufacturer are given below. The flow rate used in the tracer study was 0.05 m^3/s. Using these data, determine whether any dead volume exists in the tank.

t, s	C_T/C_{T_i}
10	0.95
50	0.78
100	0.61
150	0.47
200	0.37
300	0.22
400	0.14

SOLUTION:

1. Plot the theoretical and actual response curves for the CFSTR. The theoretical response of a CFSTR with a volume of 15 m^3 and flow rate of 0.05 m^3/s to a slug tracer input is shown in Fig. 6.16, along with the actual data. Clearly, dead volume exists because the washout rate is faster than would be expected for a CFSTR.

2. Solve for the hydraulic detention time θ_H.

$$\theta_H = \frac{15 \text{ m}^3}{0.05 \text{ m}^3/\text{s}} = 300 \text{ s}$$

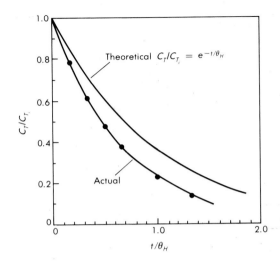

FIGURE 6.16

Actual and theoretical tracer-output response curves for Example 6.5.

3. Develop an expression for the hydraulic detention time θ_H as a function of the ratio of C_{T_i}/C_T.
   From Eq. (6.22),

$$\theta_H = t / \ln \frac{C_{T_i}}{C_T}$$

4. Calculate θ_H for each data point and determine the average detention time.
   The computed results are as follows:

t, s	θ_H, s
10	195
50	201
100	202
200	201
300	198
400	203
$\bar{\theta}_H = 200$ s	

5. Determine the dead volume.
   a. The active volume is 10 m³ (0.05 m³/s × 200 s).
   b. The dead volume is the difference between the actual and the active volume:

$$\text{Dead volume} = 15 \text{ m}^3 - 10 \text{ m}^3$$
$$= 5 \text{ m}^3$$

COMMENT

In considering multiparameter models, it is important to remember that they are strongly affected by flow. Increasing flow could result in a decrease in dead volume because of higher turbulence. A decrease in flow might increase dead volume. Similarly, temperature changes may result in stratification and changes in dead volume.

Plug-Flow Reactor Followed by Stirred-Tank Reactor

A plug-flow reactor followed by a stirred-tank reactor results in a translation in time. If a slug of tracer is injected into the input of the systems shown in Fig. 6.15(c) or 6.15(d), the result is the same, and is shown in Fig. 6.17. The slug moves through the PFR unchanged and enters the CFSTR, where it is washed out according to the expression $C_T/C_{T_i} = e^{-t/\theta_H}$. If the CFSTR is ahead of the PFR, exactly the same result occurs. Dispersion modifies the shape of the effluent curve but not the concept. When reactive materials are being monitored, the systems may differ, depending on the order of reaction, but, conceptually, the time translation provided by the PFR is the important factor.

Recycle

Recycle of effluent to the influent (or some intermediate point) is a method of accounting for backmixing. Although dispersion is a form of backmixing, other mechanisms, such as secondary currents, are often more significant. For reaction orders greater than zero, the PFR system with backmixing, shown in Fig. 6.15(e), will not perform as well as a conventional PFR because the average reactant concentration is lower, and hence the reaction rate will be lower. Letting α be the ratio of the recycled flow to the input flow results in the following steady-state mass

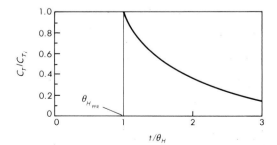

FIGURE 6.17

Response curve for a PFR followed by a CFSTR or a CFSTR followed by a PFR subjected to a slug input of tracer.

balance at point 1 in Fig. 6.15(e):

$$QC_{A_i} + \alpha QC_A = (Q + \alpha Q)C_{A_i}^* \qquad (6.26a)$$

$$\quad \text{In} \qquad \text{Out}$$

$$C_{A_i}^* = \frac{QC_{A_i} + \alpha QC_A}{Q + \alpha Q} \qquad (6.26b)$$

Because the reactor inlet concentration $C_{A_i}^*$ is less than the feed concentration C_{A_i}, the initial reaction rate is decreased by recycling for reaction orders greater than zero. For a first-order reaction the ratio of the effluent concentration C_A to $C_{A_i}^*$ and C_{A_i} is given by the following equations:

$$\frac{C_A}{C_{A_i}^*} = e^{-k\theta_H^*} \qquad (6.27)$$

where

$$\theta_H^* = \theta_H/(1 + \alpha)$$
$$\frac{C_A}{C_{A_i}} = \frac{e^{-k\theta_H/(1+\alpha)}}{1 + \alpha(1 - e^{-k\theta_H/(1+\alpha)})} \qquad (6.28)$$

The steady-state response of PFRs to recycle for a first-order reaction is shown in Fig. 6.18. Recycle rates must be quite high to have a measurable effect, but the effect is clearly negative. In water and wastewater treatment processes, recycle ratios often approach 1, but economic considerations rarely allow larger values.

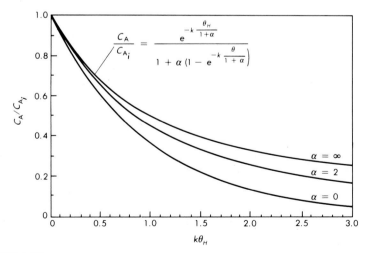

FIGURE 6.18

Response curves for a PFR with recycle for first-order reaction.

6.6

HETEROGENEOUS REACTION SYSTEMS

The reactions described up to this point have been homogeneous; that is, they occur throughout the liquid phase. Many important reactions occur at specific sites, and this fact must be considered in the analysis of the reaction system. Uptake of oxygen by bottom (benthic) organisms in a stream and adsorption of organic material on the surface of activated carbon granules are examples of heterogeneous reaction systems. First-order, second-order, saturation, or other types of models are used. Rates are reported as moles (or mass) converted per unit time per unit area, rather than per unit volume, however.

Transport of reactants to reaction sites plays an important role in controlling or limiting the rate of heterogeneous reactions. Although transport rate also affects homogeneous reaction rates [6.4, 6.7], it is an unusual situation in environmental problems. Common problems such as benthic oxygen uptake and carbon adsorption provide examples of transport rate control.

A schematic diagram of transport of oxygen to benthic organisms is shown in Fig. 6.19. Two rates are not shown—the rate of O_2 transport in the air-to-liquid interface and of O_2 transport within the benthic layer. Each of these rates is finite and can potentially control the overall reaction rate. The driving force in each case is the concentration gradient. Dispersion in the bulk liquid is usually much faster than molecular diffusion through the laminar film or into the benthic layer. Gas-liquid transfer rates are proportional to concentration gradients close to the interface. In most cases, the liquid film (near the interface) controls the oxygen transfer rate, and the most commonly used expression is derived

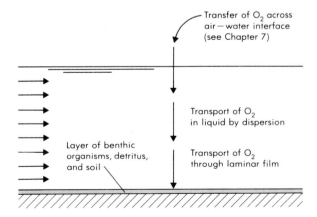

Transfer of O_2 across air—water interface (see Chapter 7)

Transport of O_2 in liquid by dispersion

Layer of benthic organisms, detritus, and soil

Transport of O_2 through laminar film

FIGURE 6.19

Oxygen transport to bottom-dwelling organisms.

from a two-film model [6.6], which is discussed further in Chapter 7.

$$N_{O_2} = K_L(C_S - C_{O_2})$$ (6.29)

where

N_{O_2} = oxygen transfer flux, g/m² · s

K_L = overall oxygen mass transfer coefficient, m/s

C_S = saturation concentration of oxygen in bulk liquid, g/m³

C_{O_2} = concentration of oxygen in bulk liquid, g/m³

Two general possibilities exist for heterogeneous reaction systems: reaction rate control and transport rate control [6.3]. If the reaction rate is slower than any of the transport rates, reactant concentrations will increase and concentration gradients will decrease. If a transport step is limiting, the reverse will occur. In shallow streams, oxygen concentrations are normally virtually constant because of turbulence. If the reaction rate is very high in the benthic layer, the O_2 concentration will drop to a very low value throughout the depth, and eventually the reaction rate r_{O_2} will equal the transfer rate N_{O_2}. When excess oxygen is available, gradients will decrease, and eventually the two rates will again be equal, but, in this case, the limiting rate will be the reaction. The two situations are illustrated in Fig. 6.20.

Adsorption of organic material on activated carbon is very often transport-rate-limited. The carbon granules are highly porous, and surface areas are of the order of 1000 m²/g. To generate such a large surface-to-mass ratio, the pores must be very small, and molecular diffusion is the principal mechanism of internal transport. Molecular diffusion coeffi-

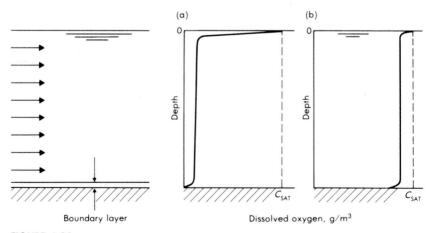

FIGURE 6.20

Examples of oxygen-rate-controlling situations in streams: (a) transport rate limitations and (b) reaction rate limitations.

cients are usually 10^{-8} to 10^{-9} m^2/s, whereas dispersion coefficients are normally 10^{-2} to 10^{-3} m^2/s. It is not surprising that internal transport is often the controlling factor in adsorption processes.

Inhibition

Inhibition of heterogeneous reactions is a common phenomenon. Several inhibition mechanisms exist and are illustrated in Fig. 6.21. In some cases, competition for the reactive site occurs. A competitive reaction, possibly undesirable, may occur, or the attachment to the reactive site may be irreversible, blocking further reactions at the sight. A noncompetitive inhibition of the reaction may occur because of the attraction of an adjacent site for a molecule that blocks the reactive site. Finally, the reactive-site structure may be changed allosterically as the result of reactions occurring some distances away. Many heterogeneous reactions

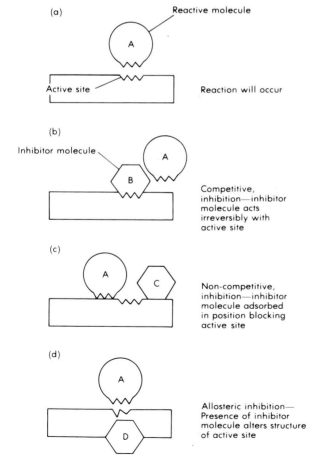

(a)

Reactive molecule

Active site

Reaction will occur

(b)

Inhibitor molecule

Competitive, inhibition—inhibitor molecule acts irreversibly with active site

(c)

Non-competitive, inhibition—inhibitor molecule adsorbed in position blocking active site

(d)

Allosteric inhibition—
Presence of inhibitor molecule alters structure of active site

FIGURE 6.21

Definition sketch showing the inhibition of heterogeneous reactions.

are catalytic, with the reaction site being on the catalyst. When irreversible binding of the inhibitor molecule occurs, the catalyst is poisoned and the reaction system fails.

Pseudohomogeneous Reactions

Pseudohomogeneous reactions are heterogeneous reactions that can be analyzed as homogeneous systems. Beds of activated carbon adsorbent are often treated as homogeneous systems with peculiar rate characteristics, because the procedure is simpler than doing a more detailed analysis. Classical analysis of benthic uptake of oxygen has treated the process as a homogeneous reaction.

KEY IDEAS, CONCEPTS, AND ISSUES

- Mathematical models developed to predict the hydraulic response of reactors used to carry out chemical and biological reactions are used to study the response of natural systems to external and internal inputs.
- Batch, complete-mix, plug-flow, cascade complete-mix, and packed-bed reactor models are used to study the response of natural systems.
- The complete-mix and plug-flow models are the ones most commonly used to study the response of natural systems.
- Combinations of complete-mix and plug-flow models have been used to study natural systems.
- The cascade complete-mix reactor model is often used to study the nonideal low regimes that exist between those described by the complete-mix and the plug-flow models.
- Plug-flow models in which the coefficient of dispersion is considered have been used extensively to model the nonideal behavior of plug-flow reactors.
- Homogeneous reactions occur simultaneously throughout the liquid phase in flowing systems. Heterogeneous reactions occur at specific sites and are usually associated with a solid surface.

DISCUSSION TOPICS AND PROBLEMS

6.1. A CFSTR is used to produce a product B from the reactant A. The stoichiometric and rate equations for the reaction are

$$2A \rightarrow B$$
$$r_A = -k[A]^2$$

For a steady-state situation write the mass-balance expressions and solve for [B] if

$$[A_i] = 5 \text{ mol/m}^3$$
$$[B_i] = 0$$
$$Q = 0.01 \text{ m}^3/\text{s}$$
$$k = 0.001 \text{ m}^3/\text{mol} \cdot \text{s}$$
$$V = 10 \text{ m}^3$$

6.2. A set of reactions involving an intermediate and two end products is to be carried out in a CFSTR under steady-state conditions. Determine the concentration of each species as a function of θ_H. Assume that $C_{A_i} = C_{A_i}$ and $C_{B_i} = C_{C_i} = C_{D_i} = 0$.

$$r_A = -k_1 C_A$$
$$r_B = 2k_1 C_A - k_2 C_B - k_3 C_B \qquad A \rightarrow B \!\!\begin{array}{c} \nearrow C \\ \searrow D \end{array}$$
$$r_C = k_2 C_B$$
$$r_D = k_3 C_B$$

6.3. In dealing with CFSTRs it is often necessary to identify approximate steady-state conditions. For example, any time a reaction is of order one or greater, steady-state conditions cannot be attained in real time. Tracer response is similar. For a step input of tracer, determine the time necessary to achieve 95 percent of the steady-state response for (a) a single CFSTR and (b) two identical CFSTRs in series.

6.4. Given a reaction $A \rightarrow B$ with $r_A = -kC_A$, determine the time t/θ_H to achieve 95 percent of the steady-state value for a step increase in influent concentration from 0 to C_{A_i} for (a) a single CFSTR and (b) two CFSTRs in series. Assume the value of $k\theta_H$ is equal to 1.0.

6.5. Estimate the reduction in bacteria during the passage of wastewater that initially contains 10^6 organisms per milliliter through three stabilization ponds that are arranged in series. The volumes of the three ponds are 10,000, 20,000, and 6000 m^3, respectively. The flow rate is 1000 m^3/d. Assume that steady-state conditions apply, that the ponds are mixed completely because of wind action, that first-order decay kinetics apply, and that the value of the reaction rate coefficient k is 1.0 d^{-1}. Would the efficiency improve if all of the ponds were equal in volume (12,000 m^3/pond)?

6.6. The concentration of ultimate BOD in a river entering the first of two lakes connected in series is equal to 20 g/m^3. If the first-order BOD rate coefficient k is 0.35 d^{-1} and each lake behaves as a CFSTR, determine the BOD_u concentration in the outlet of each lake. The steady-state river flow rate is 4000 m^3/d, and the lake volumes are 20,000 and 12,000 m^3, respectively.

6.7. A BOD removal process is carried out in three identical CFSTRs in series. The mixing mechanism transfers oxygen at a rate dependent on the oxygen deficit:

$$r_{O_2\,trans} = k_2(C_s - C_{O_2})$$

where

$r_{O_2\,trans}$ = rate of O_2 transfer, $g/m^3 \cdot d$

$\quad k_2$ = oxygen transfer coefficient, $80\ d^{-1}$

$\quad C_s$ = dissolved-oxygen saturation concentration, $8.00\ g/m^3$

The hydraulic residence time of each tank is 12 hr, the influent BOD_u to the first tank in the series is $150\ g/m^3$, and the BOD removal rate is first-order:

$$r_{BOD} = -k(BOD_u)$$

$$k = 10\ d^{-1}$$

Assuming there is no oxygen in the influent, determine the dissolved-oxygen concentration in the second tank.

6.8. When a slug of tracer is injected into a cascade of CFSTRs, a maximum value of C_T occurs at $t > 0$ for all reactors except the first. Determine the time when the peak occurs in the nth reactor in terms of θ_H.

6.9. A reaction $A + B \rightarrow C$ behaves as a first-order process with respect to A:

$$r_A = -kC_A$$

Develop a steady-state expression for the effluent concentration of B in a cascade of CFSTRs (see Fig. 6.1d) if A is nonvolatile and B is a slightly soluble gas (e.g., O_2). Assume that mass transfer of B occurs in each unit of the cascade and that all units have equal volumes.

6.10. A reactor that behaves as an ideal PFR is to be used to carry out a BOD removal reaction. The removal rate in the system is given approximately by

$$r_{BOD_u} = -\frac{k\ BOD_u}{K + BOD_u}$$

where

$$k = 0.12\ g/m^3 \cdot s$$

$$K = 30\ g/m^3$$

$$BOD_{u_i} = 150\ g/m^3$$

For a flow of $0.5\ m^3/s$ determine the reactor volume necessary to produce an effluent having $20\ g/m^3\ BOD_u$.

6.11. Explain briefly why PFR systems are more efficient, that is, require less volume, than CFSTRs in meeting specific effluent requirements.

6.12. Compare, under steady-state conditions, the volume required for (a) a CFSTR, (b) a six-CFSTR cascade, and (c) a PFR to remove 98 percent of the influent reactant A if $r_A = -kC_A$.

6.13. A reaction $A \rightarrow B$ is to be carried out in a system consisting of a PFR and a CFSTR operated at steady state. Compare the results for the two reaction expressions below for (a) PFR $\rightarrow$ CFSTR and (b) CFSTR $\rightarrow$ PFR.

System characteristics:

$$V_{PFR} = 100 \text{ m}^3$$
$$V_{CFSTR} = 1000 \text{ m}^3$$
$$Q = 0.05 \text{ m}^3/\text{s}$$
$$C_{A_i} = 150 \text{ g/m}^3$$

First-order reaction:

$$r_A = -10^{-5} C_A \text{ g/m}^3\text{-s}$$

Second-order reaction:

$$r_A = -10^{-5} C_A^2 \text{ g/m}^3/\text{s}$$

6.14. Determine the effect of recycle on the performance of a CFSTR for any type of reaction.

6.15. A reaction $A \rightarrow B$ is to be carried out in the CFSTR with recycle (see Fig. 6.15e). Determine the effect of recycle ratio α on the steady-state response. What effect does reaction order have on the response?

6.16. A reaction $A \rightarrow B$ is known to be first order with respect to A and is carried out in an ideal PFR with recycle (see Fig. 6.15e). Determine the effect of the recycle ratio α on effluent concentration of A.

6.17. A 3-km reach of stream has a nearly uniform cross section, although it passes through a number of curves. Average width is 20 m and

t, s	C_T, g/m^3	t, s	C_T, g/m^3
600	0.26	2100	1.94
800	0.55	2200	1.92
1000	0.90	2300	1.88
1200	1.25	2500	1.76
1400	1.56	2700	1.60
1600	1.78	2900	1.43
1800	1.91	3100	1.25
1900	1.94	3300	1.08
2000	1.95	3500	0.91

average depth is 1.5 m. A slug of rhodomine WT tracer is added at the top of the reach at a time when the flow rate is 30 m^3/s. The slug of tracer is added in a manner that provides rapid lateral mixing. The measured concentration 10 s after addition is 10 g/m^3. Use the data taken at the end of the reach to develop a model for flow in the river.

6.18. A river is split by an island, with part of the flow remaining in a narrow channel and the remainder flowing into a bay before returning to the main channel (see Fig. 6.22). Based on dye studies it was found that the bay behaves as a CFSTR of volume 3.14×10^6 m^3 and the channel behaves as a PFR of length 2500 m, depth 2 m, and width 30 m. The division of flow between the two channels is summarized for the period of record below.

Q, m^3/s	Q_1, m^3/s	Q_2, m^3/s
16	6	10
20	8	12
24	10	14
28	12	16
32	13	19
36	14	22
40	15	25

a. If dye is added above the junction to give an initial concentration of 10 g/m^3, determine the ideal response curve below the point the streams join. Use $Q = 28$ m^3/s and assume the 10 g/m^3 concentration extends over 280 m^3.

b. If a wastewater discharge of 4 m^3/s and 100 g/m^3 BOD$_u$ is added just above the upstream junction, determine the steady-state BOD$_u$ concentration at the lower junction. Assume a first-order removal

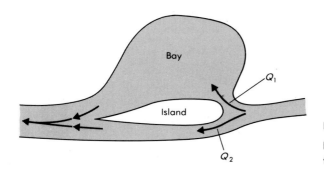

FIGURE 6.22

Definition sketch of the bay system for Problem 6.18.

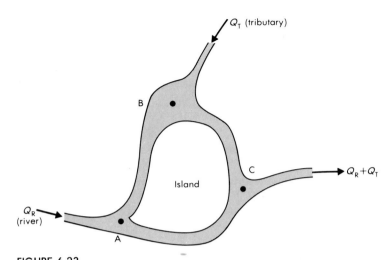

FIGURE 6.23

Definition sketch of the river system for Problem 6.19.

rate with $k = 0.2$ d^{-1} at 20°C, $T = 25$°C in both sections, and $Q = 28$ m^3/s.

c. Determine the oxygen concentration in the bay and at the lower junction for the wastewater discharge given above and a $k_2 = 0.5$ d^{-1} at 20°C in the PFR section and a $k_2 = 0.4$ d^{-1} at 20°C in the CFSTR section. Assume an upstream dissolved-oxygen concentration of 7.5 g/m^3 and a wastewater dissolved-oxygen concentration of 1.0 g/m^3. Use temperature given in part (b).

6.19. The river and tributary shown in Fig. 6.23 have flow rates of 4.0 and 0.5 m^3/s, respectively. Under the conditions given, the flow split at point A is 0.7 to Q_{AB} and 0.3 to Q_{AC}. Tracer studies have been performed, and the flow from A to B to C can be characterized as a PFR followed by CFSTR followed by a PFR, while the flow from A to C behaves as a single PFR. Volumes of these reactors are given below.

REACTOR		VOLUME, m^3
PFR	AB	15,000
	BC	10,000
	AC	30,000
CFSTR	B	135,000

For steady-state conditions, determine the concentration of reactant A at point C for the following conditions:

$$r_A = -kC_A$$
$$k = 0.5 \ d^{-1}$$
$$C_A = 50 \ g/m^3 \text{ (in tributary)}$$
$$C_A = 100 \ g/m^3 \text{ (in river)}$$

6.20. A reaction vessel with a volume of 1296 m³ and a flow rate of 0.1 m³/s is characterized hydraulically with tracers. At time = 0, 1000 g of dye is added to the influent. For the response data given below, determine the hydraulic efficiency. The hydraulic efficiency is defined as the ratio of the center of gravity (with respect to the x-axis) of the actual effluent dye distribution to the theoretical detention time.

TIME, d	DYE CONC, g/m³
0.04	Trace
0.06	1.00
0.10	0.80
0.15	0.55
0.20	0.30
0.26	0.00

REFERENCES

6.1. Bird, R. B., T. Stewart, and E. N. Lightfoot, (1962), *Transport Phenomena*, John Wiley and Sons, New York.

6.2. Davies, J. T., (1972), *Turbulence Phenomena*, Academic Press, New York.

6.3. Denbigh, K. G., and J. C. R. Turner, (1984), *Chemical Reactor Theory*, 3rd ed., Cambridge University Press, New York.

6.4. Kehrberger, G. J., J. D. Norman, E. D. Schroeder, and A. W. Busch, (1964), "BOD Progression in Soluble Substrates, VIII, Temperature Effects," *Proc. 19th Industrial Waste Conference, Purdue University, Lafayette, Ind.*

6.5. Levenspiel O., (1972), *Chemical Reaction Engineering*, 2d ed., John Wiley and Sons, New York.

6.6. Schroeder, E. D., (1977), *Water and Wastewater Treatment*, Mc-Graw-Hill Book Company, New York.

6.7. Swilley, E. L., J. O. Bryant, and A. W. Busch, (1964), "The Signifi-cance of Transport Phenomena in Wastewater Treatment Processes," *Proc. 19th Industrial Waste Conference*, Purdue University, Lafayette, Ind.

6.8. Wehner, J. F., and R. F. Wilhelm, (1958), "Boundary Conditions of Flow Reactor," *Chemical Engineering Science*, vol. 6, no. 1, p. 89.

Modeling Water Quality in the Environment

In managing water quality, quantitative predictive models are necessary because human activities and natural environmental changes (e.g., the eruption of Mount St. Helens) are continually perturbing the nation's streams, lakes, and estuaries. Effects of these perturbations must be analyzed to provide mitigating measures or to determine if such an activity is suitable for a given site. In many cases, the approach is to determine what procedures will be most effective in solving a long-term water quality problem, rather than protecting a pristine water resource.

Most models used in environmental engineering are relatively simple in concept and are used to predict extreme conditions—that is, those events occurring at low flow or maximum flow, high temperature or low temperature. Such models give "worst-case" results for the range of conditions that can reasonably be expected. Fortunately, worst-case models are conservative, and the mitigating measures developed based on these models usually result in the maintenance of a high level of water quality.

Important factors governing the movement of contaminants in the air-water-soil environment are considered in Chapter 7. These factors are required for the construction of the mathematical models used to predict water quality changes in rivers, estuaries, lakes, and groundwater basins. Each of these physical systems is dominated by a different set of characteristics and constraints. Typically, oxygen resources are considered in river and estuary models. Lake water quality is affected significantly by seasonal temperature and wind. Groundwater quality is related directly to adsorption and dissolution reactions with the minerals that make up the soil. River and estuary models are considered in Chapter 8. Lake and groundwater models are examined in Chapters 9 and 10, respectively.

7

Movement of Contaminants
in the Environment

In Part I, the physical, chemical, and biological impurities found in water
were identified, and the methods used to characterize them were discussed. As noted in Chapter 4, the distribution of many of the contaminants found in water is global in scale. The question is: Did this global
presence occur because the use of these contaminants is global or because
these contaminants are able to move through the environment? It is the
purpose of this chapter to introduce the reader to the most important
physical, chemical, and biological phenomena responsible for the movement and fate of contaminants in the environment. The principles
and concepts introduced in this chapter are applied in Chapters 8, 9,
and 10, where water quality in rivers, lakes, and groundwater aquifers
is modeled.

7.1
ENVIRONMENTAL INTERACTIONS

The biosphere can, for purposes of analysis, be divided into three major
elements: water, soil, and atmosphere. The interrelationships that occur
among these elements and humans are illustrated schematically in
Fig. 7.1. In assessing the movement of contaminants in the biosphere
it is necessary to consider (1) transport within each major element,
(2) transport to and from interfaces, and (3) transport across the interfaces of the major elements. It is also significant to note that human
inputs can occur directly into each of the three elements. In what follows,
the important phenomena controlling the movement of contaminants are
introduced.

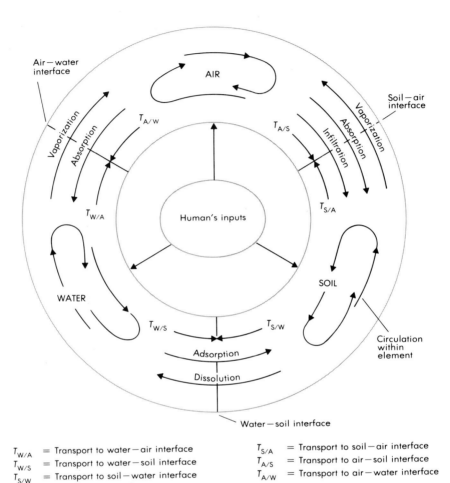

$T_{W/A}$ = Transport to water−air interface	$T_{S/A}$ = Transport to soil−air interface
$T_{W/S}$ = Transport to water−soil interface	$T_{A/S}$ = Transport to air−soil interface
$T_{S/W}$ = Transport to soil−water interface	$T_{A/W}$ = Transport to air−water interface

FIGURE 7.1

Definition sketch for the movement of contaminants in the environment. Plant and animal inputs are omitted for clarity.

7.2
LARGE-SCALE TRANSPORT OF CONTAMINANTS

The large-scale movement of contaminants in the biosphere is accomplished by air and water transport. Because this book deals primarily with water quality, the following discussion is focused on those factors affecting the movement of contaminants in surface water and groundwater. It should be noted, however, that the factors governing the movement of contaminants in the atmosphere are essentially the same.

When contaminants are released to the aquatic environment, they move with the water in streams and rivers and in groundwater aquifers. The mass balance for a contaminant in a flowing system, as given by Eq. (5.40), is

$$
\begin{array}{c}
\text{Rate of} \\
\text{accumulation of} \\
\text{mass within the} \\
\text{system boundary}
\end{array}
=
\begin{array}{c}
\text{Rate of flow} \\
\text{of mass} \\
\text{into the} \\
\text{system boundary}
\end{array}
-
\begin{array}{c}
\text{Rate of flow} \\
\text{of mass} \\
\text{out of the} \\
\text{system boundary}
\end{array}
+
\begin{array}{c}
\text{Rate of} \\
\text{generation} \\
\text{within the} \\
\text{system boundary}
\end{array}
\quad (5.40)
$$

Assuming the contaminant is nonreactive, the principal mechanisms involved in the transport of mass into and out of the system boundary are advection and hydrodynamic dispersion.

Advection

In mass transport by advection, contaminants are transported by the mean velocity of the water as it flows in an open or closed channel or through a porous medium. Mass transport due to advection can be defined as follows:

$$\text{Transport by advection} = \alpha \bar{v}_x A_x C \qquad (7.1)$$

where

$\alpha =$ porosity of medium (1.0 for water system)

$\bar{v}_x =$ average velocity in x direction, m/s

$A_x =$ cross-sectional area in x direction, m^2

$C =$ mass concentration of contaminant, g/m^3

Hydrodynamic Dispersion

The transport of contaminants by hydrodynamic dispersion (see Chapter 6) can be described for the one-dimensional case as

$$\text{Transport by dispersion} = -\alpha D_x A_x \frac{\partial C}{\partial x} \qquad (7.2)$$

where

$D_x =$ coefficient of hydrodynamic dispersion in x direction, m^2/s

$\dfrac{\partial C}{\partial x} =$ concentration gradient in x direction, $\text{g/m}^3 \cdot \text{m}$

The coefficient of hydrodynamic dispersion D_x can be expressed as the sum of two terms:

$$D_x = \bar{v}_x k_x^n + D_m \qquad (7.3)$$

where

$\bar{v}_x$ = mean velocity in x direction, m/s

k_x = characteristic property of the physical system (dispersivity), m

n = empirical constant, usually equal to 1

D_m = coefficient of molecular diffusion, m²/s

If the mean velocity is very low, the hydrodynamic dispersion is essentially equal to the molecular diffusion. Equation (7.3) is a more precise definition of the coefficient of dispersion first introduced in Eq. (6.14). However, circumstances rarely occur where the turbulent dispersion component ($\bar{v}_x k_x^n$) and the molecular diffusivity are the same order of magnitude. Because the turbulent component of dispersion is much larger than the molecular diffusivity, the latter term is rarely included in models involving hydrodynamic dispersion.

Materials Balance

Referring to Fig. 7.2, the materials balance for a nonreactive contaminant transported by advection and hydrodynamic dispersion is as follows:

$$\alpha \frac{\partial C_A}{\partial t} A_x \Delta x = \left(\alpha \bar{v}_x A_x C_A - \alpha D_x A_x \frac{\partial C_A}{\partial x} \right)\Big|_x$$

$$\text{Accumulation} = \qquad\qquad \text{Inflow}$$

$$- \left(\alpha \bar{v}_x A_x C_A - \alpha D_x A_x \frac{\partial C_A}{\partial x} \right)\Big|_{x+\Delta x}$$

$$- \qquad\qquad \text{Outflow} \qquad\qquad\qquad\qquad (7.4)$$

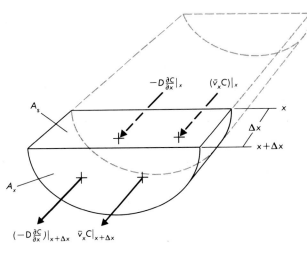

FIGURE 7.2

Definition sketch for the one-dimensional analysis of the movement of a conservative contaminant by advection and hydrodynamic dispersion in a steam.

Simplifying and taking the limit as Δx approaches zero yields

$$\frac{\partial C_A}{\partial t} = D_x \frac{\partial^2 C_A}{\partial x^2} - \bar{v}_x \frac{\partial C_A}{\partial x} \qquad (7.5)$$

Equation (7.5) is the basic expression used to describe the change in concentration of a conservative contaminant brought about by advection and hydrodynamic dispersion. Specific solutions to Eq. (7.5) will depend on the boundary conditions. Numerous solutions for Eq. (7.5) for a variety of boundary conditions may be found in Refs. [7.4] and [7.11]. The application of Eq. (7.5) is illustrated in Chapters 8 and 10, where the movement of contaminants in river and groundwater systems is modeled. The coefficient of dispersion is considered further in Refs. [7.1] and [7.4].

7.3
TRANSPORT TO AND FROM INTERFACES

Before contaminants can be transported across the interface between elements of the biosphere, they must be transported to the interface between elements. Although a detailed discussion of the various transport mechanisms is beyond the scope of this text, it is important to identify some of them. The importance of these mechanisms in bringing about changes in water quality will vary, depending on the specific contaminant.

Transport Phenomena

In the water environment, contaminants are brought to the water-air interface primarily by (1) hydraulically induced turbulence and circulation, (2) thermal- and wind-induced circulation, and (3) molecular diffusion. Transport to the water-soil interface is accomplished by (1) hydraulically, thermally, and wind-induced circulation; (2) advection and turbulent dispersion; (3) sedimentation; and (4) percolation.

In the air environment, particles such as spray droplets, dusts, aerosols, wind-blown dusts, and other contaminants are transported to the air-water and air-soil interfaces by dry deposition in the absence of rain and snow, precipitation, snowfall, and wind erosion and deposition. When contaminants are deposited on the water and ground surface they typically are not resuspended.

In the soil environment, contaminants are transported to the soil-air interface by a variety of factors, including water movement caused by advection, hydrodynamic dispersion and evaporation; capillary phenomenon; and molecular diffusion. Transport to the soil-water interface is typically by diffusion within the soil structure.

Other Transport Processes

In addition to the aforementioned transport phenomena, other processes that are important in the transport of contaminants include evapotranspiration from plants and a variety of municipal and industrial operations that bypass the various interfaces and discharge contaminants directly into each of the elements. Examples of the latter include SO_2 released from high stacks, contaminants in treated municipal and industrial wastewaters discharged to streams and groundwater aquifers, and excess pesticides and herbicides from overhead spraying. These discharges are representative of direct human input into the the three elements of the biosphere, as illustrated in Fig. 7.1.

7.4
AIR-WATER INTERFACE TRANSFERS

The transfer of material across the air-water interface occurs primarily by evaporation and absorption. Before considering the subjects of evaporation and absorption it will be helpful to review some basic physical-chemical relationships.

Useful Physical-Chemical Relationships

In considering the transfer of contaminants across the air-water interface, it will be useful to review the concepts of the distribution coefficient and vapor pressure.

The Distribution Coefficient

When a substance that is soluble in two phases is added to a system of two immiscible phases, the substance will be distributed in each in fixed proportions at a given temperature, independent of the quantity of the substance. The above is a statement of the distribution law. The ratio of the concentrations in each phase is defined as the *distribution coefficient* (or the *partition coefficient*) [7.17, 7.20]. For dilute solutions, the preceding is also a working statement of Henry's law [see Eq. (2.83)]. An identical law holds for the distribution of a solute between two immiscible liquids. The coefficient is constant only when the given solute dissolves in both solvents in the same form and no association or dissociation takes place. In practice, the distribution ratio is seldom strictly a constant. Stated mathematically, the distribution coefficient is

$$\frac{C_{X/A}}{C_{X/B}} = K_D \qquad (7.6)$$

where

$C_{X/A}$ = concentration of solute X in solvent A, g/m^3

$C_{X/B}$ = concentration of solute X in solvent B, g/m^3

K_D = distribution or partition coefficient

If the solute remains unchanged during its distribution between two solvents, the distribution coefficient can be used to calculate the efficiency of an extraction process in which a given solvent is used to extract a solute from another solvent (see Example 7.1). For the purposes of comparison and analysis, the octanol : water solvent system is used to characterize a variety of organic substances (see Section 10.4). Values of the octanol : water distribution coefficient K_{OW} are readily available in the literature [7.17]; selected values are presented in Table 10.5 in Chapter 10.

EXAMPLE 7.1

EXTRACTION OF A SOLUTE

Develop an expression that can be used to predict the performance of a single or multiple extraction process. Assume water containing x_0 g of an extractable solute is to be treated with an organic solvent. The distribution coefficient is defined as $K = C_S/C_W$ where C_S is the concentration of the solute in the organic solvent and C_W the concentration of the solute in the water.

SOLUTION:

1. Develop an expression to determine the amount of solvent remaining after one extraction.
   a. Assume the starting volumes of solvent and water are A and B, respectively.
   b. If x_1 is the amount of material remaining in the water after the first extraction, then

   $$C_W = \frac{x_1}{B}$$

   $$C_S = \frac{x_0 - x_1}{A}$$

   c. Substitute these values in the expression for the distribution coefficient and solve for x_1, the amount remaining:

   $$\frac{B(x_0 - x_1)}{Ax_1} = K$$

   $$x_1 = x_0 \left(\frac{B}{B + KA} \right)$$

2. Develop an expression for multiple extractions. The amount of the solute remaining of the second extraction is

$$x_2 = x_1 \left(\frac{B}{B + KA} \right)$$

Substituting for x_1 yields

$$x_2 = x_0 \left(\frac{B}{B + KA} \right)^2$$

For n extractions the amount remaining is

$$x_n = x_0 \left(\frac{B}{B + KA} \right)^n$$

3. The amount of material extracted is

$$x_{ext} = x_0 \left[1 - \left(\frac{B}{B + KA} \right)^n \right]$$

COMMENT

The distribution law cannot be applied with any measure of accuracy to solutions with constituent concentrations higher than about 1 N because such solutions no longer behave as ideal solutions.

Vapor Pressure

All solids and liquids possess a vapor pressure, which is a measure of the tendency of the substance to evaporate. Conceptually, the vapor pressure can be considered to be the solubility of the material in air at a given temperature. Within limits, the vapor pressure increases as the temperature increases. Thus every solid or liquid has a tendency to evaporate until the pressure of the vapor above them is equal to the equilibrium value at a given temperature [7.20, 7.22].

Starting with the Clapeyron-Clausius equation [7.7] and assuming the latent heat of vaporization is constant, the expression for the change in vapor pressure with temperature is

$$\ln p_v = C - \frac{\Delta H_v}{R} \frac{1}{T} \tag{7.7}$$

where

p_v = vapor pressure, atm

C = constant

ΔH_v = molecular latent heat of vaporization, J/mol

R = universal gas constant, J/mol · K

T = temperature, K

Equation (7.7) is generally written as

$$\ln p_v = A + \frac{B}{T} \tag{7.8}$$

where A and B are constants.

An alternative form of Eq. (7.8), known as the Antione equation [7.7, 7.17], is more commonly used:

$$\ln p_v = A - \frac{B}{T + C} \tag{7.9}$$

where

$A, B, C = $ constants

$T = $ boiling point, K

Values for A, B, and C for a wide variety of compounds may be found in Refs. [7.7], [7.16], and [7.18].

Of more specific interest is the vapor density (mass/unit volume) which represents the concentration of a given constituent in the gaseous phase [7.20]. The vapor density d_v is obtained from the universal gas law:

$$p_v V = \frac{m}{M} RT \tag{7.10}$$

$$d_v = \frac{m}{V} = \frac{p_v M}{RT} \tag{7.11}$$

where

$d_v = $ vapor density, g/L or g/m^3

$p_v = $ equilibrium vapor pressure, atm

$V = $ volume, L

$m = $ mass of substance, g

$M = $ gram molecular mass of substance, g/mol

$R = $ universal gas constant, 0.082 L · atm/mol · K

$T = $ temperature, K

The application of Eq. (7.11) is illustrated in Example 7.2.

EXAMPLE 7.2

DETERMINATION OF MASS DENSITY

Determine the vapor mass density at equilibrium of the agricultural pesticide Lindane at 30°C. The vapor pressure and molecular mass are 1.24×10^{-8} atm and 291 g/mol, respectively.

SOLUTION:
Compute the vapor density in grams per cubic meter of Lindane using Eq. (7.11).

$$d_v = \frac{p_v M}{RT}$$
$$= \frac{(1.24 \times 10^{-8} \text{ atm}) \, 291 \text{ g/mole}}{(0.082 \text{ L} \cdot \text{atm/mol} \cdot \text{K}) \, 303 \text{ K}}$$
$$= 1.45 \times 10^{-7} \text{ g/L}$$
$$= 1.45 \times 10^{-4} \text{ g/m}^3$$

COMMENT
By comparison, the vapor density for water at 30°C is about 49 g/m^3.

Volatilization

Volatilization is the process whereby liquids and solids vaporize and escape to the atmosphere. Most formulations for the vaporization process are based on (1) theoretical considerations of energy exchange, (2) empirical correlations (Fig. 7.3), and (3) various combinations of the two. The general subject of water evaporation is dealt with extensively in most texts dealing with chemical engineering [7.1, 7.22] and hydrology [7.2, 7.13].

In considering the vaporization of a contaminant from the surface of a liquid or solid, the factors that must be considered are illustrated in Fig. 7.4. As shown, vaporization is a three-step process involving (1) escape from the interface, (2) diffusion through the surface boundary

FIGURE 7.3

Class A evaporation pan equipped with hook gage, anenometer, and water temperature probe. Pan evaporation rates, once correlated, are used to predict evaporation rates from open water surfaces.

Wind blowing across surface

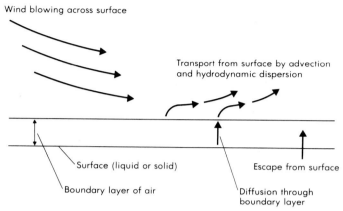

Transport from surface by advection
and hydrodynamic dispersion

Surface (liquid or solid)   Escape from surface

Boundary layer of air   Diffusion through
                                             boundary layer

FIGURE 7.4

Definition sketch for the vaporization of a pure compound.

Source: Adapted from Ref. [7.20].

layer, and (3) dispersion [7.20]. The escape from the surface depends
primarily on the vapor pressure of the contaminant at the given temperature. After the contaminant has escaped the surface, it must diffuse
outward in the quiescent (stagnant) boundary layer that is normally
present. The contaminant will be transported away from the stagnant
layer by advection and turbulent dispersion.

A simplified model that can be used to describe the escape due to
vaporization of a contaminant from a surface can be derived by considering the unit volume of a water body such as shown in Fig. 7.5. Based on
experimental observations, it has been found that the rate of mass
transfer of a contaminant is approximately proportional to the difference
between the saturation (equilibrium) and existing concentration of the
contaminant solution. Stated mathematically the rate of vaporization is

$$r_C = K(C - C_s) \qquad (7.12)$$

where

r_C = rate of mass transfer (vaporization), $g/m^2 \cdot hr$

K = coefficient of mass transfer, m/hr

C = concentration of contaminant, g/m^3

C_s = saturation concentration of contaminant, g/m^3

It should be noted that C_s is a function of the atmospheric partial
pressure. If the compound of interest is absent from the air, $C_s = 0$. Also,
volatilization only occurs if $C > C_s$. If $C < C_s$, absorption occurs.

If it is assumed that the rate term given by Eq. (7.12) is the only term
involved in the transfer of the contaminant, then the resulting materials

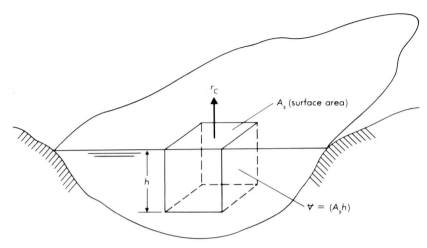

FIGURE 7.5

Definition sketch for the analysis of the vaporization of a contaminant from a water surface.

Source: Adapted from Ref. [7.19].

balance for the volume element shown in Fig. 7.5 is

$$\frac{dC}{dt}(A_s h) \;=\; 0 \;-\; 0 \;-\; r_C A_s \tag{7.13}$$

Accumulation = Inflow − Outflow − Outflow due
                                                 to evaporation

where

A_s = surface area, m^2

h = depth, m

and other terms are as defined previously.

Substituting for r_C in Eq. (7.13) and simplifying yields

$$\frac{dC}{dt} = -\frac{K}{h}(C - C_s) \tag{7.14}$$

The integrated form of Eq. (7.14), given previously as Eq. (2.85), is

$$\frac{C_t - C_s}{C_0 - C_s} = e^{-(K/h)t} \tag{7.15}$$

where

C_0 = concentration of contaminant at time $t = 0$, g/m^3

C_t = concentration of contaminant at time t, g/m^3

and other terms are as defined previously.

In the literature, the coefficient K/h in Eq. (7.15) is often reported as $K_L a$, where K_L is the liquid-film mass-transfer coefficient and a is equal to the ratio of the interface area A to the volume V. The volumetric mass-transfer coefficient $K_L a$ is often used where it is difficult to measure the interfacial contact area between the gas and the liquid. Equation (7.15) can be used to estimate the time required for a contaminant concentration to drop to a given value. Representative vaporization parameters for various chemical compounds are given in Table 7.1. The application of Eq. (7.15) is illustrated in Example 7.3.

TABLE 7.1

Evaporation Parameters for Various Compounds at 25°C

COMPOUND	MOLECULAR MASS, g/mol	SOLUBILITY IN WATER, g/m³	VAPOR PRESSURE, mmHg	K, m/hr
Alkanes				
n-octane ($C_8 H_{18}$)	114.0	0.66	14.1	0.124
2,2,4-trimethyl pentane ($C_8 H_{18}$)	114.0	2.44	49.3	0.124
Aromatics				
Benzene ($C_6 H_6$)	78.0	1780	95.2	0.144
Toluene ($C_7 H_8$)	92.0	515	28.4	0.133
o-Xylene ($C_8 H_{10}$)	106.0	175	6.6	0.123
Cumene ($C_9 H_{12}$)	120.0	50	4.6	0.119
Naphthalene ($C_{10} H_8$)	128.0	33	0.23	0.096
Biphenyl ($C_{12} H_{10}$)	154.0	7.48	0.057	0.092
Pesticides				
DDT ($C_{14} H_9 Cl_5$)	354.5	0.0012	1×10^{-7}	9.34×10^{-3}
Lindane ($C_6 H_6 Cl_6$)	291.0	7.3	9.4×10^{-6}	1.5×10^{-4}
Dieldrin ($C_{12} H_8 Cl_6 O$)	381.0	0.25	1×10^{-7}	5.33×10^{-5}
Aldrin ($C_{12} H_8 Cl_6$)	365.0	0.2	6×10^{-6}	3.72×10^3
Polychlorinated biphenyls (PCBs)				
Aroclor 1242 ($C_{12} H_7 Cl_3$)	257.5	0.24	4.06×10^{-4}	0.057
Aroclor 1248 ($C_{12} H_6 Cl_4$)	292.0	5.4×10^{-2}	4.94×10^{-4}	0.072
Aroclor 1254 ($C_{12} H_5 Cl_5$)	326.5	1.2×10^{-2}	7.71×10^{-5}	0.067
Aroclor 1260 ($C_{12} H_4 Cl_6$)	361.0	2.7×10^{-3}	4.05×10^{-5}	0.067
Other				
Mercury (Hg)	200.6	3×10^{-2}	1.3×10^{-3}	0.092

Source: Adapted from Ref. [7.14].

EXAMPLE 7.3

EVALUATION OF CONTAMINANT VAPORIZATION TIMES

Determine the time required for the concentrations of toluene and Dieldrin spilled in a shallow lake to be reduced to one-half their initial values. Assume the temperature of the water is 25°C, the average lake depth is 2.0 m, and there are no interactions between the compounds.

SOLUTION:

1. Derive an expression that can be used to determine the time required for the concentration of a contaminant to reach one-half the initial value. If the initial background level of the contaminants in the atmosphere is low, then the term C_s in Eq. (7.15) can be neglected. Assuming C_s can be neglected and substituting known values in Eq. (7.15) yields

$$\frac{0.5\, C_{A_0}}{1.0\, C_{A_0}} = e^{-(K/h)t_{1/2}}$$

$$t_{1/2} = \frac{0.69h}{K}$$

2. Using the data given in Table 7.1, determine the time required for the concentrations in the lake to reach one-half their original values.

   a. For toluene:

   $$t_{1/2} = \frac{0.69\,(2.0\text{ m})}{0.133\text{ m/hr}}$$

   $$= 10.4\text{ hr}$$

   b. For Dieldrin:

   $$t_{1/2} = \frac{0.69\,(2.0\text{ m})}{5.33 \times 10^{-5}\text{ m/hr}}$$

   $$= 25{,}891\text{ hr}$$

COMMENT

The time required for the concentration of Dieldrin to reach one-half of the initial value can be used as an argument for the development and use of agricultural chemicals that are more readily broken down in the environment.

Gas Absorption

The process whereby a gas is taken up by a liquid is known as *absorption*. The gas is absorbed by diffusing into the liquid through the air-water interface. The diffusion of a gas in a liquid is brought about by the random thermal motions of the molecules [7.5, 7.6]. In considering the

absorption of a gas two conditions must be considered: absorption in a
quiescent liquid and absorption in an agitated liquid [7.5].

Absorption in Quiescent Liquid

The diffusion of oxygen from the atmosphere in a quiescent liquid
can be illustrated schematically as shown in Fig. 7.6(a). In general, the
one-dimensional transport (flux) of a gas A across a unit area perpendicular to the direction of flow by diffusion is given by

$$\text{Transport by diffusion} = -D_m \frac{\partial C_A}{\partial z} \qquad (7.16)$$

where

D_m = coefficient of molecular diffusion, m^2/s

$\dfrac{\partial C_A}{\partial z}$ = concentration gradient in the z direction, $g/m^3 \cdot m$

The time variation of the concentration, as illustrated in Fig. 7.6(a), can

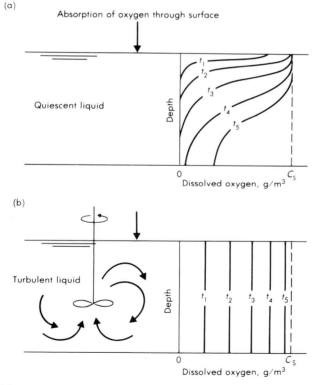

FIGURE 7.6

**Definition sketch for the absorption of a gas (oxygen) (a) in a quiescent liquid and
(b) in an agitated liquid. The gas concentration is assumed to be zero at time t_o in
both cases.**

be analyzed by writing a materials balance for the volume element, as shown in Fig. 7.7(a):

$$\frac{\partial C_A}{\partial t}\Delta z \quad = -D_m \left.\frac{\partial C_A}{\partial z}\right|_z + D_m \left.\frac{\partial C_A}{\partial z}\right|_{z+\Delta z} \qquad (7.17)$$

Accumulation =   Diffusion   Diffusion
                    in   out

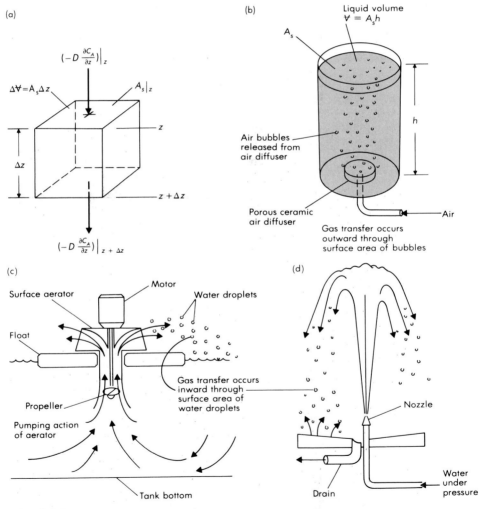

FIGURE 7.7

Definition sketch for the analysis of the absorption of a gas: (a) in a quiescent liquid; (b) from rising gas bubbles released at the bottom of a tank; (c) by water droplets produced with a mechanical aerator; and (d) by water droplets produced with a nozzle under pressure.

Taking the limit as Δz approaches zero yields

$$\frac{\partial C_A}{\partial t} = D_m \frac{\partial^2 C_A}{\partial z^2} \qquad (7.18)$$

Equation (7.18) is the basic expression used to describe the change in concentration of the gas A in the liquid with time and distance because of molecular diffusion. Solutions for Eq. (7.18) for a variety of boundary conditions may be found in Refs. [7.3] and [7.4].

Absorption in Agitated Liquid

Because the surface of most water bodies found in nature is seldom quiescent, Eq. (7.18) cannot be used to describe the gas transfer that occurs in such conditions. A simplified model that can be used to describe the transfer of a gas in an agitated liquid (see Fig. 7.6b) was proposed by Whitman in 1923 [7.5, 7.21]. The model proposed by Whitman is illustrated schematically in Fig. 7.8. As shown there, transfer of a gas must occur across the two films that are assumed to exist at the gas-liquid interface. Lightly soluble gases (e.g., O_2, N_2, and CO_2) encounter primary resistance to transfer from the liquid film, while very soluble gases (e.g., NH_3) encounter the primary resistance to transfer from the gas film. Both films offer resistance to the transfer of gases of intermediate solubility.

For gases with low solubility such as O_2, N_2, and CO_2, the interface concentration C_i is essentially equal to the saturation concentration C_s. Thus following the same line of reasoning used to develop Eq. (7.15) leads to the development of Eq. (7.19), which can be used to estimate the increase in the concentration of a gas in solution:

$$\frac{C_s - C_t}{C_s - C_0} = e^{-(K/h)t} \qquad (7.19)$$

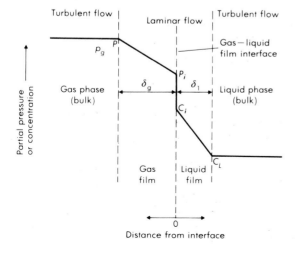

FIGURE 7.8

Definition sketch for the two-film model of the gas absorption process.

where

C_s = saturation concentration of gas in solution, g/m^3

C_t = concentration of gas in solution at time t, g/m^3

C_0 = initial concentration of gas in solution, g/m^3

K = coefficient of mass transfer, m/hr

h = depth, m

t = time, hr

Equation (7.19) is essentially the same as Eq. (7.15), with the exception that the order of the saturation concentration terms on the left-hand side of the equation is reversed. The term K/h is also known as the *reaeration coefficient* k_2. The reaeration coefficient, used in modeling the oxygen resources of streams, is considered further in Chapter 8.

7.5
WATER-SOIL INTERFACE TRANSFERS

The transfer of constituents across the water-soil interface depends primarily on adsorption-desorption and dissolution (solubilization) of constituents from the solid phase. Other important phenomena include chemical processes such as ion exchange and the precipitation of constituents from the liquid phase on the solid phase and physical processes such as sedimentation and mechanical filtration.

Adsorption

Adsorption is said to have occurred if there is an accumulation of material at the interface between two phases. A variety of adsorption systems can be defined, including the adsorption of a gas on a solid or a liquid on a solid. With respect to the movement of constituents across the water-soil interface, the adsorption of a constituent from solution on a solid (soil) is of most interest.

Factors Affecting Adsorption

Important factors that affect the adsorption process are the properties of the adsorbent (the material on which adsorption is occurring) and adsorbate (the material being adsorbed). For a more detailed discussion of these properties Refs. [7.3], [7.20], and [7.21] are recommended.

Properties of the Adsorbent
In considering the adsorption process in soils the following factors must be considered: (1) the chemical characteristics of the mineral and organic fractions of the soil or porous medium, (2) the nature of

adsorption process, whether physical or chemical or both, (3) the nature of the bonds formed between the adsorbent and the adsorbate (e.g., van der Waals physical forces of attraction, hydrophobic bonding, or hydrogen bonding), and (4) local environmental factors such as temperature and pH.

Properties of the Adsorbate

To determine whether a contaminant will be adsorbed and to what extent, the following chemical properties must be considered: (1) the pK value for the contaminant, (2) the solubility of the contaminant, and (3) the soil distribution (partition) coefficient (see discussion below).

Adsorption Process and Models

Because of the many complex interactions that exist between the adsorbent and adsorbate, the adsorption process is often conceptualized as a three-step process involving macrotransport, microtransport, and sorption. *Macrotransport* (bulk transport) involves the movement of the contaminant with the water, as described previously in Section 7.2. *Microtransport* is used to describe the diffusion of the contaminant in the quiescent liquid layer next to the soil. The nonspecific term *sorption* is used to describe the attachment of the contaminant to the surface of the soil and its subsequent movement within the soil structure.

Adsorption models used in the study of the removal of contaminants from solution are based on equilibrium conditions, even though equilibrium is rarely achieved in practice. Because these models are used to describe relationships at constant temperature, they are known as *adsorption isotherms*. To describe the adsorption of constituents from solution the models proposed by Freundlich and Langmuir are commonly used.

The Freundlich Adsorption Isotherm

The most commonly encountered model used to describe the adsorption of a material from solution was proposed by Freundlich [7.9, 7.12].

$$\frac{x}{m} = K_F C_e^{1/n} \tag{7.20}$$

where

$\quad x = $ mass of material adsorbed (adsorbate) on solid phase, g

$\quad m = $ mass of solid (adsorbent) on which adsorption is taking place, g

$\quad K_F = $ Freundlich adsorption coefficient

$\quad C_e = $ concentration of material being adsorbed remaining in solution at equilibrium, g/m^3

$\quad n = $ empirical coefficient

Although the Freundlich adsorption isotherm is an empirical expression, it has proved useful under widely varying conditions. One explanation might be that nonpolar materials in water include a wide range of organic compounds and that this range of materials requires a number of theoretical expressions that combined can be represented adequately by Eq. (7.20). The empirical coefficients in the Freundlich equation are obtained by plotting $\ln(x/m)$ versus $\ln C_e$ or (x/m) versus C_e on log-log paper.

The Langmuir Adsorption Isotherm

A theoretical adsorption model based on the assumption that the adsorbent surface is saturated when a monolayer has been adsorbed was developed by Langmuir [7.12, 7.21]. The Langmuir adsorption isotherm is given by the following expression:

$$\frac{x}{m} = \frac{abC_e}{1 + bC_e} \tag{7.21}$$

where

a = empirical constant

b = saturation coefficient, m^3/g

and other terms are as defined previously. The Langmuir adsorption isotherm is similar in form to the saturation-type rate expression [Eq. (5.16)], but the two must not be confused. One denotes a rate, the other equilibrium. The coefficients in the Langmuir equation are obtained by plotting $C_e/(x/m)$ versus C_e on arithmetic paper, with Eq. (7.21) written as follows:

$$\frac{C_e}{(x/m)} = \frac{1}{ab} + \frac{C_e}{a} \tag{7.22}$$

Selection of an appropriate adsorption model is made by fitting the adsorption isotherms to experimental data (see Example 7.4).

EXAMPLE 7.4

DETERMINATION OF ADSORPTION ISOTHERM COEFFICIENTS

Given the following results from a batch adsorption test for the removal of an organic contaminant from water with granular activated carbon, determine the appropriate absorption model and the corresponding coefficients. The liquid volume used in the batch experiments was 1 L.

MASS OF CARBON, g	RESIDUAL CONTAMINANT CONC C_e, g/L
0.0	20
0.9	13
1.7	10
4.0	6
7.0	4
10.0	3

SOLUTION:

1. Derive the values needed to plot the Freundlich and Langmuir adsorption isotherms from the batch adsorption test data.

CONTAMINANT MASS, g

C_0	C_e	x	m, g	$\dfrac{x}{m}$	$\dfrac{C_e}{(x/m)}$
20	13*	7	0.9	7.8	1.67
20	10	10	1.7	5.9	1.70
20	6	14	4.0	3.5	1.71
20	4	16	7.0	2.3	1.75
20	3	17	10.0	1.7	1.76

*13 g (13 g/L × 1.0 L)

2. Plot the Freundlich and Langmuir adsorption isotherms (Fig. 7.9).
3. Determine the adsorption isotherm coefficients.
   a. Freundlich:

$$\frac{1}{n} = \text{slope} = 1.0 \qquad n = 1.0$$

When $x/m = 1.0$, $C_e = 1.7$

$$K_F = (x/m)/C_e = 1.0/1.7$$
$$= 0.59$$

   b. Langmuir:

$$\frac{1}{a} = \text{slope} = \frac{1.76 - 1.67}{3 - 13} = -0.009$$
$$a = -111.1$$
$$\frac{1}{ab} = 1.7$$
$$b = -0.005$$

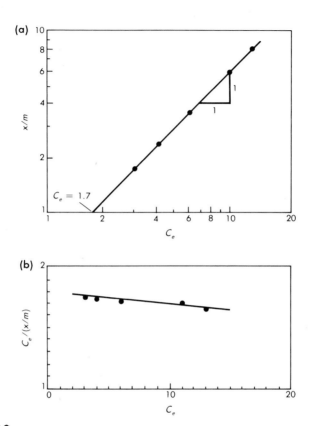

FIGURE 7.9

Functional analysis of adsorption data for Example 7.4: (a) Freundlich isotherm and (b) Langmuir isotherm.

COMMENT

Because negative coefficients have no meaning, use of the Langmuir isotherm is inappropriate.

Because the value of the exponent n in the Freundlich adsorption isotherm is equal to 1, the Freundlich coefficient K_F corresponds to the soil distribution coefficient K_{SD} (see following discussion).

The Soil Distribution Coefficient

The soil distribution coefficient K_{SD} is analogous to the distribution coefficient K_D, defined in Eq. (7.6), and is used to define the approximate distribution of a contaminant between the solid and liquid phase:

$$K_{SD} = \frac{S}{C} \tag{7.23}$$

where

$\quad K_{SD}$ = soil distribution coefficient, m^3/g

$\qquad S$ = mass of solute sorbed per unit mass of dry soil, g/g

$\qquad C$ = concentration of solute in the liquid phase, g/m^3

The soil distribution coefficient will vary with the contaminant and the characteristics of soil system, particularly the clay content [7.8]. If the slope of the Freundlich isotherm, as represented by the term $1/n$, is found to be equal to 1, then the isotherm is said to be linear and the Freundlich coefficient K_F is identical to the soil distribution coefficient K_{SD}. Application of the soil distribution coefficient in the analysis of the movement of contaminants in groundwater is illustrated in Chapter 10.

Dissolution-Precipitation

As noted in Chapter 2, most of the constituents found in natural groundwaters are derived from the dissolution of minerals with which the water has come in contact. For example, the dissolution or precipitation of calcium carbonate is controlled by the following reaction:

$$CaCO_3 + H_2CO_3 \rightleftharpoons Ca^{+2} + 2HCO_3^- \qquad (7.24)$$

Equation (7.24) is an example of the dissolution-precipitation reactions that can occur in groundwater to alter water quality. The dynamics of Eq. (7.24) and other similar equations are controlled by changes in the pressure, temperature, pH, and concentration. Additional details on the dissolution process may be found in Ref. [7.8].

7.6
SOIL-AIR INTERFACE TRANSFERS

The major factors responsible for the transfer of contaminants across the soil-air interface include evaporation (volatilization), infiltration, and sorption. Volatilization from plant surfaces was discussed previously in Section 7.3.

Volatilization

As in the case with water, the evaporation process for soil or soil-plant surfaces, termed *evapotranspiration*, is modeled on a theoretical analysis of energy exchange, empirical observations, and various combinations of the two. Evapotranspiration is considered in detail in Refs. [7.2] and [7.13]. It should be noted that evapotranspiration is the basis of the Blaney-Criddle model introduced in Chapter 1 [Eq. (1.5)].

The volatilization of contaminants from the soil surface depends on pH, temperature, soil characteristics, soil moisture, the degree of absorption, and wind movement (turbulence) over the soil surface. The approach used most commonly to model the vaporization process is the same as outlined in Section 7.4 for the air-water interface.

Infiltration

The term *infiltration* is used to describe the entry of surface water into the ground. *Percolation* is the term applied to the movement of the water through the interstices of the underlying earth. As water from rainfall, melting snow, or irrigation systems infiltrates into the soil, contaminants deposited on the surface are transported across the air-soil interface into the underlying soil. The subsequent movement of these contaminants is subject to the factors discussed previously in Sections 7.2 and 7.5. Infiltration is considered in greater detail in Refs. [7.2] and [7.13].

Sorption

The nonspecific term "sorption" is used to describe the transfer of gaseous contaminants across the soil-air interface in the absence of percolation. Vaporization, pH, temperature, soil characteristics, soil moisture, and the distribution coefficient are important variables in the sorption process. The approach outlined in Section 7.4 for absorption is used to model the process.

7.7
TRANSFORMATION OF CONTAMINANTS IN THE ENVIRONMENT

In the previous sections, the movement of contaminants in surface water and groundwater and between the air-water, water-soil, and soil-air interfaces has been examined. The purpose of this section is to consider some of the more important transformations that occur in the environment and how these transformations affect the concentration of contaminants and fate of contaminants in the environment [7.8]. Important transformations include bacterial conversions, natural decay, hydrolysis reactions, photochemical reactions, and oxidation-reduction reactions [7.18, 7.20].

Bacterial Conversion

Bacterial conversion is the most important process in the transformation of contaminants found in surface waters and in many groundwaters. The exertion of CBOD and NBOD considered previously in Chapter 2 are the most common examples of bacterial conversion encountered in water quality management. The bacterial transformation of toxic organic compounds is also of great significance but is beyond the scope of this text. References [7.10], [7.15], and [7.20] are recommended for further details on the many bacterial transformations that occur in nature.

Natural Decay

In nature, contaminants will decay for a variety of reasons, including mortality in the case of bacteria and photooxidation for certain organic constituents. Contaminant decay is often assumed to follow first-order kinetics such as $r_c = -kC$. Thus

$$C_t = C_0 e^{-kt} \tag{7.25}$$

where

C_t = concentration at time t, g/m^3
C_0 = concentration at time $t = 0$, g/m^3
k = first-order decay coefficient, d^{-1}
t = time, d

Determination of the first-order decay coefficient is based on field studies, as the number of variables that can affect it are impossible to enumerate or control. At best, k values measured in the field are lumped parameter values and should not be applied to other systems without field verification.

Hydrolysis Reactions

Hydrolysis reactions of concern are those reactions that occur between contaminants and water. Of special interest are the reactions that occur between organic compounds and water. For example, under anaerobic conditions, the hydrolysis of a starch can be represented as

$$(C_6H_{10}O_5)_n + nH_2O \rightarrow nC_6H_{12}O_6 \tag{7.26}$$

Hydrolysis reactions are important in the conversion of the organic matter in wastewater in sewers, in bottom deposits of streams and lakes, and in groundwater.

Photochemical Reactions

Solar radiation is known to trigger a variety of chemical reactions. Radiation in the near-ultraviolet (UV) and visible range is known to cause the breakdown of a variety of organic compounds. The breakdown of synthetic polymers (plastics) exposed to UV radiation is an example familiar to most. If photochemical breakdown is to occur in water, the contaminant must be located properly to absorb the solar radiation. Typically, contaminants must be present on or near the surface of water bodies or in the atmosphere.

Oxidation-Reduction Reactions

In an oxidation-reduction system, one substance is oxidized and the other is reduced by donating or accepting electrons. In natural systems, oxidation-reduction reactions are important, as they may serve to mediate other reactions with contaminants that may be present. The reaction of iron serves as an example of what can occur. The oxidation of iron (Fe^{+2}) by oxygen to iron (Fe^{+3}) can be represented as follows:

$$4Fe^{+2} + O_2 + 4H^+ \rightleftharpoons 4Fe^{+3} + 2H_2O \tag{7.27}$$

Thus, depending on the oxidation-reduction potential and the pH of the solution, iron may exist as Fe^{+2} or as Fe^{+3}. The corresponding reactions needed to predict what may occur, depending on whether Fe^{+2} or Fe^{+3} is present, are quite different.

7.8
PRACTICAL ASPECTS OF CONTAMINANT MOVEMENT

The purpose of this chapter has been to introduce the reader to the phenomena involved in the movement of contaminants in the biosphere. Out of necessity, many of the subjects were considered only briefly. It is hoped, however, that this introduction to the phenomena will serve as a preliminary guide to the subject. For a more in-depth discussion of the concepts introduced in this chapter, the book by Thibodeaux [7.19] is recommended. In exploring each of these subjects further, the lack of data and information necessary to predict water quality changes in the environment will be frustrating. It is hoped that in your work you will have an opportunity to help develop the needed data base.

KEY IDEAS, CONCEPTS, AND ISSUES

- Contaminants in nature are distributed primarily by movements of air and water.

- To study the movement of contaminants in the environment, their transport in air and water must be considered along with the interface transfers that can occur between water-soil, water-air, and soil-air elements.
- In the absence of any reactions, contaminants in surface water and groundwater are transported primarily by advection and hydrodynamic dispersion.
- The principal mechanisms governing air-water and soil-air interface transfers of contaminants are volatilization and absorption.
- The principal mechanisms governing water-soil interface transfers are adsorption and dissolution (solubilization). Other important mechanisms within the soil element include ion exchange, sedimentation, and mechanical filtration.
- Important contaminant transformations that occur in nature include bacterial conversions, natural decay, hydrolysis reactions, photochemical reactions, and oxidation-reduction reactions.

DISCUSSION TOPICS AND PROBLEMS

7.1. The distribution coefficient for iodine (the solute) in an amyl alcohol $(C_5H_{11}OH)$: water solvent system is equal to 200. If the equilibrium concentration of iodine in the alcohol is 375 g/m^3, determine the concentration of iodine in the water layer.

7.2. Using the expression developed in Example 7.1 for solvent extraction, determine the volume of solvent needed per extraction to reduce the concentration of a solute in 1 L of water from 100 to 1 mg/L if the value of the distribution coefficient is 10. Consider 1, 2, 4, and 8 extraction steps.

7.3. An industrial wastewater contains 1.50 g/m^3 of a priority pollutant that can be extracted with kerosene. At equilibrium, the ratio of pollutant concentrations in the water and the solvent, $C_K/C_W = K_D$, is 5.0. If 20 m^3 of wastewater must be treated per day, determine the number of extractions and the total solvent volume required to produce an effluent having a pollutant concentration of 0.15 g/m^3, using kerosene volumes of 0.05, 0.20, and 0.40 m^3.

7.4. Determine the distribution coefficient required to extract 90 percent of a pollutant using no more than five extractions and a total solvent volume of 5 percent of the wastewater volume.

7.5. Using the data given in Table C.1 in Appendix C, estimate the heat of vaporization for water.

7.6. Determine the vapor density of water in air at a temperature 20°C.

7.7. The wet well at a wastewater pump station receives a steady flow that is saturated with the PCB Aroclor 1254. Average temperature in the wet well is 17°C, and the air pressure is 0.97 atm. Determine the mass concentration of PCB to which workers are exposed.

7.8. The vapors of two liquids are known to be toxic when inhaled over prolonged periods. If the vapor pressure for liquid A is 100 mmHg and for liquid B is 60 mmHg, which of the two liquids represents the greater health risk?

7.9. An industrial wastewater initially saturated with toluene flows through a 0.2-m-deep open channel for a distance of 1000 m. Estimate the concentration of toluene at the end of channel, assuming ideal plug flow if the average velocity is 0.3 m/s.

7.10. Wastewater enters the channel of Problem 7.9 with an oxygen concentration of 0. At the end of the channel the oxygen concentration is 2.7 g/m^3. Estimate the value of the mass-transfer coefficient K for oxygen if the water temperature is 15°C.

7.11. A redwood-slat tray aerator is used to reduce the concentration of CO_2 in groundwater pumped from a confined aquifer. In the aerator, water to be treated is applied to the upper tray and allowed to drip to successive trays. Based on pilot plant tests, it has been found that the concentration of CO_2 can be reduced from 30 to 6 g/m^3 in 6.2 s in a six-tray aerator. The temperature at which the tests were conducted was 20°C. Determine the number of trays required to reduce the concentration of CO_2 to 3.5 g/m^3.

7.12. Solve Problem 7.11, but assume that the final concentration of CO_2 must be reduced to 2.0 g/m^3.

7.13. The CO_2 concentration of a groundwater is to be reduced using a fountain aerator (see Fig. 7.7d). Vertical jets will be used, and the water discharged from the jets will fall into an open storage reservoir. If the diameter of the droplets formed as the water exits from the jets is 5.0 mm, determine the pressure required at the nozzle to achieve a final CO_2 concentration of 4.0 g/m^3. The initial CO_2 concentration in the groundwater is 40 g/m^3 and the temperature is 15°C. Assume the coefficient of gas transfer for CO_2 is equal to 8.33×10^{-4} m/s.

7.14. Solve Problem 7.13, but assume that the diameter of the water droplets is 4.0 mm, that the temperature of the groundwater is 23°C, and that the final CO_2 concentration must be equal to 2.0 g/m^3.

7.15. Determine the mass-transfer coefficient for benzene using the following data. State any assumptions made in solving the problem.

TIME, hr	RELATIVE CONCENTRATION, C/C_0
0	1.0
1	0.170
2	0.030
3	0.006
4	0.001

7.16. A reaction between a dissolved solid A (molecular mass = 48 g/mol) and a dissolved gas B (molecular mass = 24 g/mol) is described by the first-order reaction:

A + B → products

$$r_A = -k[B]$$

where C_A and C_B are the respective mass concentrations in grams per cubic meter. Transfer of the gas from the atmosphere above the reactor is described by Henry's law, and the rate equation is

$$x_B = K_H p_B \qquad K_H = 3 \times 10^{-3} \text{ atm}^{-1}$$
$$N_B = K(C_B^* - C_B) \qquad K = 10 \text{ m/d}$$

where

N_B = mass of B transferred per unit of area-time, g/m²-d

C_B^* = equilibrium concentration of B based on Henry's law

Determine C_A and the rate r_A as a function of time (0 to 2.5 d) for a batch reactor with a volume of 1.0 m³ and a surface area of 1.0 m² if $k = 0.3$ d⁻¹, $C_{A_0} = 100$ g/m³, $C_{B_0} = 0$, and $p_B = 0.04$ atm.

7.17. The following test results were obtained for the adsorption of an organic acid from 1 L of solution with 10 g of soil. Determine the absorption isotherm that best fits the experimental test results; also determine the coefficients for the selected isotherm.

CONCENTRATION, g/m³	
Initial	Final
0.520	0.484
0.260	0.231
0.112	0.093
0.056	0.042
0.038	0.028

7.18. A landfill has adsorbed 0.3 g Aldrin per gram of soil. Desorption into rainwater percolating through the landfill follows a Freundlich isotherm with $K_F = 1.0$ and $n = 2$. Determine if the percolating water is likely to become saturated with Aldrin.

7.19. Wastewater containing 2.01 g/m^3 of pesticide percolates through a soil treatment system. Adsorption has been determined to follow a Langmuir isotherm with $a = 1.01$ and $b = 0.003$ m^3/g. Determine the mass of pesticide that will potentially accumulate in the soil if the cross-sectional area is 20 m^2 and the depth is 4 m. The dry density of the soil is approximately 2400 kg/m^3.

REFERENCES

7.1. Bird, R. B., W. E. Stewart, and E. N. Lightfoot, (1960), *Transport Phenomena*, John Wiley and Sons, New York.

7.2. Bouwer, H., (1978), *Groundwater Hydrology*, McGraw-Hill Book Company, New York.

7.3. Browman, M. G., and G. Chesters, (1977), "The Solid Water Interface," in *Fate of Pollutants in the Air and Water Environments*, Part 1, edited by I. H. Suffet, Marcel Dekker, New York.

7.4. Crank, J., (1957), *The Mathematics of Diffusion*, Oxford University Press, London.

7.5. Danckwerts, P. V., (1970), *Gas Liquid Reactions*, McGraw-Hill Book Company, New York.

7.6. Davies, J. T., and E. K. Rideal, (1963), *Interfacial Phenomena*, 2d ed., Academic Press, New York.

7.7. Dean, J. A., (ed.), (1973), *Lange's Handbook of Chemistry*, 11th ed., McGraw-Hill Book Company, New York.

7.8. Freeze, R. A., and J. A. Cherry, (1979), *Groundwater*, Prentice-Hall, Englewood Cliffs, N.J.

7.9. Freundlich, H., (1926), *Colloid and Capillary Chemistry*, Methuen & Company, London.

7.10. Higgins, I. J., and R. G. Burns, (1975), *The Chemistry and Microbiology of Pollution*, Academic Press, London.

7.11. Jacobs, M. H., (1967), *Diffusion Processes*, Springer-Verlag, New York.

7.12. Langmuir, I., (1918), "The Absorption of Gases on Plane Surfaces of Glass, Mica, and Platinum," *J. American Chemical Society*, vol. 40, no. 9, p. 1361.

7.13. Linsley, R. K., M. A. Kohler, and J. L. H. Paulhus, (1975), *Applied Hydrology*, 2d ed., McGraw-Hill Book Company, New York.

7.14. Mackay, D., and P. J. Leinonen, (1975), "Rate of Evaporation of Low-Solubility Contaminants from Water Bodies to Atmosphere," *Environmental Science and Technology*, vol. 9, no. 19, p. 1178.

7.15. Matsumura, F., and C. R. Krishna Murti, (1982), *Biodegradation of Pesticides*, Plenum Press, New York.

7.16. Ohe, S., (1976), *Computer Aided Data Book of Vapor Pressure*, Data Book Publishing Company, Tokyo.

7.17. *Organic Chemical Contaminants in Groundwater: Transport and Removal*, (1981), AWWA Seminar Proceedings, Report No. 20156, AWWA, Denver.

7.18. *Selected Values of Properties of Chemical Compounds*, (1966), Manufacturing Chemists Association Research Project, Thermodynamic Research Center, Department of Chemistry, Texas A&M University, College Station.

7.19. Thibodeaux, L. J., (1979), *Chemodynamics: Environmental Movement of Chemicals in Air, Water, and Soil*, John Wiley and Sons, New York.

7.20. Tinsley, I. J., (1979), *Chemical Concepts in Pollutant Behavior*, John Wiley and Sons, New York.

7.21. Weber, W. J. Jr., (1972), *Physicochemical Processes for Water Quality Control*, Wiley Interscience, New York.

7.22. Whitwell, J. C., and R. K. Toner, (1973), *Conservation of Mass and Energy*, McGraw-Hill Book Company, New York.

8

Water Quality in Rivers and Estuaries and Near Ocean Outfalls

A major aspect of water quality management involves the modeling of water quality changes in streams, rivers, estuaries, and oceans subject to both natural and anthropogenic inputs. The development of water quality models involves the application of materials balances and kinetic expressions to describe the response of the physical system. By modeling river and estuarine systems it is possible to assess the assimilative capacity of these systems, and thus to predict the impacts of proposed developments and natural occurrences. It is, therefore, the purpose of this chapter to introduce the reader to the methodology involved in the development of water quality models and to illustrate how such models are applied.

8.1

MIXING OF POLLUTANTS AND RIVER WATER

In most models used in the analysis of the effects of pollutant discharges on rivers, complete mixing at the discharge point is assumed. Complete mixing is rarely the case, but for steady inputs the approach is conservative and reasonable, considering the difficulty of determining parameters and coefficients. Slug discharges, such as might result from an industrial spill or wrecking of a barge, present another case, however. A complete analysis is beyond the scope of this text, but estimates of mixing distances are useful in making initial decisions.

Conditions in particular streams, such as bends, depth, and flow rate, strongly affect the mixing distance. Fischer et al. [8.8] suggest that the length of travel L_m required for complete mixing of a centerline point

337

discharge is given by Eq. (8.1):

$$L_m = 0.03 \; uw^2/D_T \qquad\qquad (8.1)$$

where

   L_m = length of travel to point of complete mixing, m

   u = average advective velocity, m/s

   w = stream width, m

   D_T = transverse dispersion coefficient, m²/s

A point discharge at the stream edge will require a mixing length four times as great because the transverse distance is twice as large.

   Estimation of D_T is difficult in slow-moving rivers, but it can be approximated to within ±50 percent using Eq. (8.2) [8.8]:

$$D_T = 0.2 \, \overline{H} u^* \qquad\qquad (8.2)$$

where

   $\overline{H}$ = average depth of flow, m

   u^* = shear velocity, m/s

   $\quad = \sqrt{gR_H S}$

   g = acceleration due to gravity, 9.81 m/s²

   R_H = hydraulic radius, m

   $\quad = \dfrac{\text{cross-sectional area of flow, m}^2}{\text{wetted perimeter, m}}$

   S = hydraulic gradient, m/m

   Buoyant plumes, such as discharges from power stations, are slow to mix with rivers because of the difference in density. Most such plumes "float" on the surface for considerable distances until heat exchange with the atmosphere and river and turbulent mixing reduce the temperature-density differential. Temperature differentials for power stations are usually similar to those for municipal discharges, but flows are generally much greater. Once-through cooling water flows may be 20 m³/s per 1000 MW or greater, for example, constituting a good-sized river in themselves. Analysis of the mixing characteristics of cooling water discharges usually requires physical modeling.

8.2
SIMPLE RIVER MODELS

One of the earliest mathematical water quality models was developed for the Ohio River by H. S. Streeter and E. B. Phelps [8.26]. This model, usually referred to as the *Streeter-Phelps model*, is used to predict the oxygen deficit in a river resulting from the discharge of a waste. It is the prototype river model in a number of ways. The Streeter-Phelps oxygen-sag model, in its simplest form, relates the rate of change of the oxygen deficit with distance to the respective spatial rates of deoxygenation and reoxygenation. The model is derived for steady-state conditions, but is usually presented in the literature as a non-steady-state model because ideal plug-flow conditions are assumed, and, as noted in Chapter 6, the equations for a batch reactor and a PFR at steady state are identical.

Deoxygenation in Rivers

The oxygen in rivers and streams is depleted by (1) the bacterial oxidation of the suspended and dissolved organic matter discharged to them from both natural and anthropogenic sources and by (2) the oxygen demand of sludge and benthic deposits. Depletion of oxygen by benthic deposits is considered in a subsequent section.

The concept of biological oxidation of organic matter was introduced in Chapter 2. The amount of oxygen required to stabilize a waste is normally measured by the BOD test; BOD is, therefore, the primary source of oxygen depletion or utilization in a waterway. The rate of deoxygenation r_{O_2} is

$$r_{O_2} = -kL \qquad (8.3)$$

where

r_{O_2} = rate of deoxygenation, $g/m^3 \cdot d$

k = first-order reaction-rate constant, d^{-1}

L = ultimate BOD remaining at point in question, g/m^3

Assuming ideal plug flow, a mass balance performed on a differential section of river yields

$$\frac{dL}{d\theta_H} = -kL \qquad (8.4)$$

As noted in the BOD discussion in Chapter 2, the integrated form of the above equation is by Eq. (8.5):

$$L = L_i e^{-k\theta_H} \qquad (8.5)$$

where

L_i = ultimate BOD at the point of discharge, g/m^3

θ_H = hydraulic detention time, d

Thus the rate of deoxygenation obtained by combining Eqs. (8.5) and (8.3) is

$$r_{O_2} = -kL_i e^{-k\theta_H} \qquad (8.6)$$

Inorganic and organic material, too heavy to remain in suspension, will settle out, forming a sludge deposit or benthic layer on the bottom of the stream. Sludge deposits in the bottoms of slow-moving rivers can exert a significant oxygen demand on the overlying river water. Although most of the sludge will be undergoing anaerobic decomposition, which is a relatively slow process, aerobic decomposition can take place at the interface between the sludge and the flowing water.

Rates of deposition and scour vary with the velocity of flow and turbulence of the river. At times, sedimentation may reduce the BOD load in the river water if settleable solids are discharged or if coagulation of colloidal matter takes place. At other times, scour will increase the BOD load by returning these particles to the river water. The benthic load must be assessed for each reach of river to determine its importance in the total oxygen balance. In many rivers, it is a factor that can be eliminated from consideration.

Reoxygenation in Rivers

Aside from the oxygen contained in the waters of the tributary rivers, surface drainage, and groundwater inflow, the sources of oxygen replenishment in river water are absorption from the atmosphere (see Chapter 7) and photosynthesis of aquatic plants and algae. In modeling the oxygen resources of rivers, the term *reaeration* is used to describe the atmospheric absorption of oxygen. The rate of atmospheric reaeration is proportional to the dissolved-oxygen deficiency. As noted in Chapter 7, the rate of reaeration r_R can be expressed as:

$$r_R = k_2 (C_s - C_{O_2})$$

(8.7)

where

r_R = rate of reaeration, $g/m^3 \cdot d$

k_2 = reaeration constant, d^{-1}

C_s = dissolved-oxygen saturation concentration, g/m^3

C_{O_2} = dissolved-oxygen concentration, g/m^3

The reaeration constant can be estimated by determining the characteristics of the stream and using one of the many empirical formulas that have been proposed over the past 50 years [8.6, 8.16, 8.20, 8.26]. Equation (8.8) is a generalized formula proposed for natural streams [8.20]:

$$k_2 = \frac{294(D_L u)^{1/2}}{\overline{H}^{3/2}}$$

(8.8)

where

D_L = coefficient of molecular diffusion for oxygen, m^2/d

u = mean stream velocity, m/s

$\overline{H}$ = average depth of flow, m

The variation of the coefficient of molecular diffusion with temperature can be approximated with the following expression:

$$D_{L_T} = D_{L_{20}} \times 1.037^{(T-20)}$$

(8.9)

where

D_{L_T} = coefficient of molecular diffusion for oxygen at temperature T, m^2/d

$D_{L_{20}}$ = coefficient of molecular diffusion for oxygen at 20°C, m^2/d
= 1.760×10^{-4} m^2/d

1.037 = temperature coefficient

T = temperature of river water, °C

An empirical expression for k_2, developed from studies in the Tennessee Valley area of rivers that varied in depth from 0.6 to 3.4 m and in velocity from 0.5 to 1.5 m/s, is [8.6]

$$k_2 = \frac{5.23\, u}{\overline{H}^{1.67}}$$

(8.10)

where

k_2 = reaeration constant, d^{-1}

and other terms are as defined previously. Typical values of k_2 are given in Table 8.1. Note that the temperature correction coefficient is given as a footnote in Table 8.1. The coefficient is used in the same manner as the BOD temperature correction coefficient defined in Eq. (2.79).

TABLE 8.1
Typical Reaeration Constants for Various Water Bodies

WATER BODY	RANGES OF k_2 AT 20°C* (base e), d^{-1}
Small ponds and backwaters	0.10–0.23
Sluggish streams and large lakes	0.23–0.35
Large streams of low velocity	0.35–0.46
Large streams of normal velocity	0.46–0.69
Swift streams	0.69–1.15
Rapids and waterfalls	> 1.15

*For other temperatures use $k_{2_T} = k_{2_{20}} 1.024^{T-20}$.
Source: Adapted in part from Refs. [8.18] and [8.24].

Development of the Oxygen-Sag Model

In most river analyses, it is assumed that wastes that are discharged to the river are distributed evenly over the cross section of the river. While the assumption of uniform distribution may be far from the truth in the immediate vicinity of the outlet, the validity of the assumption improves in most cases as the wastes proceed downstream (see Section 8.1). It can also be assumed that no mixing occurs along the axis of the river, which is reasonable provided the river is not extremely turbulent.

Initial Mixing

If it is assumed that the river and waste are mixed completely at the point of discharge, then the concentration of a constituent in the river-waste mixture at $x = 0$ is given by

$$C_0 = \frac{Q_r C_r + q_w C_w}{Q_r + q_w} \qquad (8.11)$$

where

$C_0 =$ initial concentration of constituent at point of discharge, g/m^3

$Q_r =$ river flow rate, m^3/s

$C_r =$ concentration of constituent in river before mixing, g/m^3

$q_w =$ wastewater flow rate, m^3/s

$C_w =$ concentration of constituent in wastewater, g/m^3

In many small rivers the wastewater flow makes up a major fraction of the total flow. For example, Leeds, England, a city of approximately 500,000, discharges a dry-weather flow of about 1.4 m^3/s to the river Aire, which has a width of about 20 m, an average flow of 12.7 m^3/s, and a minimum flow of 1.5 m^3/s. In this situation, and particularly at low flow, mixing conditions would certainly be good, and the Streeter-Phelps model is probably very appropriate. By comparison, Cincinnati, Ohio, also with a population of approximately 500,000, discharges to the Ohio River, which has an average flow of 3100 m^3/s, with a range between 300 and 8000 m^3/s, and a width of well over 100 m. Mixing conditions would be poor in the Ohio River unless a diffuser was used to provide distribution across the channel.

Change in Oxygen Resources

The change in the oxygen resources of a river can be modeled by assuming that the river (Fig. 8.1) is essentially a plug-flow reactor (Fig. 8.2). Over any incremental volume, the following materials balance can be written for the concentration of oxygen in the stream (neglecting hydrodynamic dispersion):

$$\frac{\partial C_{O_2}}{\partial t} \Delta V = QC_{O_2}|_x - QC_{O_2}|_{x+\Delta x} + \quad r_{O_2}\Delta V \quad + \quad r_R \Delta V \qquad (8.12)$$

Accumulation = Inflow − Outflow + Deoxygenation + Reoxygenation

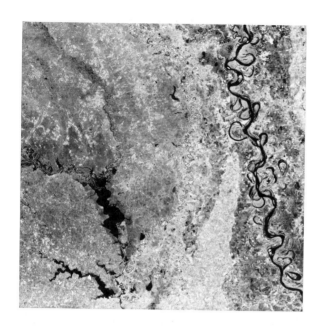

FIGURE 8.1

Landsat view of the Mississippi River near Lake Village, Arkansas. (Landsat 1, 2, & 3 Coverage Index, path 25, row 37, MSS band 7.)

where

$$C_{O_2} = \text{concentration of oxygen in river, g/m}^3$$
$$Q = \text{volumetric flow rate in river, m}^3/\text{s}$$
$$V = \text{volume, m}^3$$

and other terms are as defined previously. Substituting for r_{O_2} and r_R from Eqs. (8.3) and (8.7), collecting terms, and taking the limit as Δx

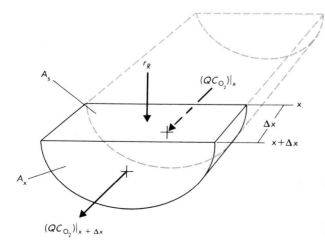

FIGURE 8.2

Definition sketch for the plug-flow model used in river analysis.

approaches zero yields

$$\frac{\partial C_{O_2}}{\partial t} = -Q\frac{\partial C_{O_2}}{\partial V} - kL + k_2(C_s - C_{O_2}) \tag{8.13}$$

If steady-state conditions are assumed ($\partial C_{O_2}/\partial t = 0$), the above expression can be simplified to

$$0 = -Q\frac{dC_{O_2}}{dV} - kL + k_2(C_s - C_{O_2}) \tag{8.14}$$

Substituting θ_H for V/Q, Eq. (8.14) becomes

$$\frac{dC_{O_2}}{d\theta_H} = -kL + k_2(C_s - C_{O_2}) \tag{8.15}$$

If the oxygen deficit D_{O_2} is defined as

$$D_{O_2} = (C_s - C_{O_2}) \tag{8.16}$$

then the change in the deficit with residence time is equal to

$$\frac{dD_{O_2}}{d\theta_H} = -\frac{dC_{O_2}}{d\theta_H} \tag{8.17}$$

Using the above relationships, Eq. (8.15), written in terms of the oxygen deficit, is

$$\frac{dD_{O_2}}{d\theta_H} = kL - k_2 D_{O_2} \tag{8.18}$$

Substituting for L from Eq. (8.5) and rearranging yields

$$\frac{dD_{O_2}}{d\theta_H} + k_2 D_{O_2} = kL_i e^{-k\theta_H} \tag{8.19}$$

When $\theta_H = 0$, $D_{O_2} = D_i$, and the integrated form of Eq. (8.19) is given by Eq. (8.20) (see Appendix I):

$$D_{O_2} = \frac{kL_i}{k_2 - k}(e^{-k\theta_H} - e^{-k_2\theta_H}) + D_i e^{-k_2\theta_H} \tag{8.20}$$

where

D_i = initial oxygen deficit at the point of waste discharge, g/m^3

Equation (8.20) is the classic Streeter-Phelps oxygen-sag equation, which is most commonly used to model the oxygen resources of a river. It must be used with caution, however, because it applies to channels of uniform cross section where effects of algae and sludge deposits are negligible.

The graphical representation of the Streeter-Phelps oxygen-sag equation is shown in Fig. 8.3. Active biological decomposition begins immediately after discharge. This decomposition utilizes oxygen. Because

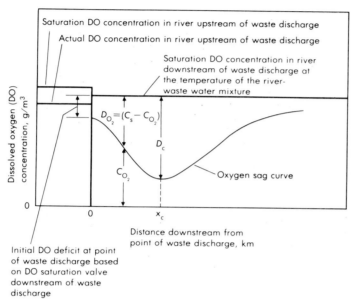

FIGURE 8.3

Characteristic oxygen-sag curve obtained using the Streeter-Phelps equation.

atmospheric reaeration is proportional to the dissolved-oxygen deficit, its rate will increase with increasing deficit. Finally a critical point is reached at which the rate of oxygen utilized for waste decomposition equals the rate of atmospheric reaeration, and the minimum dissolved-oxygen concentration occurs. This critical point is at x_c in Fig. 8.3. Downstream from this point, the rate of reaeration is greater than the rate of utilization, and the dissolved oxygen begins to increase. Eventually, the stream will show no effects of the waste discharge. This is the phenomenon of *natural stream purification* [8.16, 8.24, 8.29].

Critical Dissolved-Oxygen Deficit

The critical or maximum dissolved-oxygen deficit D_c at the point x_c can be determined by setting $dD_{O_2}/d\theta_H$ in Eq. (8.19) to zero. When this is done the critical deficit is given by

$$D_c = \frac{k}{k_2} L_i e^{-k\theta_H^*} \tag{8.21}$$

where θ_H^* is the flow time required to reach the critical point. The value of θ_H^* can be determined by differentiating Eq. (8.20) with respect to θ_H and setting $dD_{O_2}/d\theta_H$ equal to zero:

$$\theta_H^* = \frac{1}{k_2 - k} \ln\left[\frac{k_2}{k} \left(1 - \frac{D_i(k_2 - k)}{kL_i} \right) \right] \tag{8.22}$$

The distance x_c is equal to

$$x_c = \theta_H^* u \tag{8.23}$$

where u is the velocity of flow in the river.

The use of the Streeter-Phelps equation in determining the dissolved-oxygen concentration (e.g., oxygen-sag curve) in a river is illustrated in Example 8.1. Additional details on the oxygen-sag method of analysis may be found in Refs. [8.13], [8.19], [8.20], [8.22], [8.24], and [8.29]. Additional information on the biology and ecology of streams may be found in Refs. [8.14] and [8.15].

EXAMPLE 8.1

OXYGEN-SAG ANALYSIS FOR A RIVER

It is proposed to discharge wastewater to a small river having a minimum flow of 4 m^3/s (50-yr return period). The allowable BOD_u that may be discharged is to be based on maintaining a minimum dissolved-oxygen concentration of 5.5 g/m^3. Based on laboratory studies, the k value for the wastewater is 0.2 d^{-1}. The wastewater flow will not exceed 1.5 m^3/s. The temperature of the river at low flow is 25°C, and the estimated value of k_2 is 0.41 d^{-1} at 25°C. Determine the allowable effluent BOD_u in the river at the point of discharge and in the waste discharge for an initial dissolved-oxygen concentration in the river of 7.5 g/m^3 after mixing. Assume a wastewater temperature of 25°C.

SOLUTION:

1. Determine the initial deficit and the deficit at the critical point in the oxygen-sag curve. At 25°C, $C_S = 8.2$ g/m^3 (see Appendix F).
   a. The initial deficit is

   $$D_i = 8.2 - 7.5 = 0.7 \text{ g/m}^3$$

   b. The deficit at the critical point is

   $$D_C = 8.2 - 5.5 = 2.7 \text{ g/m}^3$$

2. Adjust k for temperature, using a θ value of 1.056 (see Chapter 2). At 25°C,

   $$k = 0.2(1.056)^{25-20}$$
   $$= 0.26 \text{ d}^{-1}$$

3. Find the ultimate BOD at the point of discharge and the critical time. To find the required values, Eqs. (8.21) and (8.22) must be solved simultaneously, using a trial-and-error procedure. The ap-

propriate equations and substituted values are

a. Equation (8.21):

$$D_c = \frac{k}{k_2}\left(L_i e^{-k\theta_H^*}\right)$$

$$2.7 = \frac{0.26}{0.41}L_i e^{-0.26\theta_H^*}$$

b. Equation (8.22):

$$\theta_H^* = \frac{1}{k_2 - k}\ln\left[\frac{k_2}{k}\left(1 - \frac{D_i(k_2 - k)}{kL_i}\right)\right]$$

$$\theta_H^* = \frac{1}{0.41 - 0.26}\ln\left[\frac{0.41}{0.26}\left(1 - \frac{0.7(0.41 - 0.26)}{0.26L_i}\right)\right]$$

c. Substitute assumed values of θ_H^* in Eq. (8.21) and solve for L_i. Substitute computed values of L_i in Eq. (8.22) and solve for θ_H^*. Repeat until θ_H^* values match.

θ_H^*, d (assumed)	θ_H^*, d (computed)
2.0	2.65
2.25	2.67
2.50	2.69
2.70	2.71[†]
2.75	2.72
3.00	2.74

[†] Use 2.71.

d. Compute the ultimate BOD of the point of discharge using Eq. (8.21) (step 3a).

$$L_i = 8.59 \text{ g/m}^3$$

4. Determine the ultimate BOD of the wastewater.

Assuming the BOD upstream of the discharge is negligible, a materials balance can be made to calculate the effluent BOD_u:

$$q_w\left(BOD_{u_w}\right) + Q_r(0) = (Q_r + q_w)(8.59)$$

$$BOD_{u_w} = \frac{5.5}{1.5}(8.59) = 31.5 \text{ g/m}^3$$

Other Oxygen Sources and Sinks

Other oxygen sources and sinks (reactions leading to the consumption, or depletion, of oxygen) in rivers include photosynthesis, respiration of photosynthetic organisms, and benthic oxygen demand. Including expressions for these sources and sinks requires a considerable amount of field data. Photosynthetic contributions of oxygen occur only during daylight hours and are quite seasonal in nature. The primary contributors are algae, but the actual species present may vary a great deal. At night, algae continue to respire but do not produce oxygen, thus becoming an oxygen sink. Highly eutrophic waters may range in dissolved oxygen from supersaturated during hot, sunny days to anaerobic at night, although such conditions are unusual in flowing streams.

Photosynthesis

Photosynthesis by phytoplankton, particularly algae, is a major oxygen source in lakes and slow-moving streams. Photosynthetic organisms, like other autotrophic groups, utilize CO_2 as a carbon source. As a result, under rapid growth conditions, the carbonate equilibrium may be altered also. Because nitrogen is assimilated in the growth process, nitrogen concentrations (as $NH_3 - N$ or $NO_3^- - N$) may be very low in systems having high phytoplankton populations.

Net phytoplankton growth rate r_p is usually described as the sum of the specific growth, respiration, settling, and predation rates. Growth and respiration rates are usually measured in controlled laboratory systems and then extrapolated to field conditions. Settling rates are not significant in rivers but may be a factor in lakes. Predation by zooplankton and fish has been considered in a number of studies but is difficult to measure in the field.

Photosynthetic oxygen production is a function of the algae concentration (often measured as chlorophyll a), water depth and temperature, and light intensity and duration. Because there is more incident

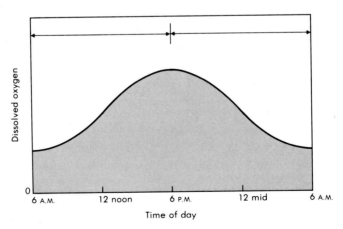

FIGURE 8.4

Diurnal variation of dissolved oxygen in water containing large algal populations. Under extreme conditions, dissolved oxygen may decrease to zero.

radiation when the sun is high in the sky than when it is near the horizon, the rate of photosynthesis is assumed to be sinusoidal. Respiration, on the other hand, is assumed to be constant because it does not depend on light radiation. Where large populations of algae are present, a diurnal variation in dissolved-oxygen concentration occurs, as shown in Fig. 8.4. An excellent summary of a number of models that have been proposed for these relationships is presented in Ref. [8.28].

In streams the principal algae mass is attached to bottom materials, particularly rocks. This allows the algal culture to extract nutrients from the passing water. Not surprisingly, the photosynthetic contribution to the oxygen balance is greatest in clear, shallow streams. Lakes are generally deeper than streams, and light extinction usually prevents significant contributions from attached algae. Suspended algal concentrations of 10 g/m^3 (dry mass) are considered large and contribute significantly to the oxygen balance.

Deoxygenation Due to Sludge Deposits

Large benthic oxygen demands are quite often localized to short reaches, where sludge accumulates. In well-managed rivers, organic sludges should not accumulate to an extent where demand is a problem. Many rivers have been subjected to uncontrolled municipal and industrial discharges for long periods, however, and deposits exist. In cases where removing the deposits is impractical, prediction of the effects on the oxygen balance is necessary.

A second source of benthic oxygen demand is from the growth of attached organisms such as the filamentous bacteria often released in wastewater discharges. Control of these growths is not always possible, and evaluation of their effects is necessary.

Benthic oxygen demand is usually measured by extracting small sections of bottom material for laboratory evaluation, although in-situ measurements have been made [8.25]. Zero-order models are the norm, and the use of Eqs. (8.24) and (8.25) and of the typical values given in Table 8.2 is recommended [8.27]:

$$r_B = \frac{r_B^*}{\overline{H}} \tag{8.24}$$

where

$r_B^* =$ uptake rate, $g/m^2 \cdot d$

$r_B =$ uptake rate, $g/m^3 \cdot d$

$\overline{H} =$ average depth, m

The temperature correction for r_B is

$$r_{B_T} = r_{B_{20}}(1.065)^{T-20} \tag{8.25}$$

where 1.065 is the temperature coefficient.

TABLE 8.2

Typical Benthic Oxygen Demand Rates at 20°C

BOTTOM TYPE	TYPICAL VALUE OF r_B^*, $g/m^2 \cdot d$
Filamentous bacteria ($10 \ g/m^2$)	-7
Municipal sewage sludge near outfall	-4
Municipal sewage sludge, "aged,"	
downstream of outfall	-1.5
Estuarine mud	-1.5
Sandy bottom	-0.5
Mineral soils	-0.07

Source: From Ref. [8.27].

The Modified Oxygen-Sag Equation

Estimation of photosynthesis and respiration is far more difficult than estimation of benthic demand. Temperature, nutrient concentration, sunlight, and turbidity are important coupled variables. Aquatic plants as well as algae are often significant contributors. Although a large body of literature has been developed, satisfactory general expressions are not available. Reported values of photosynthetic oxygen production r_{AP} (g O_2/m^2 surface area $\cdot$ d) range from 9 $g/m^2 \cdot d$ for the Truckee River in California, a shallow, fast-moving, unshaded mountain stream, to 0.3 $g/m^2 \cdot d$ for parts of the Neuse River system of North Carolina [8.27]. Reported respiration rate values (averaged over 24 hr) vary between 0.5 and 10 $g/m^2 \cdot d$.

Photosynthetic, respiratory, and benthic rates are generally added to Eq. (8.15) as zero-order (constant) terms. Performing a mass balance for the dissolved-oxygen concentration, as in the development of Eq. (8.13), the following steady-state differential equation is derived:

$$\frac{dC_{O_2}}{d\theta_H} = -kL + k_2\left(C_s - C_{O_2}\right) + r_{AP} + r_{AR} + r_B \qquad (8.26)$$

where

$r_{AP} =$ rate of algal photosynthesis, $g/m^3 \cdot d$

$r_{AR} =$ rate of algal respiration, $g/m^3 \cdot d$

$r_B =$ rate of benthic demand, $g/m^3 \cdot d$

Note that $r_{AP} > 0$ and r_{AR} and $r_B < 0$. Expressed in terms of the oxygen deficit, the integrated form of Eq. (8.26) is given by Eq. (8.27):

$$D_{O_2} = \frac{kL_i}{k_2 - k}\left(e^{-k\theta_H} - e^{-k_2\theta_H}\right) + D_i e^{-k_2\theta_H}$$

$$- \frac{r_{AP} + r_{AR} + r_B}{k_2}\left(1 - e^{-k_2\theta_H}\right) \qquad (8.27)$$

The use of this equation requires the evaluation of many parameters—in particular, r_{AP}, r_{AR}, and r_B. The magnitude of the effects of algae and sludge deposits on the oxygen economy of a river can be determined only from detailed testing and analysis of the river in question (see Example 8.2).

EXAMPLE 8.2

OXYGEN SAG ANALYSIS INCLUDING BENTHIC DEMAND

Determine the effect of a benthic demand on the oxygen concentrations in the stream of Example 8.1. Assume $r_B^* = -0.8/\text{m}^2 \cdot \text{d}$ and $\overline{H} = 3$ m.

SOLUTION:

1. When the benthic oxygen demand is considered, Eqs. (8.21) and (8.22) must be modified as given below:

$$\theta_H^* = -\frac{1}{k} \ln \left[\frac{k_2 D_c + r_B}{kL_i} \right]$$

$$\theta_H^* = \frac{1}{k_2 - k_1} \ln \left[\frac{k_2}{k} - \frac{k_2 D_i (k_2 - k) + r_B (k_2 - k)}{k^2 L_i} \right]$$

2. Substituting the values from Example 8.1 and $r_B = (-0.8 \text{ g/m}^2 \cdot \text{d})/(3 \text{ m}) = 0.27 \text{ g/m}^3 \cdot \text{d}$ in the modified equations, the corresponding values of θ_H^* and L_i found by trial and error are

$$\theta_H^* = 3.08 \text{ d} \ (2.71 \text{ d in Example 8.1})$$

$$L_i = 6.95 \text{ g/m}^3 \ (8.59 \text{ g/m}^3 \text{ in Example 8.1})$$

COMMENT

By comparing the above values to those obtained in Example 8.1 it can be seen that the benthic demand results in an increase of the distance to the critical sag point and a decrease in the allowable BOD input.

8.3
TIDAL RIVERS AND ESTUARIES

Tidal action results in water moving upstream during portions of the tidal cycle, even though the net flow over the cycle is downstream. An illustration is provided by considering the movement of a slug of tracer injected into a tidal stream at the point in time when the net velocity changes in direction from upstream to downstream (high-water slack). The dye concentration distribution will become increasingly dispersed as

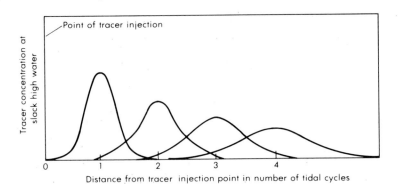

FIGURE 8.5

Tracer dispersion in tidal flow.

the tidal movement carries the dye downstream. If the concentration is measured at high-water slack during each cycle, the concentration profiles will be of the type shown in Fig. 8.5. Dye distribution will be nonsymmetrical, with the longer tail downstream. In simple estuaries, one-dimensional dispersion is the dominant factor. For tidal rivers and estuaries a U will be used to denote the net downstream velocity.

If a mass balance on a reactant B is made and time is averaged over the tidal cycle to give a pseudo-steady-state expression, Eq. (8.28) results:

$$D \frac{d}{dx}\left(\frac{dC_B}{dx} \right) - \frac{d}{dx}(UC_B) + r_B + S_B = 0 \tag{8.28}$$

where S_B represents a distributed source of B, and the expression is an extension of Eq. (6.16a). It is important to remember that D here represents the coefficient of dispersion, not the DO_2 deficit. For constant values of A and U and a first-order reaction, Eq. (8.28) reduces to Eq. (6.16b). It should be noted that the solution to Eq. (6.16b) given in Eq. (6.21) is not applicable to tidal rivers and estuaries. O'Connor [8.19] has developed a solution that can be used to obtain the concentration profile upstream and downstream of the peak, occurring at high-water slack for situations where $S_B = 0$ and $r_B = -kC_B$.

Upstream ($x < 0$):

$$\frac{C_B}{C_{B_i}} = \exp(j_1 x) \tag{8.29}$$

Downstream ($x > 0$):

$$\frac{C_B}{C_{B_i}} = \exp(j_2 x) \tag{8.30}$$

where

$$j_1 = \frac{U}{2D}\left(1 + \sqrt{1 + \frac{4kD}{U^2}}\right) \tag{8.31}$$

$$j_2 = \frac{U}{2D}\left(1 - \sqrt{1 + \frac{4kD}{U^2}}\right) \tag{8.32}$$

U = net downstream velocity calculated from freshwater
    flow, m/s

D = coefficient of dispersion, m^2/s

The value of C_B at $x = 0$ is affected by dispersion. O'Connor [8.19] gives the C_{B_i} value as

$$C_{B_i} = \frac{W}{Q\sqrt{1 + \frac{4kD}{U^2}}} \tag{8.33}$$

where

W = mass of reactant B entering per unit time, g/s

Q = freshwater flow rate, m^3/s

and other terms are as defined previously.

The time-averaged results of applying the steady-state model are probably the best that can be obtained. Changes occur throughout the tidal cycle, but variations in coefficient values are very difficult to predict adequately. There is also a spatial averaging involved in the determination of the coefficients that makes incorporation of non-steady-state terms an exercise in futility. It must be remembered that values of the velocity, velocity gradient, and tracer concentration are taken at particular points in time over long distances and are averaged to provide values for the models.

The Tidal Dispersion Coefficient

Determination of the tidal dispersion coefficient can be accomplished theoretically or experimentally. The following theoretical expression has been proposed [8.10]:

$$D = 63\, nU_T R_H^{5/6} \tag{8.34}$$

where

D = tidal dispersion coefficient, m^2/s

n = Manning's roughness coefficient

U_T = maximum tidal velocity, m/s

R_H = hydraulic radius, m

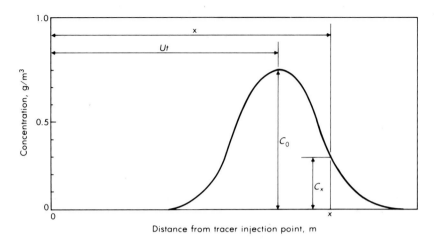

FIGURE 8.6

Definition sketch for the determination of the tidal dispersion coefficient from an instantaneous dye release.

Experimental methods of measuring D include instantaneous dye release and salinity intrusion. When instantaneous dye release is used, the dye concentration distribution is measured at a time after the release, as shown in Fig. 8.6. The peak of the dye has moved a distance Ut from the release point where U is the net freshwater stream velocity. Using the peak dye concentration as a reference, the concentration of dye at other points (x) is measured, and the value of D is determined as the average result, using Eq. (8.35):

$$\ln\frac{C_x}{C_0} = -\frac{(x - Ut)^2}{4Dt} \tag{8.35}$$

As before, measurements should be made at a consistent point in the tidal cycle, and the normal point is high-water slack.

Where salinity intrusion occurs, the salt concentration of the water can be used as a tracer to determine the value of D. For a conservative material such as salt, the generation rate is equal to zero, and the solution of Eq. (8.28) is

$$\ln\frac{C}{C_0} = \frac{U}{D}x \tag{8.36}$$

where x is the distance upstream from the sea (m) and is therefore < 0.

Values of D reported in the literature are listed in Table 8.3. Most of the values are less than 300 m²/s, but the range is quite large in all cases. Thus models suffer from a good deal of insensitivity. The effect of error in estimating parameter values is shown in Figs. 8.7 and 8.8, where a

TABLE 8.3

Tidal Dispersion Coefficients for Selected Estuaries

ESTUARY	D, m^2/s	REFERENCE
Upper Delaware	60–210	8.19
Lower Delaware	210–330	8.19
Upper Potomac	6–18	8.12
Middle Potomac	18–180	8.12
Lower Potomac	180–300	8.11
Waccasassa	60–81	8.28
New York Harbor	300–720	8.13
James	270–330	8.19
Hudson	240	8.19
Severn (England), summer	54–120	8.2
Severn, winter	123–530	8.2
Thames, low flow	54–84	8.2
Thames, high flow	330	8.2

range of values of U and D are used in the solution of Eqs. (8.29) to (8.33). Sensitivity to parameter variation can also be seen in these figures.

Two-Variable Systems

Coupled systems are of primary concern in water quality management. The previous example of dissolved-oxygen–BOD coupling is the most commonly encountered coupled system, but nutrient-algae and other

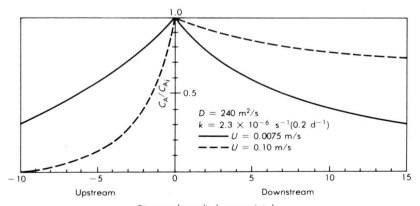

FIGURE 8.7

Distribution of reactant around discharge point for a steady-state dispersion model of a tidal estuary with a first-order reaction with variable freshwater velocities.

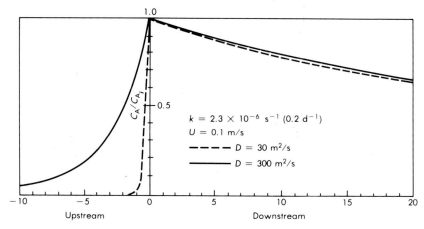

FIGURE 8.8

Distribution of reactant around discharge point for a steady-state dispersion model of a tidal estuary with a first-order reaction with variable tidal dispersion coefficients.

systems are important also. For the dissolved-oxygen–BOD problem, it is again easier to solve for the dissolved-oxygen deficit D_{O_2}. For one-dimensional steady-state conditions, Eq. (8.28) is written as

$$D\frac{d^2 D_{O_2}}{dx^2} - U\frac{dC_D}{dx} - k_2 D_{O_2} + kL = 0 \tag{8.37}$$

$D_{O_2} = 0$ at $x = -\infty$

$D_{O_2} = 0$ at $x = +\infty$

When BOD is the reactant in Eqs. (8.29), (8.30), and (8.33), O'Connor [8.19] has developed the following solutions for the dissolved-oxygen deficit:

Upstream ($x < 0$):

$$D_{O_2} = \frac{k}{k_2 - k}\frac{W}{Q}\left[\frac{\exp\frac{U}{2D}(1 + j_3)x}{j_3} - \frac{\exp\frac{U}{2D}(1 + j_4)x}{j_4}\right] \tag{8.38}$$

Downstream ($x > 0$):

$$D_{O_2} = \frac{k}{k_2 - k}\frac{W}{Q}\left[\frac{\exp\frac{U}{2D}(1 - j_3)x}{j_3} - \frac{\exp\frac{U}{2D}(1 - j_4)x}{j_4}\right] \tag{8.39}$$

where

$$j_3 = \sqrt{1 + \frac{4kD}{U^2}} \tag{8.40}$$

$$j_4 = \sqrt{1 + \frac{4k_2 D}{U^2}} \tag{8.41}$$

$W =$ mass of BOD entering per unit time, g/s

$Q =$ freshwater flow rate, m³/s

and other terms are as defined previously. Example solutions for two velocity conditions are shown in Fig. 8.9.

Multidimensional Systems

A *complex system* might be defined as any system that does not fit a simple one-dimensional model [see Eq. (8.35)]. Such a definition would not depend on the first-order model because numerical solutions can be generated for more sophisticated kinetics quite easily. The primary reason for complexity can be seen in the abscissa of Fig. 8.8. Predicted changes occur over tens of kilometers. Most estuaries change in geometry, and a distance of 10 km may result in a large variation in D and U. Most estuaries are characterized by transverse and vertical mixing due to wind, geometry, or stratification. Where these effects are not dominant,

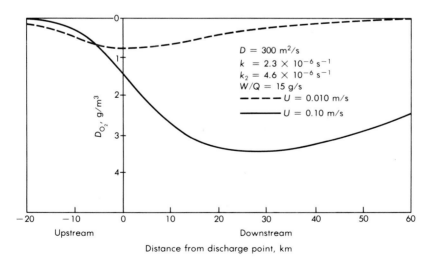

FIGURE 8.9

Oxygen deficit predicted for two freshwater velocities in a tidal estuary. Note that the greater upstream deficit is associated with the lower velocity.

one-dimensional models can be used; otherwise more complex grid models must often be used.

Two approaches are used in breaking estuaries into segments. The continuous-solution approach utilizes the dispersion model [Eq. (8.35)] and forces the deficits (D_{O_2}) to be equal at the boundaries. A CFSTR cascade is used to model the segmented system in the finite-section approach.

Continuous-Solution Method

A general mathematical solution to Eq. (8.37) is

$$D_{O_2} = G\exp(g_1 x) + H\exp(g_2 x) + \frac{k}{k_2 - k}\left[I\exp(j_1 x) + J\exp(j_2 x)\right]$$

(8.42)

where

$G, H, I, J = \text{constants}$

$$g_1 = \frac{U}{2D}\left(1 + \sqrt{1 + \frac{4k_2 D}{U^2}}\right)$$

(8.43)

$$g_2 = \frac{U}{2D}\left(1 - \sqrt{1 + \frac{4k_2 D}{U^2}}\right)$$

(8.44)

and other terms are as defined previously.

Because conditions may vary in the various segments, the oxygen transfer coefficient k_2 may change and should be double-subscripted; the dispersion coefficient and velocity may also vary. The general equation can then be written where the subscript j refers to the segment number:

$$D_{O_{2_j}} = G_j \exp(g_{1_j} x) + H_j \exp(g_{2_j} x)$$

$$+ \frac{k_j}{k_{2_j} - k_j}\left[I_j \exp(j_{1_j} x) + J_j \exp(j_{2_j} x)\right]$$

(8.45)

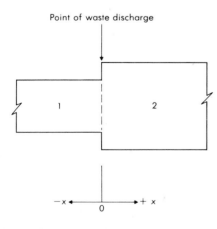

Point of waste discharge

$-x \longleftarrow \quad \longrightarrow + x$

0

FIGURE 8.10

Definition sketch for a two-segment estuary.

Equation (8.45) must be solved at the boundary values to determine the coefficient (G_j, H_j, I_j, J_j) values. For any system with more than two segments, the problem is quite tedious. Matrix methods are useful, but the major problem is generating boundary values and evaluating terms in the intermediate segments. Consideration of the two-segment system shown in Fig. 8.10 provides ample explanation.

Appropriate boundary conditions are

At $x = -\infty$

$$D_{O_{21}} = 0$$

At $x = \infty$

$$D_{O_{22}} = 0$$

At $x = 0$:

$$D_{O_{21}} = D_{O_{22}}$$

Writing a materials balance at $x = 0$ results in

$$\left| Q_1 D_{O_{21}} - D_1 A_1 \frac{dD_{O_{21}}}{dX_1} \right|_{x=0} = \left| Q_2 D_{O_{22}} - D_2 A_2 \frac{dD_{O_{22}}}{dx} \right|_{x=0}$$

mass in   mass out

The four coefficients—H_1, J_1, G_2, I_2—can be shown to be equal to zero by application of the boundary conditions at $x = \pm\infty$, leaving four coefficients—G_1, I_1, H_2, J_2—to be evaluated. Applying the third boundary condition results in Eq. (8.46):

$$D_{O_{20}} = G_1 + \frac{k}{k_{21} - k} I_1 = H_2 + \frac{k}{k_{22} - k} J_2 \qquad (8.46a)$$

$$G_1 - H_2 = \frac{k}{k_{22} - k} J_2 - \frac{k}{k_{21} - k} I_1 \qquad (8.46b)$$

The fourth boundary condition can be applied by evaluating Eq. (8.45) and its derivative at $x = 0$ and substituting into the materials balance.

$$(Q_1 - D_1 A_1 g_{11}) G_1 + \frac{k}{k_{21} - k} (Q_1 - D_1 A_1 j_{11}) I_1 =$$

$$(Q_2 - D_2 A_2 g_{22}) H_2 + \frac{k}{k_{22} - k} (Q_2 - D_2 A_2 j_{22}) J_2 \qquad (8.47)$$

Writing Eqs. (8.46) and (8.47) in matrix form:

$$\begin{bmatrix} a_{11} & a_{12} \\ a_{21} & a_{22} \end{bmatrix} \begin{bmatrix} G_1 \\ H_2 \end{bmatrix} = \begin{bmatrix} b_{11} & b_{12} \\ b_{21} & b_{22} \end{bmatrix} \begin{bmatrix} I_1 \\ J_2 \end{bmatrix} \qquad (8.48)$$

where

$$a_{11} = 1 \qquad\qquad\qquad a_{12} = -1$$

$$a_{21} = Q_1 - D_1 A_1 g_{11} \qquad a_{22} = -(Q_2 - D_2 A_2 g_{22})$$

$$b_{11} = \frac{-k}{k_{21} - k} \qquad\qquad b_{12} = \frac{k}{k_{22} - k}$$

$$b_{21} = \frac{-k}{k_{21} - k}(Q_1 - D_1 A_1 j_{11}) \qquad b_{22} = \frac{k}{k_{22} - k}(Q_2 - D_2 A_2 j_{22})$$

The values of I_1 and J_2 can be obtained independently from a first-order decay model, because the decay term (BOD) is independent of oxygen concentration. Development of the expressions for I_1 and J_2 using the alternative method is identical to the above analysis, but the expressions are simpler and result in Eq. (8.49):

$$J_2 = I_1 = \frac{W}{D_1 A_1 j_{11} - Q_1 - D_2 A_2 j_{22} + Q_2} \qquad (8.49)$$

Finite-Section Analysis

Finite-section analysis is similar to the analysis discussed in Example 6.3. The input-output terms must include a tidal mixing expression as well as the flow term. Thomann [8.27] uses an average length l of the adjacent section to form a pseudo concentration gradient:

$$\frac{dC}{dx} = \frac{C_n - C_{n-1}}{l_{n-1}} \qquad (8.50)$$

where

$C = $ concentration of contaminant, g/m^3

$x = $ distance, m

$l = $ length of adjacent section, m

The concentration gradient is then used to describe input (or output due to tidal mixing), and the mass balance on the nth section is given by Eq. (8.51):

$$\underbrace{\frac{dc}{dt} V_n}_{\text{accumulation}} = \underbrace{\left[(QC)_{n-1} - D_{n-1} A_{n-1} \left(\frac{C_n - C_{n-1}}{l_{n-1}} \right) \right]}_{\text{inflow}}$$

$$\underbrace{- \left[(QC)_n - D_n A_n \left(\frac{C_{n+1} - C_n}{l_n} \right) \right]}_{\text{outflow}} + \underbrace{W_n + r V_n}_{\text{generation}} \qquad (8.51)$$

where

C = concentration of contaminant, g/m^3

V = volume, m^3

D = coefficient of dispersion, m^2/s

A = cross-sectional area, m^2

l = length of adjacent section, m

W_n = waste loads, g/s

r = rate of reaction, g/m$^3 \cdot$ s

As in the continuous-solution method, the non-steady-state term is eliminated and Eq. (8.51) is reduced to an algebraic expression. The advantage of finite-section analysis is the ability to insert waste loads (W_n) at particular sections. In continuous-solution analysis, loads must

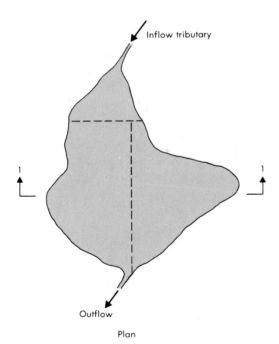

Inflow tributary

Outflow

Plan

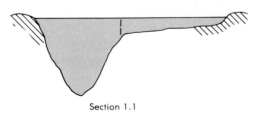

Section 1.1

FIGURE 8.11

Plan and elevation of a lake characterized by distinct shallow and deep regions. Dashed lines show possible sections for multidimensional analysis.

be either distributed or at the boundaries. Two- or three-dimensional modeling is also possible using finite sections, but the complexity of natural systems and the difficulty of obtaining data for verification make multidimensional modeling extremely difficult.

In cases where stratification or depth variation is sharp, as shown in Fig. 8.11, multidimensional models are feasible, however. A second type of multidimensional problem is shown in Fig. 8.12, where a river delta results in multiple channels. Each channel can be set up as a finite section, and because physical measurements can be taken, the modeling effort can be more precise.

Link-Node Models

Link-node models are a more flexible form of the finite-section model. One, the "dynamic estuary model," was developed by Water Resources Engineers, Inc., during the 1960s and uses "pipes" between the CFSTRs, which make up the nodes [8.21]. Incorporation of the pipes allows a

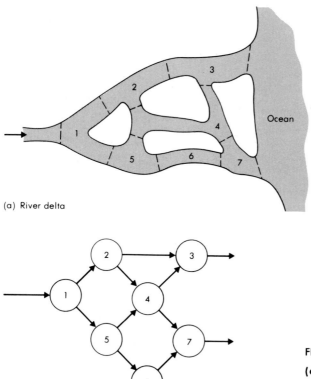

(a) River delta

(b) Equivalent multidimensional network.

FIGURE 8.12

(a) River delta and (b) resulting multidimensional network of a model composed of finite element sections.

degree of control over the flow into each section. Reactions are allowed to take place only in the CFSTRs. Thus a structured response to flow variation can be built into the system.

8.4
OCEAN OUTFALLS

Communities located on the ocean often discharge wastewater directly to the sea through submerged outfalls (Fig. 8.13). In most cases, a suitable receiving stream is unavailable, and the purpose of the ocean outfall installation is to provide enough dilution of the previously treated wastewater to prevent deterioration of the environment near the discharge point and the surfacing of recognizable waste materials. In the past, the idea of treating wastewater prior to ocean disposal was considered quite novel, or perhaps bizarre. The ocean was considered an infinite sink, and it was inconceivable that relatively small flows of wastewater could damage the marine environment. But closed beaches and dead zones near outfalls have led to more careful consideration of the issue. Although U.S. Environmental Protection Agency regulations do not favor ocean disposal, discharge to the ocean has been allowed where it can be shown that the effects of such a discharge will be minimal and an acceptable level of treatment is provided.

Thermal effects are of particular interest in outfall design. One reason is that discharges generally are lower in density than the receiving

FIGURE 8.13

Ocean outfall from a small California coastal community. The outfall was exposed because of beach erosion.

waters and tend to be buoyant. A second reason is that local warming of the environment changes the ecological makeup of the discharge area. Heated effluents, such as cooling water from power plants, can result in greatly increased growth rates and changes in species makeup. Because productivity of the region is increased, this effect is sometimes referred to as *thermal enrichment* as well as *thermal pollution*. In either case, a decision must be made about the desirability of the change.

Ocean Outfall Design

Submarine outfalls are designed to provide maximum initial mixing or dilution and to ensure that pollutants do not surface or reach the shore in measurable quantities. Formation of sludge deposits on the ocean floor and related changes in floor ecology have only recently been considered a problem. Recent studies of the ocean floor have resulted in concern about long-term effects on important marine species, including kelp. Quite possibly, the solution will be to prohibit discharge of sludges and require a higher degree of treatment prior to discharge.

An ocean outfall consists of the pipe carrying the effluent to the discharge area and the end section, or diffuser (Fig. 8.14). Among the more common types of diffusers in use are those with side ports and risers (Fig. 8.15). Side-port diffusers have a number of holes on each side of the final section. Port spacing is close enough to allow the discharge to be considered slot jets [8.3]. Risers are used where the pipe is in a trench,

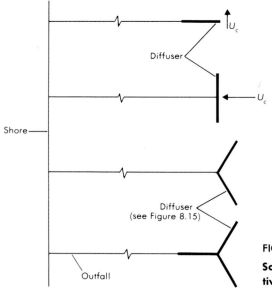

FIGURE 8.14

Schematic of ocean outfall with alternative diffuser arrangements.

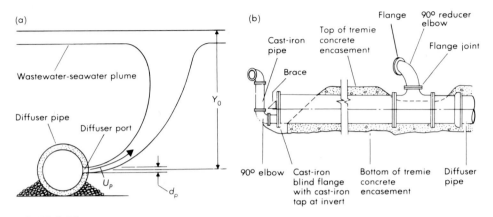

FIGURE 8.15

Typical ports used in ocean outfall diffusers: (a) side ports built in diffuser section and (b) riser ports built in diffuser section.

is surrounded by rock or coral formations, or must be heavily ballasted along its length. Horizontal discharge is achieved by using elbows or side ports at the top of the riser. Use of risers has two disadvantages: The construction process is more expensive, and, of more importance, the chance of damage by anchors, fishing equipment, or dredges is real. Ports are placed so that they are not directly opposite each other. Such placement of the ports has advantages with respect to pipe strength and minimizing jet mixing from the two sides. In the past, the diameter of the diffuser discharge ports varied from 75 to 150 mm. Recently, with improvements in pretreatment, port diameters as small as 20 mm have been used.

Ocean Outfall Operation

The operation of an ocean outfall can be described as follows. Because the density of wastewater is lighter than that of seawater, wastewater (or treated effluent) discharged from a diffuser port usually forms a buoyant plume (Fig. 8.16a). As the plume rises toward the surface, seawater is entrained until a stable, somewhat homogeneous wastewater-seawater mixture (often called a *wastewater field*) develops, usually at or near the surface. As seawater is entrained in the buoyant plume, the concentration of a contaminant will be reduced because of dilution. The dilution achieved as the wastewater rises to the surface is known as *initial dilution*. It should be noted that if the degree of stratification in the ocean at the point of discharge is sufficient, a submerged field will develop (Fig. 8.16b).

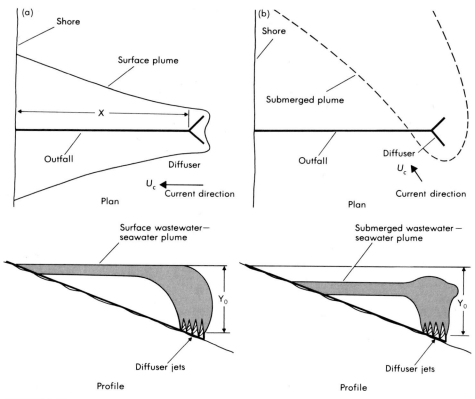

FIGURE 8.16

Definition sketch for the formation and movement of a wastewater-séawater plume from an ocean outfall: (a) surface plume and (b) submerged plume.

Once a field starts to develop, it will begin to move with the ocean currents, as shown in Fig. 8.16. The field spreads as it moves away from the point of discharge because of turbulent hydrodynamic dispersion (also known as *eddy diffusion*). As the field spreads, the concentration of a contaminant is further reduced. The contaminant reduction achieved as the plume spreads is termed *dilution due to dispersion*.

In addition to the initial dilution and the dilution due to dispersion, the concentration of a reactive contaminant can also be diluted by *natural decay* from chemical and biological reactions and physical means. The most common cause of decay is that attributed to the natural action of bacteria and other microorganisms. Thus in determining whether the concentration of a contaminant will be acceptable at some distance from the point of discharge, three factors must be considered: initial dilution,

dilution due to dispersion, and decay. Each of these factors is examined separately in the following discussion.

Initial Dilution

Initial dilution resulting from the jet discharge is a function of the relative density of the wastewater and the seawater, jet diameter, jet velocity, current velocity, and depth of discharge. Two methods are used to estimate the initial dilution. They are the current flux and the buoyant plume methods, which are considered below.

Current Flux Method

In moderate currents, the initial dilution S_1 can be estimated by Eq. (8.52):

$$S_1 = \frac{L_d h_0 U_c}{Q} \tag{8.52}$$

where

S_1 = dilution factor due to initial dilution

L_d = length of diffuser, m

h_0 = depth of water over diffuser referenced to top of diffuser
or ocean bottom at diffuser location, m

U_c = current velocity over diffuser, m/s

Q = wastewater flow rate, m^3/s

As written, Eq. (8.52) represents the concentration that would exist if the water flowing over the diffuser in a given time period were to mix completely with the wastewater discharged from the outfall during the same time period. Because complete mixing seldom occurs, many designers will use an effective mixing depth equal to 0.50 to 0.25 times the value of h_0.

Buoyant Plume Method

For stagnant water Abraham [8.1] developed a correlation between the dilution ratios achieved and the diameter of the port, the depth of discharge, and the densimetric Froude number. At low-current velocities, the initial correlation developed by Abraham, as shown in Fig. 8.17, can be used to estimate the initial dilution. The densimetric Froude number N_{DF} can be computed using Eq. (8.53):

$$N_{DF} = \frac{U_p}{\sqrt{\left(\dfrac{\rho_{sw} - \rho_{ww}}{\rho_{sw}}\right) g d_p}} \tag{8.53}$$

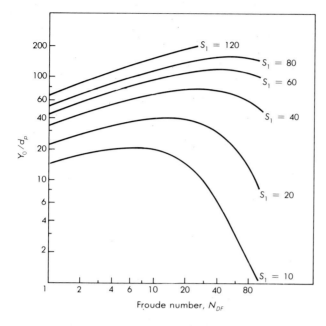

FIGURE 8.17

Initial mixing: S_1 in a turbulent jet.

Source: From Ref. [8.1].

where

N_{DF} = densimetric Froude number, unitless

U_p = velocity of fluid leaving discharge port, m/s

ρ_{sw} = density of seawater, kg/m³

ρ_{ww} = density of wastewater, kg/m³

g = acceleration due to gravity, 9.81 m/s²

d_p = diameter of discharge port, m

As noted above, the curves presented in Fig. 8.17 are based on experimental results with quasistagnant systems. The term *low-current velocities* is difficult to define, but representative values are in the range of 0.5 m/s. In practice, both methods should be applied and the lower S_1 value used (see Example 8.3).

EXAMPLE 8.3

DETERMINATION OF INITIAL DILUTION FROM SUBMARINE OUTFALL

An outfall diffuser is located at a depth of 20 m in an area having a minimum current velocity of 0.2 m/s. Seawater density is 1030 kg/m³, and the wastewater density is 980 kg/m³. Wastewater is discharged through 10 ports, each 0.10 m in diameter, facing opposite directions and spaced 2 m apart. The wastewater flow rate is 0.3 m³/s. Estimate the initial dilution using the two approaches defined by Eq. (8.52) and Fig. 8.17. Use an effective depth of 0.5 h_0 in applying Eq. (8.52).

SOLUTION:

1. Determine the initial dilution using the current flux approach [Eq. (8.52)] with an effective mixing depth of 0.5 h_0. Assume the effective length of the diffuser is 10 m (20 ports × 2.0 m/port).

$$S_1 = \frac{L_d h_0 U_c}{Q}$$

$$= \frac{(20 \text{ m})(20 \text{ m} \times 0.5)(0.2 \text{ m/s})}{0.3 \text{ m}^3/\text{s}}$$

$$= 133$$

2. Determine the initial dilution using the buoyant plume method.
   a. Compute Y_0/d_p where Y_0 is the depth of water over the diffuser referenced to the centerline of the diffuser discharge ports, d_p. For the purposes of this problem assume Y_0 is equal to 20 m.

   $$Y_0/d_p = \frac{20 \text{ m}}{0.10 \text{ m}} = 200$$

   b. Compute the diffuser port discharge velocity:

   $$A_p = \pi(0.10)^2/4$$
   $$= 0.0079 \text{ m}^2$$
   $$U_p = \frac{0.3 \text{ m}^3/\text{s}}{(10)(0.0079 \text{ m}^2)}$$
   $$= 3.80 \text{ m/s}$$

   c. Compute the densimetric Froude number:

   $$N_{DF} = \frac{3.80}{\sqrt{\left(\frac{1030 - 980}{1030}\right)(9.81)(0.10)}}$$

   $$= 17.4$$

   d. Determine S_1 using Fig. 8.17 and the values of Y_0/d_p and N_{DF} determined above.

   $$S_1 \approx 120 \text{ (from Fig. 8.17)}$$

COMMENT

In general, the lower of two values is used as an estimate of the initial dilution.

Dilution due to Dispersion

Dispersion in the buoyant plume has been modeled using a number of approaches [8.3, 8.7, 8.9]. The model proposed by Brooks [8.3] is based on the assumptions that the discharge behaves as a slot jet, that vertical and longitudinal mixing are much less than horizontal mixing, and that the initial dispersion processes have equalized the density of the wastewater and ambient seawater (see Fig. 8.16a). Brooks used the following exponential equation to model the lateral dispersion coefficient:

$$D_L = \alpha L_x^n \tag{8.54}$$

where

D_L = coefficient of lateral dispersion, m²/s

α = empirical coefficient

L_x = width of waste plume, m

n = exponent

Values of α reported in the literature vary over a considerable range. For an n value equal to 4/3, values for α range from 0.0015 to 0.049 cm$^{2/3}$/s. Pearson [8.23] has suggested a value of 0.01 cm$^{2/3}$/s. Assuming the plume begins as a slot, the initial value of the coefficient is computed using the effective length for the width of the plume.

The dilution due to dispersion, S_2, is obtained by solving for the concentration C_x at a distance x from the diffuser and dividing the computed value into the initial concentration of the waste after initial dilution. The dilution due to dispersion, S_2, and the plume width L_x are given in Eqs. (8.55) and (8.56), respectively [8.3, 8.4]:

$$S_2 = \frac{C_0}{C_x} = \frac{1}{\mathrm{erf}\sqrt{\dfrac{3/2}{[1 + \frac{2}{3}\beta(x/b)]^3 - 1}}} \tag{8.55}$$

$$L_x = b\left(1 + \frac{2}{3}\frac{x}{b}\right)^{3/2} \tag{8.56}$$

where

S_2 = dilution factor due to dispersion, C_0/C_x

C_0 = concentration of wastewater after initial dilution, g/m³

C_x = concentration of wastewater at centerline of plume at a distance x from the diffuser, g/m³

erf(x) = error function (x) (see Appendix J)

$\beta = 12 D_L/U_c b$, dimensionless number

D_L = coefficient of lateral dispersion, m²/s

U_c = critical current velocity, m/s

b = effective length of diffuser

$x =$ distance along centerline of plume measured from the diffuser, m

$L_x =$ effective width of plume at a distance x from the diffuser (see Fig. 8.16a), m

A nomograph for the solution of Eq. (8.55) is presented in Fig. 8.18 [8.5]. Effective diffuser length is the length at right angles to the current. Because current direction varies with time, this can be an important factor. Often, to maximize dispersion, diffusers are laid at angles other than 90° to the shore, or in Y or V shapes (see Fig. 8.14).

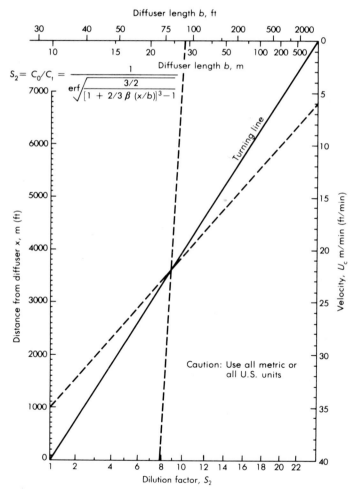

FIGURE 8.18

Nomograph solution for Eq. (8.55).

Source: From Ref. [8.5].

Because plumes tend to rise because of their buoyancy, as well as to move downcurrent, the actual flow path distance is generally greater than x. Methods of estimating the actual distance traveled are available, but a more conservative approach is to use the downcurrent distance. In general, the expressions presented for dilution provide conservative estimates of dilution, but it should be clear that the values generated are only estimates. Many simplifying assumptions have been made in the derivations, including the massive assumption that conditions are steady in the marine environment.

Dilution due to Decay

A third factor, in addition to initial dilution and dispersion, that is important in reducing the concentration of a waste once discharged in an ocean environment is decay. Depending on the contaminant under study, decay can be produced by natural causes; by photochemical reactions; by hydrolysis reactions; and by physical means such as flocculation, precipitation, and sedimentation. As noted in Chapter 7, decay is often modeled as a first-order function:

$$C_t = C_0 e^{-kt} \tag{7.25}$$

In an ocean outfall system the time term would be related to distance along the centerline of the plume. The application of Eq. (7.25) is considered in Example 8.4.

EXAMPLE 8.4

REDUCTION OF BACTERIA WITH AN OCEAN OUTFALL

An ocean outfall diffuser is located at a water depth of 25 m at a distance of 5000 m from a public beach and is at an angle of 45° to the water's edge. The total wastewater discharged through six 0.15-m-diameter diffuser ports spaced at 3 m is 0.3 m³/s. The coliform bacteria concentration in the discharge is $10^5/mL$, and the first-order death rate in seawater of coliforms is approximately 0.3 hr^{-1}. Determine the maximum concentration of coliform bacteria reaching the beach if the velocity of the onshore current is 0.5 m/s. Assume $\rho_{ww} = 980$ kg/m³ and $\rho_{sw} = 1030$ kg/m³.

SOLUTION:

1. Determine the initial dilution.
    a. Using the current flux approach [Eq. (8.52)] with an effective mixing depth of 0.5 h_0:
    Assume the effective length of the diffuser, based on port spacing, is 18 m (6 ports × 3 m/port):

$$S_1 = \frac{(18 \text{ m} \times \sin 45°)(25 \text{ m} \times 0.5)(0.5 \text{ m/s})}{0.3 \text{ m}^3/\text{s}} = 265$$

b. Using the buoyant plume method:
   (1) Compute Y_0/d_p

   $$Y_0/d_p = 25 \text{ m}/0.15 = 167$$

   (2) Compute the diffuser port discharge velocity:

   $$A_p = \pi(0.15)^2/4$$
   $$= 0.018 \text{ m}^2$$
   $$U_p = \frac{0.3 \text{ m}^3/\text{s}}{(6)(0.018 \text{ m}^2)}$$
   $$= 2.78 \text{ m/s}$$

   (3) Compute the densimetric Froude number:

   $$N_{DF} = \frac{2.78 \text{ m/s}}{\sqrt{\left(\dfrac{1030 - 980}{1030}\right)(9.81)(0.15)}} = 10.4$$

   (4) Determine S_1 using Fig. 8.17:

   $$S_1 \approx 120$$

c. Use a dilution factor of 120, the lower of the two values for S_1.

2. Determine the dilution due to dispersion. The dilution due to dispersion is found using Fig. 8.18. Note that the effective diffuser length b is 12.7 m (18 m $\times$ sin 45°):

$$S_2 = 17$$

3. Determine dilution due to decay. The decrease in concentration due to death of coliform bacteria (C_0/C) will be labeled S_3. Inverting Eq. (7.25) given above yields

$$S_3 = \frac{C_0}{C} = e^{kt}$$

$$t = \frac{5000 \text{ m}}{0.5 \text{ m/s}} = 10,000 \text{ s}$$

$$= 2.78 \text{ hr}$$
$$S_3 = e^{(0.3)(2.78)} = 2.3$$

4. Determine maximum coliform concentration at the beach. The maximum concentration at the beach, taking into account all of the dilution factors tending to reduce the concentration of coliform organisms, is

$$C_b = C_i/(S_1 S_2 S_3)$$
$$= (10^5/\text{mL})/(120 \times 17 \times 2.3)$$
$$= 21.3/\text{mL}$$

8.5

EFFECTS OF WATER RESOURCES MANAGEMENT FACILITIES

While the focus of this chapter has been on the water quality changes that occur in rivers and estuaries, it is equally important to note that facilities constructed for the management of rivers can also have a significant effect on water quality.

Unintended Changes due to Construction of Facilities

Quite often the effects of constructing management facilities on rivers for purposes of flood control or land reclamation are not recognized in the design phase, but experience has provided many lessons. Sediment carried by rivers is removed in the relatively quiet waters of reservoirs, resulting in the gradual filling of the reservoir, clearer water downstream, and a loss of gravels needed for fish spawning beds. Thus the character of a river is changed below a dam as well as above. Before filling, reservoirs are stripped of vegetation because of the effect of decaying organic matter on water quality and the potential navigation hazards of trees. The soil disturbance caused by the stripping results in release of nutrients and in more rapid eutrophication.

Similar problems occur when development takes place on lakeshores. Nutrients are released and washed into the lake, and the eutrophication process is accelerated. Marina construction often results in production of excellent conditions for algal growth and eventually in green rather than blue lagoons. Even more severe problems occur with construction of small artificial lakes. Often these lakes are in arid regions, and the lake is an attractive feature of a subdivision or other development. Because water is in short supply and quite often must be purchased, the rapid turnover necessary to prevent algal blooms is impossible. Runoff from houses built along the shore enters the lake, and in some cases septic tank leachate drains into the lake; both such drainages accelerate eutrophication.

Aswan High Dam: A Case Study

In recent years the need for agricultural development in developing nations has placed great stress on available water resources, and many large facilities have been constructed to provide more water. Because the focus of these projects is on providing water for irrigation, water quality effects and relationships are often overlooked. An example is provided by the construction of the Aswan High Dam and Lake Nasser in Egypt [8.21]. This great project increased that nation's potentially arable land from 2.8 million to 3.6 million ha. The hydroelectric capacity of the project is 10 billion kWh per annum, a particularly valuable asset as fossil fuels become increasingly expensive. Filling and utilization of Lake

Nasser began in 1964. By 1970 Egypt was self-sufficient in wheat and an exporter of rice. The fish yield of Lake Nasser was up to 5000 tonnes annually. Optimists believe yields could be considerably greater, although the long-term outlook is uncertain.

Balanced against these benefits are a number of negative results of the project. Virtually all sediment and the accompanying nutrients have been trapped behind the dam since 1964. The nutrient loss has destroyed the Mediterranean sardine fishery (18,000 tonnes in 1964, 500 tonnes in 1969). Five Nile Delta lakes have also become less productive. Salt water is penetrating the delta, and currently irrigable areas are being lost. The decrease in sediment deposition along the delta shoreline has resulted in a recession of several millimeters per year. The deposition of 50 to 100 million tonnes of nutrient-laden silt in Lake Nasser has greatly decreased the nutrients in irrigation water and spawned a large fertilizer industry that consumes a significant fraction of the dam's hydroelectric power. Nile mud was previously a source of material for brick making, but the raw material now is deposited in the reservoir. Finally, and perhaps most importantly, control of flows below the dam has led to an increase in schistosomiasis infection.

Judging the net value of a project such as the Aswan High Dam is impossible. It is clear that many of the negative factors of the project were either not considered or ignored in the design process, and the actual benefit to Egypt will be far less than predicted. Mitigation of the project's impacts will be difficult, particularly because mitigating measures were not designed into the original system. Perhaps the most disturbing impact is the direct effect on people in terms of the increased incidence of shistosomiasis and the loss of age-old delta lands.

KEY IDEAS, CONCEPTS, AND ISSUES

- The Streeter-Phelps oxygen-sag model is commonly used to model the oxygen resources in a river.

- In the simple form of the Streeter-Phelps model, the rate of deoxygenation is balanced by atmospheric reaeration. If the rate of deoxygenation exceeds the rate of reaeration, low-dissolved-oxygen or anaerobic conditions can develop in the river.

- Deoxygenation in rivers is brought about by bacterial oxidation of suspended and dissolved organic matter and by the oxygen demand of benthic deposits. The latter demand is not considered in the conventional Streeter-Phelps analysis.

- Reoxygenation (reaeration) is brought about by the absorption of atmospheric oxygen. The rate of reaeration is a function of the average river velocity and depth.

- Other factors that are often considered in the development of oxygen models for rivers include algal photosynthesis, algal respiration, and benthic demands.
- In evaluating the oxygen resources of tidal rivers and estuaries, dispersion caused by tidal mixing is an important factor.
- Three factors are involved in the dilution of a contaminant discharged to the ocean environment through an outfall: initial dilution, dispersion, and decay.
- In an ocean outfall system initial dilution is brought about by the mixing of wastewater and seawater as the wastewater is discharged from a diffuser. Dilution by dispersion is brought about by the spreading of the wastewater-seawater plume as it moves away from the diffuser. Dilution by decay results from several reactions, including those of natural death, photochemical oxidation, and hydrolysis.

DISCUSSION TOPICS AND PROBLEMS

8.1. Compare the O'Connor and Dobbins [Eq. (8.8)] and Churchill et al. [Eq. (8.10)] expressions for k_2, the reaeration coefficient, by plotting the value of k_2 as a function of the velocity u at 20°C. Use three depths: 0.5, 2.0, and 4.0 m. Discuss the possible reasons for the observed differences. How would you choose a value for a stream that had not been experimentally evaluated?

8.2. A wastewater discharge to a small stream has the following characteristics:

	SEASON	
ITEM	Winter	Summer
Discharge		
q_w, m³/s	0.1	0.1
T, °C	21	21
BOD_5, g/m³	100	100
DO, g/m³	0.0	0.0
$k_{20°C}$, d⁻¹	0.2	0.2
Stream		
Q, m³/s	3	1.5
T, °C	12	25
BOD_5, g/m³	0.0	0.0
$DO_{deficit}$, g/m³	0.0	0.0
k_T, d⁻¹	0.08	0.26
k_{2_T}, d⁻¹	0.55	0.40

Determine the greatest dissolved-oxygen deficit and the minimum dissolved-oxygen concentration downstream.

8.3. A wastewater flow of 1.0 m^3/s is discharged into a river flowing at 49 m^3/s. Temperature of the river downstream of the discharge is 15°C. The ultimate BOD of the discharge is 150 g/m^3, and the 20°C k value is 0.23 d^{-1}. The dissolved-oxygen concentration just downstream of the discharge is 8.0 g/m^3, and the measured (at 15°C) k_2 value is 0.15 d^{-1}. Determine (a) the minimum dissolved-oxygen concentration downstream of the discharge and (b) the travel time to the sag point. Assume $Cl^- = 0$ g/m^3 and upstream $BOD = 0$.

8.4. The Knostrop Sewage Treatment works of Leeds, England, was designed for a dry-weather flow of 120,000 m^3/d of domestic and industrial wastewater. Discharge from the works is to the river Aire, which has a dry-weather flow of 2.55 m^3/s. Assuming that photosynthesis is negligible and that the benthic demand is typical of municipal wastewater downstream of outfalls, determine the critical stream dissolved-oxygen concentration as a function of effluent BOD and two stream temperatures (use $T = 10$ and 20°C). Prepare a plot of the critical stream dissolved oxygen as a function of the effluent BOD. Assume the dissolved-oxygen deficit upstream of the outfall is 2.0 g/m^3, the value of k is 0.2 d^{-1}, and the dissolved oxygen in the discharge is 3.0 g/m^3. River depth can be taken as 1.5 m, and width is approximately 20 m.

8.5. A small stream ($q = 15$ m^3/s) flows into a larger river ($Q = 300$ m^3/s). The larger river is 300 m wide and has an average depth of 6 m and a hydraulic gradient of 0.001 m/m. Determine the distance required for complete mixing to be achieved. How can this answer be correlated with the Streeter-Phelps model?

8.6. A discharge from a large wastewater treatment plant is released near the bank of a large river. The discharge contains several grams per cubic meter of chlorine, used as a disinfectant, which is highly toxic to fish. Is this type of discharge better than a discharge (a) in the center of the river? (b) at several points across the river's width?

8.7. Using the curve shown in Fig. 8.6, determine the coefficient of tidal diffusion. Assume the following data are applicable: $U = 0.1$ m/s, $t = 21,600$ s, and $x = 3,200$ m.

8.8. Verify the correctness of the curves shown in Fig. 8.7.

8.9. Verify the correctness of the curves shown in Fig. 8.8.

8.10. Verify the correctness of the curves shown in Fig. 8.9.

8.11. A series of salinity measurements has been made on a 5-km section of an estuary. The section can be considered a uniform cross section, and during the period of interest, the freshwater flow rate was 13 m^3/s and the freshwater velocity was 0.13 m/s. Estimate the tidal dispersion coefficient from the data given below. Chloride concentration at the end of the estuary is 18.2 kg/m³. Samples were taken at high-water slack.

DISTANCE FROM MOUTH, km	CHLORIDE CONC, kg/m³
3	3.64
4	2.32
5	1.39
6	0.73
7	0.47
8	0.29

8.12. For the estuary of Problem 8.11, determine the effect of a wastewater discharge at a point 7000 m from the mouth. The discharge contains 80 g/m³ BOD_u and has a flow rate of 1 m³/s. Assume $k = 0.2$ d^{-1} and $k_2 = 0.45$ d^{-1}. Temperature is 20°C.

8.13. The estuary shown in Fig. 8.19(a) can be described by the CFSTR system of Fig. 8.19(b). Average values of the system parameters are

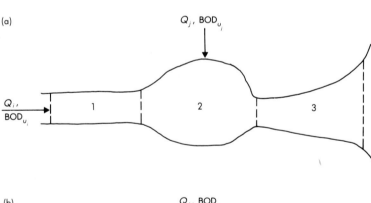

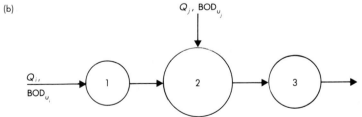

FIGURE 8.19

Diagram and schematic representation of the estuary in Problem 8.13.

given below for the normal temperature of 15°C. Volumes are for high-water slack.

$$k = 0.14 \text{ d}^{-1} \qquad\qquad V_2 = 3.5 \times 10^7 \text{ m}^3$$
$$k_{21} = 0.35 \text{ d}^{-1} \qquad\qquad V_3 = 3 \times 10^7 \text{ m}^3$$
$$k_{22} = 0.25 \text{ d}^{-1} \qquad\qquad \bar{A}_1 = 165 \text{ m}^2$$
$$k_{23} = 0.40 \text{ d}^{-1} \qquad\qquad \bar{A}_2 = 330 \text{ m}^2$$
$$D_1 = 150 \text{ m}^2/\text{s} \qquad\qquad \bar{A}_3 = 350 \text{ m}^2$$
$$D_2 = 75 \text{ m}^2/\text{s} \qquad\qquad \text{BOD}_{ui} = 10 \text{ g/m}^3$$
$$D_3 = 300 \text{ m}^2/\text{s} \qquad\qquad \text{BOD}_{uj} = 250 \text{ g/m}^3$$
$$Q_i = 55 \text{ m}^3/\text{s} \qquad\qquad C_{O_{2_i}} = 7.64 \text{ g/m}^3$$
$$Q_j = 2 \text{ m}^3/\text{s} \qquad\qquad C_{O_{2_j}} = 0.0 \text{ g/m}^3$$
$$V_1 = 1 \times 10^7 \text{ m}^3$$

Salinity = 25 ppt (parts per thousand)

Determine the dissolved-oxygen concentration at high-water slack, assuming salinity is constant in the three segments but very low upstream and dispersion above 1 is negligible.

8.14. The San Francisco Bay system can be viewed crudely as an ideal CFSTR. Consider the following data on the bay:

Surface area $= 1.24 \times 10^9 \text{ m}^2$

Average depth at low tide $= 6$ m

Average tidal rise of 1.5 m, twice a day

Freshwater discharge, 6-mon winter-spring average $= 1000 \text{ m}^3/\text{s}$

Freshwater discharge, 6-mon summer-fall average $= 200 \text{ m}^3/\text{s}$

a. Ignoring the influence of the tides, what are the mean hydraulic detention times under winter-spring and summer-fall conditions?
b. If the tides are included, what are the mean hydraulic detention times?
c. If the wastewater discharge averages 60 m^3/s from all sources around the bay and if these discharges contain a tracer at a concentration of 100 g/m^3, what is the average tracer concentration in water flowing out the Golden Gate? (Courtesy of Jim Hunt.)

8.15. A treated wastewater is to be discharged through a diffuser 30 m long and having 20 ports, each 0.10 m in diameter. Maximum flow rate will be 0.6 m^3/s. The diffuser will be set parallel with the beach because onshore currents are relatively common. If the ocean floor slope averages 1.0 percent, determine the length of outfall necessary to provide a dilution factor of 400 in an onshore current of 0.15 m/s.

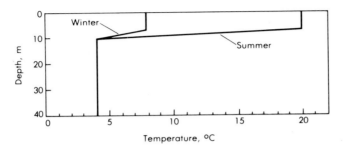

FIGURE 8.20

Temperature versus depth data for Problem 8.17.

Assume wastewater and seawater densities are 990 and 1020 kg/m³, respectively.

8.16. Wastewater effluent is to be discharged through a diffuser 50 m long having 30 ports, each 0.125 m in diameter. The maximum flow rate will be 1 m³/s. The diffuser is set parallel to the beach because of an onshore current of 0.15 m/s. If the ocean floor slope is 1.5 percent, determine the length of the outfall necessary to provide a dilution factor of 200. Assume a wastewater density of 990 kg/m³ and a seawater density of 1020 kg/m³. (Courtesy of Jim Hunt.)

8.17. Coastal waters undergo density stratification caused by temperature gradients. The density stratification is seasonally dependent, as shown in Fig. 8.20. The following densities are known for seawater:

TEMPERATURE, °C	DENSITY, kg/m³
4	1027.00
8	1026.51
12	1025.83
16	1025.00
20	1024.02

Wastewater effluent at a density of 1000.00 kg/m³ is discharged at a depth of 30 m and undergoes a 50-fold dilution (49 parts of seawater to 1 part wastewater. Assuming no further dilution occurs, will the diluted wastewater effluent rise above the thermocline during summer or winter? (Courtesy of Jim Hunt.)

REFERENCES

8.1. Abraham, G., (1963), *Jet Diffusion in Stagnant, Ambient Fluid*, Delf Hydraulics Laboratory Publication 29.

8.2. Bowden, K. F., (1967), "Circulation and Diffusion," in *Estuaries*, edited by G. H. Lauff, AAAS Publication No. 85, Washington, D.C.

8.3. Brooks, N. H., (1960), "Diffusion of Sewage Effluent in an Ocean Current," *Proc. First International Conference on Waste Disposal in the Marine Environment*, University of California, Berkeley.

8.4. _____ (1973), "Dispersion in Hydrologic and Coastal Environments," Environmental Protection Agency Report 660/3-73-010, Washington, D.C.

8.5. Burchett, M. E., G. Tchobanoglous, and A. J. Burdoin, (1967), "A Practical Approach to Submarine Outfall Calculation," *Public Works*, vol. 98, no. 5, p. 95.

8.6. Churchill, M. A., H. L. Elmore, and R. A. Buckingham, (1962), "Prediction of Stream Reaeration Rates," *J. Sanitary Eng. Division*, ASCE, vol. 88, SA4, p. 1.

8.7. Fan, L. N., and N. H. Brooks, (1969), *Numerical Solutions of Turbulent Buoyant Jet Problems*, Technical Report KH-R-18, W. M. Keck Laboratory of Hydraulics and Water Resources, California Institute of Technology, Pasadena.

8.8. Fischer, H. B., E. J. List, R. C. Y. Koh, J. Imberger, and N. H. Brooks, (1979), *Mixing in Inland and Coastal Waters*, Academic Press, New York.

8.9. Grace, R. A., (1978), *Marine Outfall Systems*, Prentice-Hall, Englewood Cliffs, N.J.

8.10. Harleman, D. R. F., (1964), "The Significance of Longitudinal Dispersion in the Analysis of Pollution in Estuaries," *Proc. 2d International Conference on Water Pollution Research*, Tokyo, Pergamon Press, New York.

8.11. Hetling, L. J., (1968), "Simulation of Chloride Concentrations in the Potomac Estuary," FWPCA Tech. Paper No. 12, Middle Atlantic Region, U.S. Dept. of Interior.

8.12. _____ and R. L. O'Connell, (1966), "A Study of Tidal Dispersion in the Potomac River," *Water Resources Research*, vol. 2, no. 4, p. 825.

8.13. Hydroscience, Inc., (1968), Mathematical Models for Water Quality for the Hudson-Champlain and Metropolitan Coastal Water Pollution Control Project, Leonia, N.J.

8.14. Hynes, H. B. N., (1971), *The Biology of Polluted Waters*, University of Toronto Press, Toronto.

8.15. _____ (1970), *The Ecology of Running Waters*, University of Toronto Press, Toronto.

8.16. Klein, L., (1959, 1962, 1966), *River Pollution, 1: Chemical Analysis, 2: Causes and Effects, 3: Control*, Butterworths, London.

8.17. Koh, R. C. Y., and N. H. Brooks, (1975), "Fluid Mechanics of Waste Disposal in the Ocean," *Annual Review of Fluid Mechanics*, vol. 7, p. 187.

8.18. Metcalf and Eddy, Inc., (1979), *Wastewater Engineering*, 2d ed., revised by G. Tchobanoglous, McGraw-Hill Book Company, New York.

8.19. O'Connor, D. J., (1960), "Oxygen Balance of an Estuary," *J. Sanitary Engineering Division*, ASCE, vol. 86, SA3, p. 35.

8.20. _____ and W. E. Dobbins, (1956), "Mechanism of Reaeration in Natural Streams," *J. Sanitary Engineering Division*, ASCE, vol. 82, SA6, p. 1115.

8.21. Oglesby, R. T., C. A. Carlson, and J. A. McCann (eds.), (1972), *River Ecology and Man*, Academic Press, New York.

8.22. Orlob, G. T. (ed.), (1983), *Mathematical Modeling of Water Quality*, Wiley Interscience, New York.

8.23. Pearson, E. A. (ed.), (1960), *Waste Disposal in the Marine Environment*, Pergamon Press, New York.

8.24. Phelps, E. B., (1944), *Stream Sanitation*, John Wiley and Sons, New York.

8.25. Stein, J. E., and J. G. Denison, (1966), "In Situ Benthal Oxygen Demand of Cellulosic Fibers," *Proc. 3rd International Conference on Water Pollution Research*, Munich.

8.26. Streeter, H. W., and E. B. Phelps, (1925), "A Study of the Pollution and Natural Purification of the Ohio River, III. Factors Concerned in the Phenomena of Oxidation and Reaeration," U.S. Public Health Service, Bulletin No. 146.

8.27. Thomann, R. V., (1972), *Systems Analysis and Water Quality Management*, Environmental Research Applications, Inc., New York.

8.28. University of Florida, (1964), *A Study of Estuarine Pollution Problems Progress Report II*, Department of Civil Engineering, Gainesville.

8.29. Velz, C. J., (1970), *Applied Stream Sanitation*, Wiley Interscience, New York.

9

Water Quality in Lakes and Reservoirs

Water quality in lakes and reservoirs is subject to the natural degradation processes of eutrophication and the impacts of societal development, which greatly speed eutrophication. Lakes are classified as oligotrophic, mesotrophic, or eutrophic, depending on the availability of nutrients and the productivity of the water in terms of living material. Progression from the low-productivity oligotrophic state through the mesotrophic and into the eutrophic condition is a normal aging process that results from recycling and accumulation of nutrients over long periods of time. Thus oligotrophic lakes such as Lake Tahoe and Lake Superior will eventually be eutrophic no matter what type of management is undertaken. Human activities increase the rate of eutrophication, often by several orders of magnitude, and a process that would take tens of thousands of years if left to nature may occur in a century or less with the aid of wastewater discharges and agricultural runoff.

Oligotrophic lakes are usually more aesthetically pleasing than eutrophic lakes, and the water quality is generally more suitable for a wider range of uses. Highly eutrophic bodies often are subject to algal blooms, rapid increases in algae populations that are unsightly, wash up on beaches to decay in odoriferous splendor, and attach to swimmers, boats, and fishing equipment. Fish populations are generally greater in eutrophic than in oligotrophic lakes, and the numbers of so-called trash fish are much greater.

9.1
HYDRAULIC BEHAVIOR OF LAKES AND RESERVOIRS

Lakes and reservoirs behave as nonideal (arbitrary flow) reactors with long hydraulic residence times. Even small lakes typically have residence times in the tens of years. Such systems are little affected by flow, and the

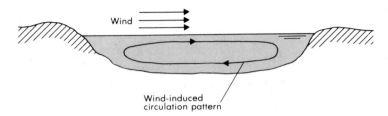

FIGURE 9.1

Circulation resulting from the wind blowing over the surface of a shallow lake.

major perturbations are caused by wind and temperature. Wind causes circulatory motion, as shown in Fig. 9.1. This circulation is enough to keep small, unstratified lakes and impoundments used for wastewater treatment (Fig. 9.2) nominally well mixed, and CFSTR models are often satisfactory for water quality descriptions of these systems (see Example 9.1).

Large and particularly deep lakes often have circulation cells form and behave as systems of CFSTRs in parallel or series. The simple system shown in Fig. 9.3 would result from some type of density stratification. Circulation in the lower cell would be weak, but, because the residence time is long, behavior would still be as a CFSTR. A point of interest is that the water surface will remain virtually flat and the density gradient line between the two circulation cells will be tilted because of the necessity of balancing the surface stress with a hydrostatic pressure gradient. An excellent description of these processes is given in Ref. [9.3].

FIGURE 9.2

Shallow oxidation ponds used for the treatment of wastewater (see Chapter 14).

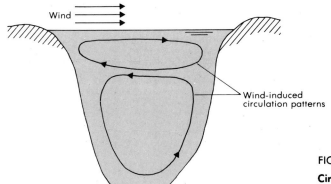

FIGURE 9.3

Circulation in deep lake with two wind-generated cells.

EXAMPLE 9.1

ANALYSIS OF A LAKE WITH A CFSTR MODEL

The contents of a tanker truck carrying a liquid organic waste are spilled accidentally into a small lake. The resulting initial concentration of the waste in the lake is 100 mg/L. The volume of the lake is 10^5m^3. A stream that flows into and out of the lake has a flow rate of 1000 m^3/d. If the organic waste in solution undergoes first-order photochemical decay $(r_C = -kC)$ with a k value of 0.005 d^{-1}, determine the time required for the concentration of the waste in the lake to be reduced to 5 percent of the initial value.

SOLUTION:

1. Write a materials balance for the lake assuming it can be modeled as a CFSTR.

$$\frac{dC}{dt} \mathcal{V} = QC_i - QC + (-kC)\mathcal{V}$$

accumulation = inflow − outflow + generation

2. Solve the materials balance for C for the boundary condition that $C = C_0$ at $t = 0$. Note that $C_i = 0$ because the flow into the lake does not contain any waste.

$$\int_{C_0}^{C} \frac{dC}{C} = -\left(\frac{1}{\theta_H} + k\right)\int_{0}^{t} dt$$

$$\ln \frac{C}{C_0} = -\left(\frac{1}{\theta_H} + k\right)t$$

3. Substitute known values and solve for t.

$$C = 0.05C_0$$

$$\theta_H = 10^5 \ \text{m}^3/1000 \ \text{m}^3/\text{d}$$

$$= 100 \ \text{d}$$

$$\ln \frac{0.05C_0}{C_0} = -\left(\frac{1}{100} + 0.005\right)t$$

$$t = -3.00/(0.015)$$

$$= 200 \ \text{d}$$

COMMENT

The situation described in this problem has occurred a number of times. When a spill occurs, usually there is no information on the hydrodynamics of the lake. In such a situation, the approach used in this problem can be applied to obtain a conservative estimate of the time required for the concentration to be reduced to a given level. If the flow in the lake is known to be nonideal, a less conservative estimate can be made by assuming the lake can be modeled as two or more CSFTRs in series.

9.2

EFFECTS OF PHYSICAL PROCESSES ON WATER QUALITY

Water quality in lakes and reservoirs is more often related to eutrophication and temperature than to organic material, BOD, and oxygen deficit. Obviously, oxygen-related factors are coupled to eutrophication and temperature, but usually these are secondary effects. Small lakes are sometimes heavily polluted and dissolved-oxygen levels may drop to low values. Removal of the organic discharges quickly changes the conditions, however. Increased eutrophication rates are much more difficult to control because nutrients are recycled. For example, nitrogen added to a lake is assimilated by algae or aquatic plants. When the organisms die, the bulk of the nitrogen is released and is available for assimilation by new cells. Thus nutrients accumulate in the system.

Mixing in Reservoirs and Lakes

Conceptualization of mixing in reservoirs and lakes must be dominated by the fact that residence times are long. Because systems with long residence times are not significantly affected by entrance and exit flow effects, the primary factors controlling mixing are wind and temperature. Temperature variation and density variation may be considered synony-

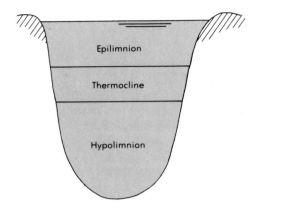 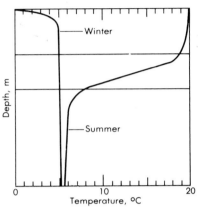

FIGURE 9.4

Typical temperature profile in a thermally stratified lake.

mous in most cases, and because temperature is simple to measure, it is used as a fundamental parameter.

Lakes gain and lose energy through the surface because of shear forces from wind, solar heating, and radiant cooling. In warm weather, vertical convection currents are formed because of differential cooling and heating during the day and night. Gradually, the water at lower levels becomes significantly cooler and more dense than that at the surface, and convective forces are damped out except in a surface layer called the *epilimnion*, which may be a few meters deep. Although the epilimnion is well mixed, the lower layer, or *hypolimnion*, is weakly mixed and usually has distinct gradients in nutrient and oxygen concentrations near the bottom. Between the epilimnion and the hypolimnion is the thermocline, a layer of varying depth having a sharp temperature gradient. The result is a temperature profile similar to that shown in Fig. 9.4. In cold weather, the surface layer may cool below 4°C and a stably stratified profile will result.

Turnover

In very cold climates and in lakes with considerable turbidity, the epilimnion temperatures may approach 4°C, whilie the hypolimnion temperatures are lower than 4°C. Density of the epilimnion will be greater than that of the hypolimnion, and the system becomes unstable. Small perturbations, as from wind shear, result in a turnover of the lake contents, and for a period the lake is completely mixed. After mixing, the entire contents of the lake will be less than 4°C, and the lake will restratify with the colder water near the surface. Surface freezing may occur after this "fall turnover." A "spring turnover" also occurs as the water warms on the surface, the maximum density develops as temperatures approach 4°C, and instability develops.

Turbidity is related to the formation of the unstable conditions because solar energy is adsorbed by particles near the surface. Heating and cooling cycles are restricted to the region near the surface, and the variations become more severe. Lake Tahoe on the California-Nevada border has very clear water, with significant light penetration to 35 m. As eutrophication progresses, the growth of algae will increase the turbidity, and the potential both for turnover and surface freezing will become greater.

Turnover affects water quality in two ways: (1) by changes in nutrient and temperature distribution and (2) by movement of bottom materials throughout the volume. Quite often, nutrient materials accumulate in the lower depths, either as sediment or because biological activity is lower. When these materials are brought to the surface with sunlight and higher temperatures and oxygen concentrations, eutrophication rates are increased. Thus the effect of a turnover on an oligotrophic lake such as Tahoe would be very great.

Stratification and Temperature Distribution

Temperature distribution is an important variable in any lake for ecological reasons. In multipurpose reservoirs, however, temperature distribution is also important from the design-operation viewpoint. Many reservoirs are designed so that released water can be of a particular temperature to meet requirements of migrating fish, spawning grounds, or seed germination. Stratification allows the reservoir to be operated with respect to temperature, and destratification (turnover) destroys this ability to control the system. Normally, stratification develops over a period of a few weeks, and the times of year that instabilities develop are not critical. Recognition of the effects of destratification must be made in designing facilities, however.

Stratification and Dissolved-Oxygen Distribution

When stratification occurs in a relatively shallow lake, reservoir, or impoundment, it is often possible that the oxygen resources will be affected because of algal growth, bacterial activity, and the oxygen demands of the bottom muds. An example of the situation that can develop in a shallow impoundment is illustrated in Fig. 9.5. In referring to that figure, it can be seen that if water is withdrawn at a depth greater

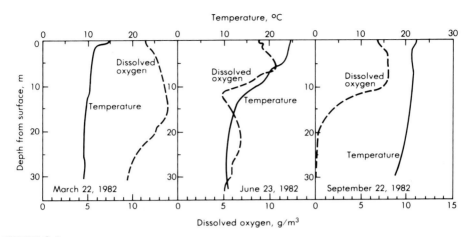

FIGURE 9.5

Typical temperature and dissolved-oxygen profiles for a small impoundment—Lake Mendocino near Ukiah, Calif.

Source: U.S. Department of Interior Geological Survey.

than about 12 m during the month of September it can have a serious effect on the downstream fisheries.

9.3

MODELING OF LAKES AND RESERVOIRS

Modeling of lakes and reservoirs is extremely complex. Heat transfer from the surface requires careful evaluation of the wind conditions and ambient temperature variations. Convective mixing is coupled to wind-generated mixing, and in stratified lakes the level of stratification is important. Such modeling always requires sophisticated analysis that is well beyond the scope of an introductory course. In addition, such problems require extensive field data for their evaluation. Modeling approaches can be divided into two general categories, one-dimensional and multidimensional. One-dimensional models work best in describing small to medium-sized, well-mixed bodies. Typically the length of the major axis will be less than 50 km.

Shallow, weakly stratified, well-stratified, and/or very large bodies must be modeled using two or three spatial dimensions plus time. Solution of such models is beyond the scope of this text, but the general concepts are straightforward, and a few simple approaches will be discussed later in the chapter. A more complete introduction is given in Refs. [9.5] and [9.6].

One-Dimensional Models

As stated above, well-stratified lakes are most suited to the application of one-dimensional models. It has been suggested that the densimetric Froude number could be used as a criterion of stability and hence of stratification [9.6]:

$$N_{DF} = \frac{U}{\sqrt{\dfrac{\Delta\rho}{\rho_0} gd}} \tag{9.1}$$

where

N_{DF} = densimetric Froude number, unitless
Q = flow rate, m^3/s
U = average flow-through velocity Q/bd, m/s
b = average width, m
d = average depth, m
$\Delta\rho$ = change in mass density over depth d, kg/m^3
ρ_0 = reference density, kg/m^3
g = gravitational constant, 9.81 m/s^2

Well-stratified impoundments have $N_{DF} \ll 1/\pi$; weak stratification is indicated when $0.1 < N_{DF} < 1$; and fully mixed systems have $N_{DF} > 1$. Several examples taken from Ref. [9.5], are given in Table 9.1. The application of Eq. (9.1) is illustrated in Example 9.2.

TABLE 9.1

Stratification and the Densimetric Froude Numbers of Several Impoundments

IMPOUNDMENT	LOCATION	LENGTH, km	N_{DF}	CONDITION OF STRATIFICATION
Wells Reservoir	Washington	46	3.8	Fully mixed
Lake Roosevelt	Washington	200	0.46	Weak
Cayuga Lake	New York	60	0.015	Strong
Hungry Horse Reservoir	Montana	47	0.0026	Strong
Ross Lake	Washington	32	0.0042	Strong
Lake Päijänne	Finland	120	0.0004	Strong

Source: From Ref. [9.5].

EXAMPLE 9.2

ESTIMATION OF LAKE STRATIFICATION

A lake has average length, depth, and width of 20, 0.05, and 4 km, respectively. The discharge is 500 m³/s. Mass density of the water during the summer varies from 998 kg/m³ at the surface to 1025 kg/m³ at the bottom. Estimate the stability of the lake using $\rho_0 = 1000$ kg/m³.

SOLUTION:

1. Determine the average flow-through velocity U.

$$U = Q/bd$$
$$= (500 \text{ m}^3/\text{s})/(4000 \text{ m})(50 \text{ m})$$
$$= 0.0025 \text{ m/s}$$

2. Determine the densimetric Froude number using Eq. (9.1).

$$N_{DF} = \frac{U}{\sqrt{\dfrac{\Delta\rho}{\rho_0}gd}}$$

$$= \frac{0.0025}{\sqrt{\dfrac{(1025 - 998)}{1000}(9.81)(50)}}$$

$$= \frac{0.0025}{\sqrt{(0.027)(9.81)(50)}}$$

$$= 6.87 \times 10^{-4}$$

3. According to the criteria set forth for use with Eq. 9.1, the lake is strongly stratified.

Vertical Modeling

A typical conceptual model of a stratified impoundment is shown in Fig. 9.6. Three vertical sections are shown, but any number are possible. Temperature gradients, distributed inflows and outflows to and from aquifers, and outlet level are all factors in determining the number of sections used. The primary discharge is from the middle segment in Fig. 9.6, which would be a common practice in a reservoir. In fact, many reservoirs are designed for withdrawal at multiple levels, and model development must take this into consideration in defining the section thickness.

A materials balance is made on each section to yield a set of equations describing the impoundment. Each of the terms can be a

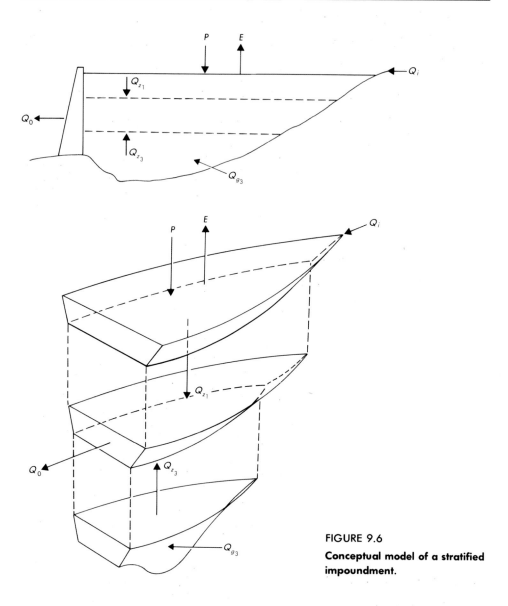

FIGURE 9.6

Conceptual model of a stratified impoundment.

function of a set of parameters. For example, surface inflow Q_i would be a function of season, antecedent precipitation, and recent hydrologic history. Evaporation E would be a function of season, surface and air temperatures, and wind speed. Precipitation P is generally a seasonal function. Flows between sections (Q_{z_1}, Q_{z_3}) are determined by internal hydraulic factors, density gradients, and external contributions and extractions. The outflow Q_{0_2} from a reservoir is usually controlled, but outflow from a natural impoundment would be a function of exit head.

In this latter case outflow rate can be estimated using an overall mass balance and weir-type relationships.

Describing mass movement within a stratified impoundment, or development of a stratified model, requires materials balances on vertical sections:

$$\frac{\partial V_i}{\partial t} = Q_{i_i} + Q_{z_{i-1}} + Q_{z_{i+1}} + P - E - Q_{z_i} - Q_{0_i} \pm Q_{g_i} \qquad (9.2)$$

Accumulation = \qquad\qquad Inflow \qquad - \qquad Outflow \qquad $\pm$ \qquad Inflow/ outflow

where

V_i = volume of ith section, m^3

Q_{i_i} = surface inflow rate to ith layer, m^3/t

Q_z = z-axis flow rates between sections, m^3/t

P = precipitation rate, m^3/t

E = evaporation rate, m^3/t

Q_{0_i} = withdrawal rate, m^3/t

Q_{g_i} = groundwater inflow or outflow rate, m^3/t

For the three-section system of Fig. 9.6 application of Eq. (9.2) results in Eqs. (9.3), (9.4), and (9.5):

$$\frac{\partial V_1}{\partial t} = Q_{i_1} + P - E - Q_{z_1} \qquad (9.3)$$

$$\frac{\partial V_2}{\partial t} = Q_{z_1} + Q_{z_3} - Q_{0_2} \qquad (9.4)$$

$$\frac{\partial V_3}{\partial t} = Q_{g_3} - Q_{z_3} \qquad (9.5)$$

Solution of the materials balance equations requires descriptions of each term. As noted above, the external terms can be described through conventional hydrologic engineering methods.

Description of internal flows between the sections requires a heat energy balance of the following general form:

$$\frac{\partial H_i}{\partial t} = (h_i - h_0 + h_{sz})_i - (h_{w_i} - h_{w_{i+1}}) - (h_{d_i} - h_{d_{i+1}}) \qquad (9.6)$$

where

$H_i = c_p \rho V_i T_i$, the heat content of the ith section, J

c_p = specific heat, J/kg · K

ρ = mass density, kg/m^3

$h_{i_i} = c_p \rho Q_{i_i} T_{i_i}$, the heat inflow, J/s

$h_{0_i} = c \rho Q_{0_i} T_{0_i}$, the heat outflow, J/s

$h_{sz_i} = \int_z^{z+\Delta z} q_{sz} a_z \, dz$, the direct insulation, J/s

$q_{sz} = (1 - \beta)q_{sn}e^{-\zeta z}$, the solar radiation flux, $J/m^2 \cdot s$

β = ratio of absorbed to incoming radiation

ζ = bulk extinction coefficient, m^{-1}

a_z = area of horizontal plaue at depth z, m^2

$h_{w_i} = c_p \rho Q_{z_i} T_{z_i}$, advected heat, J/s

$h_{d_i} = c_p \rho D_z a_z \, \partial T / \partial z$, heat dispersion, J/s

D_z = dispersion coefficient, m^2/s

Among the parameters of Eq. (9.6), estimation of the solar radiation penetrating the surface and the vertical dispersion coefficient is difficult. Considerable success has been achieved in predicting temperature profiles, however, as shown in Fig. 9.7.

Multidimensional Models

Discussion of multidimensional models is beyond the scope of this book, but it is important to point out that considerable work has been done in this area and that much success has been achieved. Most current modeling efforts utilize finite-element models for horizontal layers. Mass and energy transport between layers is described using classical hydrodynamic approaches. Usually models are restricted to two layers, the hypolimnion and the epilimnion. Impressive results of such modeling efforts are shown in Fig. 9.8. Lake Ontario is a large, complex body of water, and the remarkable precision attained by Simons [9.6] is a significant achievement. It should be noted that the results shown in Fig. 9.8 required a great deal of computer time in addition to a great deal of effort in development and calibration of the equations.

Ecological Modeling

Lakes and reservoirs contain a wide variety of living organisms ranging from bacteria to large fish. Usually the categories of phytoplankton (bacteria, algae, and floating plants), zooplankton (protozoans and small animals), small fish, and large fish are used. Bottom-dwelling plants and animals (such as crustaceans) may be included as separate categories in some models. Phytoplankton growth results from the availability of organics and nutrients in the case of bacteria and of nutrients and sunlight in the case of the algae. Zooplankton and small fish feed on the phytoplankton, and large fish feed on the small fish. A conceptual model of the ecological system can be constructed as shown in Fig. 9.9.

In lakes the predominant phytoplankton species are photosynthetic (blue-green and green algae). Consequently, the phytoplankton population tends to be concentrated at the surface. Major parameters in predicting phytoplankton concentration are sunlight time, temperature,

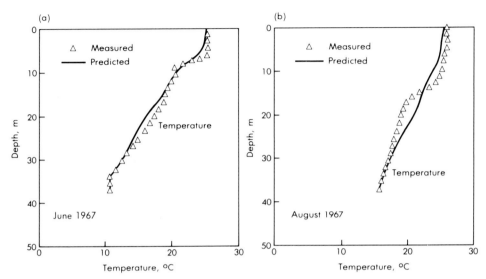

FIGURE 9.7

Predicted and measured temperatures in Lake Hartwell, Ga.

Source: Adapted from Ref. [9.3].

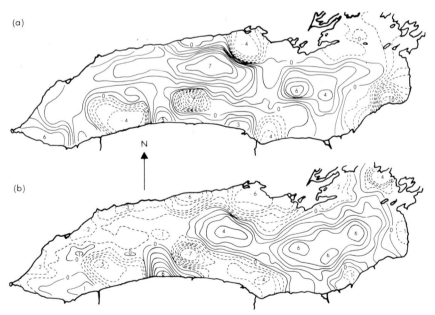

FIGURE 9.8

(a) Observed and (b) computed temperature changes (°C) in Lake Ontario, August 2 to 5, 1972, corresponding to the second layer of the model (between 10 and 20 m).

Source: From Ref. [9.6], courtesy of the American Meteorological Society.

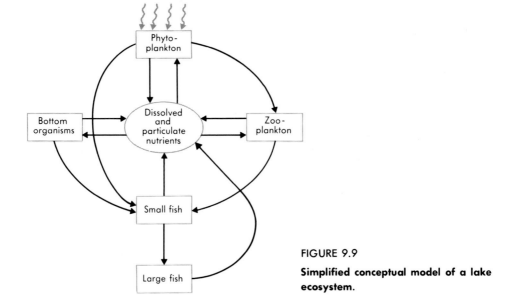

FIGURE 9.9

Simplified conceptual model of a lake ecosystem.

and turbidity (which affects the extinction coefficient). Because turbidity increases with phytoplankton growth, a form of feedback control exists. Phytoplankton models are often broken into submodels describing particular groups (e.g., diatoms) that have different growth responses [9.2, 9.4].

Zooplankton feed on phytoplankton and can be expected to be concentrated near the surface also. Growth rates are dependent upon temperature and the amount of phytoplankton available. In most cases a first-order relationship is assumed. The rate coefficient is approximated using laboratory experiments, and a final value is obtained through model calibration.

Fish and benthic organisms are treated in much the same way as the zooplankton. Usually fish populations do not play a dominant role in water quality characteristics, and the purpose of their inclusion is to predict the converse effect, i.e., that of water quality changes on fish.

Because of the importance of nutrients and temperature as a driving force in phytoplankton growth, ecological models must be coupled to hydrodynamic lake models. Nutrients such as nitrogen and phosphorus generally result from wastewater discharges or agricultural runoff. Movement of nutrients through the lake is described using mass balances, with the advective terms taken from the hydrodynamic models. Generally lakes are broken into a group of compartments over which homogeneity is assumed. Because phytoplankton populations decrease sharply with

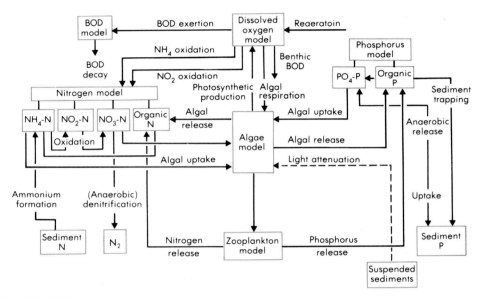

FIGURE 9.10

Processes considered in the development of a complex limnological model.

Source: From Ref. [9.1].

depth, the upper layer is shallow. The complexity required in ecological models is illustrated in Fig. 9.10.

9.4

SIGNIFICANCE OF LAKE AND RESERVOIR MODELING

The extreme difficulty of lake and reservoir modeling forces the question of why it must be done. Real-time control is not possible for a number of reasons: Response times of lakes are long, many variables are uncontrollable or immeasurable, and natural factors such as weather play an important role. Models can be used to answer three very important questions, however: Why has the current situation developed? What is likely to happen to lake water quality and organism populations in the future? What will happen if the conditions are changed by a new discharge, removal of a present discharge, or a change in discharge quality? The answers to these questions allow the public to make plans and choices. Because lakes are valuable resources, maintenance of lake water quality is of major significance (Fig. 9.11).

A second important use of lake and reservoir models is directly related to engineering design. Facilities such as cooling ponds, marinas, and artificial impoundments for subdivisions all have particular water

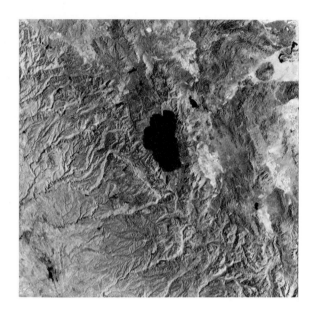

FIGURE 9.11

Landsat view of Lake Tahoe and surrounding area. Lake Tahoe, located on the border between California and Nevada, is noted for its unusual water clarity (greater than 30 m). (Landsat 1, 2, & 3 Coverage Index, path 46, row 33, MSS band 7.)

quality requirements. These water bodies are generally small and shallow, and the modeling is less complex and costly. The operation of such facilities can be improved through model development and use.

KEY IDEAS, CONCEPTS, AND ISSUES

- Lakes are described as being oligotrophic, mesotrophic, or eutrophic, depending on the availability of nutrients and the productivity of the water in terms of living matter.

- Lakes and reservoirs behave as nonideal reactors with long residence times.

- Water quality in lakes is more often a function related to eutrophication and temperature than to the presence of organic material, BOD, and oxygen deficit.

- Lake turnover affects water quality in two ways: (1) nutrients and temperature distributions are altered and (2) bottom materials (e.g., nutrients) are circulated throughout the volume.

- Lake and reservoir models can be divided into two general categories —namely, one-dimensional and multidimensional.

- Description of the internal flows between sections in a stratified lake model is based on a heat energy balance.

- Ecological models are useful in sensitivity studies in which alternative impacts are to be evaluated.

> • Although lakes are difficult to control on a real-time basis, mathe-
> matical models can be used to study the long-term impacts of
> variables that are uncontrollable or immeasurable and of natural
> factors such as weather.

DISCUSSION TOPICS AND PROBLEMS

9.1. Determine the time required for the waste concentration in the
discharge from the lake in Example 9.1 to be reduced to 5 percent of
the initial value if it is assumed that the hydraulics of the lake can be
modeled as two equal-volume CFSTRs in series and that all of the
waste is deposited into the first of the two reactors.

9.2. The lake shown in Fig. 9.12 has a total volume of 10^8 m³. A series of
tracer studies were conducted to determine the hydraulic characteristics of the lake. Slugs of tracer (1000 kg) were dumped at points A

(a)

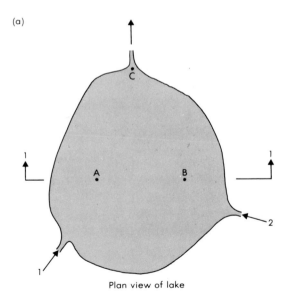

Plan view of lake

(b)

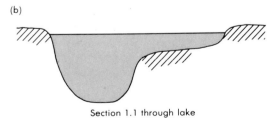

Section 1.1 through lake

FIGURE 9.12

Definition sketch for Problem 9.2.

and B in two sequential experiments. In a third experiment, 1000 kg of tracer was distributed over the entire surface of the lake. Concentration measurements were made at point C. Results of the experiments are given below. Average values of Q_1 and Q_2 are 0.95 and 0.50 m³/s, respectively.

TIME, d	TRACER CONCENTRATION AT C, mg/m³		
	Uniform distribution	Release at A	Release at B
50	9.4	14.0	0.9
100	8.8	12.9	1.6
150	8.3	11.4	2.2
200	7.8	10.7	2.8
250	7.3	9.4	3.2
300	6.9	8.6	3.6
350	6.4	7.7	3.8

a. Develop a hydraulic model for the lake.
b. Discuss the factors that would control algal biomass production in the lake and where problems are likely to be the greatest in space and time.

9.3. A reservoir is located on a small river at an elevation having a mean atmospheric pressure of 745 mmHg. The reservoir has a full volume of 10^8 m³, but varies between 10^7 and 10^8 m³ during the year. A waste discharge enters the river at a point just upstream of the reservoir so that all BOD exertion occurs in the reservoir and the dissolved oxygen concentration in the river remains near saturation. The characteristics of the reservoir and the waste discharge and river are given below. Using the given data, estimate the oxygen deficit of the reservoir during each quarter. Assume that $k = 0.2$ d^{-1} at 20 °C, $k_2 = 0.35$ d^{-1} at 20 °C, and that all withdrawals are from the epilimnion. For the purposes of analysis, assume the reservoir BOD

Reservoir Data

TIME PERIOD	$\bar{Q}_i^*$, m₃/s	$\bar{Q}_0$, m³s	$V_s^\dagger$, m³ × 10⁷	$V_e^\dagger$, m³ × 10⁷	THERMO-CLINE DEPTH, m	EPILIMNION VOL, m³ × 10⁷	$T_E^\ddagger$, °C	$T_H^\ddagger$, °C
July–Sept.	0.20	3.40	9.45	6.93	30–14	4.60–2.08	20	15
Oct.–Dec.	1.80	1.00	6.93	7.56	14–9	2.08–1.29	14	13
Jan.–Mar.	3.60	0.50	7.56	9.97	— —	— —	10	10
Apr.–June	0.80	1.50	9.97	9.42	33–29	5.00–4.45	14	12

*Combined river-wastewater flow.
† Volume at start and end of period.
‡ Temperature of epilimnion and hypolimnion.

and oxygen deficit are equal to zero on July 1. Also, if the incoming river-waste flow sinks to the hypolimnion, assume no interaction occurs within the epilimnion.

Waste Discharge and River Data

TIME PERIOD	BOD_u LOADING, kg/d	TEMP. OF RIVER – WASTEWATER MIXTURE °C
July–Sept.	7000	19
Oct.–Dec.	6000	12
Jan.–Mar.	6000	10
Apr.–June	7000	15

9.4. A small farm reservoir behaves as two identical CFSTRs in series. Flow into the reservoir consists of runoff from pasture land having an average BOD_5 of 50 g/m³ and an average flow rate of 0.5 m³/s. Volume of the reservoir is 2.16×10^6 m³. If $k = 0.2$ d^{-1} and $k_2 = 0.35$ d^{-1} at 20°C, determine the dissolved-oxygen concentration if the water temperature is 20°C, the inflow is anaerobic, and steady–state conditions exist.

9.5. A stratified reservoir can be described adequately by two volumetric prisms, as shown in Fig. 9.13. Destratification occurs for a two-week period in March and a second two-week period in November each year. During the stratified periods liquid and nutrient exchange between the two segments is negligible. Nitrogen enters the lake with

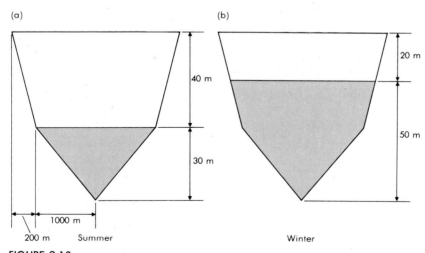

FIGURE 9.13

Definition sketch for Problem 9.5.

agricultural and natural runoff and from a wastewater discharge into a tributary. Using the data given below, estimate the increase in the mass of nitrogen in the lake over a two-year period. Surface area of the lake is 12×10^6 m^2.

	INFLOW, m^3/d			N, g/m^3		
SEASON	Ag.	Trib.	Nat.	Ag.	Trib.	Nat.
Summer	1000	15,000	1300	25	2	10
Winter	—	42,000	3800	—	1	10

9.6. The lake of Problem 9.5 has average epilimnion and hypolimnion temperature values during winter and summer of 6 and 3°C and 12 and 6°C, respectively. Determine the relative degree of stratification.

9.7. A small lake with a volume equal to 500,000 m^3 is approximately completely mixed. The lake is used as a fishing pond by local residents, but pesticide-containing runoff from local farms may cause damage to the fish population. Natural flow through the system is 1000 m^3/d and agricultural runoff is 200 m^3/d for 200 d. Pesticide concentration in the runoff is 0.1 g/m^3. When will the concentration increase to the 0.02 g/m^3 level known to be toxic to the fish?

9.8. Consider the water quality problem in a large lake surrounded by a city. The lake has an area of 11×10^6 m^2, a volume of 3.6×10^8 m^3, and an average depth of 33 m. During a six-month period the lake is thermally stratified, with the epilimnion being 10 m deep. A population of 200,000 persons live in the drainage basin of the lake, and wastewater treatment plant effluents are discharged into the lake. Assume that the wastewater effluent enters the stagnant hypolimnion during the six months of summer stratification. What level of BOD removal is required to prevent DO depletion by one-half during stratification? Assume a BOD$_5$ loading of 90 grams per person per day (untreated) and an initial hypolimnion DO concentration of 11.3 g/m^3. (Courtesy of Jim Hunt.)

9.9. Consider now the problem of algae production and decay from wastewater effluents discharged into the lake of Problem 9.8. The stoichiometric equation for algae production is given by

$$106CO_2 + 90H_2O + 16HNO_3 + H_3PO_4 + \text{light} \rightarrow$$
$$C_{106}H_{199}O_{45}N_{16}P_1 + 309/2O_2$$

Assume phosphorus is the growth-limiting nutrient, all wastewater effluent phosphorus produces algae stoichiometrically, and all algae produced settle into the hypolimnion and then decay causing a BOD exertion. Determine the necessary phosphorus treatment required to prevent hypolimnion dissolved oxygen from reaching one-half the initial value. Assume a daily PO_4^{-3} loading of 5 g of phosphorus per person and an initial bottom DO concentration of 11.3 g/m^3. (Courtesy of Jim Hunt.)

9.10. A large reservoir (volume = 60×10^9 m^3) has been designed with 20 percent of its volume for sediment storage. If the influent streams have a total flow rate of 55×10^6 m^3/d and an average suspended solids concentration of 350 g/m^3, determine the expected life of the reservoir. Assume an average sediment density of 1400 kg/m^3.

9.11. In tropical countries a serious endemic waterborne disease, onchocerciasis, is associated with rapidly flowing water. Installation of a dam and reservoir will eliminate onchocerciasis (often called river blindness) but may lead to even more serious problems. Use reference materials in your campus library to determine the important relationships that are involved.

9.12. Multipurpose reservoirs are often designed with several water quality objectives as basic design criteria. Discuss briefly the factors associated with reservoir design and operation for the following water uses: (a) irrigation, (b) spawning, (c) siltation below the dam, (d) gravel production, (e) cold-water game fish, and (f) warm-water game fish.

REFERENCES

9.1. Baca, R. G., and R. C. Arnett, (1976), "A Finite Element Water Quality Model for Eutrophic Lakes" in *Finite Elements in Water Resources, 4*, edited by W. G. Gray, G. Pinder, and C. Brebbia, Devon Pentech, Plymouth.

9.2. Canale, R. P. (ed.), (1976), *Modeling Biochemical Processes in Aquatic Ecosystems*, Ann Arbor Science Publishers, Ann Arbor.

9.3. Fontane, D. G., and J. P. Bohan, (1974), *Richard B. Russell Lake Water Quality Investigation*, Waterways Experiment Station Technical Report H-74-14, Vicksburg, Miss.

9.4. Jorgensen, S. E., (1983), "Ecological Modeling of Lakes" in *Mathematical Modeling of Water Quality in Lakes and Reservoirs*, edited by G. T. Orlob, Wiley Interscience, New York.

9.5. Orlob, G. T., (1983), "One-Dimensional Models for Simulation of Water Quality in Lakes and Reservoirs" in *Mathematical Modeling of Water Quality*, edited by G. T. Orlob, Wiley Interscience, New York.

9.6. Simons, T. J., (1975), "Verification of Numerical Models of Lake Ontario: Part II. Stratified Circulation and Temperature Changes," *J. Physical Oceanography*, vol. 5, no. 1, p. 98.

10

Water Quality in Groundwater Systems

The primary source of drinking water in the United States is groundwater. The importance of groundwater is evident when it is considered that about 80 percent of all public water supplies are obtained from groundwater resources [10.8]. In the period from 1950 through 1975 agricultural usage of groundwater increased from 13.2 to 41.6 billion liters per day. It is estimated that the total withdrawal of groundwater for irrigation, industry, and public and rural supplies amounted to 310 billion liters per day in 1975 [10.8, 10.17].

In the past, the quality of most groundwaters was not a major concern. Treatment usually consisted of chlorination and, if necessary, the removal of iron and manganese and other specific constituents. But within the last 10 years it has become apparent that because of human activities many of the nation's groundwater aquifers have been contaminated by compounds other than those present in the natural environment. As noted previously, many of the compounds that have been found in groundwater are known to be carcinogenic and/or mutagenic. Although their concentrations might be minute, the presence of these compounds is clearly a serious threat to the nation's groundwater resources.

It is the purpose of this chapter to introduce readers to (1) the sources of contaminants, (2) the general aspects of groundwater movement, (3) the movement of nonreactive constituents in groundwater, and (4) the movement of reactive components in groundwater and then to discuss some of the practical problems involved in the study of the movement of contaminants and the methods used to improve the quality of contaminated groundwaters.

10.1

SOURCES OF CONTAMINANTS IN GROUNDWATER

Natural groundwaters contain a variety of chemical species and gases in solution. As noted in Chapter 2, most of the ionic species in groundwater are derived from the surrounding soil and rocks. Soil bacteria, present in all groundwater aquifers (water-bearing strata), may also be present in

TABLE 10.1

Potential Sources of Groundwater Contaminants

SOURCE	POSSIBLE CONTAMINANTS
Accidental spills	Various inorganic and organic chemicals
Acid rain	Oxides of sulfur (SO_x) and nitrogen (NO_x)
Agricultural activities	Fertilizers, pesticides, herbicides, and fumigants
Animal feedlots	Organic matter, nitrogen, and phosphorus
Deicing of roads	Chlorides, sodium, and calcium
Deep-well injection of wastes	Variety of inorganic and organic compounds, radioactive materials, and radionuclides
Hazardous waste disposal sites	Various inorganic compounds (particularly heavy metals) and organic compounds (e.g., pesticides and priority pollutants)
Industrial-liquid-waste storage ponds and lagoons	Heavy metals and various cleaning solvents and degreasing compounds
Landfills, industrial	Wide variety of inorganic and organic compounds
Landfills, municipal	Heavy metals, gases, organic compounds, and inorganic compounds (e.g., calcium, chlorides, and sodium)
Land disposal of liquid and semisolid industrial wastes	Organic compounds, heavy metals, and various cleaning solvents and degreasers
Land disposal of municipal wastewater and waste sludges	Organic compounds, inorganic compounds, heavy metals, microbiological contaminants, etc.
Mining	Minerals and acid mine drainage
Rainfall	Chloride, sulfate, organic compounds, etc.
Saltwater intrusion	Inorganic salts
Septic-tank leaching fields or beds (soil absorption areas)	Organic matter, nitrogen, phosphorus, bacteria, etc.
Storage tanks, underground	Organic cleaning and degreasing compounds, petroleum products, and other hazardous wastes

Source: Adapted in part from Refs. [10.8] and [10.12].

FIGURE 10.1

Surface pond used without authorization for the disposal of cleaning solvents and oils at a small industrial facility.

the groundwater, depending on the size of the interstices of the porous medium.

In addition to the natural sources of contaminants, human activities have also contributed to the buildup of the contaminants in groundwaters. Potential sources of groundwater contaminants are reported in Table 10.1. When the range of possible sources of contamination is considered, it is little wonder that cases of groundwater contamination have been reported in all parts of the United States. A typical example of an unauthorized discharge from a small industrial activity is illustrated in Fig. 10.1.

10.2
GROUNDWATER MOVEMENT

The subject of groundwater movement is quite complex. For the purposes of this chapter, the movement of groundwater will be considered in its simplest forms to illustrate the principles involved. More complex groundwater models will be discussed to provide a broader perspective of the field of groundwater modeling.

Darcy's Law

The basic empirical relationship governing the movement of groundwater through a porous medium is known as Darcy's law [10.2, 10.5, 10.6, 10.16]. First reported in 1856 by the French engineer Henry Darcy, the law which bears his name is written as

$$v = -K\frac{dh}{dl} \tag{10.1}$$

where

v = superficial flow velocity, m/s

K = hydraulic conductivity, m/s

$\dfrac{dh}{dl}$ = hydraulic gradient, m/m

The minus sign in Darcy's law arises from the fact that the head loss is negative. The hydraulic conductivity K (in meters per second) is dependent on the properties of the porous medium and the fluid [10.4]:

$$K = Cd^2\frac{\gamma}{\mu} \tag{10.2}$$

where

C = constant of proportionality, unitless

d = grain size of porous medium, m

γ = specific weight of water, kN/m^3

μ = dynamic viscosity of water, N $\cdot$ s/m^2

The term Cd^2 is solely a function of the porous medium and is often identified as the *specific or intrinsic permeability* [10.4]. Representative values for the hydraulic conductivity are reported in Table 10.2.

The discharge through a cross-sectional area of an aquifer is given by

$$Q = -AK\frac{dh}{dl} \tag{10.3}$$

TABLE 10.2

Typical Values for the Hydraulic Conductivity for Various Soil Types

SOIL TYPE	K, m/s
Clays	$< 10^{-8}$
Peat	10^{-8} to 10^{-7}
Silt	10^{-7} to 10^{-6}
Loam	10^{-7} to 10^{-5}
Very fine sands	10^{-6} to 10^{-5}
Fine sands	10^{-5} to 10^{-4}
Coarse sands	10^{-4} to 10^{-3}
Sand with gravel	10^{-3} to 10^{-2}
Gravels	$> 10^{-2}$

Source: Adapted from Refs. [10.2], [10.5], and [10.16].

where

Q = hydraulic discharge, m³/s

A = cross-sectional area, m²

and other terms are as defined previously. It should be noted that the application of Eq. (10.3) is based on the assumption that the porous medium is homogenous and is saturated with water. The application of Eq. (10.3) is illustrated in Example 10.1.

EXAMPLE 10.1

DETERMINATION OF TIME OF FLOW THROUGH A POROUS MEDIUM

Determine the time of flow of water from the upper to the lower lake shown in Fig. 10.2. If a contaminant that moves with the water undergoes a first-order decay with time ($r_c = -kC$), estimate the concentration discharged to the lower lake. Assume the following data are applicable:

Cross-sectional area $A = 50$ m²
Head loss $h = 40$ m
Length of flow path $L = 2000$ m
Hydraulic conductivity $K = 10^{-3}$ m/s
Porosity of porous medium $\alpha = 0.4$
Initial concentration of contaminant $C_0 = 100.0$ g/m³
First-order decay constant $k = 0.001$ d⁻¹ (base e)

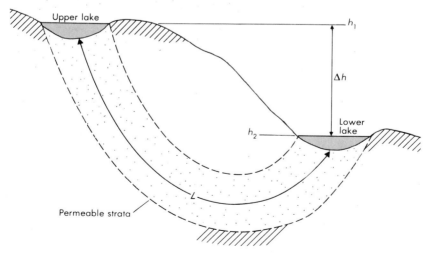

FIGURE 10.2

Definition sketch for Example 10.1: Two lakes connected by a subsurface permeable strata.

SOLUTION:

1. Determine the superficial and true pore velocities of flow through the porous medium.
   a. Superficial velocity

   $$v = -K\frac{dh}{dl}$$

   $$= -(10^{-3} \text{ m/s})\frac{-40 \text{ m}}{2000 \text{ m}} = 2.0 \times 10^{-5} \text{ m/s}$$

   b. Pore velocity

   $$v_p = v/\alpha$$

   $$= 2.0 \times 10^{-5}/0.4 = 5.0 \times 10^{-5} \text{ m/s}$$

2. Determine the time of travel.

   $$t = \frac{2000 \text{ m}}{5.0 \times 10^{-5} \text{ m/s}}$$

   $$= 4 \times 10^{7} \text{ s}$$

   $$= 463 \text{ d}$$

3. Determine the concentration discharged to the lower lake.
   a. For a first-order decay, the appropriate equation is

   $$C = C_0 e^{-kt} \qquad [\text{Eq. (7.25)}]$$

   b. The concentration discharged to the lower lake is

   $$C = (100.0 \text{ g/m}^3) e^{-0.001(463)}$$

   $$= 62.9 \text{ g/m}^3$$

Dupuit-Forchheimer Theory of Free-Surface Flow

The movement of contaminants in groundwater in unconfined systems bounded by a free surface is usually treated as a steady-state flow problem, and a simple Dupuit-Forchheimer analysis is often used. The Dupuit-Forchheimer analysis is based on the assumption that streamlines are virtually horizontal (potential lines are vertical) and that the slope of the hydraulic gradeline is the same as the free-surface slope [10.2, 10.5, 10.7, 10.16]. Although the model is for unconfined aquifers, with suitable assumptions it can be extended to confined aquifers.

Analysis of Groundwater Flow

For a small section of the aquifer of Fig. 10.3 (considering flow in only two directions—x and z), the materials balance expression for the

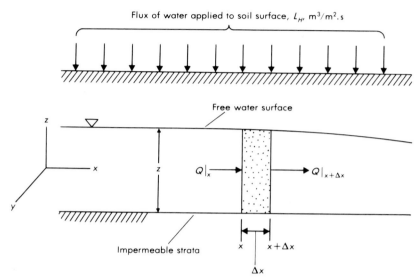

FIGURE 10.3

Definition sketch for a Dupuit-Forchheimer analysis of free-surface groundwater flow subject to surface recharge.

small section subject to surface recharge is

$$\frac{\partial(\Delta V)}{\partial t} = Q|_x - Q|_{x+\Delta x} + L_H y \Delta x \tag{10.4}$$

accumulation = inflow − outflow + inflow due to recharge

where

ΔV = differential volume element, m^3

Q = flow rate, m^3/s

L_H = uniform flux of water added through soil surface, m^3/m$^2 \cdot$ s

y = width of aquifer section subject to recharge, m

Assuming steady-state conditions exist, collecting terms, dividing by $y\Delta x$, and taking the limit as Δx approaches zero results in Eq. (10.5):

$$-\frac{d\left(\dfrac{Q_x}{y}\right)}{dx} + L_H = 0 \tag{10.5}$$

Because the depth z of the unconfined aquifer varies with distance x, $Q_x = v_x z y$ and Eq. (10.5) can be rewritten as

$$-\frac{d(zv_x)}{dx} + L_H = 0 \tag{10.6}$$

Substituting Darcy's law $\left(v_x = -K \dfrac{dz}{dx} \right)$ for v_x and taking the derivative yields

$$\frac{K}{2} \left(\frac{d^2 z^2}{dx^2} \right) + L_H = 0 \tag{10.7}$$

The solution to Eq. (10.7) is dependent upon the boundary conditions. If L_H is uniform over a large distance or if there is a plane of symmetry, the solution to Eq. (10.7) is

$$z = \left(z_0^2 - \frac{L_H}{K} x^2 \right)^{1/2} \tag{10.8}$$

where

z = groundwater depth under the recharge section at distance x, m

z_0 = groundwater depth where $x = 0$, m

Clearly Eq. (10.8) has limitations as a model. The aquifer depth cannot drop to zero because v_x would become infinite. In practice, the

(a)

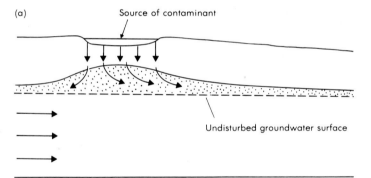

System where water containing contaminant does not mix with the groundwater

(b)

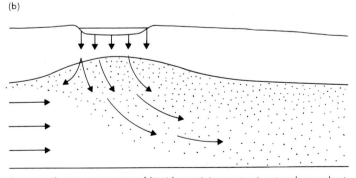

System with extensive mixing of liquid containing contaminant and groundwater

FIGURE 10.4

Two models of groundwater mixing with applied surface water.

model is used in a different manner. The undisturbed groundwater depth is used as a reference, and the change in depth due to flow is determined. Values of L_H are generally quite small, and groundwater depths are usually large enough to allow use of the equation over considerable areas. As an example, a heavy storm might produce 50 mm of precipitation in a 6-hr period. Assuming this precipitation entered the groundwater at a uniform rate, L_H would be 2.3×10^{-6} m/s. The result would be a local increase in the groundwater table due to storage, but a number of storms would need to occur to make this increase significant.

Mixing Models

When contaminants are discharged from the surface, two conditions can exist when those contaminants reach the groundwater. Two extreme situations are shown in Fig. 10.4. The zero-mixing model is the worst case in terms of dilution and water quality effects at points where the aquifer surfaces. The complete-mixing model is the worst case where trace elements and toxic constituents are involved, because the entire aquifer is contaminated. Clearly if the groundwater were to be treated to remove a given contaminant, the zero- or no-mix situation is desirable.

EXAMPLE 10.2

EFFECT OF LAND WASTEWATER TREATMENT SYSTEM ON GROUNDWATER LEVEL

The 50-m-wide land wastewater treatment system depicted in Fig. 10.5 will discharge to a groundwater aquifer. The depth of the groundwater is 50 m and the slope of the free surface is 0.01 m/m. The treatment system

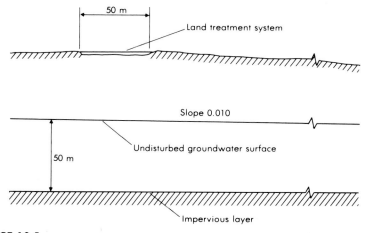

FIGURE 10.5

Definition sketch for the land treatment system of Example 10.2.

will be loaded at a rate of 35 mm/d, and some mounding is expected. Initially flow will be in both the $+x$ and $-x$ directions, but because a sink does not exist in the $-x$ direction, the steady-state flow will be entirely toward a distant stream. Soil in the aquifer is classified as fine sand with a hydraulic conductivity K equal to 10^{-3} m/s. For these conditions, determine the depth of water above the normal groundwater at the upstream and downstream boundaries of the land treatment system.

SOLUTION:

1. For the system described, the zero-mixing model (see Fig. 10.4) is appropriate.

2. It is necessary to divide the system into two sections, one under the land treatment system and the other between the land treatment system and the stream.

   a. For the section under the land treatment system Eq. (10.8) is written as

$$z_e = \left(z_0^2 - \frac{L_H}{K} w^2 \right)^{1/2}$$

   where

   $z_e =$ groundwater depth at treatment system downstream boundary, m

   $z_0 =$ groundwater depth at treatment system upstream boundary, m

   $w =$ treatment system width in the x direction, m

   b. For the constant-flow section between the edge of the treatment system and the stream the depth of flow will be approximately constant at steady-state. For this condition, Eq. (10.6) reduces to

$$\frac{dv_x}{dx} = 0$$

   Integrating the above expression yields

   $v_x =$ constant

   But from Darcy's law v_x is equal to

$$v_x = -K \frac{dz}{dx}$$

   c. Assuming a steady flow, a constant depth, and a hydraulic gradeline of $dz/dx = -0.01$ m/m, the velocity of flow in the section between the edge of the treatment system and the stream is

$$v_x = -K \frac{dz}{dx} = -\left(10^{-3} \text{ m/s}\right)\left(-0.01 \text{ m/m}\right) = 10^{-5} \text{ m/s}$$

2. Determine the depth of flow above the undisturbed aquifer. Assume that the applied water flows through a layer $z_e - 50$ m deep which has the same hydraulic gradient as the groundwater.

a. The depth of flow at the downstream boundary z_e is

$$z_e - 50 \text{ m} = \frac{L_H w}{v_x}$$

where $z_e - 50$ m is the depth above the undisturbed aquifer. Substitute and solve for z_e.

$$z_e - 50 \text{ m} = \frac{(0.035 \text{ m/d})(50 \text{ m})}{(10^{-5} \text{ m/s})(86{,}400 \text{ s/d})}$$

$$= 2.025 \text{ m}$$
$$z_e = 52.025 \text{ m}$$

b. The depth of flow of the upstream boundary z_0 is

$$z_0 = \left(z_e^2 + \frac{L_H}{K} w^2 \right)^{1/2}$$

$$= \left[(52.025)^2 + \frac{(0.035)(50)^2}{(10^{-3})86{,}400} \right]^{1/2}$$

$$= 52.035 \text{ m}$$

COMMENT:

The relatively high increase in groundwater depth results from the assumption that there is no flow in the y direction.

10.3

TRANSPORT OF NONREACTIVE CONTAMINANTS IN GROUNDWATER

In the previous section the movement of groundwater was considered without allowance for any contaminants that may be present. In this present section the movement of nonreactive contaminants is considered in homogeneous aquifer systems without and with dispersion.

Simplified Mixing Model

The simplest approach that can be used to assess the concentration of a contaminant that may exist in a groundwater is to assume that complete mixing occurs to some depth in the underlying groundwater. For example, in Fig. 10.4(a) if no intermixing occurs, the concentration of the nonreactive constituent in the mounded water is the same as the influent value. If, on the other hand, it is assumed that some intermixing occurs, then the concentration will be reduced. The determination of the nitrate concentration in a groundwater well located down gradient from a rapid infiltration basin is illustrated in Example 10.3.

EXAMPLE 10.3

ESTIMATION OF EFFECT OF MIXING DEPTH ON NITRATE CONCENTRATION IN GROUNDWATER WELL

Determine the effect of the mixing depth on the maximum concentration of nitrate that would be expected in a well downstream from treated effluent spreading basins such as shown in Fig. 10.6. Assume the discharge of treated effluent per 100 m is 1000 m^3/d. The treated wastewater contains 15.0 mg/L of nitrate ($NO_3^- - N$). Assume the hydraulic conductivity of the aquifer is 10^{-5} m/s and that the slope of the upstream groundwater surface is -0.018 m/m. The background level on nitrate in the groundwater is 0.5 mg $NO_3^- - N/L$.

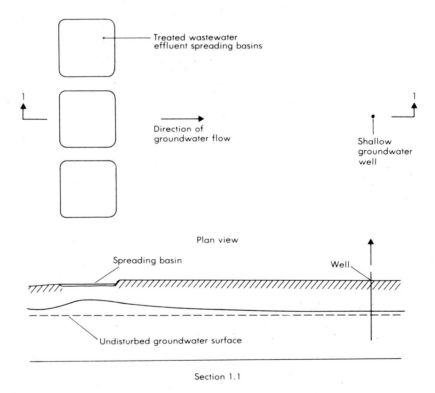

Plan view

Section 1.1

FIGURE 10.6

Definition sketch for Example 10.3:

SOLUTION:

1. Determine the groundwater flow as a function of the mixing depth. The groundwater flow per 100 m of width can be estimated using Eq. (10.3):

$$Q = -(100 \text{ m} \times h)(10^{-5} \text{ m/s})(-0.018 \text{ m/m})(86{,}400 \text{ s/d})$$

where h is the mixing depth in meters.

h, m	Q, m³/d
50	78
100	155
150	233
200	311
250	389

2. Determine the maximum concentration of nitrate in the water well as a function of mixing depth. To determine the maximum concentration of nitrate, assume the total extent of the spreading basins is such that the effects of lateral spreading can be neglected.

   a. For the above condition, the appropriate steady-state mass balance is

$$N_m(Q_g + Q_{ww}) = Q_g N_g + Q_{ww} N_{ww}$$

| Outflow combined | = | Inflow groundwater | + | Inflow wastewater |

   where

   N_m = maximum concentration of nitrate in well water, g/m³
   Q_g = flow of groundwater, m³/d
   Q_{ww} = flow of wastewater, m³/d
   N_g = concentration of nitrate in groundwater, g/m³
   N_{ww} = concentration of nitrate in wastewater, g/m³

   b. Substituting known values and solving for N_m yields

h, m	Q_g, m³/d	Q_{ww}, m³/d	N_m, mg/L
0	0	1000	15.0
50	78	1000	14.0
100	155	1000	13.1
150	233	1000	12.3
200	311	1000	11.6
250	389	1000	10.9

FIGURE 10.7

Typical spreading basins used for the disposal of treated wastewater effluent.

COMMENT

In practice, the concentration of nitrate in the well would be less than the computed values because of denitrification in the soil. Where only one or two spreading basins are used at one time, such as shown in Fig. 10.7, lateral spreading would tend to reduce further the nitrate concentration. Further details on this subject may be found in Ref. [10.14].

Transport by Advection and Dispersion

The transport of a nonreactive contaminant in a groundwater aquifer occurs by advection and dispersion. Performing a materials balance on a section of the aquifer shown in Fig. 10.8 yields

$$\frac{\partial C}{\partial t} \alpha \, \Delta V = \left(\alpha \bar{v}_x A_x C - \alpha D_x A_x \frac{\partial C}{\partial x} \right)\bigg|_x - \left(\alpha \bar{v}_x A_x C - \alpha D_x A_x \frac{\partial C}{\partial x} \right)\bigg|_{x+\Delta x}$$

Accumulation =   Inflow   −   Outflow

where

C = concentration of nonreactive contaminant, g/m^3

α = porosity

V = volume, m^3

$\bar{v}_x$ = average fluid velocity in x direction, m/s

A_x = cross-sectional area in x direction, m^2

D_x = coefficient of hydrodynamic dispersion in x direction, m^2/s

Simplifying the above equation and taking the limit as Δx approaches zero yields Eq. (10.9), which is the same as Eq. (7.5):

$$\frac{\partial C}{\partial t} = D_x \frac{\partial^2 C}{\partial x^2} - \bar{v}_x \frac{\partial C}{\partial x} \tag{10.9}$$

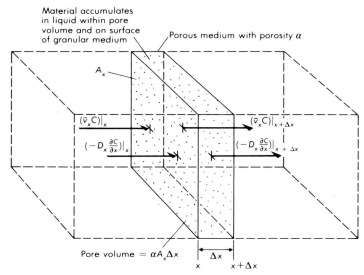

FIGURE 10.8

Definition sketch for the one-dimensional analysis of the movement of a conservative contaminant in a groundwater aquifer.

Solutions for Eq. (10.9) for various boundary conditions may be found in Ref. [10.3]. For a constant step function input of contaminant C_0, the following boundary conditions would apply:

$$C(x,0) = 0 \qquad x \geq 0$$
$$C(0,t) = C_0 \qquad t \geq 0$$
$$C(\infty,t) = 0 \qquad t \geq 0$$

For these boundary conditions the following solution has been developed [10.11]:

$$\frac{C}{C_0} = \frac{1}{2}\left[\operatorname{erfc}\left(\frac{x - \bar{v}_x t}{2\sqrt{D_x t}}\right) + \exp\left(\frac{\bar{v}_x x}{D_x}\right)\operatorname{erfc}\left(\frac{x + \bar{v}_x t}{2\sqrt{D_x t}}\right)\right] \qquad (10.10)$$

where

C = concentration of contaminant at distance x, g/m

C_0 = initial concentration of contaminant, g/m^3

erfc = complementary error function (see Appendix J)

x = distance, m

$\bar{v}_x$ = average velocity in x direction, m/s

t = time, s

D_x = coefficient of hydrodynamic dispersion in x direction, m^2/s

Tabulated values for the error function and the complementary error function are given in Appendix J.

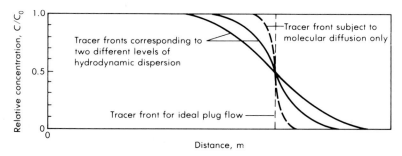

FIGURE 10.9

Typical breakthrough curve for a contaminant subject to hydrodynamic dispersion.

Equation (10.10) can be used to predict the breakthrough curve for a contaminant subject to hydrodynamic dispersion (Fig. 10.9). When the distance or time is large, the second term on the right-hand side of Eq. (10.10) can be neglected (see Example 10.4). At very low velocities, the hydrodynamic dispersion is essentially equal to the molecular diffusion [see Eq. (7.3)]. At high velocities the hydrodynamic dispersion is equal to a constant times the average velocity, and mechanical dispersion is most important. Typical values for the hydrodynamic dispersion coefficient and the molecular diffusion coefficient are reported in Tables 10.3 and 10.4, respectively. In most groundwater aquifers composed of a granular medium, molecular diffusion is not very important. In a fractured aquifer, mechanical dispersion is important. Ultimately, as discussed in Section 10.5, prediction of the movement of contaminants is complicated by the local geology and the lack of field data.

TABLE 10.3

Typical Values for the Coefficient of Hydrodynamic Dispersion

TYPE OF GEOLOGICAL DEPOSIT	GROUNDWATER VELOCITY, m/s	D, m^2/s
Coarse materials	10^{-4}	2×10^{-8}
	10^{-3}	1.5×10^{-6}
Fine materials	10^{-5}	1×10^{-9}
	10^{-4}	3×10^{-9}
	10^{-3}	1×10^{-8}

Source: Adapted from Refs. [10.2], [10.5], and [10.6].

TABLE 10.4

Typical Coefficients of Molecular Diffusion

CONSTITUENT	TYPE OF GEOLOGIC DEPOSIT	D_m, m^2/s
Nonreactive chemical species	Fine-grained clayey material	1×10^{-11} to 10^{-10}
	Course-grained unconsolidated material	1×10^{-10} to $< 2 \times 10^{-9}$
Nonreactive chemical species in water		1×10^{-9} to 5×10^{-9}

Source: Adapted from Ref. [10.6].

EXAMPLE 10.4

DETERMINATION OF CONTAMINANT BREAKTHROUGH

A conservative contaminant is accidentally discharged to the upper lake shown in Fig. 10.2. If the contents of the lake are assumed to be mixed completely, estimate the times required for the concentrations entering the lower lake to reach 10 and 90 percent of the concentration in the upper lake. Assume that the data from Problem 10.1 are applicable and that the value of the coefficient of hydrodynamic dispersion is equal to 10^{-6} m^2/s. Note that there is an assumption in the boundary conditions that the concentration in the upper lake remains approximately constant.

SOLUTION:

1. Because the distance involved (2000 m) is large, assume the right-hand term in Eq. (10.10) can be neglected. The resulting equation is

$$\frac{C}{C_0} \simeq \frac{1}{2}\left[\text{erfc}\left(\frac{x - \bar{v}_x t}{2\sqrt{D_x t}}\right)\right]$$

2. Set up a computation table and solve for C/C_0 for various travel times. Use 1-d time increments before and after the time of travel determined in Example 10.1 (1157 d).

3. The times required for 10 and 90 percent of the original concentration to be reached are as follows:

   a. For $C/C_0 = 0.1$

   $t = 460$ d

   b. For $C/C_0 = 0.9$

   $t = 465$ d

TRAVEL TIME, d	β^*	erf(β)[†]	erfc(β)[‡]	C/C_0
459	1.359[§]	0.94	0.06	0.030
460	1.015	0.85	0.15	0.075
461	0.672	0.66	0.34	0.170
462	0.329	0.36	0.64	0.320
463	− 0.013	− 0.01	1.01	0.505
464	− 0.354	− 0.38	1.38	0.690
465	− 0.694	− 0.67	1.67	0.835
466	− 1.034	− 0.86	1.86	0.930
467	− 1.373	− 0.95	1.95	0.975

$$*\beta = \left(\frac{x - \bar{v}_x t}{2\sqrt{D_x t}} \right)$$

[†] Values for erf(β) are from Appendix J.
[‡] erfc(β) = 1 − erf(β)

$$§\beta = \frac{2000 \ \text{m} - (5 \times 10^{-5} \ \text{m/s})(459 \ \text{d})(86,400 \ \text{s/d})}{2\sqrt{(10^{-6} \ \text{m}^2/\text{s})(459 \ \text{d})(86,400 \ \text{s/d})}}$$

COMMENT

The results obtained with Eq. (10.10) are highly dependent on the value of the coefficient of hydrodynamic dispersion. As the value of the coefficient of hydrodynamic dispersion approaches the value of the coefficient of molecular diffusion (i.e., 10^{-10} m^2/s), the travel times determined in step 3 are essentially the same as the theoretical travel time.

10.4
TRANSPORT OF REACTIVE CONTAMINANTS IN GROUNDWATER

Contaminants moving into the groundwater may be adsorbed on particle surfaces or may enter into chemical and biochemical reactions as well as be carried along with the groundwater flow. For example, phosphate, which reacts with multivalent cations in soil, is usually not a migrating chemical species. Where land wastewater treatment systems are used, such as in Example 10.2, the adsorption sites will become saturated, and a front of phosphate-saturated soil will gradually progress through the system until the entire flow field is saturated (Fig. 10.10). Ammonium ion is quickly absorbed on clay particle surfaces where it is oxidized to NO_2^- and NO_3^- by nitrifying bacteria. The very soluble oxidation products may be assimilated by plants or bacteria in the soil or carried into the groundwater.

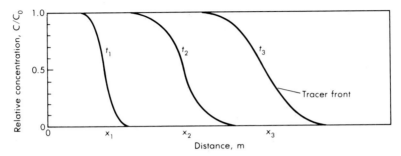

FIGURE 10.10

Definition sketch for the movement of a saturated contaminant front in a ground-water aquifer.

Some biochemical transformations of NO_3^- occur in the groundwater, particularly if organics are present, but accumulation of NO_3^- is a common occurrence, especially where heavy agricultural fertilization is practiced. Heavy metal ions are strongly adsorbed on clay surfaces, and thus they present a problem similar to that of phosphate: A saturated front can progress through the soil. A major environmental problem evolves from assimilation of heavy metals by plants.

Since 1970, identification of groundwaters polluted by organic materials has greatly increased. Solvents such as trichloroethane (TCE), herbicides, soil fumigants such as dibromochloropentane (DBCP), and pesticides such as Lindane and Dieldrin are being found on a regular basis. Sources include manufacturers' container storage locations, dumps, and points of use (see Table 10.1).

Transport with Reaction

Performing a mass balance for the transport of a reactive contaminant in a porous medium results in the following equation, which is essentially the same as Eq. (10.9), with the exception of the rate term that has been added to describe the particular reaction(s) in question:

$$\frac{\partial C}{\partial t} = D_x \frac{\partial^2 C}{\partial x^2} - \bar{v}_x \frac{\partial C}{\partial x} + r_c \qquad (10.11)$$

where

r_c = reaction rate, $g/m^3 \cdot s$

and other terms are as defined previously in the development of Eq. (10.9).

The reaction rate may be governed by adsorption, hydrolysis reactions, precipitation, or a combination of factors. Although some closed-form solutions exist, most solutions to Eq. (10.11) are numerical [10.1, 10.13, 10.15].

Transport with Adsorption

Adsorption is the most common way in which contaminants are removed from the groundwater as it moves through a porous medium. If hydrodynamic dispersion is neglected, the materials balance (see Fig. 10.8) for a contaminant subject to adsorption in a groundwater aquifer is

$$\frac{\partial S}{\partial t} \rho_b \Delta V + \alpha \frac{\partial C}{\partial t} \Delta V = \alpha \bar{v}_x A_x C|_x - \alpha \bar{v}_x A_x C|_{x+\Delta x}$$

$$\begin{array}{ccccc} \text{Accum} & & \text{Accum} & & \\ \text{on solid} & + & \text{in liquid} & = \text{Inflow} & - & \text{Outflow} \\ \text{phase} & & \text{phase} & & \end{array}$$

where

S = mass of solute (contaminant) sorbed per unit mass of dry soil, g/g

ρ_b = bulk density of soil, g/m^3

and other terms are as defined previously. Substituting $(K_{SD}) \partial C / \partial t$ for $\partial S / \partial t$ [see Eq. (7.23)] and taking the limit as Δx approaches zero yields

$$-\bar{v}_x \frac{\partial C}{\partial x} = \left(1 + \frac{\rho_b}{\alpha} K_{SD}\right) \frac{\partial C}{\partial t} \tag{10.12}$$

where

K_{SD} = soil distribution coefficient, m^3/g

and other terms are as defined previously.

Where the partitioning of the contaminant between the soil and the groundwater can be described adequately by the soil distribution coefficient K_{SD} (see Chapter 7), the retardation of the contaminant front relative to the liquid can be described with the following relationship [10.4, 10.7, 10.12]:

$$\frac{\bar{v}_x}{\bar{v}_{xc}} = \left(1 + \frac{\rho_b}{\alpha} K_{SD}\right) \tag{10.13}$$

where

$\bar{v}_x$ = average velocity of groundwater, m/s

$\bar{v}_{xc}$ = average velocity of the $C/C_0 = 0.5$ point of the retarded contaminant concentration profile, m/s

and other terms are as described previously.

If it is assumed that α for most aquifers varies from 0.2 to 0.4 and that the corresponding values for ρ_b are approximately 1.6 to 2.1 × 10^6

g/m^3, then Eq. (10.13) can be written as [10.7]

$$\frac{\bar{v}_x}{\bar{v}_{xc}} = \left(1 + 4 \times 10^6 \, K_{SD}\right) \text{ to } \left(1 + 10 \times 10^6 \, K_{SD}\right) \tag{10.14}$$

If the value of K_{SD} is equal to zero, the contaminant is nonreactive and no retardation occurs. If the value of K_{SD} is greater than about 10^{-4} the contaminant is essentially immobile. The value of K_{SD} can be estimated by using the following expression [10.10]:

$$K_{SD} = 6.3 \times 10^{-7} \, f_{OC}(K_{OW}) \tag{10.15}$$

where

f_{OC} = fraction of organic carbon in the soil, g/g

K_{OW} = octanol : water distribution coefficient

Selected values for K_{OW} for various organic compounds are reported in Table 10.5. Additional K_{OW} values for a variety of compounds can be found in Ref. [10.8].

Transport with Diffusion

Groundwater flow is laminar, and vertical mixing does not occur to a great extent. Thus contaminants added to the groundwater travel vertically by molecular diffusion, a very slow process. This situation partially accounts for the surprising differences often found in the quality of water from different wells in the same aquifer. An idea of the relative transport velocities can be obtained from consideration of a simplified model of the transport of a reactive material into an undisturbed aquifer. Assuming that a one-dimensional model is satisfactory, a materials balance on the

TABLE 10.5

**Octanol : Water Distribution Coefficients
for Selected Organic Compounds**

COMPOUND	EMPIRICAL FORMULA	log K_{OW}
Benzene	C_6H_6	2.01
Carbon tetrachloride	CCl_4	2.72
Chlorobenzene	C_5H_5Cl	2.49
Chloroform	$CHCl_3$	1.95
DDT	$C_{14}H_9Cl_5$	4.98
Phenol	C_6H_6O	1.49

Source: Adapted from Ref. [10.7].

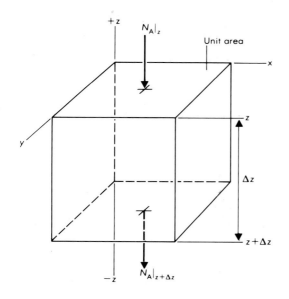

FIGURE 10.11

Definition sketch for the flux of a contaminant into the underlying groundwater.

volume element shown in Fig. 10.11 can be used to develop Eq. (10.16):

$$\frac{\partial C_A}{\partial t} \Delta z = N_A|_z - N_A|_{z+\Delta z} + r_A \Delta z \tag{10.16}$$

Accumulation = Inflow − Outflow + Generation

where

$\quad C_A$ = concentration of contaminant A, mol/m^3

$\quad z$ = distance in z direction, m

$\quad N_A|_z$ = molar flux of A in the z direction, mol/m$^2 \cdot$ s

$\quad r_A$ = rate of reaction, mol/m$^3 \cdot$ s

Assuming that steady-state conditions are reached, collecting terms, dividing by Δz, and taking the limit as Δz approaches zero,

$$-\frac{dN_A}{dz} + r_A = 0 \tag{10.17}$$

At steady state the molar flux is given by Fick's first law:

$$N_A|_z = -D \frac{dC_A}{dz} \tag{10.18}$$

Substituting for N_A in Eq. (10.17) yields

$$D \frac{d^2 C_A}{dz^2} + r_A = 0 \tag{10.19}$$

For a zero-order reaction, $r_A = -k$ and Eq. (10.19) becomes

$$D\frac{d^2C_A}{dz^2} - k = 0 \tag{10.20}$$

where the boundary conditions are

$$z = 0 \qquad C_A = C_{A_0}$$
$$z = z^* \qquad C_A = 0$$

The term z^* represents the distance at which the concentration of A in the aquifer is equal to zero.

Equation (10.20) can be integrated to give

$$C_A = \frac{k}{2D}z^2 + Az + B \tag{10.21}$$

where A and B are the constants of integration.

Setting $z = 0$ results in $B = C_{A_0}$. The second integration constant A is evaluated by noting that at steady state, the flux at the interface $N_A|_{z=0}$ must equal the total amount of A reacting, Vk. For a unit area $V = z^*$:

$$N_A|_{z=0} = -D\frac{dC_A}{dz}\bigg|_{z=0} = kz^*$$

Differentiating Eq. (10.21) with respect to z yields

$$\frac{dC_A}{dz} = \frac{k}{D}z + A$$

Setting the above two expressions for dC_A/dz equal at $z = 0$ and solving for A:

$$A = -\frac{k}{D}z^*$$

Substituting for A and B in Eq. (10.21) results in

$$C_A = \frac{k}{2D}z^2 - \frac{k}{D}z^*z + C_{A_0} \tag{10.22}$$

When $z = z^*$, $C_A = 0$, and z^* is equal to

$$z^* = \left(\frac{2C_{A_0}D}{k}\right)^{1/2} \tag{10.23}$$

$$C_A = \frac{1}{2}\frac{k}{D}z^2 - \left(2C_{A_0}\frac{k}{D}\right)^{1/2}z + C_{A_0} \tag{10.24}$$

The application of Eqs. (10.23) and (10.24) is illustrated in Example 10.5.

EXAMPLE 10.5

MOVEMENT OF REACTIVE MATERIAL INTO GROUNDWATER BY DIFFUSION

A metal ion is removed in groundwater by an approximately zero-order precipitation reaction. If $k = 2.3 \times 10^{-8}$ mol/m$^3 \cdot$s and $D = 10^{-8}$ m^2/s, estimate the vertical concentration profile resulting from an interface concentration of 0.7 mol/m^3.

SOLUTION:

1. Determine the value z^* using Eq. (10.23).

$$z^* = \left(\frac{2C_{A_0}D}{k} \right)^{1/2}$$

$$z^* = \left[2\frac{0.7(10^{-8})}{2.3 \times 10^{-8}} \right]^{1/2} = 0.78 \text{ m}$$

Therefore the ion does not penetrate a great distance.

2. Using Eq. (10.24) determine the concentration of the metal C_A as a function of the vertical distance.

$$C_A = \frac{2.3 \times 10^{-8}}{2 \times 10^{-8}} z^2 - \left[2(0.7)\frac{2.3 \times 10^{-8}}{1 \times 10^{-8}} \right]^{1/2} z + (0.7)$$

z, m	0.01	0.02	0.04	0.08	0.12	0.16	0.20	0.30	0.40	0.60
C_A, mol/m^3	0.68	0.66	0.63	0.56	0.50	0.44	0.39	0.27	0.17	0.04

COMMENT

It is important to note that, as illustrated by this example, many contaminants do not move significantly into the groundwater aquifer by molecular diffusion, especially where a reaction is involved.

10.5

PRACTICAL CONSIDERATIONS IN GROUNDWATER QUALITY MANAGEMENT

In the preceding sections, the movement of contaminants in groundwater has been described assuming uniform flow conditions and the availability of data on the porous medium, the molecular diffusion coefficient, and the distribution coefficient. Unfortunately, ideal flow conditions seldom exist, and usable data are usually not available in most situations. Some of the complicating factors encountered in the study and modeling of the movement of contaminants in groundwater are considered below. Some control measures and treatment techniques used to contain and restore contaminated aquifers are also considered.

Nonuniform Field Conditions

One of the most difficult problems and perhaps the most commonly encountered in the field is the nonuniform nature of most groundwater aquifers and lack of information on what is actually under the ground at a specific study location.

Two examples of the types of conditions that can be encountered in the field are illustrated in Figs. 10.12 and 10.13. In Fig. 10.12 a dense slug of the contaminant is moving in an opposite direction to the more dilute contaminant. In Fig. 10.13 the seepage from a waste disposal site has been intercepted by a clay lens. In both of these cases the only way to detect such conditions is to drill a sufficient number of observation wells.

Lack of Contaminant and Site Data

In situations where one or more contaminants are known to have entered a groundwater, it is important to have data available on their solubility, density, viscosity, volatility, and stability (reactivity) [10.9]. Data on the soil distribution coefficients for the various contaminants will also be needed. Unfortunately the amount of such data available is limited, usually to a specific site where a problem has developed or to sites where long-term studies have been conducted. At this time, if the movement of a contaminant is to be modeled so that corrective measures can be taken, special field and laboratory studies will have to be conducted.

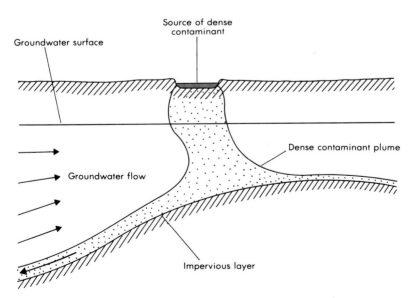

FIGURE 10.12

Movement of a slug of a dense contaminant from a surface source.

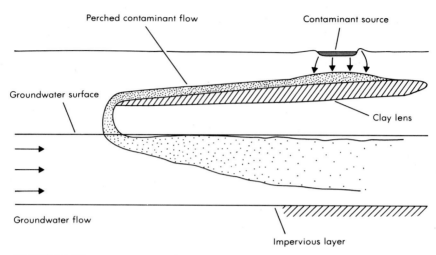

FIGURE 10.13

Movement of a contaminant from a surface source intercepted by a clay lens.

The most common methods used to obtain the data needed to define the movement of contaminants are pilot-scale column or thin-layer leaching studies. In conducting such studies it must be recognized that it is difficult to reproduce natural conditions in the laboratory. However, if the results of laboratory studies can be correlated to the results of field studies, it is often possible to make reasonable predictions.

Control Measures for Contaminated Aquifers

The types of measures that may be undertaken to control the contamination of an underground aquifer will depend on the nature and source of the contaminant(s). Different measures will be required if the source of the contaminant is continuous or a one-time input (Fig. 10.14). For both a continuous and a one-time input, it will be necessary to trace the movement of the contaminant plume from the source. Typically, the plume is traced by drilling wells that can be sampled at various depths. Chloride concentration data, collected from monitoring wells, were used to develop the chloride plumes shown in Fig. 10.15. Assuming the source can be identified and the contaminant plume can be traced, the next question is what to do about it. In general, cleanup and control measures fall into three categories: containment, removal, and treatment [10.9].

Containment Options

Containment involves the control of the spread of the contaminants. Common methods of containment include surface sealing, the instal-

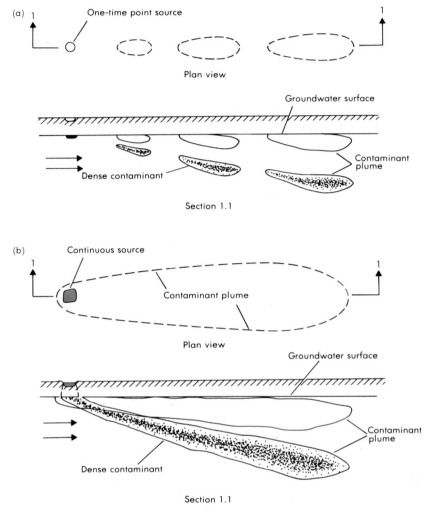

FIGURE 10.14

Generalized movement of a contaminant plume for contaminants of different densities from a surface source: (a) one-time point source discharge and (b) continuous-source discharge.

lation of physical barriers, and the use of hydraulic barriers. Surface sealing involves sealing the surface to prevent infiltration and, in some cases, subsurface percolation. Physical and hydraulic barriers are used to limit the spread of the contaminant plume. Slurry walls, collection trenches, sheet piling, and grout curtains have been used as physical barriers with varying success (Fig. 10.16). Physical barriers work best where there is an impermeable underlying strata. The effectiveness of

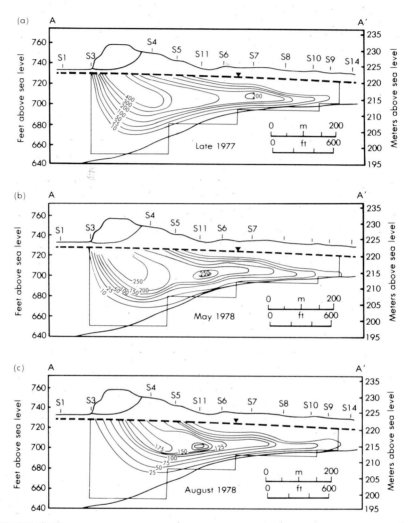

FIGURE 10.15

Measured chloride plume under a well-monitored landfill site. (a) Late 1977, (b) May 1978, and (c) August 1978.   Source: Courtesy of Waterloo Research Institute [10.18].

physical barriers beyond depths of about 40 m is uncertain. Hydraulic barriers are formed by combinations of pumping and injection wells (Fig. 10.17).

Removal Options

Contaminated sites can be cleaned up by removing surface liquid and sludge sources by pumping and dredging. Contaminated soil that is

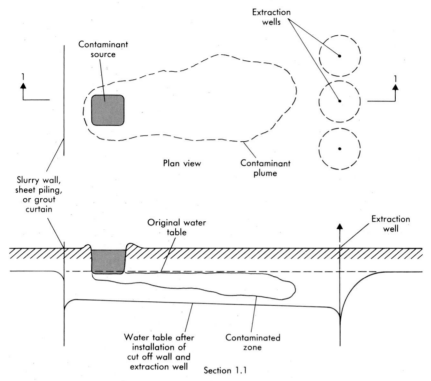

FIGURE 10.16

Definition sketch for the use of slurry walls, grout curtains, or sheet piling; surface sealing; and extraction wells to control the spread of a contaminant from a surface waste disposal site.

accessible may also have to be excavated. Contaminated groundwater can best be removed for treatment using drains or by pumping with horizontal or vertical wells (Fig. 10.17). Where treatment is needed, one of the most common approaches is to reinject the treated groundwater. If an alternative source of water is available it can be used for injection. Treated groundwater, not reinjected, is usually discharged to streams or rivers on a temporary basis.

Treatment Options

Wastes and groundwater removed from contaminated sites may have to be treated before disposal or reuse. Immobilization, separation, conversion, concentration, or thermal destruction are some of the methods used for the treatment of groundwater contaminants (see Table 11.4). Treatment methods are considered further in Part IV.

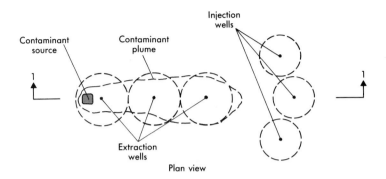

Plan view

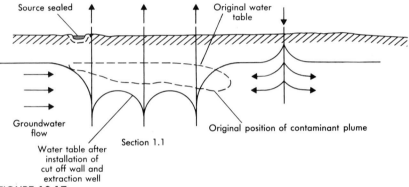

FIGURE 10.17

Definition sketch for the use of extraction and injection wells to control the spread of a contaminant plume from a surface waste disposal site.

KEY IDEAS, CONCEPTS, AND ISSUES

- Groundwater is the primary source of drinking water in the United States.

- Groundwater in various parts of the country has become contaminated by such a wide variety of anthropogenic contaminants that generalizations about the movement and effects of a given contaminant are inappropriate.

- The movement of groundwater can be described by Darcy's law.

- Nonreactive contaminants are transported in groundwater aquifers by advection and hydrodynamic dispersion.

- In a homogeneous porous medium, the vertical transport of a contaminant by molecular diffusion is an extremely slow process.

- The movement of contaminants in groundwater is retarded by a variety of reactions, the most notable being adsorption.
- The distribution coefficient can be used to assess the potential movement of a contaminant in a groundwater aquifer.
- In practice, tracing the movement of contaminants in groundwater is complicated and expensive because of the nonuniformity of most groundwater aquifers and the lack of usable field data.

DISCUSSION TOPICS AND PROBLEMS

10.1. Estimate the effect that a 10°C rise in temperature would have on the percolation rate of leachate from a sanitary landfill. Use temperature increments of 0–10, 20–30, 40–50, and 60–70 °C.

10.2. The discharge from a constant-head permeameter is 3.7 mm/s at 20°C. What will the discharge be at 7 and at 33°C?

10.3. For some reason, a lined hazardous-waste landfill was located on top of a confined aquifer. Situated approximately 500 m away is a community water supply well which is pumped at a rate of 10^4 m³/d. If an earthquake occurs which splits open the liner and pollutes the aquifer, how much time is available to shut down the well and avoid withdrawing contaminated water? Assume the following aquifer characteristics are applicable: hydraulic conductivity = 50 m/d, porosity = 0.2, aquifer thickness = 5 m, and aquifer width = 50 m. (Courtesy of Jim Hunt.)

10.4. Estimate the maximum flow of leachate from the bottom of a landfill lined with a 0.2 m thick clay layer. Assume the groundwater table is coincident with the bottom of the clay layer and that the depth of leachate in the landfill is 0.5 m above the clay layer. The hydraulic conductivity of the clay material is 5.0×10^{-8} m/s.

10.5. Estimate the thickness of a clay layer that would be required to limit the flow of leachate from the bottom of a landfill to a value of 1.0 mm/d. Assume the hydraulic conductivity of the clay material is 5×10^{-8} m/s and that the surface of the groundwater is 0.68 m below the surface of the leachate collected in the bottom of the landfill.

10.6. Demonstrate that Eq. (10.8) can be derived from Eq. (10.7).

10.7. Consider a vertical clay barrier of thickness w separating contaminated groundwater from clean groundwater (Fig. 10.18). The clay material has a hydraulic conductivity K. The contaminated

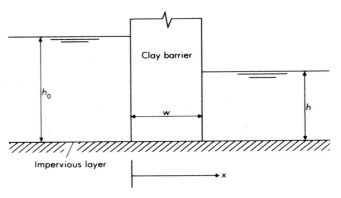

FIGURE 10.18
Definition sketch for Problem 10.7.

groundwater has a water table depth h_0 that is greater than the water table depth h in the clean groundwater.

a. Given that the steady-state differential equation for variable water depth in an unconfined aquifer is

$$\frac{d^2h^2}{dx^2} = 0$$

solve for the water table depth as a function of the horizontal distance x inside the clay barrier.

b. Derive an expression for the steady-state flow rate through the barrier if it has a length L. (Courtesy of Jim Hunt.)

10.8. You are asked to place a grassy area on top of a building. Soil with a hydraulic conductivity K is placed on top of a flat concrete roof, and water is applied at a constant rate of I (depth per time) above the grass surface. Derive the following steady-state differential equation for the water table elevation h as a function of horizontal distance x. Assume one-dimensional flow is occurring and that drains are installed at the edge of the system. (Courtesy of Jim Hunt.)

$$\frac{d^2h^2}{dx^2} + \frac{2I}{K} = 0$$

10.9. For the grassy roof of Problem 10.8, consider soil having a K value equal to 10 m/d, a rainfall rate I equal to 0.01 m/d, and a grassy area width of 25 m. An impermeable wall exists at x equal to 0 and the water table depth is not to exceed 0.3 m at the right-hand boundary. For these conditions:

a. What are the two boundary conditions that must be satisfied by the differential equation derived in Problem 10.8?

b. What is the depth of the water table as a function of horizontal position in the grassy area?

c. What is the depth of soil required to prevent the water table from reaching the surface? (Courtesy of Jim Hunt.)

10.10. Determine the maximum application rate of treated effluent in Example 10.3 if the nitrate concentration in the downstream well is not to exceed 10 mg/L when the mixing depth is assumed to be 175 m. In practice, the mixing depth can be assumed to be approximately equal to the depth of the well.

10.11. Treated wastewater is discharged to the ground using percolation ponds that drain to an aquifer immediately below the ponds that is 1000 m wide and 40 m deep. The aquifer feeds a stream 6000 m from and 40 m below the percolation ponds. Hydraulic conductivity of the aquifer is 0.0001 m/s. The source of water in the aquifer is a spring a short distance above the ponds, and the water is nearly saturated with oxygen (8.0 g/m^3).

a. Determine the travel time from the ponds to the stream assuming a porosity of 0.3 and complete mixing.

b. Assuming that the treated wastewater is distributed uniformly across the aquifer, determine the maximum effluent BOD concentration allowable from the 0.008 m^3/s capacity plant if the groundwater discharged to the stream is to contain 2.0 mg/L BOD. Assume $k = 0.05 \text{ d}^{-1}$.

c. If the BOD can be reduced to zero, what effect will an $NH_3 - N$ concentration of 30 g/m^3 have on the water quality? Assume that nitrification is a first-order reaction with respect to N and the rate coefficient, k_N is 0.05 d^{-1}.

10.12. A percolation pond 200 m wide is located at a point 40 m above an unconfined aquifer. The aquifer is 100 m deep and has an average hydraulic gradient of 0.002. The hydraulic conductivity is 10^{-3} m/s. Determine the increase in groundwater depth for a loading of 20 mm/d.

10.13. Prepare a sketch illustrating the movement of a contaminant front (see Fig. 10.10) for a conservative contaminant subject to advection and hydrodynamic dispersion. Plot the contaminant curves at distances of 1000, 5000 and 10,000 m from the source of the contaminant. Assume the coefficient of hydrodynamic dispersion is equal to 10^{-5} m^2/s and the velocity of the groundwater is 2×10^{-5} m/s.

10.14. Solve Problem 10.13 for distances of 4000, 8000 and 12,000 m from the source.

10.15. Using Eq. (10.10), determine the relative concentration (C/C_0) of a contaminant as a function of depth (i.e., distance from the source) brought about molecular diffusion alone. Assume the coefficient of molecular diffusion is 1.0×10^{-10} m^2/s. Plot curves for 10, 100, 1000, and 10,000 years on log-log paper for concentration ratios varying 1.0 to 0.01 and for depths from 0.1 to 100 m.

10.16. Solve Problem 10.15 for a coefficient of molecular diffusion of 1.0×10^{-8} m^2/s.

10.17. Determine the relative movement of a contaminant front in an aquifer subject to the following conditions. The K_{OW} value for the contaminant is 100, the porosity of the aquifer is 0.2, the bulk density of the aquifer material is 2.0×10^6 g/m^3, and the fraction of organic carbon in the soil is 0.15.

10.18. Solve Problem 10.17 for phenol and DDT. What conclusion can be derived from the results of these computations?

REFERENCES

10.1. Bachmat, Y., J. Bredehoeft, D. Holtz, and S. Sebastian, (1980), *Groundwater Management: The Use of Numerical Models*, Water Resources Monograph Series, vol. 5, American Geophysical Union, Washington, D.C.

10.2. Bouwer, H., (1978), *Groundwater Hydrology*, McGraw-Hill Book Company, New York.

10.3. Crank, J., (1957), *The Mathematics of Diffusion*, Oxford University Press, London.

10.4. Davis, S. N., and R. J. M. DeWiest, (1966), *Hydrogeology*, John Wiley and Sons, New York.

10.5. De Wiest, R. J. M., (1965), *Geohydrology*, John Wiley and Sons, New York.

10.6. Freeze, R. A., and J. A. Cherry, (1979), *Groundwater*, Prentice-Hall, Englewood Cliffs, N.J.

10.7. Hansch, C., and A. Leo, (1979), *Substitute Constants for Correlation Analysis in Chemistry and Biology*, John Wiley and Sons, New York.

10.8. Hess, A. F., J. E. Dyksen, and H. J. Dunn, (1983), "Groundwater Contamination: Challenge of the 80's," *Water Technology*, vol. 6, no. 7, p. 40.

10.9. Jennings, A. A., and R. L. Sholar, (1984), "Hazardous Waste Disposal Network Analysis," *J. Environmental Engineering Division*, ASCE, vol. 110, no. 2, p. 325.

10.10. Karickhoff, S. W., D. S. Brown, and T. A. Scott, (1979), "Sorption of Hydrophobic Pollutants on Natural Sediments," *Water Research*, vol. 13, no. 3, p. 241.

10.11. Ogata, A., (1970), "Theory of Dispersion in a Granular Medium," U.S. Geological Survey Professional Paper 411-I.

10.12. *Organic Chemical Contaminants in Groundwater: Transport and Removal*, (1981), AWWA Seminar Proceedings, Report No. 20156.

10.13. Pickens, J. F., and W. C. Lennox, (1976), "Numerical Simulation of Waste Movement in Steady Groundwater Flow Systems," *Water Resources Research*, vol. 12, no. 2, p. 171.

10.14. *Process Design Manual Land Treatment of Municipal Wastewater*, (1981), Technology Transfer, USEPA 625/1-81-013, Cincinnati.

10.15. Sykes, J. F., S. B. Pahwa, R. B. Lantz, and D. S. Ward, (1982), "Numerical Simulation of Flow and Contaminant Migration at an Extensively Monitored Landfill," *Water Resources Research*, vol. 18, no. 6, p. 1687.

10.16. Todd, D. K., (1980), *Groundwater Hydrology*, 2d ed., John Wiley and Sons, New York.

10.17. _____, (1983), *Ground-Water Resources of the United States*, Premier Press, Berkeley, Calif.

10.18. Waterloo Research Institute, (1982), "Canadian Forces Base Borden Sanitary Landfill Research Program," vol. 1, University of Waterloo, Waterloo, Ontario.

PART IV

Modification of Water Quality

Water quality characteristics can be changed in a number of ways and for a number of reasons. The effects of wastewater discharges and of impoundments were discussed in Part III. Methods of modifying water quality for specific purposes (e.g., for treating drinking water and for meeting wastewater discharge requirements) are discussed in this part. The two most common water quality modification problems are (1) providing suitable domestic water supplies and (2) treating the used domestic water (wastewater) in preparation for discharge to receiving waters. Similar problems exist with industrial installations, although certain water quality requirements for water supply and discharges, such as those for the electronics industry, may be much more stringent than for municipal systems.

The development of treatment systems for water and wastewater requires a fundamental understanding of physical, chemical, and biological phenomena on which the various treatment units and processes are based. Flow sheets for the commonly used physical, chemical, and biological treatment operations and processes, in both potable water production and wastewater treatment, are developed in Chapter 11. The physical, chemical, and biological treatment methods used in water supply and wastewater treatment are introduced and discussed in Chapters 12, 13, and 14. The synthesis of treatment systems using the unit operations and processes introduced in Chapters 11 through 14 is considered in Chapter 15.

Readers are reminded of the statement in the preface to this text that only an introduction to water quality modification is presented in this volume. More detailed theoretical analyses and descriptions of the various treatment methods are presented in the second volume of this series. The hydraulic design of water and wastewater facilities is considered in the third volume.

11

Introduction to Water and Wastewater Treatment

Over the years, a variety of methods have been developed for the treatment of water and wastewater. In most situations, a combination or sequence of methods will be needed. The specific sequence required will depend on the quality of the untreated water or wastewater and the desired quality of the product. Although treating water is relatively inexpensive on a per-cubic-meter basis, there is little opportunity to modify water quality directly in most natural systems such as streams, lakes, and groundwaters because of the large volumes involved. Rather, what is done is to treat the water used for public water supplies and to treat wastewater before it is returned to the environment in engineered systems.

It is the purpose of this chapter to present an overview of the various methods and means used for the treatment of water and wastewater. Topics to be considered in this chapter include (1) the classification of treatment methods, (2) the various methods used for the treatment of specific contaminants, (3) the configuration of typical water and wastewater treatment plants, and (4) the techniques used to size the individual treatment facilities. The physical, chemical, and biological treatment methods introduced in this chapter are considered in detail in Chapters 12, 13, and 14, respectively. The synthesis of the various treatment methods into complete treatment plants is considered in Chapter 15.

11.1

CLASSIFICATION OF TREATMENT METHODS

The contaminants in water and wastewater are removed by physical, chemical, and biological means. The specific methods are classified as physical unit operations, chemical unit processes, and biological unit processes. Although several of these operations and processes are combined in most treatment systems, they are usually considered separately —a practice followed in this text. By considering each group separately, it is possible to examine the fundamental principles involved apart from their application in the treatment of water and wastewater.

Physical Unit Operations

Treatment operations in which change is brought about through the application of physical forces are classified as physical unit operations. Typical unit operations include screening, mixing, gas transfer, sedimentation, and filtration. Physical operations are considered in detail in Chapter 12.

Chemical Unit Processes

Treatment processes in which the removal or treatment of contaminants is brought about by the addition of chemicals or by chemical reactions are classified as chemical unit processes. Chemical precipitation and disinfection are two important examples. Chemical processes are considered in Chapter 13.

Biological Unit Processes

Treatment processes in which the removal of contaminants is brought about by biological means are classified as biological unit processes. The activated sludge process used for the treatment of the organic matter in wastewater is perhaps the best-known example. Biological processes are considered in Chapter 14.

11.2
TREATMENT METHODS FOR WATER, WASTEWATER, AND CONTAMINATED WATERS

In Part I, a wide variety of contaminants that may be found in water and wastewater were identified. Contaminants that may have to be removed from groundwater, surface water, and wastewater to meet specific water quality objectives are identified in Table 11.1. Because of their importance, treatment methods for contaminants of anthropogenic origin are also considered.

Water Treatment Methods

The most important objective of water treatment is to produce a water that is biologically and chemically safe for human consumption. Quality

TABLE 11.1

Typical Contaminants Found in Various Waters that May Need to Be Removed to Meet Specific Water Quality Objectives*

CLASS	TYPICAL CONTAMINANTS FOUND IN		
	Groundwater	Surface water	Wastewater
Floating and suspended materials	None	Branches, leaves, algal mats, soil particles	Wood, rags, paper, grit, food wastes, feces
Colloidal materials	Microorganisms, trace organic and inorganic constituents[†]	Clay, silt, organic materials, pathogenic organisms, algae, other microorganisms	Food wastes, feces, pathogenic bacteria, other microorganisms, silt
Dissolved materials	Iron and manganese, hardness ions, inorganic salts, trace organic compounds	Organic compounds, tannic acids, hardness ions, inorganic salts	Organic compounds (e.g., BOD) nutrients, heavy metals, inorganic salts
Dissolved gases	Carbon dioxide, hydrogen sulfide	[‡]	Ammonia, hydrogen sulfide, methane
Immiscible liquids	[§]	Oils and greases	Oils and greases

*Specific water quality objectives may be related to drinking water standards, industrial use requirements, effluent discharge requirements, or agricultural reuse.

[†] Typically of anthropogenic origin.

[‡] Gas supersaturation may have to be reduced if surface water is to be used in fish hatcheries.

[§] Unusual in natural groundwater aquifers.

requirements similar to those for domestic use will generally apply for most industrial users. In some cases, such as in the manufacture of printed circuits, even higher quality requirements may have to be met.

The principal contaminants found in water and the unit operations and processes used for their removal are summarized in Table 11.2. As noted there, commonly used water treatment methods are either physical operations or chemical processes. Biological processes are not used because appreciable amounts of organic matter are not present in most natural waters and biological processes are not suitable in situations where contaminant concentrations are low. In general, effluents from biological treatment processes do not meet source standards for domestic water supplies. However, many community water supplies contain treated effluents from upstream wastewater discharges (e.g., the New Orleans water supply contains discharges from Minneapolis–St. Paul and all other communities in the Mississippi watershed), but dilution and the assimilative capacity of the receiving water are sufficient to make the mixture acceptable as a water supply source.

Wastewater Treatment Methods

The principal objective of wastewater treatment is to produce an effluent that can be discharged without causing serious environmental impacts. The principal contaminants found in wastewater and the unit operations and processes used for their removal are summarized in Table 11.3. Processes and operations used in wastewater treatment are similar to those used in water treatment, except for biological methods. The principal use of biological treatment is for the removal of easily biodegradable organic compounds, although biological processes are also used for removal of nitrogen and phosphorus in some situations. A large number of biological process configurations are in use, and several combine physical operations and chemical and biological processes within the same unit.

Contaminated Groundwater Treatment Methods

Serious health and environmental hazards exist in a wide variety of chemicals of anthropogenic origin that have been found in both surface waters and groundwaters. The principal objectives to be met in treating contaminated groundwaters are to eliminate the health hazards and to restore, to the extent possible, the quality of the groundwater. Treatment methods for contaminated groundwaters are listed in Table 11.4.

TABLE 11.2

**Unit Operations, Processes, and Treatment Systems
Used to Remove the Major Contaminants Found in Water**

CONTAMINANT	UNIT OPERATION, UNIT PROCESS, OR TREATMENT SYSTEM	CLASSIFICATION*
Pathogenic organisms	Chlorination	C
	Ozonation	C
Turbidity and suspended matter	Screening	P
	Sedimentation	P
	Filtration	P
	Coagulation/flocculation/ sedimentation/filtration	C/P/P/P
Color	Adsorption	P/C
	Ion exchange	C
	Coagulation/flocculation/ sedimentation/filtration	C/P/P/P
Tastes and odors	Oxidation (aeration)	P/C
	Adsorption	P
	Chemical oxidation	C
Organic matter	Adsorption	P
	Ion exchange	C
	Ozonation	C
	Coagulation/flocculation/ sedimentation/filtration	C/P/P/P
Hardness ions, $Ca^{+2} + Mg^{+2}$	Chemical precipitation	C
	Ion exchange	C
Dissolved gases	Aeration	P
	Vacuum deaeration	P
	Chlorination	C
	Ion exchange	C
Heavy metals	Chemical precipitation	C
	Ion exchange	C
Iron and manganese	Ion exchange	C
	Oxidation/precipitation/filtration	C/P/P
Dissolved solids	Reverse osmosis	P
	Distillation	P

*C = chemical, P = physical.

TABLE 11.3

Unit Operations, Processes, and Treatment Systems
Used to Remove the Major Contaminants Found in Wastewater

CONTAMINANT	UNIT OPERATION, UNIT PROCESS, OR TREATMENT SYSTEM	CLASSIFICATION*
Suspended solids	Screening and comminution	P
	Sedimentation	P
	Flotation	P
	Filtration	P
	Coagulation/sedimentation	C/P
	Land treatment	P
Biodegradable organics	Activated sludge	B
	Trickling filters	B
	Rotating biological contactors	B
	Aerated lagoons	B
	Oxidation ponds	B
	Intermittent sand filtration	P/B
	Land treatment	B/C/P
	Physical/chemical	P/C
Pathogens	Chlorination	C
	Ozonation	C
	Land treatment	P
Nutrients:		
Nitrogen	Suspended-growth nitrification and denitrification	B
	Fixed-film nitrification and denitrification	B
	Ammonia stripping	C/P
	Ion exchange	C
	Breakpoint chlorination	C
	Land treatment	B/C/P
Phosphorus	Metal salt coagulation/sedimentation	C/P
	Lime coagulation/sedimentation	C/P
	Biological/chemical phosphorus removal	B/C
	Land treatment	C/P
Refractory organics	Carbon adsorption	P
	Tertiary ozonation	C
	Land treatment systems	P
Heavy metals	Chemical precipitation	C
	Ion exchange	C
	Land treatment	C/P
Dissolved inorganic solids	Ion exchange	C
	Reverse osmosis	P
	Electrodialysis	C

*B = biological, C = chemical, P = physical.

TABLE 11.4

**Unit Operations, Processes, and Treatment Systems Used to
Remove Anthropogenic Contaminants Found in Groundwater**

CONTAMINANT	UNIT OPERATION, UNIT PROCESS, OR TREATMENT SYSTEM	CLASSIFICATION*
Acids and alkalies	Membrane separation/reuse	P
	Neutralization	C
Biodegradable organics	Activated sludge	B
	Aerobic filtration	B
	Aerobic ponds	B
	Anaerobic digestion	B
Heavy metals	Adsorption (traces)	P
	Chemical reduction	C
	Gravimetric separation	P
	Precipitation	C
Inorganic ions and salts	Adsorption (traces)	P
	Encapsulation/fixation	P
	Gravimetric separation	P
	Landfill	P
	Membrane separation/reuse	P
	Precipitation	C
Organometallic compounds	Adsorption (traces)	P
	Chemical oxidation/reduction	C
	Gravimetric separation	P
	Precipitation	C
Reactive ions and compounds	Chemical oxidation/reduction	C
	Dissolving	C
	Encapsulation/fixation	P
	Hydrolysis	C
Refractory organics	Chemical oxidation	C
	Incineration	C
	Solvent extraction/incineration	C
	Solvent extraction/reuse	C
Solvents and oils	Acid/caustic stripping	C
	Distillation	P
	Evaporation	P
	Filtration	P
	Gravimetric separation	P
	Incineration	C
	Steam stripping	P

*B = biological, C = chemical, P = physical.
Source: Adapted from Ref. [11.4].

11.3

TREATMENT PROCESS FLOW SHEETS

Depending on the contaminants to be removed, an almost limitless number of process combinations can be developed using the unit operations and processes reported in Tables 11.2, 11.3, and 11.4. The term "flow sheet" is used to describe a particular combination of unit operations and processes used to achieve a specific treatment objective. Apart from the analysis of the technical feasibility of the individual treatment methods, as discussed in Chapter 15, the exact flow-sheet configuration will depend on factors such as (1) the client's needs, (2) the designer's past experience, (3) regulatory agency policies on the application of specific treatment methods, (4) the availability of equipment suppliers, (5) what use can be made of existing facilities, (6) the availability of qualified operating personnel, (7) initial construction costs, and (8) future operation and maintenance costs. Conventional flow sheets for the treatment of water and wastewater are presented and discussed below.

Water Treatment

The treatment required for a water depends on the source of supply. For example, the treatment of water from a remote mountain catchment area may involve screening, filtration, and disinfection, whereas the treatment for river water may involve screening, coagulation/flocculation, sedimentation, filtration, and disinfection. Groundwater from deep wells may require no treatment at all. Depending on the local geology, both groundwater and surface supplies may require softening for hardness removal. A flow sheet for the treatment of a groundwater containing iron and manganese is shown in Fig. 11.1. As shown, chlorine and atmospheric oxygen are used to oxidize the iron and manganese. After sedimentation, granular-medium filtration is used to remove any residual oxidation products.

A typical flow sheet for the complete treatment of river water is shown in Fig. 11.2. As shown, the bar racks and screens are used to remove large debris. The coagulation/flocculation, sedimentation, and filtration treatment steps are used for the removal of turbidity and pathogenic organisms. Chemical coagulation and flocculation are used to produce a flocculant precipitate that enmeshes the colloidal and suspended solids in the water. The precipitate is removed by gravity sedimentation. Filtration is used to remove any residual solids remaining after sedimentation. Disinfection is used for the control of pathogenic organisms. After disinfection, the treated water is often stored in a clear well prior to being pumped into the distribution system. The storage lagoons are used for the treatment of the materials removed from the raw water and the materials added to bring about treatment.

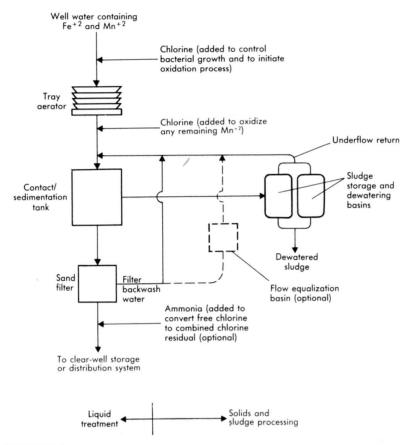

FIGURE 11.1

Typical flow sheet for the removal of iron and manganese from a groundwater.

Wastewater Treatment

The treatment required for a wastewater will depend on the effluent discharge requirements. For example, where an ocean discharge is used, removal of large debris by screens and of settleable solids by sedimentation may be the only treatment steps that are required. Where treated effluent is to be discharged to an inland stream, complete treatment may be required. Discharges to environmentally sensitive lakes, streams, and estuaries may require additional treatment to remove specific constituents.

In the literature, treatment schemes are often identified as primary, secondary, or advanced (also known as tertiary). In *primary treatment*, a portion of the suspended solids and organic matter is removed from the wastewater. This removal is usually accomplished with physical oper-

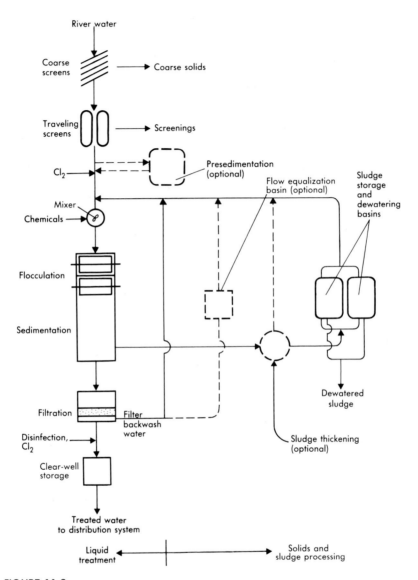

FIGURE 11.2

Typical flow sheet for the treatment of river water for domestic use.

ations such as screening and sedimentation. The effluent from primary treatment will ordinarily contain considerable organic material and will have a relatively high BOD. The further treatment of the effluent from primary treatment to remove the residual organic matter and suspended material is known as *secondary treatment*. In general, biological processes employing microorganisms are used to accomplish secondary treatment.

The effluent from secondary treatment usually has little BOD_5 and suspended solids and may contain several milligrams per liter of dissolved oxygen. When required for water reuse or for the control of eutrophication in receiving waters, *advanced (tertiary) treatment* is used for the removal of suspended and dissolved materials remaining after secondary treatment.

The choice of a set of treatment methods depends on several factors, including discharge permits and available disposal facilities. Actually, the distinction between primary, secondary, and advanced treatment is rather arbitrary, as many modern treatment methods incorporate physical, chemical, and biological processes in the same operation. A more rational approach would be to drop these arbitrary distinctions and to focus instead on the optimum combinations of operations and processes that must be used to achieve the required treatment objectives. For enforcement purposes, the Environmental Protection Agency has established standards for secondary treatment (see Table 11.5 and Chapter 4).

Typical flow sheets for the treatment of wastewater are presented in Figs. 11.3 and 11.4. The flow sheets shown in Fig. 11.3 are used for small communities, whereas the flow sheet shown in Fig. 11.4 is for a larger community. In the flow sheets shown in Fig. 11.3, the large solids in the

TABLE 11.5

Definition of Secondary Treatment as Established by the U.S. Environmental Protection Agency

CHARACTERISTIC OF DISCHARGE	UNIT OF MEASUREMENT	AVERAGE MONTHLY CONCENTRATION	AVERAGE WEEKLY CONCENTRATION
BOD_5	mg/L	30*[†]	45[†]
Suspended solids[‡]	mg/L	30*[†]	45[†]
Hydrogen-ion concentration	pH units	6.0–9.0[§]	

*Or, in no case more than 15 percent of influent value.

[†]Arithmetic mean.

[‡]Treatment plants with stabilization ponds and flows $< 7570 \text{ m}^3/\text{d}$ (2 Mgal/d) are exempt.

[§]Continuous, only enforced if caused by industrial wastewater or in-plant treatment.

Note: On November 16, 1983, EPA proposed allowing authorities to substitute carbonaceous biochemical oxygen demand (CBOD) for biochemical oxygen demand (BOD) when establishing secondary treatment limits for publicly owned treatment works. In addition, EPA proposed some revised secondary treatment requirements. Under the new requirements, communities using trickling filters or wastewater treatment ponds would have to achieve on a consistent basis at least 65 percent removal of BOD and suspended solids and would have to attain a minimum level of effluent quality of 45 mg/L for BOD and suspended solids as measured over a 30-d period. Further easing of the limits would be allowed when a state could show that trickling filters and ponds could not achieve the 45 mg/L level because of local climatic and geographic conditions, but under no circumstances could adjustments cause an adverse effect on water quality.

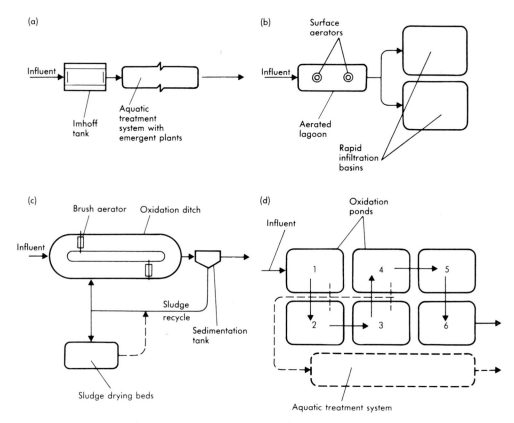

FIGURE 11.3

Flow sheets for typical wastewater treatment plants for small communities: (a) Imhoff tank followed by an aquatic treatment system; (b) aerated lagoon followed by rapid infiltration basins; (c) oxidation ditch with separate sedimentation tank and sludge drying beds; and (d) oxidation ponds with a direct discharge or followed by an aquatic treatment system. Disinfection may be required with flow sheets (a), (c), and (d).

incoming wastewater are screened or reduced in size by comminution. In Fig. 11.3(a) and 11.3(d), the wastewater, after pretreatment in either an Imhoff tank or in wastewater oxidation ponds, is discharged to an aquatic marsh for final treatment. In Fig. 11.3(b) the wastewater is disposed of by percolation into the soil after biological treatment. In Fig. 11.3(c) a suspended-growth biological process is used for treatment. Under current regulations, the final effluent from flow sheets in Fig. 11.3(a), (c) and (d) would have to be disinfected before discharge to the environment.

In larger communities (Fig. 11.4), racks are used to remove large debris and the size of coarse solids is reduced by comminution. The use of grit-removal facilities will depend on the nature and condition of the

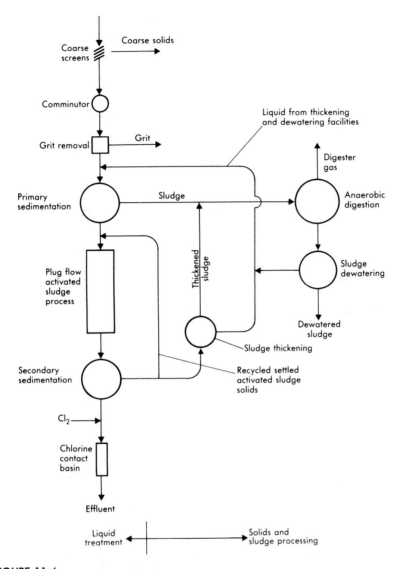

FIGURE 11.4

Flow sheet for a typical wastewater treatment plant for a large community. The plant employs the activated sludge process for biological treatment.

incoming sewers. Flow metering may precede or follow grit removal. In some plants, flow metering is accomplished at the effluent end of the plant. Settleable solids are removed in the primary sedimentation tank. Biological treatment is used to remove colloidal and dissolved organic matter and any residual solids escaping the primary sedimentation tank. The secondary sedimentation tank is an integral part of the biological

treatment process. Disinfection is used to control the discharge of pathogenic microorganisms. Dechlorination is used to remove any residual
chlorine before discharge into ecologically sensitive areas.

Contaminated Groundwater Treatment

In Chapter 10, it was noted that the cleanup of a contaminated groundwater could be accomplished by pumping out the contaminated water,
treating it, and reinjecting the treated water back into the groundwater
aquifer. A flow sheet for the treatment of a groundwater contaminated

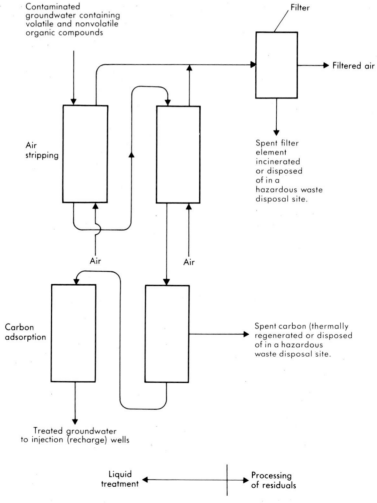

FIGURE 11.5

**Typical flow sheet for the treatment of a groundwater contaminated with volatile
and nonvolatile organic compounds of anthropogenic origin.**

with volatile and nonvolatile organic compounds is presented in Fig. 11.5. As shown, gas (air) stripping is used to remove the volatile organics, and activated carbon is used for the removal of the nonvolatile organic compounds by adsorption. The air from the stripping tower containing the organic compound removed from the groundwater is passed through a filter before being discharged. The filter medium must be disposed of at a hazardous-waste disposal site. Where the concentration of the contaminant is sufficient to support microbial activity, selected strains of microorganisms have been used for treatment. In the future, as the field of genetic engineering develops, it is anticipated that enzymes and microbial cultures capable of treating a wide range of compounds will be developed.

11.4
SLUDGE PROCESSING

In recent times, disposal of the material removed from water and wastewater (including the material added to bring about treatment) has become one of the most difficult problems in the implementation of any treatment system. The material removed by treatment is usually identified as *sludge*. Typical methods that have been used for the processing of sludge from water and wastewater treatment plants are considered below.

Sludge from Water Treatment Plants

In general, because coagulant and filter-backwash sludges are quite different, different methods of disposal should be used. In most new water filtration plants, and in many modified older water treatment plants, filter backwash is returned to the raw-water intake (see Fig. 11.2). Using this method, the only sludge that must be processed is the sludge removed from the settling basin. In smaller treatment plants, storage basins are used to equalize the backwash return-water flow rate. The capacity of the equalization basins should be such that the rate at which backwash water is returned to the treatment process is equal to or less than 10 percent of the average plant filtration rate. The combined sludge removed from the settling basin is usually disposed of in lagoons. Because lagoons are used both for storage and drying a minimum of two lagoons is required. Drying is usually completed within six months.

In very large treatment plants, where space is at a premium, coagulant sludge may be thickened before drying. In some cases, it has been possible to reuse the coagulant sludge for the manufacture of bricks and building blocks. Some small treatment plants discharge combined coagulant sludge to municipal sewers.

Sludge from Wastewater Treatment Plants

In small wastewater treatment plants without sludge digestion facilities, sludge is disposed of in lagoons or on sludge drying beds, as shown in Fig. 11.3. In intermediate and large plants, biological sludge is usually thickened before anaerobic sludge digestion (see Fig. 11.4). Digested sludge is either dewatered and trucked away for disposal or applied to storage lagoons or drying beds for dewatering. Dewatered sludge from storage lagoons or drying beds is usually disposed of by landfilling or land spreading.

Sludge from Contaminated Groundwater Treatment

Waste materials removed from contaminated groundwaters are usually concentrated prior to disposal in specially designated hazardous-waste disposal sites. In some cases, the waste materials can be incinerated or pyrolyzed.

11.5
IMPLEMENTATION OF TREATMENT METHODS

In both water and wastewater treatment, the physical implementation of the unit operations and processes identified in Tables 11.2, 11.3, and 11.4 is accomplished in specially designed tanks or other appropriate facilities in which the pertinent operational and environmental variables can be controlled. The operational and environmental variables are controlled so that treatment can be accomplished at accelerated rates as compared with the rates that would be expected in natural systems. In theory, the size and configuration of the tankage should depend on the kinetics governing the specific operation and process and the treatment objectives to be met. However, based on past experience, a variety of empirical design parameters have been developed. These parameters are commonly used for the design of unit operations and processes in routine applications. In some cases, because of the unusual nature of some wastes, pilot-plant testing has been necessary to define design parameters.

Kinetic Analysis

Where chemical and biological processes are to be used, it is now common practice to determine the required tank volume on the basis of a materials balance analysis in which the applicable reaction and/or process kinetics controlling the process are considered. The method of approach followed in the analysis of the various treatment operations and processes is exactly the same as outlined in Chapters 5 and 6.

Use of Design Parameters

Over the years, various design parameters and criteria have been developed from practical experience for the design of conventional water and wastewater treatment operations and processes. Typical design parameters are based on detention time, surface loading rates, and mass loading rates. The most commonly used parameters are defined below.

1. Hydraulic detention time, θ_H:

$$\theta_H = \frac{\text{volume } V, \text{ m}^3}{\text{flow rate } Q, \text{ m}^3/\text{d}} \qquad (11.1)$$

2. Hydraulic surface loading rate, $S_{Q/A}$:

$$S_{Q/A} = \frac{\text{flow rate } Q, \text{ m}^3/\text{d}}{\text{surface area } A, \text{ m}^2} \qquad (11.2)$$

3. Mass surface loading rate, $S_{M/A}$:

$$S_{M/A} = \frac{\text{mass of material applied, kg/d}}{\text{surface area } A, \text{ m}^2} \qquad (11.3)$$

4. Mass per volume loading rate, $V_{M/V}$:

$$V_{M/V} = \frac{\text{mass of material applied, kg/d}}{\text{volume } V, \text{ m}^3} \qquad (11.4)$$

5. Mass per mass loading rate, $M_{M/M}$:

$$M_{M/M} = \frac{\text{mass of material applied, kg/d}}{\text{mass of material in system, kg}} \qquad (11.5)$$

The application of some of these expressions is illustrated in Example 11.1.

EXAMPLE 11.1

DESIGN OF A PRIMARY SEDIMENTATION TANK

Design a circular sedimentation tank with a hydraulic detention time of 2.5 hr and a surface overflow rate of 40 m³/m² · d for a town with a population of 20,000 persons. Assume the wastewater flow is 450 L/capita · d. Determine the tank depth and diameter, assuming that the sludge removal mechanism to be used is available in 0.5-m-diameter increments and in depth increments of 0.2 m.

SOLUTION:

1. Determine the required diameter of the sedimentation tank.
   a. The surface area is

   $$A_s = \frac{20,000 \times 450 \text{ L/capita} \cdot \text{d}}{1000 \text{ L/m}^3 \times 40 \text{ m}^3/\text{m}^2 \cdot \text{d}}$$

   $$= 225 \text{ m}^2$$

   b. The tank diameter is

   $$\text{Dia} = \left(\frac{4 \times 225}{\pi}\right)^{1/2}$$

   $$= 16.9 \text{ m}$$

   Use 17 m.

2. Determine the required depth.
   a. The volume is

   $$V = \frac{(20,000 \times 450 \text{ L/capita} \cdot \text{d})(2.5/24)}{1,000 \text{ L/m}^3}$$

   $$= 938 \text{ m}^3$$

   b. The depth is

   $$d = \frac{938 \text{ m}^3}{225}$$

   $$= 4.17$$

   Use 4.2 m.

COMMENT

An allowance of 5 to 10 percent of the volume is usually made for the sludge removal equipment.

Pilot-Plant Studies

For most applications, the use of conventional design parameters and criteria has proved to be adequate. However, where the characteristics of the water or wastewater to be treated are unusual or where new process designs are proposed to improve performance or to reduce costs, the conduct of pilot-plant studies has proved to be of great value. Pilot-plant testing involves the construction of a smaller- (pilot-) scale plant of the proposed facilities. The pilot facilities are used to obtain kinetic and performance data and operating experience that will be used in evaluating the economic feasibility of the full-scale plant and in its design. A typical example of a pilot plant designed to assess the feasibility of an aquatic treatment system using water hyacinths, and including several pretreatment options, is shown in Fig. 11.6.

FIGURE 11.6

City of San Diego aquatic treatment pilot plant. One of three pretreatment options following primary sedimentation, the pulsed-bed filter, is shown in the foreground. Aquatic treatment system using water hyacinths is shown in the background.

KEY IDEAS, CONCEPTS, AND ISSUES

- The contaminants in water and wastewater are removed by physical, chemical, and biological means.

- Specific treatment methods are classified as physical unit operations, chemical unit processes, and biological unit processes.

- The term "flow sheet" is used to describe a particular combination of unit operations and processes used to meet a specific treatment objective.

- The unit operations and processes used in a particular flow sheet will depend on the contaminants to be removed and the degree of treatment that is required.

- The disposal of the material removed from water and wastewater (including the material added to bring about treatment) is one of the most difficult problems in the development of any treatment system.

- Implementation of unit operations and processes in treatment process flow sheets is based on kinetic analysis, design parameters developed from experience, and results of pilot-plant studies.

DISCUSSION TOPICS AND PROBLEMS

11.1. A water source chosen for a community has the following character-
       istics:

CHARACTERISTIC	VALUE
Ca^{+2}, g/m^3	100.00
Mg^{+2}, g/m^3	40.00
Cl^-, g/m^3	177.50
HCO_3^-, g/m^3	200.70
Turbidity, NTU	0.2
Suspended solids, g/m^3	0.01

Suggest a treatment process flow sheet based on Table 11.2 for
treatment of this water.

11.2. A groundwater containing TCE and chloroform from a chemical-
       wastes landfill is used as the principal water supply for a community.
       Concentrations of the contaminants are in the 0.1 to 1.0 g/m^3 range
       (well above acceptable limits). Suggest a method of mitigating the
       problem.

11.3. Secondary wastewater treatment is often defined as a system that can
       be used to remove 85 percent of the influent BOD and suspended
       solids. Briefly discuss the merit of this definition and the use of
       removal percentages in defining discharge requirements in general.

11.4. Using a list of priority pollutants provided by your instructor de-
       termine those that can be found in the pesticide/herbicide section of
       your local supermarket or garden supply store. Which of these are
       known or suspected carcinogens?

11.5. In Table 11.5 it is noted that effluent BOD and suspended solids
       from secondary treatment systems should each average less than 30
       g/m^3 but in no case should exceed 15 percent of the influent value.
       What are the advantages and disadvantages of this type of specifica-
       tion? Consider factors such as sewerage system maintenance, instal-
       lation of household garbage-disposal units, and conservation of
       water. Would the same requirements be appropriate in Germany,
       where water use is approximately half that of the United States?

11.6. Write a brief (500- to 1000-word) paper examining the equity of
       forcing downstream water users to remove materials discharged from
       upstream communities (e.g., Cincinnati, Ohio, treats water contain-
       ing materials discharged by communities between Cincinnati and
       Pittsburg).

11.7. Few water and wastewater systems have constant demands. Influent
       flows are often subject to variations in the range of one-half to five

times annual average flow. How can these variations be accommodated, and what significant differences would you anticipate between water and wastewater treatment?

11.8. Determine the hydraulic detention time and overflow (hydraulic surface loading) rate in a 30-m-diameter by 4-m-deep sedimentation tank for a flow rate of 12,000 m^3/d.

11.9. What diameter and sidewall depth is needed for a circular clarifier for a flow of 20,000 m^3/d to achieve an overflow (hydraulic surface loading) rate of 20 $m^3/m^2 \cdot d$ and a detention time of 3.0 hr?

11.10. Solve Problem 11.9, but assume that a rectangular settling tank with a length-to-width ratio of 3 to 1 is to be used.

11.11. A water treatment plant has a sedimentation tank with a volume of 1400 m^3, a water depth of 3.0 m, and a length-to-width ratio of 2.5 to 1. If the plant flow rate is 9000 m^3/d determine (a) the detention time in hours, (b) the surface loading rate, and (c) the horizontal flow-through velocity.

11.12. The influent flow and BOD_5 to a wastewater treatment plant are 4000 m^3/d and 250 g/m^3, respectively. What is the plant loading expressed in kilograms per day?

11.13. If 250 kg of chlorine is used to treat 125,000 m^3 of water, what is the dosage in milligrams per liter?

11.14. The recommended chlorine dosage for effluent disinfection is 4.0 mg/L. What amount of chlorine (in kilograms per day) is required for flows of (a) 5000 m^3/d, (b) 200 L/min, and (c) 0.5 m^3/s?

11.15. A domestic wastewater contains 250 g/m^3 of suspended solids. If 65 percent of the solids are removed during primary sedimentation, what volume of solids (sludge) would result from a flow of 4000 m^3/d if the concentration of solids in the sludge is 5.5 percent?

REFERENCES

11.1. Borchardt, J. A., W. J. Redman, G. E. Jones, and R. T. Sprague (eds.), (1981), *Sludge and Its Ultimate Disposal*, Ann Arbor Science Publishers, Ann Arbor, Mich.

11.2. *Code of Federal Regulations*, (1973), Title 40, Part 35, Appendix A.

11.3. Eckenfelder, W. W. Jr., (1980), *Principles of Water Quality Management*, CBI Publishing Company, Boston.

11.4. Edwards, R. E., N. A. Speed, and D. E. Verwoert, (1983), "Cleanup of Chemically Contaminated Sites," *Chemical Engineering*, Vol. 90, no. 4, p. 73.

11.5.   Metcalf & Eddy, Inc., (1979), *Wastewater Engineering: Treatment, Disposal, Reuse*, 2d ed., revised by G. Tchobanoglous, McGraw-Hill Book Company, New York.

11.6.   Reynolds, T. D., (1982), *Unit Operations and Processes in Environmental Engineering*, Brooks/Cole Engineering Division of Wadsworth, Monterey, Calif.

11.7.   Sanks, R. L. (ed.), (1978), *Water Treatment Plant Design for the Practicing Engineer*, Ann Arbor Science Publishing, Ann Arbor, Mich.

11.8.   Schroeder, E. D., (1977), *Water and Wastewater Treatment*, McGraw-Hill Book Company, New York.

11.9.   Tchobanoglous, G., (1975), "Wastewater Treatment for Small Communities" in *Water Pollution Control in Low Density Areas*, edited by W. J. Jewell and R. Swan, University of Vermont, Hanover, N.H.

11.10. Weber, W. J. Jr., (1972), *Physicochemical Processes for Water Quality Control*, Wiley Interscience, New York, 1972.

12

~~~~~~

Physical Treatment Methods

Physical treatment methods, as noted in Chapter 11, depend on the application of physical forces to bring about a change in water quality. Typical physical treatment methods used in water and wastewater treatment systems include screening, comminution, aeration, mixing, flocculation, gravity sedimentation, filtration, adsorption, gas stripping, and reverse osmosis. The objective in this chapter is to present the principles on which these methods are based. The methods and the rationale for selecting particular treatment methods and for synthesizing them into treatment systems are presented in Chapter 15.

12.1
APPLICATIONS OF PHYSICAL TREATMENT

Historically, physical treatment methods were used to alter or remove the particulate matter present in water and wastewater. While the alteration or removal of particulate matter remains the primary function of these methods, soluble materials are also now removed with physical treatment methods. Examples include the adsorption of organic materials on activated carbon, stripping of dissolved gases such as H_2S or NH_3, and removal of dissolved organics and inorganics by reverse osmosis. In considering the various physical treatment methods, it is important to remember that, with the exception of comminution, a by-product is formed in each application and disposal of the by-product (e.g., sludge and filter backwash) is difficult and expensive. Because of the importance of sludge in the development of water and wastewater management systems, a separate section on solids processing is included in this chapter.

12.2

SCREENING

Screening devices are used at water and wastewater treatment plants as protective devices for pumps and other equipment. Large materials such as branches, sticks, bottles, balls, and rocks can be excluded by bar racks or coarse screens. Smaller objects, including fish and twigs, are removed by fine screens. The presence of sticks and branches in rivers is not surprising, but a much wider variety of materials must be handled. Larger items such as bedsprings and furniture are sometimes thrown in rivers and may become entangled on intake structures. A tremendous variety of household items find their way into sewers and streams. Many of these items would be expected to clog toilets or household sewers; examples include false teeth, shoes, shirts, and crockery.

Applications of Screening

Screens of all types are used in water quality management facilities. Screens, also known as *bar racks*, are located at water intakes from rivers and lakes to prevent the entry of logs and other large debris (Fig. 12.1a). The clear spacing between bars composing the rack varies from 25 to 75 mm. A trolley and hoist is often installed to remove logs and other materials that may become lodged against the screen.

Traveling-belt screens of wire mesh (Fig. 12.1b) with openings varying in size from 5 to 25 mm are often used following bar racks to prevent small fish from being drawn into the intake. In some locations, microstrainers with apertures varying in size from 15 to 64 μm are used as a pretreatment step for the purpose of removing planktonic organisms (Fig. 12.1d). Microstrainers are used to pretreat water from lakes and storage reservoirs.

Screens (bar racks) are often provided at large pump stations for the purpose of removing rags and other coarse materials that can foul and damage pump impellers. Screens are installed at storm-water overflow structures to prevent the discharge of coarse materials to the aquatic environment. At wastewater treatment plants screens are used to remove materials that can foul pumps, aeration devices, sludge-scraping mechanisms, and weirs and channels. Examples of bar racks are shown in Fig. 12.2(a), (b), and (c). The bar rack in Fig. 12.2(a) is cleaned manually, whereas the solids accumulated on the bar racks shown in Figs. 12.2(b) and (c) are cleaned mechanically. A fine-wedge-wire screen used to remove smaller particles in wastewater is shown in Fig. 12.2(d).

(a)											(b)

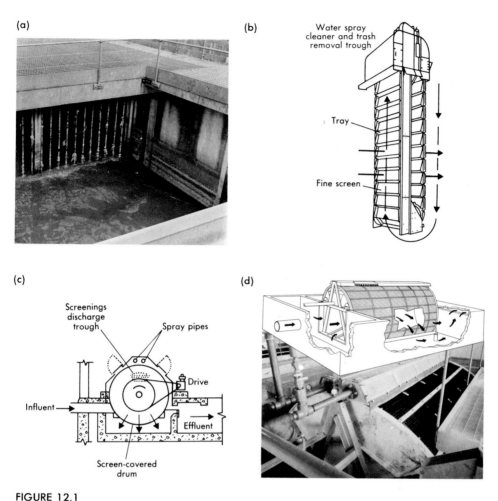

FIGURE 12.1

Typical screens used in water supply systems: (a) intake bar rack; (b) traveling screen; (c) rotary drum screen; and (d) microstrainer.

Source: Part (d) courtesy of Envirex, Inc., a Rexnord Company.

Design and Operation of Screens

Screen installations should be designed so that deposition of solids does not occur in the upstream channel and so that the buildup of head loss is minimized. An approach velocity of 0.5 m/s at average flow will minimize the deposition of material. To limit the head loss, the clear spacing between bars is made equal to the cross-sectional area of the approach channel. Also, continual cleaning is needed to limit the loss of head.

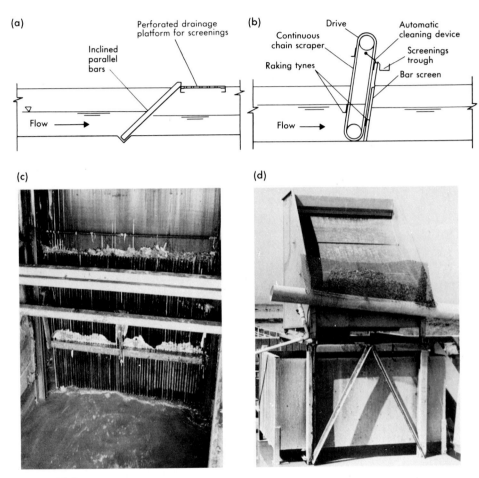

FIGURE 12.2

Typical screens used in wastewater treatment systems: (a) manually cleaned bar rack at entrance to treatment plant, (b) bar rack equipped with mechanical cleaning device, (c) photo of bar rack equipped with mechanical cleaning device, and (d) fine-wedge-wire fixed parabolic screen.

Disposal of Screenings

The quantity of screenings at water treatment facilities will vary with the time of year and the flow conditions in the river. Screenings are usually disposed of in a landfill. At wastewater treatment plants the quantity of screenings from coarse screens (bar racks) varies from 3.5 to 35 $m^3/10^6$ m^3 of wastewater, with a value of about 15 $m^3/10^6$ m^3 being typical [12.35, 12.36]. For fine screens, the quantity of screenings can be estimated as $(500/d)$ m^3/10^6 m^3 of wastewater where d is the screen opening, expressed in millimeters, for openings from 5 to 100 mm. Screenings from wastewater treatment plants are removed for disposal in landfills or ground up in comminutors or in grinder pumps and returned to the flow.

12.3
COMMINUTION

Large suspended materials entering wastewater treatment plants that are flexible enough to pass racks or coarse screens are often chopped or ground to a size that will not result in damage to the pumps. A number of devices are available from manufacturers; some of these devices are combination screens and comminutors (Fig. 12.3). Head loss through comminutors is usually between 50 and 100 mm, depending on the unit size and flow rate. Information on expected head losses is usually obtained from the manufacturers. Further, because these units are supplied complete, no detailed design is necessary.

(a)

(b)

FIGURE 12.3

Typical comminutors used at wastewater treatment plants: (a) in line circular screen with internal rotating cutting heads and (b) bar rack equipped with circular cutting head. The cutting head, normally out of the channel, is lowered periodically to comminute solids that have accumulated on the bars.

12.4

AERATION

The basic concepts necessary for discussion of aeration in water and wastewater treatment have been presented in Chapter 2 (gas-liquid equilibrium and Henry's law); Chapter 6 (oxygen transfer as a pseudohomogeneous reaction process); Chapter 7 (air/water interface transfers and gas absorption), where the two-film model was developed (see Fig. 7.8); and Chapter 8 (models of rivers and estuaries). Further application of the concepts will be presented in this chapter in the discussion of gas stripping from liquids (Section 12.10) and in Chapter 14 when aerobic biological treatment is discussed. The purpose of this section is to develop the conceptual basis for aeration as a unit operation in water supply and wastewater management.

Aeration Methods

The principal aeration devices (Fig. 12.4) used in water supply and wastewater systems may be classified as (1) waterfall aerators, (2) diffused-air (or pure oxygen) aeration systems, and (3) mechanical aerators [12.1, 12.4, 12.36, 12.56, 12.57].

Waterfall Aerators

Spray aerators, cascade aerators, and multiple-tray aerators are the principal types of waterfall aerators in common use. In spray aerators, water is forced into the air through a nozzle, as in a fountain. In the cascade aerator, water is spread out and allowed to flow in a thin sheet over a series of cascades or other obstructions to induce turbulence, to mix the absorbed gas in the water, and to expose new water surfaces to the atmosphere. In the multiple-tray aerator, water is applied to the upper of a series of trays filled with a coarse medium such as coke, stone, or ceramic balls. The water is aerated as it flows over the medium in the tray and as it falls from tray to tray.

Diffused-Air Aeration Systems

In diffused-air aeration systems, air (or pure oxygen) is introduced into the liquid being aerated in the form of bubbles which typically rise through the liquid [12.36, 12.56]. The size of the bubbles varies from coarse to extremely fine, depending on the specific aeration device. Common aeration devices include porous diffusers, nonporous diffusers, and U-tube aerators. Most porous diffusers are small-orifice devices constructed of silicon dioxide or aluminum oxide held together in a porous mass with a ceramic binder. A wide variety of nonporous diffusers are available, including jet, valved-orifice, and turbulence disk

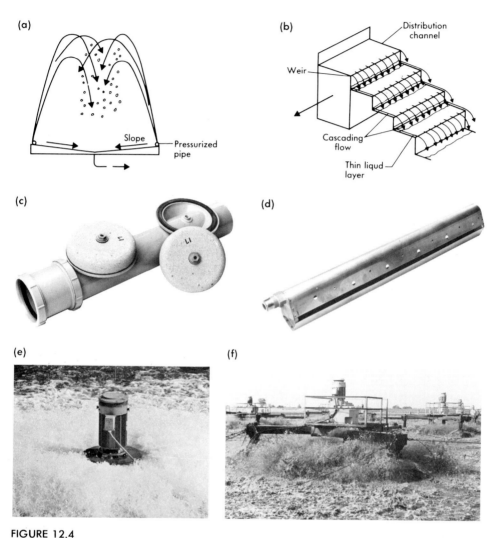

FIGURE 12.4

Typical aeration devices used in water and wastewater treatment: (a) spray and (b) cascade waterfall aerators; (c) Norton ceramic dome fine-bubble and (d) Sanitaire coarse-bubble (see also Fig. 14.4a) diffused-air aerators; and (e) high-speed and (f) slow-speed mechanical aerators.

aerators. Nonporous diffusers were developed to overcome some of the clogging problems experienced with fine-bubble porous diffusers and to reduce the capital cost of the aeration equipment. Typically nonporous diffusers are less efficient than porous diffusers. In the U-tube aerator, the partial pressure of the gas can be increased by passing water containing diffused-air bubbles through a depth of 10 to 20 m. As the partial pressure is increased the saturation concentration in the liquid is increased.

Mechanical Aerators

Mechanical aeration is achieved by breaking up the water surface mechanically. By producing a large air-water interface the transfer of oxygen from the atmosphere is enhanced. Both vertical- and horizontal-shaft aerators are in common use. In vertical-shaft aerators, aeration is achieved by exposing water droplets to the atmosphere, by surface turbulence, and by air entrainment. In horizontal-shaft aerators, a rotor or cage is used to aerate the water through surface turbulence and air entrainment. In addition to aeration, horizontal pumping is also achieved with rotor aerators.

The turbine aerator is a hybrid system involving the use of a turbine impeller (mixer) and an air source. Air discharged below the impeller is broken into fine bubbles and dispersed throughout the tank by the pumping action of the impeller.

Applications of Aeration

The principal applications of aeration in water quality management are related to the addition and removal of gases from water.

Aeration is used in water supply systems to reduce the concentration of volatile taste- and odor-causing substances such as hydrogen sulfide, for the removal of carbon dioxide from groundwaters, and for oxidation of ions such as iron and manganese. Sprays, cascades, multiple trays, bubble diffusers, and surface turbines are the most commonly used aeration devices. Examples of a number of these devices were shown previously in Fig. 7.7.

In wastewater management systems, aeration is used in sewers and treatment plant head works to control hydrogen sulfide formation, to strip volatile organic and inorganic compounds from wastewater, to supply oxygen for aerobic biological wastewater treatment, to raise treated effluent dissolved-oxygen concentrations to levels acceptable for discharge, and, in some cases, to increase the oxygen concentration of receiving waters. The latter case should be viewed as an emergency procedure and not as a solution to an environmental quality problem.

The most commonly used aeration devices are diffused-bubble systems and surface turbines. Oxygen transfer in attached biological films (see Section 14.6) is similar to that in cascade or tray aerators, but these systems are not ordinarily designed on the basis of oxygen-transfer limitations. Energy requirements for aeration in wastewater management are a function of the wastewater characteristics and consequently vary considerably from location to location. Aeration is usually the single largest energy sink in a wastewater treatment system.

Predicting Oxygen Transfer Rates in Aeration Systems

Prediction of oxygen transfer rates in aeration systems is nearly always based on an oxygen rate model such as given in Eq. (6.29):

$$r_{O_2} = K_L a \left(C_S - C_{O_2} \right)$$

where

r_{O_2} = rate of oxygen mass transfer, $g/m^3 \cdot s$

$K_L a$ = overall oxygen mass transfer coefficient, s^{-1}

C_S = saturation concentration of oxygen in bulk liquid, g/m^3

C_{O_2} = concentration of oxygen in bulk liquid, g/m^3

The overall oxygen mass transfer coefficient is usually determined in full-scale facilities or in test facilities such as shown in Fig. 12.5. If pilot-scale facilities are used to determine $K_L a$ values, scale-up must be considered [12.45]. Determination of the $K_L a$ value from test data is illustrated subsequently in Example 12.1.

Factors Affecting Oxygen Transfer

The oxygen mass transfer coefficient $K_L a$ is a function of temperature, intensity of mixing (and hence of the type of aeration device used and the geometry of the mixing chamber), and constituents in the water [12.17, 12.36, 12.49]. Temperature effects are treated in the same manner as they were treated in establishing the BOD rate coefficient—by using

FIGURE 12.5

Typical test tank used for testing the performance of floating surface aerators.

an exponential function to approximate the van't Hoff-Arrhenius relationship:

$$K_L a_{(T)} = K_L a_{(20°C)} \theta^{T-20} \qquad (12.1)$$

where

$K_L a_{(T)}$ = oxygen mass transfer coefficient at temperature T, s^{-1}

$K_L a_{(20°C)}$ = oxygen mass transfer coefficient at 20°C, s^{-1}

Reported values for θ vary with the test conditions. Typical θ values are in the range of 1.015 to 1.040 [12.1, 12.17]. A θ value of 1.024 is typical for both diffusion and mechanical aeration devices [12.17].

Effects of mixing intensity and tank geometry are difficult to deal with on a theoretical basis but must be considered in the design process because aeration devices are often chosen on the basis of efficiency. Efficiency is strongly related to the $K_L a$ value associated with a given aeration unit. In most cases an aeration device is rated for a range of operating conditions using tap water having a low total-dissolved-solids concentration. A correction factor α is used to estimate the $K_L a$ value in the actual system:

$$\alpha = \frac{K_L a(\text{wastewater})}{K_L a(\text{tap water})} \qquad (12.2)$$

where α is the correction factor. Values of α vary with the type of aeration device, the basin geometry, the degree of mixing, and the wastewater characteristics. Values of α varying from about 0.3 to 1.2 have been reported [12.49]. Typical values for diffused and mechanical aeration equipment are in the range of 0.4 to 0.8 and 0.6 to 1.2, respectively. If the basin geometry in which the aeration device is to be used is significantly different from that used to test the device, great care must be exercised in selecting an appropriate α value.

A third correction factor, β, is used to correct the test-system oxygen transfer rate for differences in oxygen solubility due to constituents in the water such as salts, particulates, and surface active substances:

$$\beta = \frac{C_S(\text{wastewater})}{C_S(\text{tap water})} \qquad (12.3)$$

Values of β vary from about 0.7 to 0.98. A β value of 0.95 is commonly used for wastewater. Because the determination of β is within the capability of most wastewater treatment plant laboratories, experimental verification of assumed values is recommended.

Application of Correction Factors

The application of the correction factors cited above can be illustrated by considering the equation used to predict field oxygen-transfer rates for mechanical surface aerators (see Fig. 12.4) based on measurements made in experimental test facilities [12.1, 12.17, 12.36]:

$$OTR_f = SOTR\left(\frac{\beta C_S - C_W}{C_{S_{20}}}\right)\theta^{T-20}(\alpha) \tag{12.4}$$

where

OTR_f = actual oxygen transfer rate under field-operating conditions in a respiring system, kg O_2/kW·h

$SOTR$ = standardized oxygen transfer rate under test conditions at 20°C and zero dissolved oxygen, kg O_2/kW·h

C_S = oxygen saturation concentration for tap water at field-operating conditions, g/m^3

C_W = operating oxygen concentration in wastewater, g/m^3

$C_{S_{20}}$ = oxygen saturation concentration for tap water at 20°C, g/m^3

and other terms are as defined previously.

With diffused-air aeration systems, the C_S values in Eq. (12.4) must be corrected to account for the higher than atmospheric oxygen saturation concentrations achieved in the reactor due to the release of air at the reactor bottom. To use Eq. (12.4) for diffused-air aeration systems, the value of C_S is taken to be the average dissolved oxygen concentration attained at infinite time [12.1]. Practical methods for determining the appropriate value of C_S for such systems are discussed in detail in Ref. [12.1]. Additional details on the determination of $K_L a$ values may be found in Refs. [12.1], [12.17], and [12.56]; Example 12.1 provides an appropriate exercise.

EXAMPLE 12.1

DETERMINATION OF $K_L a$ VALUES FROM AERATOR TEST DATA

The following data have been obtained from aeration tests conducted using a surface aerator in tap water and wastewater at 15°C. Using the given data, determine the $K_L a$ values expressed in hours for water and

TIME, min	CONCENTRATION, g/m^3	
	Tap water	Waste-water
0	0.0	0.0
20	3.1	2.1
40	5.4	3.7
60	6.8	5.0
80	7.9	6.1
100	8.6	6.9

wastewater and the alpha value at 20°C. Both the tap water and wastewater tests were conducted in the same tank. Assume the value of C_S is the same for water and wastewater.

SOLUTION:

1. Use Eq. (2.85) to determine the $K_L a$ values.
   a. Equation (2.85) is

   $$\frac{C_s - C_t}{C_s - C_0} = e^{-(K_L a)t}$$

   where

   > C_s = dissolved-oxygen saturation concentration, g/m^3
   > C_t = dissolved-oxygen concentration at time t, g/m^3
   > C_0 = initial dissolved-oxygen concentration, g/m^3
   > $K_L a$ = overall mass transfer coefficient, min^{-1}
   > t = time, min

   b. If the negative logarithm of both sides of Eq. (2.85) is taken, then the value of $K_L a$ can be determined by plotting $-\ln(C_s - C_t/ C_s - C_0)$ versus t.

   $$-\ln\left(\frac{C_s - C_t}{C_s - C_0}\right) = (K_L a)t$$

2. Prepare the aerator test data for plotting.
   a. The dissolved-oxygen saturation value at 15°C is 10.1 g/m^3. (See Appendix F.)
   b. The modified test data are

TIME, min	$-\ln\left(\dfrac{C_s - C_t}{C_s - C_0}\right)$	
	Tap water	Waste-water
0	0.00	0.00
20	0.37*	0.24
40	0.78	0.46
60	1.11	0.67
80	1.51	0.92
100	1.90	1.14

   $$* -\ln\left(\frac{10.1 - 3.1}{10.1 - 0.0}\right) = 0.37$$

3. Plot the data developed in step 2 and determine the $K_L a$ values at 15°C.
   a. The aeration test data are plotted in Fig. 12.6.

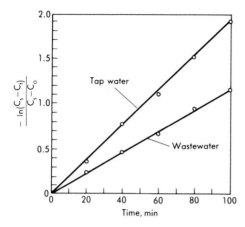

FIGURE 12.6

Functional plot of aerator test data used to determine K_La values for Example 12.1.

b. The K_La value for tap water is equal to the slope of the curve plotted in Fig. 12.6:

$$K_La = \frac{1.90 \ (60 \ \text{min}/\text{hr})}{100 \ \text{min}}$$
$$= 1.14 \ \text{hr}^{-1}$$

c. The K_La value for wastewater is

$$K_La = \frac{1.14 \ (60 \ \text{min}/\text{hr})}{100 \ \text{min}}$$
$$= 0.68 \ \text{hr}^{-1}$$

4. Determine the K_La values at 20°C. Assume $\theta = 1.024$.
   a. For tap water:

$$K_La_{(20°C)} = 1.14 \ \text{hr}^{-1}(1.024)^{20-15}$$
$$= 1.28 \ \text{hr}^{-1}$$

   b. For wastewater:

$$K_La_{(20°C)} = 0.68 \ \text{hr}^{-1}(1.024)^{20-15}$$
$$= 0.77 \ \text{hr}^{-1}$$

5. Determine the alpha value.

$$\alpha = \frac{K_La(\text{wastewater})}{K_La(\text{tap water})}$$
$$= \frac{0.77 \ \text{hr}^{-1}}{1.28 \ \text{hr}^{-1}} = 0.60$$

COMMENT

In many reference texts, the value of $K_L a$ will be determined by plotting the term $(C_s - C_t)$ versus t on log-arithmetic paper. This method of plotting can be used, as the term $(C_s - C_0)$ in the denominator of Eq. (2.85) is a constant. When this method of analysis is used, the slope of the line which corresponds to the $K_L a$ value must be multiplied by 2.303 to obtain the appropriate value of $K_L a$ for use with Eq. (2.85) expressed in the natural log form. In this example, the water and wastewater dissolved-oxygen saturation concentration was the same, but this is not normally the case.

Performance of Aerators

Evaluation of the performance of aerators or aeration systems is usually based on the number of kilograms of oxygen transferred per kilowatt

TABLE 12.1

Typical Performance Data for Various Aeration Devices

AERATION DEVICE	OXYGEN TRANSFER EFFICIENCY,* %	OXYGEN TRANSFER RATE, kg O_2/kW · h	
		SOTR*	OTR$_f$
Waterfall aerators[†]			
Spray		0.1–0.5	0.1–0.5
Cascade		0.1–0.5	0.1–0.5
Tray		0.2–0.6	0.2–0.6
Diffused-gas aerators			
Porous diffuser			
Fine bubble	20–45	4.0–4.5	1.0–2.0
Medium bubble	8–25	2.0–2.5	0.8–1.3
Coarse bubble	6–20	1.4–1.6	0.6–1.4
Turbine-Sparger		1.4–2.5	0.8–1.4
Static tube	6–20	1.5–1.6	0.7–0.9
Jet	10–25	1.7–2.4	0.7–1.4
Mechanical aerators			
Low-speed surface		1.2–2.4	0.7–1.3
Low-speed with draft tube		1.2–2.4	0.7–1.4
High-speed floating		1.2–2.4	0.7–1.3
Rotor brush		1.2–2.4	0.7–1.3

*The range of clean water transfer efficiencies and SOTR values is based on depths of 3 and 8 m respectively at a mixing intensity of about 25 kW/10^3 m³.
[†] Used for CO_2 removal (stripping) in water treatment as well as oxygen transfer.
Source: Adapted in part from Refs. [12.1] and [12.35].

hour (kg $O_2/kW \cdot h$). Representative data for the principal classes of aeration devices are presented in Table 12.1.

12.5
MIXING

Mixing is perhaps the most universal of all processing operations [12.23]. Mixing is needed where one substance must be mixed completely with another, where a uniform concentration or temperature must be maintained in a process reactor, where it is desired to break up thermal stratification, and in a variety of similar applications.

Common applications of mixing in water treatment include the mixing of coagulant chemicals with the water to be treated and the addition of chlorine for disinfection (see discussion of these topics in Chapter 13). In wastewater treatment, one of the most important applications of mixing is in biological treatment, where the contents of the reactor must be mixed continuously for effective treatment to occur (see discussion of biological treatment in Chapter 14).

Types of Mixers

The types of mixing devices or methods in water and wastewater systems may be classified as (1) turbine or paddle mixers, (2) propeller mixers, (3) pneumatic mixing, (4) hydraulic mixing, and (5) in-line hydraulic and static mixing. Representative examples of these mixers or methods of mixing are illustrated in Fig. 12.7.

Turbine (paddle) and propeller mixers are used most often for the mixing of coagulating chemicals in water treatment plants. Pneumatic mixing is used in transfer channels between treatment units in both water and wastewater treatment. Hydraulic mixing is often used for the addition of chlorine in both water and wastewater treatment facilities. The in-line mixer shown in Fig. 12.7(e) is an extremely effective device for the addition of coagulating chemicals in pipelines.

Theory of Mixing

Although gases in a confined container mix rapidly by molecular diffusion, in most liquids mixing by natural molecular diffusion is a slow process [12.23]. To hasten the process, mechanical energy is added to the fluid, usually with some type of rotating agitator. When mechanical energy is added to a fluid, eddy currents are generated as a result of the velocity gradients created within the fluid. Eddy currents are important in bringing about mixing. Further, the size of the eddy (sometimes defined as the *scale* of the phenomenon) is important. Particles smaller

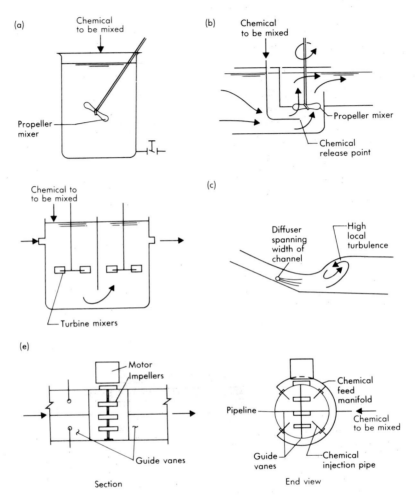

FIGURE 12.7

Typical methods and devices used for mixing in water and wastewater treatment systems in batch process: (a) using a propeller mixer; (b) using a propeller mixer in a continuous-flow process; (c) using a turbine mixer; (d) using a hydraulic jump; and (e) using an in-line mixer.

than the eddy size will move in a fixed position to each other within the eddy, and complete mixing is not achieved. This mixing phenomenon can be observed in a coffee cup or teacup. By carefully stirring the contents of the cup it is easy to get the entire fluid mass moving (vortexing) in a circular pattern without any intermixing.

Degree of Mixing

The degree of mixing in a fluid is a function of the magnitude of the eddy currents or turbulence formed within the liquid and the forces in the fluid tending to dampen the formation of eddy currents or turbulence. The degree of mixing can be defined as [12.23, 12.38]

$$D = \frac{\text{driving force}}{\text{resistance}} \tag{12.5}$$

where D is the degree of mixing, the driving force is composed of the forces producing eddy currents or turbulence, and the resistance is composed of the forces tending to dampen the formation of eddy currents or turbulence.

When the entire fluid in a confined reactor is in a turbulent flow regime, a high degree of mixing can be expected. The mechanical energy required to bring about turbulent conditions has been found to be a function of the agitator geometry, the reactor geometry, and the physical properties of the fluid(s) being mixed.

Power Requirements for Mixing

On the basis of inertial and viscous forces, the following mathematical relationship has been developed using the techniques of dimensional analysis [12.23, 12.33, 12.38, 12.43]:

$$N_p = \frac{p}{\rho n^3 d_i^5} \tag{12.6}$$

where

N_p = power number, dimensionless
p = power requirement, W
ρ = mass density of fluid, kg/m³
n = rotational speed, r/s
d_i = diameter of mixer impeller, m

For purposes of analyzing the power requirements for mixing, a standard tank configuration has been defined as shown in Fig. 12.8. For a standard tank without baffles, it can be shown that the mixing operation can be characterized with a power function defined as follows [12.23]:

$$\phi = \frac{N_p}{(N_F)^y} = k(N_R)^x \tag{12.7}$$

where

ϕ = power function, dimensionless
N_F = Froude number, $n^2 d_i/g$
k = arbitrary constant
N_R = Reynolds number, $n\rho d_i^2/\mu$
y, x = experimental constants

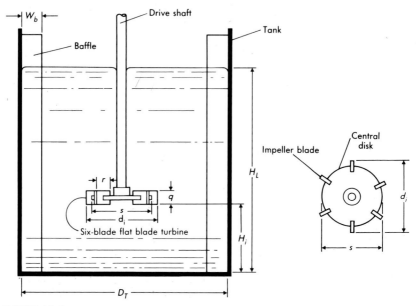

FIGURE 12.8

Standard tank configuration used in the analysis of mixer performance.

Source: Adapted from Ref. [12.23].

Notes: 1. The agitator is a six-blade flat turbine impeller
   2. Impeller diameter, $d_i = 1/3$ tank diameter
   3. Impeller height from bottom, $H_i = 1.0$ impeller diameter
   4. Impeller blade width, $q = 1/5$ impeller diameter
   5. Impeller blade length, $r = 1/4$ impeller diameter
   6. Length of impeller blade mounted on the central disk $= r/2 = 1/8$ impeller diameter
   7. Liquid height, $H_L = 1.0$ tank diameter
   8. Number of baffles $= 4$ mounted vertically at tank wall and extending from the tank bottom to above the liquid surface
   9. Baffle width, $W_b = 1/10$ tank diameter
  10. Central disk diameter, $s = 1/4$ tank diameter

and other terms are as defined previously. For a tank with baffles in which there is no liquid vortexing, y is equal to zero and

$$\phi = N_p = k(N_R)^x \tag{12.8}$$

The power function curve for the standard tank shown in Fig. 12.8 is given in Fig. 12.9. In the laminar region, it has been found that x is equal

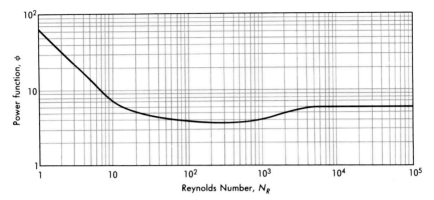

FIGURE 12.9

Mixing power function curve for standard tank configuration shown in Fig. 12.8.

Source: Adapted from Ref. [12.23].

to -1, so that

$$p = \left(\rho n^3 d_i^5\right) k \left(\frac{n\rho d_i^2}{\mu}\right)^{-1}$$

$$= k\mu n^2 d_i^3 \tag{12.9}$$

For fully turbulent flow, the value of the exponent x is equal to zero and $p = k\rho n^3 d_i^5$. Values of k for a variety of mixers for laminar and turbulent flow conditions are given in Table 12.2. The application of Eq. (12.6) is illustrated in Example 12.2.

TABLE 12.2

Values of k Used for Computing Power Requirements for Mixing

TYPE OF IMPELLER	LAMINAR RANGE	TURBULENT RANGE
Propeller, square pitch, 3 blades	41.00	0.32
Propeller, pitch of two, 3 blades	43.5	1.0
Turbine, 6 flat blades	71.0	6.30
Turbine, 6 curved blades	70.0	4.80
Fan turbine, 6 blades	70.0	1.65
Turbine, 6 arrowhead blades	71.0	4.00
Flat paddle, 6 blades	36.5	1.70
Shrouded turbine, 2 curved blades	97.5	1.08
Shrouded turbine with stator (no baffles)	172.5	1.12

Source: From Ref. [12.43].

EXAMPLE 12.2

POWER REQUIREMENTS FOR MIXING

Determine the power requirements for 3-m-diameter, six-blade flat-blade turbine impeller mixer running at 15 r/min in a 10-m-diameter mixing tank of standard configuration. Also compute the power per thousand cubic meters of volume. Assume the fluid being mixed is water, the water depth is 9.0 m, and the temperature is 15°C.

SOLUTION:

1. Determine the Reynolds number.

$$N_R = \frac{n \rho d_i^2}{\mu}$$

$$= \frac{(15/60 \text{ r/s}) \, 999.1 \text{ kg/m}^3 \, (3 \text{ m})^2}{1.139 \text{ N} \cdot \text{s/m}^2}$$

$$= 1974$$

2. Determine the power requirement.
   a. From Fig. 12.9, the power function for a Reynolds number of 1974 is equal to 4.6.
   b. Compute the power requirement using the modified form of Eq. (12.6).

$$p = \phi \rho n^3 d_i^5$$

$$= 4.6(999.1 \text{ kg/m}^3) \, (0.25 \text{ r/s})^3 \, (3 \text{ m})^5$$

$$= 17{,}450 \text{ W}$$

3. Determine the power per unit volume.
   a. The volume of the mixing tank is

$$V = (\pi/4)(10 \text{ m})^2 (9 \text{ m})$$

$$= 706.5 \text{ m}^3$$

$$= 0.71 \times 10^3 \text{ m}^3$$

   b. The power per thousand cubic meters of tank volume is

$$P = \frac{17.45 \text{ kW}}{0.71 \times 10^3 \text{ m}^3}$$

$$= 24.5 \text{ kW}/10^3 \text{ m}^3$$

COMMENT

The power input per unit volume is a rough measure of mixing effectiveness, based on the reasoning that more input power creates greater turbulence, and greater turbulence leads to better mixing [12.23, 12.38]. The typical power requirement for complete mixing in biological treatment units varies from about 10 to 30 $\text{kW}/10^3 \text{ m}^3$.

The power required for mixing depends on the chemicals being mixed, the type of mixer, the geometry of the mixing chamber, and the characteristics of the liquid. Conventional rapid-mixing chambers in water treatment have detention times varying from 30 to 60 s, with power requirements varying from 50 to 200 kW/10^3 m^3. The corresponding velocity gradients (see Section 12.6) are in the range of 300 to 1000 s^{-1}. For in-line mechanical mixers, the detention time is on the order of 0.5 to 1.0 s, with a power requirement of about 100 kW/10^3 m^3. The corresponding velocity gradients are up to 3500 s^{-1} [12.56, 12.57]. In wastewater treatment the power required for complete mixing varies from about 10 to 30 kW/10^3 m^3 of tank volume [12.35, 12.36].

12.6
FLOCCULATION

Flocculation is the treatment operation in which particle collisions are brought about hydrodynamically, typically using rotating paddles (Fig. 12.10a and b). When coagulated particles in water and wastewater collide they tend to aggregate, and thus flocculation results in the growth of larger particles (Fig. 12.10c). Because of viscous drag, large particles settle faster than small particles, and therefore aggregation is highly desirable. Flocculation is most important in water treatment, where the primary particles of interest are chemically destabilized (coagulated) colloids. The process of coagulation is discussed in Chapter 13. These particles would gradually aggregate through perikinetic flocculation brought about by Brownian motion, but the process is very slow. The rate can be greatly enhanced by providing velocity gradients that bring about orthokinetic flocculation. A number of stochastic and kinetic models for the flocculation process have been proposed [12.21, 12.25, 12.48, 12.51]. In practice, variables such as the characteristics of a particular water, the coagulant used, pH, and temperature are important, however. Experimental studies are often helpful as part of the design process [12.45].

Power Input in Flocculation

Orthokinetic flocculation is enhanced by the production of velocity gradients in the liquid. Particles moving at different speeds are more likely to collide and combine into larger particles. For this reason flocculation operations are generally characterized by the mean velocity gradient. In practice, the mean velocity gradient is a function of the mixing power input, and units are designed on this basis. It must be remembered that mean velocity gradients used in design are average values and that a great deal of point-to-point variation is to be expected.

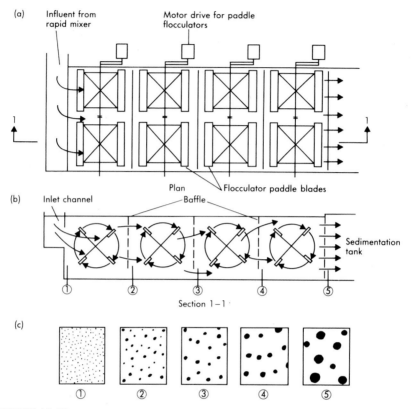

FIGURE 12.10

Definition sketches for the flocculation process: (a) plan view of a four-compartment flocculator. Note: Each set of flocculator paddles is powered by a separate motor drive so that the power input to each compartment can be varied. (b) Section 1-1 through the flocculator. (c) Particle size at indicated location through flocculator.

The mean velocity gradient is related to the power input through the shear stress τ on an element of fluid, as shown in Fig. 12.11 and Eqs. (12.10) through (12.15):

$$\tau = \mu \frac{\overline{dv}}{dy} \tag{12.10}$$

where

τ = shear stress, Pa (N/m^2)

μ = dynamic viscosity, N · s/m^2

v = fluid velocity, m/s

$\dfrac{\overline{dv}}{dy}$ = mean velocity gradient, s^{-1}

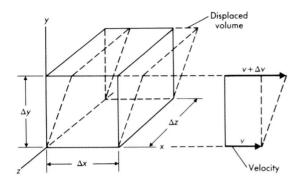

FIGURE 12.11

Fluid element subjected to differential velocity.

The power input into the fluid element can be written in terms of the mean velocity gradient $\overline{dv/dy}$:

$$p = \tau \Delta x \, \Delta y \Delta z \frac{\overline{dv}}{dy} \tag{12.11}$$

where p is the power input in watts.

The power input per unit volume is

$$P = \tau \frac{\Delta x \, \Delta y \Delta z}{\Delta x \, \Delta y \Delta z} \frac{\overline{dv}}{dy}$$

$$= \tau \frac{\overline{dv}}{dy} \tag{12.12}$$

where P is the power input per unit volume in watts per cubic meter. By substituting Eq. (12.10) into Eq. (12.12), the power per unit volume can be defined in terms of the shear stress and the mean velocity gradient:

$$P = \mu \left(\frac{\overline{dv}}{dy} \right)^2 \tag{12.13}$$

$$\frac{\overline{dv}}{dy} = \left(\frac{P}{\mu} \right)^{1/2} \tag{12.14}$$

In the literature dealing with flocculation the letter G is used to denote the mean velocity gradient [12.9, 12.18], and Eq. (12.13) is written as

$$P = \mu G^2 \tag{12.15}$$

The mean velocity gradient can now be defined as

$$G = \sqrt{\frac{P}{\mu}} = \sqrt{\frac{p}{\mu V}} \tag{12.16}$$

where V is the volume and the other terms are as defined previously.

Flocculation Systems

In practice, flocculation involving the dissipation of energy in a fluid can be accomplished mechanically using paddles or turbine mixers (Figs.

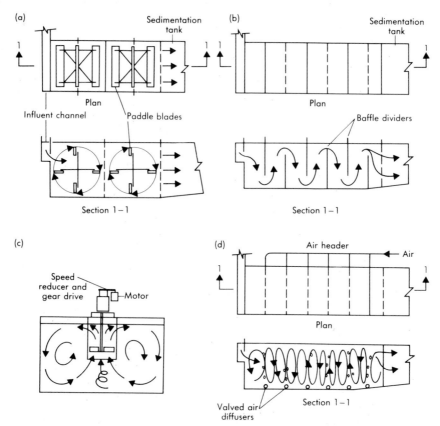

FIGURE 12.12

**Representative flocculation reactors; (a) paddle flocculator; (b) over-and-under baffled floccu-
lator; (c) turbine flocculator; and (d) air flocculator.**

12.12a and c), hydraulically in baffled chambers (Fig. 12.12b), and by air
injection (Fig. 12.12d).

The power input for an over-and-under baffle such as shown in Fig.
12.12(b) can be estimated using Eq. (12.17):

$$p = \rho g Q h \tag{12.17}$$

where

ρ = density of fluid, kg/m^3

g = acceleration due to gravity, 9.81 m/s^2

Q = fluid flow rate, m^3/s

h = head loss, m

The corresponding value for the velocity gradient is

$$G = \sqrt{\frac{\rho g Q h}{\mu V}} = \sqrt{\frac{gh}{\nu \theta_H}} \tag{12.18}$$

where

$\nu =$ kinematic viscosity, m^2/s

$\theta_H =$ hydraulic detention time V/Q, s

Power input in a mechanical paddle system (Fig. 12.13) can be related to drag on flocculator paddles and therefore to motor output:

$$F_D = C_D A \rho_w \frac{v^2}{2} \tag{12.19}$$

$$p = F_D v = C_D A \rho_w \frac{v^3}{2} \tag{12.20}$$

where

$F_D =$ drag force, N

$C_D =$ coefficient of drag (1.8 for flat blades)

$A =$ total cross-sectional area of flocculator paddles, m^2

$\rho_w =$ fluid density, kg/m^3

$v =$ fluid velocity, m/s

Using a motor-flocculator efficiency E_f that accounts for losses in the gear box and U-joints, the mean velocity gradient G can be written in terms of the rated motor-power output p_r.

$$G = \sqrt{\frac{C_D A \rho_w v^3}{2V\mu}} = \sqrt{\frac{E_f p_r}{V\mu}} \tag{12.21}$$

FIGURE 12.13

Flocculation tank equipped with paddle flocculators.

The chemical reactions of the coagulation process are very rapid and must be carried out in a rapid mixing unit to avoid undesirable side reactions. Initial flocculation steps also take place in the rapid-mix unit. As the floc particles grow in size, the turbulence of a rapid-mix unit breaks down the larger aggregates. To optimize the development of floc particles, a separate flocculation unit having lower velocity gradients (20 to 50 s^{-1}) and detention times (V/Q) of 30 to 60 min is used following the rapid mix unit.

Pilot testing of flocculation for design should include the rapid- and slow-mix steps. In the slow-mix step a range of velocity gradients should be used. Performance is judged by the clarified effluent quality using a standard settling test. The most common test unit in use, called a *jar test*, is a bench-scale system using 1-L reactors (see Fig. 13.13). Data from jar tests are useful, but accurate scale-up is an unreasonable expectation.

EXAMPLE 12.3

FLOCCULATION POWER REQUIREMENTS

Average flow in a water treatment operation is 20,000 m^3/d. It is desired to design a flocculation unit with a 30-min hydraulic residence time and a mean velocity gradient of 40 s^{-1}. Motor-drive unit efficiencies of 60 percent can be expected. Estimate the required motor size.

SOLUTION:

1. The expression for power as derived from Eq. (12.21) is

$$G = \sqrt{\frac{E_f P_r}{V \mu}}$$

$$P_r = \frac{V \mu G^2}{E_f}$$

2. Substitute known values and solve for P_r.

$$V = 20,000 \text{ m}^3/\text{d} \times \frac{30 \text{ min}}{1440 \text{ min/d}} = 417 \text{ m}^3$$

$$\mu \simeq 10^{-3} \text{ kg/m} \cdot \text{s}$$

$$E_f = 0.60$$

$$G = 40 \text{ s}^{-1}$$

$$P_r = \frac{(417 \text{ m}^3)(10^{-3} \text{ kg/m} \cdot \text{s})}{(0.60)} (40 \text{ s}^{-1})^2$$

$$= 1112 \text{ W}$$

12.7

SEDIMENTATION

Sedimentation is the unit operation in which suspended materials are removed from the liquid phase by gravity settling. Historically, sedimentation has been, and continues to be, the most common treatment method used in both water and wastewater treatment systems.

Applications of Sedimentation

Common applications of sedimentation in water treatment include (1) pretreatment of surface waters prior to conventional water treatment, (2) settling of coagulated and flocculated waters prior to filtration (Fig. 12.14a), (3) settling of coagulated flocculated waters in chemical water softening, and (4) settling of waters treated for iron and manganese removal [12.41].

In wastewater treatment, the principal uses of sedimentation are for the removal of grit and other coarse solids, of suspended solids before biological treatment (Fig. 12.14b), and of the biological solids produced during biological treatment.

Sedimentation Theory

Conceptually, sedimentation operations are quite simple in both water and wastewater treatment. In the idealized system, particles move horizontally with the flow and vertically under gravitational forces (Fig. 12.15a). Interaction between particles does not occur in ideal discrete particle setting. Thus, referring to Fig. 12.15(b), the following force

(a)   (b)

FIGURE 12.14

Typical sedimentation tanks used for removal of (a) flocculant suspensions in water treatment and (b) suspended material in untreated wastewater.

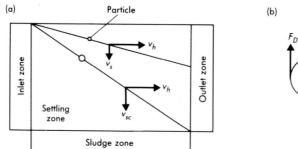

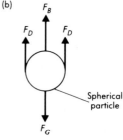

FIGURE 12.15

Definition sketch for a spherical particle settling in (a) an idealized rectangular horizontal-flow settling tank and (b) a fluid subject to the forces of buoyancy, gravity, and drag.

balance can be written for a discrete particle that is settling:

$$m_p \frac{dv_s}{dt} = F_G - F_B - F_D \qquad (12.22)$$

where

m_p = mass of settling particle, kg
v_s = particle settling velocity, m/s
F_G = gravitational force, N
F_B = buoyant force, N
F_D = drag force, N

The net gravitational force is given by

$$F_G - F_B = (\rho_p - \rho_w) g V_p \qquad (12.23)$$

where

ρ_p = density of particle, kg/m^3
ρ_w = density of water, kg/m^3
g = acceleration due to gravity, m/s^2
V_p = volume of particle ($\pi d_p^3/6$), m^3
d_p = diameter of particle, m

The drag force is a function of the cross-sectional area of the particle, the settling velocity of the particle, the liquid density, and the coefficient of drag:

$$F_D = C_D A_p \rho_w \frac{v_s^2}{2} \qquad (12.24)$$

where

C_D = coefficient of drag

A_p = cross-sectional area of particle ($\pi d_p^2/4$), m^2

ρ_w = density of water, kg/m^3

v_s = particle settling velocity, m/s

For spherical particles, the coefficient of drag can be estimated using the following relationship:

$$C_D = \frac{24}{N_R} + \frac{3}{\sqrt{N_R}} + 0.34 \tag{12.25}$$

where

N_R = Reynolds number, dimensionless

$$= \frac{v_s d_p \rho_w}{\mu}$$

μ = liquid viscosity, kg/m · s

and other terms are as defined previously.

In the ideal system, the terminal settling velocity is attained quickly, and the acceleration term can be assumed to be negligible. Thus, Eq. (12.22) can be rewritten as

$$F_G - F_B = F_D \tag{12.26}$$

Substituting Eq. (12.23) for $F_G - F_B$ and Eq. (12.24) for F_D and solving for the settling velocity v_s yields

$$v_s = \sqrt{\frac{4}{3} \frac{g(\rho_p - \rho_w) d_p}{C_D \rho_w}} \tag{12.27}$$

where d_p is the diameter of the particle, in meters, and the other terms are as defined previously. When $N_R < 0.3$, the first term of Eq. (12.25) predominates, and the discrete particle settling rate becomes (Stokes' law)

$$v_s = \frac{g(\rho_p - \rho_w) d_p^2}{18\mu} \tag{12.28}$$

where

ρ_p = particle density, kg/m^3

μ = liquid viscosity, kg/m · s

Particle densities in water and wastewater treatment vary considerably, but fall into distinct bands. Organic materials (bacteria, food particles, and fecal particles) usually have densities in the 1030 to 1100 kg/m^3 range. Chemical flocs produced in precipitation reactions have densities in the 1400 to 2000 kg/m³ range, and mineral particles usually have densities around 2500 kg/m³. The viscosity of water is about 0.001 kg/m · s at 20°C, but varies considerably with temperature (see Appendix C). The application of the above equations is illustrated in Example 12.4.

EXAMPLE 12.4

DETERMINATION OF TERMINAL SETTLING VELOCITY

Determine the terminal settling velocity for a sand particle with an average diameter of 0.5 mm and a density of 2600 kg/m³ settling in water at 20°C. Density and viscosity values may be found in Appendix C.

SOLUTION:

1. Determine the terminal settling velocity using Stokes' law, Eq. (12.28).

$$v_s = \frac{g(\rho_p - \rho_w)d_p^2}{18\mu}$$

$$= \frac{9.81 \text{ m/s}^2(2600 - 998.2)\text{kg/m}^3(5 \times 10^{-4} \text{ m})^2}{18 \times 1.002 \times 10^{-3} \text{ N} \cdot \text{s/m}^2}$$

$$= 0.22 \text{ m/s}$$

2. Check the Reynolds number.

$$N_R = \frac{\phi v_s d_p \rho_w}{\mu}$$

where ϕ is a shape factor to account for the irregularities of the sand. Using a value of 0.85 for ϕ the Reynolds number is:

$$N_R = \frac{0.85(0.22 \text{ m/s})(5 \times 10^{-4} \text{ m})(998.2 \text{ kg/m}^3)}{1.002 \times 10^{-3} \text{ N} \cdot \text{s/m}^2}$$

$$= 93.2$$

3. Because the Reynolds number computed in step 2 is greater than 0.3, Eq. (12.27) must be used to solve for the settling velocity.
   a. Solve for the coefficient of drag using Eq. (12.25) and the above value as the first estimate of N_R.

$$C_D = \frac{24}{N_R} + \frac{3}{\sqrt{N_R}} + 0.34$$

$$= \frac{24}{93.2} + \frac{3}{\sqrt{93.2}} + 0.34$$

$$= 0.91$$

b. Solve for the velocity using Eq. (12.27).

$$v_s^2 = \frac{(4/3)(9.81 \text{ m/s}^2)(2600 - 998.2)\text{kg/m}^3(5 \times 10^{-4} \text{ m})}{0.91(998.2 \text{ kg/m}^3)}$$

$v_s = 0.11 \text{ m/s}$

4. Repeat steps 2 and 3 to check the accuracy of the velocity computed in step 3b.

   a. Use the velocity computed in step 3b to estimate the new Reynolds number:

   $$N_R = \frac{0.85(0.11)(5 \times 10^{-4})(998.2)}{1.002 \times 10^{-3}}$$

   $= 46.5$

   b. Compute the new C_D value:

   $$C_D = \frac{24}{46.5} + \frac{3}{\sqrt{46.5}} + 0.34$$

   $= 1.30$

   c. Compute the new velocity:

   $$v_s^2 = \frac{(4/3)(9.81)(2600 - 998.2)(5 \times 10^{-4})}{1.30(998.2)}$$

   $= 0.0081 \text{ m}^2/\text{s}^2$

   $v_s = 0.09 \text{ m/s}$

5. Repeating steps 2 and 3 using the data from step 4 yields

   $N_R = 38.1$

   $C_D = 1.46$

   $v_s = 0.085 \text{ m/s}$

6. Repeating steps 2 and 3 using the data from step 5 yields

   $N_R = 36.0$

   $C_D = 1.51$

   $v_s = 0.083 \text{ m/s}$

   The above estimate of the velocity is sufficiently accurate.

Overflow Rate in Settling Tanks

Overflow rate is a standard design parameter that is derived from discrete-particle settling analysis. All sediment suspensions have a mixture of particle sizes, and thus have particles with a range of settling velocities. If it is assumed that a homogeneous inlet zone exists in a sedimentation tank (see Fig. 12.15a), then the same particle size distribu-

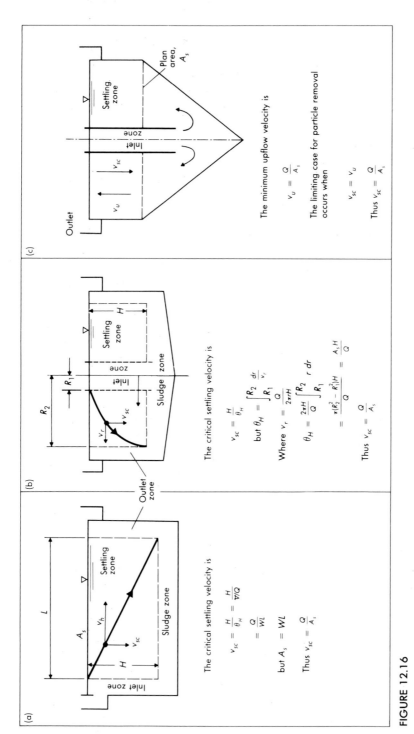

The critical settling velocity is

$$v_{sc} = \frac{H}{\theta_H} = \frac{H}{H/Q}$$

$$= \frac{Q}{WL}$$

but $A_s = WL$

Thus $v_{sc} = \frac{Q}{A_s}$

The critical settling velocity is

$$v_{sc} = \frac{H}{\theta_H}$$

but $\theta_H = \int_{R_1}^{R_2} \frac{dr}{v_r}$

Where $v_r = \frac{Q}{2\pi r H}$

$$\theta_H = \frac{2\pi H}{Q} \int_{R_1}^{R_2} r \, dr$$

$$= \frac{\pi (R_2^2 - R_1^2) H}{Q} = \frac{A_s H}{Q}$$

Thus $v_{sc} = \frac{Q}{A_s}$

The minimum upflow velocity is

$$v_u = \frac{Q}{A_s}$$

The limiting case for particle removal occurs when

$$v_{sc} = v_u$$

Thus $v_{sc} = \frac{Q}{A_s}$

FIGURE 12.16

Definition sketch for idealized settling in (a) a rectangular horizontal-flow, (b) a circular radial-flow, and (c) an upflow sedimentation tank.   Source: Adapted from Ref. [12.4].

tion enters at all depths. All particles having a settling velocity greater than v_{sc} will be removed, no matter where they enter, but particles having a velocity $v_s < v_{sc}$ will be removed in the proportion v_s/v_{sc}. Thus the critical settling velocity becomes a useful design parameter. Manipulating v_{sc} in terms of other variables results in the overflow rate (OFR), Q/A_s.

$$v_{sc} = \frac{H}{\theta_H} = \frac{H}{V/Q} = \frac{Q}{WL} = \frac{Q}{A_s} \qquad (12.29)$$

where

v_{sc} = critical settling velocity, m/s

H = tank depth, m

θ_H = hydraulic detention time, s

Q = volumetric flow rate, m³/s

W = tank width, m

L = tank length, m

A_s = surface area, m²

The development of the relationship defined by Eq. (12.29) for the three common types of sedimentation tanks used for water and wastewater treatment is given in Fig. 12.16.

Typical overflow rates used in practice are given in Table 12.3. The relationship between overflow rates and removal efficiency is illustrated

TABLE 12.3

Typical Overflow Rates Used for the Design of Sedimentation Tanks in Water and Wastewater Treatment Plants

TYPE OF OPERATION	OVERFLOW RATE, m/d	
	Range	Typical*
Water treatment		
Alum coagulation		
Turbidity removal	40–60	48
Color removal	35–45	40
Lime softening		
Low magnesium	60–110	80
High magnesium	50–90	65
Wastewater treatment		
Primary treatment		
Primary only	30–60	40
Primary with waste activated sludge return	22–40	30
Secondary treatment		
Activated sludge		
(excluding extended aeration)	16–32	24
Activated sludge		
(following extended aeration)	10–24	16
Trickling filter	16–30	24

*The typical value is based on average plant flow; for peak flow use twice the typical value.

in Example 12.5. It is important to remember that the overflow rate is a
particle settling velocity, not a fluid velocity.

EXAMPLE 12.5

ANALYSIS OF DISCRETE PARTICLE SETTLING

A suspension of particles having an average density of 1400 kg/m³ is to
be treated in a tank designed for an overflow rate of 25 m³/m² · d. For
the particle size distribution given, determine the fraction F_r removed.

EFFECTIVE DIAMETER, mm	MASS FRACTION
0.010	0.10
0.015	0.20
0.020	0.25
0.025	0.15
0.030	0.05
0.035	0.15
0.040	0.05
0.045	0.05

SOLUTION:

1. Apply Eq. (12.27) to determine the particle velocity for each size and
   use the overflow rate to determine the fraction removed in each
   group.

$$v_s = \sqrt{\frac{4}{3} \frac{(9.81)(400) d_p}{C_D(1000)}}$$

$$= 2.29 \left(\frac{d_p}{C_D}\right)^{1/2} \text{ m/s}$$

$$v_{sc} = \frac{25 \text{ m}^3/\text{m}^2 \cdot \text{d}}{86,400 \text{ s/d}} = 2.89 \times 10^{-4} \text{ m/s}$$

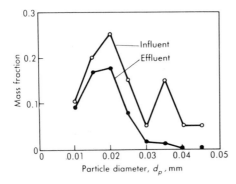

FIGURE 12.17

**Influent and effluent particle size
distribution for Example 12.5.**

When $N_R > 0.3$ the problem must be solved by trial and error. For this problem the values of N_R are all less than 0.3, so Eq. (12.28) can be used.

d_p, mm	MASS FRACTION	STOKES' LAW VELOCITY, v_s, m/s	N_R	v_s/v_{sc}	MASS FRACTION REMOVED	MASS FRACTION REMAINING
0.010	0.10	2.18×10^{-5}	2.21×10^{-4}	0.075	0.008	0.092
0.015	0.20	4.90×10^{-5}	7.41×10^{-4}	0.169	0.034	0.166
0.020	0.25	8.72×10^{-5}	1.71×10^{-3}	0.301	0.075	0.175
0.025	0.15	13.62×10^{-5}	3.4×10^{-3}	0.471	0.071	0.079
0.030	0.05	19.62×10^{-5}	5.8×10^{-3}	0.678	0.034	0.016
0.035	0.15	26.71×10^{-5}	9.4×10^{-3}	0.923	0.138	0.012
0.040	0.05	34.88×10^{-5}	1.4×10^{-2}	0.205	0.050	0.000
0.045	0.05	44.15×10^{-5}	2.0×10^{-2}	1.526	0.050	0.000
Total	1.00				0.460	0.540

2. Plot the influent and effluent particle size distribution.

The distributions are plotted in Fig. 12.17. As shown there, the particle size distributions for the influent and effluent are quite different.

Practical Sedimentation Systems

The ideal sedimentation systems described above have a number of major differences from actual systems. First, the assumption that particles do not interact is incorrect; in fact, several types of interaction occur. Second, flow patterns in real work tanks cause considerable deviation from the ideal conditions.

Particle Interaction

Three types of interaction occur: (1) flocculation in which particles collide and adhere and particle growth results; (2) fluidic, in which particles are so close together that flow is restricted and the particles move as a block; and (3) mechanical, in which particles physically interact, particularly in a compressive mode, water is squeezed out of the flocs, and particle volume may decrease (Fig. 12.18). These three types of interaction result in three classes of sedimentation (Types II, III, and IV) other than discrete and are often referred to by class or as flocculant, hindered, and compression (or thickening), respectively. *Flocculant settling* occurs only when chemical or biological flocs form in the system. This commonly occurs in water treatment plants. Mineral suspensions and primary sedimentation in wastewater treatment plants are best characterized as discrete-particle sedimentation systems.

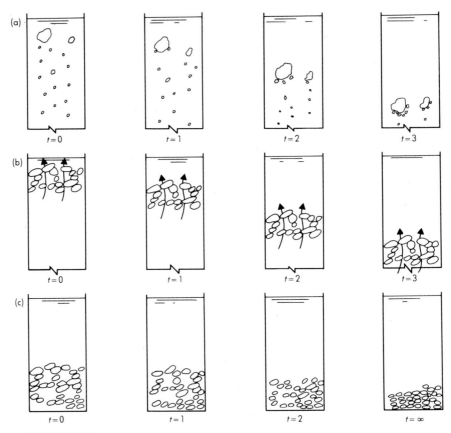

FIGURE 12.18

Nonideal particle interactions in sedimentation systems: (a) flocculation in which particle size increases during sedimentation; (b) fluidic in which fluid flow through the interstices controls the settling rate; and (c) mechanical in which the particle volume is reduced through consolidation.

Because particle growth occurs in flocculant sedimentation, the settling velocities increase, and removal efficiencies are better than predicted from discrete-particle analysis. *Hindered settling* is characterized by formation of a sharp, clear water–solids-suspension interface that settles at a characteristic rate. It is the interface settling rate that is used to determine the maximum allowable OFR in such systems. Because hindered settling rates are considerably lower than individual particle settling rates, design for hindered conditions is generally unadvisable except where solids thickening is desired.

Methods of analysis for Types II, III, and IV sedimentation are beyond the scope of this volume. Details may be found in Refs. [12.16], [12.35], [12.41], and [12.50].

Flow Patterns

Sedimentation tanks in use are quite different from the ideal unit of Fig. 12.15(a). The inlet is usually a weir followed by a baffle and as such behaves more like a slot than a uniform entrance. Effluent is collected in launders suspended in the tank at empirically located points. Thus the inlet and exit conditions are quite different from the ideal configuration. Typical rectangular and circular tanks are shown in Fig. 12.19, and typical dimensions are given in Table 12.4.

Within a tank, circulation cells form and result in dead volume and short circuiting. The most common causes of circulation-cell formation are wind, temperature, and small density differences (Fig. 12.20). Density differences not related to temperature usually are caused by the presence of a high concentration of suspended solids or a combination of suspended and total dissolved solids where brackish water infiltrates into sewers. Circulation cells are quite stable, and considerable loss of solids removal efficiency occurs because of their presence. Over the past 10 years, considerable effort has been made to evaluate the flow characteristics of sedimentation tanks and to arrange internal baffles to improve flow patterns [12.32, 12.44]. In general, to improve the performance of

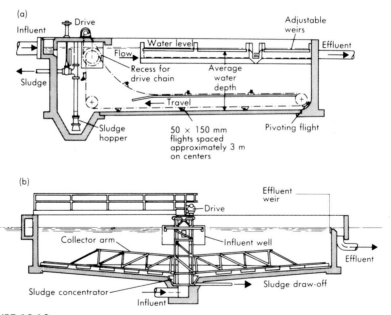

FIGURE 12.19

Typical sedimentation tanks: (a) rectangular (longitudinal section without skimmer) and (b) circular radial flow.

Source: Courtesy of Walker Process.

TABLE 12.4

Typical Dimensions of Sedimentation Tanks

	VALUE	
	Range	Typical
Rectangular		
Depth, m	3–5	3.5
Length, m	15–90	25–40
Width, m	3–24	6–10
Circular		
Depth, m	3–5	4.5
Diameter, m	4–60	12–45
Bottom slope, mm/m	60–160	80

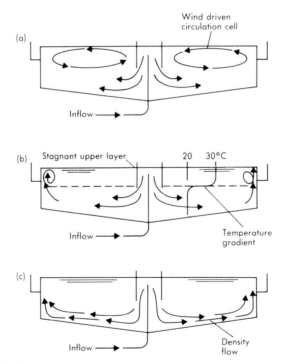

FIGURE 12.20

Nonideal conditions in circular sedimentation tanks: (a) formation of wind-driven circulation cells, (b) thermal stratification, and (c) density currents. To improve the effluent quality, the discharge weir should be placed in the dead zone.

both rectangular and circular tanks subject to cell formation, it is recommended that the effluent weirs be placed in the tank dead zone.

Sludge Handling

Solids removed in sedimentation tanks are initially stored on the tank bottom, and the storage volume is part of the design. Most settling tanks constructed today incorporate mechanical sludge-removal systems that collect the settled material with traveling scrapers called *flights* (see Fig. 12.19). The sludge is scraped into hoppers and pumped to a disposal system. Sludge disposal is costly and in wastewater treatment often accounts for more than 50 percent of the total treatment expense. Sludge disposal is considered in Section 12.12.

Sludge concentrations resulting from sedimentation typically vary from somewhat less than 1 percent (10,000 g/m^3) for poor biological wastewater treatment sludges to greater than 10 percent (100,000 g/m^3) for water treatment sludges on a dry-mass basis. At higher concentrations, pumping becomes a problem because the material is non-Newtonian. Volumes of sludge produced vary considerably with the source of water or wastewater; typical values are given in Section 12.12.

Tube and Lamella Settlers

Tube and lamella settlers are sedimentation units designed to follow ideal theory more closely. Constructed of bundles of tubes or plates set at selected angles to the horizontal, these settlers have a very short settling distance, and circulation is damped because of the small size of the tubes (Fig. 12.21). Typical tube sizes are 25 to 50 mm, and when the imposed angle is greater than 40° solids tend to slide out because of gravitational force. Tube settlers have proved to be efficient units and can be retrofitted in older tanks or designed as part of new tanks [12.14, 12.15, 12.55]. The major drawback is a tendency to clog because of the accumulation of biological growths and grease.

The critical settling velocity for a tube settler is defined in the same manner as is that of ideal sedimentation tanks.

$$v_{ct} = \frac{h}{\theta_H} = \frac{hu}{L} \tag{12.30a}$$

$$= v_{sc} \cos\theta \tag{12.30b}$$

where

v_{ct} = critical settling velocity for tube settler, m/s

h = cross-sectional distance through tube, m

θ_H = detention time in tube settler, s

u = fluid velocity through tube, m/s

L = length of tube settler, m

v_{sc} = discrete settling velocity of critical particle, m/s

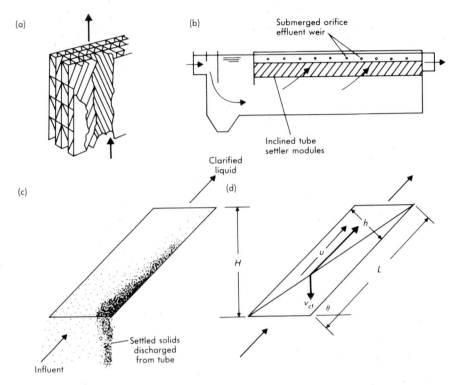

FIGURE 12.21

Definition sketch for a tube settler: (a) module of inclined tubes; (b) tubes installed in a rectangular sedimentation tank; (c) operation; and (d) inclined tube details.

Upflow Sludge-Blanket Clarifier

In some water treatment plants, the flocculation and clarification steps are carried out in a single unit known as an *upflow sludge-blanket* or *solids-contact clarifier* (Fig. 12.22; also Fig. 12.16c). In most combined units, flow enters in the center where it is mixed with the coagulating chemical and then flocculated. The flocculated solids form a sludge blanket through which the water must flow in an upward direction. Clarified water is withdrawn at the top through submerged orifices. Sludge is continually withdrawn from the bottom of the unit to maintain the desired sludge-blanket level [12.4, 12.24].

Flotation Thickening

Sedimentation systems allow particles having densities greater than water to settle out. The same particles can be attached to small gas bubbles and brought to the surface through flotation. Two types of flotation systems

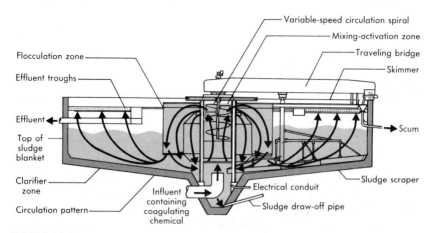

FIGURE 12.22

Typical upflow sludge-blanket clarifier.

Source: Courtesy of the Passavant Corporation.

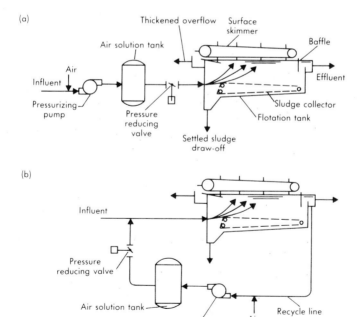

FIGURE 12.23

Schematic of dissolved-air flotation systems: (a) pressurized full flow without recycle and (b) pressurized recycle.

are in use: dissolved air and fine bubble. Two dissolved-air flotation systems are in common use (Fig. 12.23). In the first, the entire influent stream is pressurized, whereas in the second a recycle stream is pressurized. In either case, when the pressurized water is released in the flotation unit under atmospheric conditions small gas bubbles are formed. In turn, the small bubbles attach to the solids, and the result is solids flotation. The thickened solids are skimmed from the surface. In fine-bubble flotation, small bubbles are introduced through fine-pore diffusers, and the result is essentially the same as in dissolved-air flotation. Maintenance of the fine-pore diffusers offsets the savings associated with not needing a pressurizing pump and air-solution tank system.

To date, dissolved-air systems have predominated in wastewater treatment applications, although flotation is still not commonly used. Flotation system design is generally based on experience factors and pilot-plant modeling. A reliable method of predicting process performance on the basis of sludge characteristics has not been developed.

12.8
FILTRATION

The filtration process is used for two main purposes: suspended-solids removal and thickening. The discussion in this section is limited to the removal of suspended solids. Thickening, normally considered a sludge processing technique, is discussed in Section 12.12. Historically, filtration has been used as a polishing step in the treatment of water. In this application, filters are used to remove any residual suspended solids that are not removed in the sedimentation process. In wastewater treatment, filtration is used in a similar application, typically following complete physical-chemical or biological treatment.

Types of Filters

Two general types of filters are in common use: granular-medium filters (Fig. 12.24a) and surface filters (Fig. 12.24b) [12.26]. Most granular-medium filters are simply beds composed of various types of granular media. In a granular-medium filter, the removal of suspended material occurs within and on the surface of the filter bed. Surface filters usually involve the removal of suspended material by means of straining a thin membrane or straining surface. The polycarbonate filters used to determine the concentration of suspended solids in water (see Chapter 2) and a kitchen colander are typical examples of the latter. Because of the scope of this text, the following discussion will be limited to granular-medium filters. Details on surface filters may be found in Ref. [12.6].

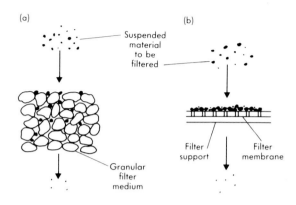

(a)

Suspended
material
to be
filtered

Granular
filter
medium

(b)

Filter
support

Filter
membrane

FIGURE 12.24

**Definition sketch for different types
of filters: (a) granular-medium and
(b) surface filters.**

Description of a Granular-Medium Filter

To introduce the subject of granular-medium filtration, it is appropriate
to describe the physical features of a rapid granular-medium filter, the
filtration process in which suspended material is removed from the liquid,
and the backwash process that is used to clean the filter. Other types of
granular-medium filters are discussed in the following sections.

Physical Features

The general features of a rapid granular-medium filter are illustrated
in Fig. 12.25(a). As shown, the filtering medium (sand in this case) is
supported on a gravel layer, which, in turn, rests on the filter underdrain
system. Filtered water, collected in the underdrain, is discharged to a
storage reservoir or to the distribution system. The underdrain system is
also used to reverse the flow to backwash the filter. The water to be
filtered enters the filter from an inlet channel. The hydraulic control of
the filter is described in a subsequent section.

Filtration Process

During filtration, water or wastewater containing suspended matter
is applied to the top of the filter bed (Fig. 12.25a). As the water filters
through the granular medium, the suspended matter in the fluid is
removed by a variety of mechanisms, including mechanical and chance
contact straining, impaction, interception, adsorption, flocculation, and
sedimentation (Fig. 12.26). With the passage of time, as material accumulates within the interstices of the granular medium, the head loss
through the filter starts to build up beyond the initial value, as shown in
Fig. 12.27. At some point the operating head loss reaches some predetermined head loss value, and the filter must be cleaned.

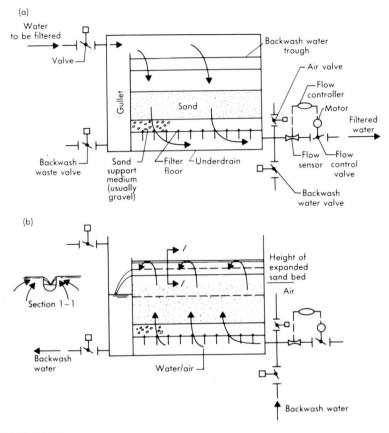

FIGURE 12.25

Definition sketch for a rapid granular-medium (sand) filter: (a) during operation and (b) during backwashing.

The terminal head loss is a controlling variable, but so too is the effluent quality. Toward the end of a filter run, as the interstices of the granular medium become filled, the suspended-solids removal efficiency is reduced, and the amount of suspended matter in the filter effluent starts to increase. When a preselected value of effluent quality is reached, the filter must be cleaned. Ideally, the time required for the head loss buildup to reach the preselected terminal value should correspond to the time when the suspended solids in the effluent reach the preselected terminal value for acceptable quality.

Backwash Process

When either the terminal head loss or quality values are reached, the filter must be cleaned. Most granular-medium filters are cleaned by

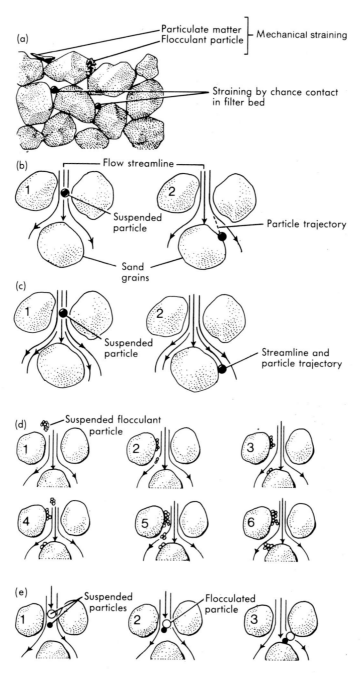

FIGURE 12.26

**Definition sketch for the removal of suspended matter in a granular-medium filter.
(a) By straining, (b) By sedimentation and inertial impaction, (c) By interception,
(d) By adhesion, and (e) By flocculation**

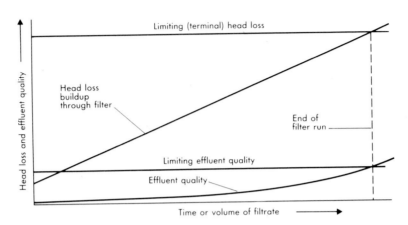

FIGURE 12.27

Definition sketch for the length of a filter run based on head loss and effluent quality (usually measured as turbidity).

reversing the flow through the filter bed (see Fig. 12.25b). Filtered water is pumped through the bed at a rate sufficient to partially expand the bed. The suspended matter arrested within the filter is removed by the shear forces created by backwash water as it moves up through the bed [12.3, 12.5, 12.11]. Additional cleaning is achieved as the particles of the filtering medium abrade against each other. The suspended solids removed from the filter are removed with the wash water in the wash-water troughs. Backwash hydraulics are considered in the section dealing with the design of granular-medium filters.

Other Granular-Medium Filters

While the granular-medium filter shown in Fig. 12.25(a) is the most commonly used, a variety of filter configurations have been used (Fig. 12.28). Among them are the deep-bed filter, the pulsed-bed filter (developed by D. S. Ross), the multimedia filter, and the slow sand filter.

Deep-Bed Filters

The deep-bed filter (Fig. 12.28a and b) is essentially the same as the conventional filter, but the depth of the filter bed and the size of the filtering media are considerably greater than the corresponding values in a conventional filter.

Pulsed-Bed Filters

In the pulsed-bed filter (Fig. 12.28c) a shallow layer of sand (25 mm) rests on a screened underdrain. Unlike all other granular-medium filters, the underdrain system below the filtering medium is open to the atmo-

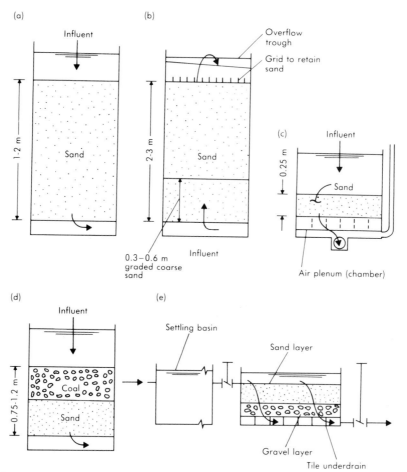

FIGURE 12.28

Types of rapid granular-medium filter beds: (a) single-medium downflow deep-bed filter; (b) single-medium upflow deep-bed filter with surface grid to retain sand (the sand between the grids forms an arch during the filtration process); (c) pulsed-bed filter with underdrain system open to the atmosphere; (d) dual-media downflow filter, and (e) slow sand filter.

sphere. In the pulsed-bed filter, as the head loss builds up the filter is pulsed with air by the action of the backwash pumps. Filtered water is pumped back into the underdrain system, and the air trapped in the air plenum (chamber) below the sand bed is pushed up through the bed to loosen and redistribute the solids accumulated at the water-sand inter-face. Because of the redistribution of solids that occurs within the bed at the end of the pulse cycle, the rate of head-loss buildup is decreased, and the filter run length is increased. The effect of the pulse action on the buildup of head loss is discussed in the section dealing with hydraulic control.

Multimedia Filters

These filters are essentially the same as the conventional single-medium filter, with the exception that two (sand and anthracite) or three (garnet, sand, and anthracite) media are used (Fig. 12.28d).

Slow Sand Filters

The first filter used for a public water supply was a slow sand filter (Fig. 12.28e) [12.22]. As shown, influent enters the filter, passes through the sand, and is collected in the underdrain. Suspended matter is removed in the surface mat that forms on the filter and in the interstices of the upper layers of sand. (Early German investigators called the surface mat that formed on slow sand filters a *Schmutzdecke*.) When the filter becomes clogged, the filter bed is drained, the upper layer of sand is removed, new sand is added, and the filter is put back into operation. Filtration rates used are in the range of 2 to 5 L/m² · min. The head loss in slow-sand filters varies from about 0.05 m initially to a final value of about 1.25 m, depending on the design of the filter.

Hydraulic Control of Granular-Medium Filters

Although a wide variety of hydraulic control schemes has been used in the design of filtration systems, the two most commonly used can be categorized as (1) constant head-constant rate or variable head-constant rate and (2) either constant-head or variable-head declining rate [12.5, 12.7].

Constant-Rate Filtration

In the past, constant-rate filtration was used most commonly. In this method of operation, the total operating head on the filter is fixed and the flow through the filter is controlled at a constant rate. The rate is controlled with a throttling valve on the discharge side of the filter (Fig. 12.29a).

A variation that is often used is to allow the water level to build up over the surface of the filter bed while maintaining a constant rate of flow through the filter (Fig. 12.29b). Constant flow to the individual filters is maintained by means of a weir. With the pulsed-bed filter (Fig. 12.29c), constant flow is achieved by pumping or by the use of an influent flow-splitting weir.

Declining-Rate Filtration

In declining-rate filtration, as the name implies, the rate of flow though the filter is allowed to decline as the rate of head loss builds up with time. Declining-rate filtration systems are either influent-controlled

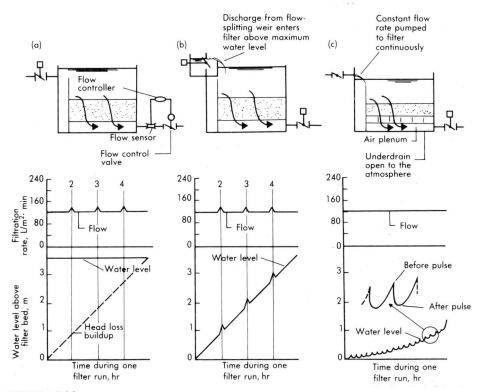

FIGURE 12.29

Definition sketch for constant-rate filtration: (a) fixed head-constant rate; (b) variable head-constant rate; (c) variable head-constant rate with pulsed-bed filter. Note: Curves for filters in (a) and (b) are for the operation of one filter in a bank of four filters. The numbers represent the filter that is backwashing during the filter run. In practice, the time before backwashing will not be the same for all of the filters.

or effluent-controlled. With influent control, a large influent channel or header is used to provide water to the individual filters, each of which is equipped with an isolation valve (Fig. 12.30). During operation the water level is essentially the same in all of the filters because the large header acts as a reservoir. Thus, the flow to each filter is equal to the flow that can be processed by each filter at a given moment. When a filter is taken out of service for backwashing, the level on the remaining filters rises slightly to maintain the same rate of flow through the system.

When a bank of filters that has been out of service is brought back into service, the operating water level is below the wash-water trough, and the filters operate as constant-rate filters, the flow rate being controlled by the settling of the influent control valves. Because extremely high filtration rates are possible when the filters are clean, a flow-restricting device (usually an orifice) is placed in the filter effluent-discharge piping.

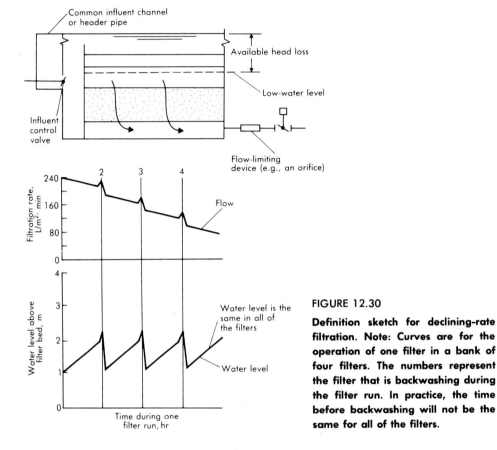

FIGURE 12.30

Definition sketch for declining-rate filtration. Note: Curves are for the operation of one filter in a bank of four filters. The numbers represent the filter that is backwashing during the filter run. In practice, the time before backwashing will not be the same for all of the filters.

When the water level in the filter is above the level of the wash-water troughs, the filters operate as variable-head declining-rate filters. In general, the clean-water head loss through the piping system, filtering media, and underdrains is about 0.80 to 1.20 m, so that the actual operating low-water level is above the wash-water troughs.

Modeling of the Filtration Process

During the past 60 years considerable effort has been devoted to the modeling of the filtration process. The models fall into two general categories: those models used to predict the clean-water head loss through a filter bed and those models used to predict the performance of filters for the removal of suspended solids.

Clean-Water Head Loss

Mathematical models of clean-water flow through a porous medium are similar to those developed for other flows through porous media, such

TABLE 12.5

Formulas Governing the Flow of Clean Water Through a Granular Medium

FORMULA	DEFINITION OF TERMS
Carmen-Kozeny [12.10, 12.29]	C = coefficient of compactness (600 to 1200)

Carmen-Kozeny [12.10, 12.29]

$$h = \frac{f}{\phi}\frac{1-\alpha}{\alpha^3}\frac{L}{d}\frac{v^2}{g} \quad (12.31a)$$

$$h = \frac{1}{\phi}\frac{1-\alpha}{\alpha^3}\frac{Lv^2}{g}\sum f\frac{p}{d_g} \quad (12.31b)$$

$$f = 150\frac{1-\alpha}{N_R} + 1.75 \quad (12.32)$$

$$N_R = \frac{\phi\, dv\,\rho}{\mu} \quad (12.33)$$

Fair-Hatch [12.19]

$$h = k\nu S^2\frac{(1-\alpha)^2}{\alpha^3}\frac{L}{d^2}\frac{v}{g} \quad (12.34a)$$

$$h = k\nu\frac{(1-\alpha)^2}{\alpha^3}(Lv)\left(\frac{S}{\phi}\right)^2\sum\frac{p}{d_g^2} \quad (12.34b)$$

Rose [12.42]

$$h = \frac{1.067}{\phi}C_d\frac{1}{\alpha^4}\frac{L}{d}\frac{v^2}{g} \quad (12.35a)$$

$$h = \frac{1.067}{\phi}\frac{Lv^2}{\alpha^4 g}\sum C_d\frac{p}{d_g} \quad (12.35b)$$

$$C_d = \frac{24}{N_R} + \frac{3}{\sqrt{N_R}} + 0.34 \quad (12.25)$$

Hazen [12.22]

$$h = \frac{1}{C}\frac{60}{1.8T+42}\frac{L}{d_{10}^2}v \quad (12.36)$$

DEFINITION OF TERMS

C = coefficient of compactness (600 to 1200)
C_d = coefficient of drag
d = grain diameter, m
d_g = geometric mean diameter between sieve sizes d_1 and d_2, $\sqrt{d_1 d_2}$
d_{10} = effective grain diameter, mm
f = friction factor
g = acceleration due to gravity, 9.81 m/s^2
h = head loss, m
k = filtration constant, 5 based on sieve openings, 6 based on size of separation
L = depth of filter bed, m
N_R = Reynolds number
p = percent of particles (based on mass) within adjacent sieve sizes
S = shape factor (varies between 6.0 and 7.7)
T = temperature, °C
v = filtration velocity, m/s
v = filtration velocity, m/d [in Eq. (12.36)]

α = porosity
μ = viscosity, N · s/m^2
ν = kinematic viscosity, m^2/s
ρ = density, kg/m^3
ϕ = shape factor

as that involving groundwater (see Chapter 10). Representative formulas that have been developed to predict the clean-water head loss are summarized in Table 12.5 [12.10, 12.14, 12.22, 12.29, 12.42]. Computation of the clean-water head loss through a filter is illustrated in Example 12.6.

EXAMPLE 12.6

DETERMINATION OF CLEAN-WATER HEAD LOSS IN A GRANULAR-MEDIUM FILTER

Determine the clean-water head loss in a filter bed composed of 0.3 m of uniform anthracite (with an average size of 1.6 mm) placed over 0.3-m layer of uniform sand (with an average size of 0.5 mm) for a filtration rate of 160 L/m^2 · min. Assume that the operating temperature is 20°C. Use the Fair-Hatch equation [Eq. (12.34a)] given in Table 12.5 for computing the head loss.

SOLUTION:

1. Determine the head loss through the anthracite layer using Eq. (12.34a) (Table 12.5).

$$h = kvS^2 \frac{(1-\alpha)^2}{\alpha^3} \frac{L}{d^2} \frac{v}{g}$$

   $k = 6$ (see Table 12.5)

   $v = 1.003 \times 10^{-6}$ m^2/s

   $S = 6.0$ (see Table 12.5)

   $\alpha = 0.4$ (assumed)

   $L = 0.3$ m

   $d = 1.6 \times 10^{-3}$ m

   $v = 160$ L/m^2 · min $= 2.67 \times 10^{-3}$ m/s

   $g = 9.81$ m/s^2

$$h = 6(1.003 \times 10^{-6})(6^2) \frac{(1-0.4)^2}{(0.4)^3} \frac{0.30}{(1.6 \times 10^{-3})^2} \frac{2.67 \times 10^{-3}}{9.81}$$

   $= 0.039$ m

2. Determine the head loss through the sand layer.

   $k = 6$ (see Table 12.5)

   $v = 1.003 \times 10^{-6}$ m^2/s

   $S = 6$ (see Table 12.5)

   $\alpha = 0.4$ (assumed)

   $L = 0.3$

   $d = 0.5 \times 10^{-3}$ m

   $v = 2.67 \times 10^{-3}$ m/s

$$g = 9.81 \text{ m/s}^2$$

$$h = 6(1.003 \times 10^{-6})(6^2)\frac{(1-0.4)^2}{(0.4)^3}\frac{0.30}{(0.5 \times 10^{-3})^2}\frac{2.67 \times 10^{-3}}{9.81}$$

$$= 0.40 \text{ m}$$

3. The total head loss H_T is

$$H_T = 0.039 \text{ m} + 0.40 \text{ m}$$

$$= 0.44 \text{ m}$$

Filtration Process

The mathematical characterization of the time-space removal of suspended solids can be modeled by considering a mass balance on volume element shown in Fig. 12.31. The basic mass balance for the volume element is

$$\frac{\partial q}{\partial t}\Delta V + \bar{\alpha}(t)\frac{\partial C}{\partial t}\Delta V = QC|_x - QC|_{x+\Delta x} + 0 \qquad (12.37)$$

Accum   +   Accum   = Inflow −   Outflow   + Generation
of solids   of solids
in non-active   in active
pore space   pore space

where

$\partial q/\partial t$ = change in quantity of solids deposited within the filter with time, $\text{g/m}^3 \cdot \text{s}$

$\bar{\alpha}(t)$ = average porosity (variable with time)

$\partial C/\partial t$ = change in average concentration of solids in pore space with time, $\text{g/m}^3 \cdot \text{s}$

ΔV = differential volume, m^3

Q = filtration rate, m^3/s

C = concentration of suspended solids, g/m^3

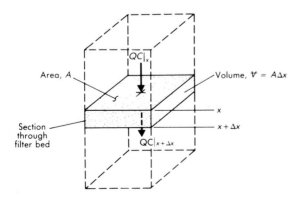

FIGURE 12.31

Definition sketch for the analysis of the filtration process.

Substituting $A \Delta x$ for ΔV and Av for Q where v is the filtration velocity (in cubic meters per square meter-second, or $m^3/m^2 \cdot s$), and taking the limit as Δx approaches zero results in

$$-v \frac{\partial C}{\partial x} = \frac{\partial q}{\partial t} + \bar{\alpha}(t) \frac{\partial C}{\partial t} \qquad (12.38)$$

The first term represents the difference between the mass of suspended solids entering and leaving the section; the second term represents the time rate of change in the mass of suspended solids present on the solid portion of the filter; and the third term represents the time rate of change in the suspended-solids concentration in the fluid portion of the filter volume. In a flowing process, the quantity of fluid contained within the bed is usually small compared with the volume of liquid passing through the bed. In this case, the materials balance becomes

$$-v \frac{\partial C}{\partial x} = \frac{\partial q}{\partial t} \qquad (12.39)$$

This equation is the one most commonly found in the literature dealing with filtration theory [12.11, 12.26, 12.27, 12.52].

To solve Eq. (12.39), an additional independent equation is required. The most direct approach is to derive a relationship that can be used to describe the change in concentration of suspended matter with distance, such as

$$\frac{\partial C}{\partial x} = \phi(V_1, V_2, V_3, \ldots) \qquad (12.40)$$

in which V_1, V_2, and V_3 are the variables governing the removal of suspended matter from solution. The change in concentration of suspended solids with distance that is observed when filtering chemically treated water and treated secondary effluent is illustrated in Fig. 12.32.

An alternative approach is to develop a complementary equation in which the pertinent process variables are related to the amount of material removed within the filter at various depths. In equation form, this may be written as

$$\frac{\partial q}{\partial t} = \phi(V_1, V_2, V_3, \ldots) \qquad (12.41)$$

Both approaches will be found in the literature [12.26]. Despite the best efforts of the research community, no general equation is available that can be used to model the filtration process. For this reason, past experience and the results of pilot-plant tests are used to establish design parameters.

Development of Head Loss During Filtration

In the past, the most commonly used approach to determine the head loss in a clogged filter was to compute it with a modified form of the equations used to evaluate the clean-water head loss (see Table 12.5). In

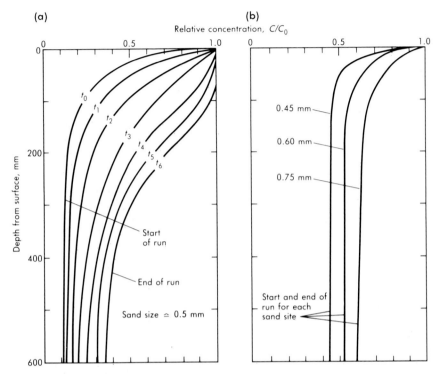

FIGURE 12.32

Normalized suspended-solids removal curves as a function of depth and time as observed during (a) the filtration of a pretreated water (coagulation/flocculation/sedimentation) and (b) the filtration of settled effluent from an activated sludge treatment process. In both instances, the filtration rate is about 80 L/m² · m.

all cases, the difficulty encountered in using these equations is that the porosity must be estimated for varous degrees of clogging. Unfortunately, the complexity of this approach renders most of these formulations useless or, at best, extremely difficult to use.

An alternative approach is to relate the development of head loss to the amount of material removed by the filter. The head loss would then be computed using the expression

$$H_t = H_0 + \sum_{i=1}^{n} (h_i)_t \tag{12.42}$$

where

H_t = total head loss at time t, m

H_0 = total initial clean-water head loss, m

$(h_i)_t$ = head loss in the ith layer of the filter at time t, m

From an evaluation of the incremental head-loss curves for uniform sand and anthracite when filtering treated wastewater effluent, the buildup of head loss in an individual layer of the filter was found to be related to the amount of material contained within the layer [12.52]. The form of the resulting equation for head loss in the ith layer is

$$(h_i)_t = a(q_i)_t^b \qquad (12.43)$$

where

$(q_i)_t$ = cumulative amount of material deposited in the ith layer at time t, mg/cm^3

a, b = constants

In this equation, it is assumed that the buildup of head loss is a function only of the amount of material removed. This assumption is only valid for wastewater effluent filtration. Representative data for uniform sand for treated wastewater effluent are presented in Fig. 12.33. The determination of the buildup of head loss during the filtration process using the data presented in that figure is illustrated in Example 12.7.

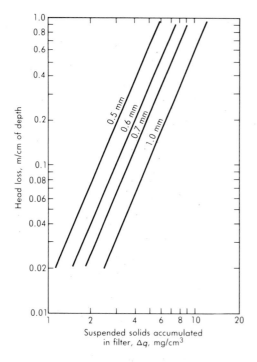

FIGURE 12.33

Head-loss buildup as a function of medium size and the amount of material removed within the filter bed for treated wastewater effluent.

EXAMPLE 12.7

ESTIMATION OF FILTER RUN LENGTH

Treated wastewater effluent is to be filtered using a shallow (600 mm) single-medium granular-medium filter. Uniform sand with a size of 0.60 mm is to be used as the filtering medium. If the average suspended-solids concentration in the treated effluent is 25.0 g/m³, the filtration rate is 80 L/m² · min, and the total available head loss is fixed at 2.5 m, estimate the length of filter run using the data given in Figs. 12.32(b) and 12.33.

SOLUTION:

1. To determine the head-loss buildup, Eq. (12.40) is rewritten as follows:

$$-v\frac{\Delta C}{\Delta x} = \frac{\Delta q}{\Delta t}$$

$$-v\frac{\Delta C}{\Delta x}(\Delta t) = \Delta q$$

where

$v =$ filtration velocity, L/cm² · min

$\Delta C = (C_{x-1} - C_x)$, mg/L

$\Delta x = (x_{x-1} - x_x)$, cm

$\Delta t = (t_2 - t_1)$, min

$\Delta q = (q_2 - q_1)$, mg/cm³

The head loss is determined by computing the value of Δq at different depths in the filter after various time intervals and finding the corresponding head loss from Fig. 12.33.

2. Set up a computation table and determine the buildup of head loss. (The explanation of the entries in the table is presented below.)

   a. The values in the first two columns are obtained from Fig. 12.32(b). The size of the depth increments selected for analysis is dictated by the shape of the normalized suspended-solids removal curve. If the curve changes gradually, larger size increments can be used.

   b. The value of C_x in column (3) is obtained by multiplying the value in column (2) by 25 mg/L, the influent suspended-solids concentration.

   c. The value of ΔC in column (4) is the difference between the suspended solids at the indicated depths.

   d. The value of Δq in columns (5), (7), and (9) is computed as follows:

$$\Delta q = -v\frac{\Delta C}{\Delta x}(\Delta t)$$

				RUN LENGTH, hr					
DEPTH				2		4		6	
x, cm (1)	C_x/C_0 (2)	C_x mg/L (3)	ΔC,* mg/L (4)	Δq, mg/cm³ (5)	Δh, m (6)	Δq mg/cm³ (7)	Δh, m (8)	Δq, mg/cm³ (9)	Δh, m (10)
0	1.00	25.0							
			6.2	3.0	0.22	6.0	1.12	8.9	3.00
2	0.75	18.8							
			2.8	0.9	0.02	1.8	0.09	2.7	0.27
5	0.64	16.0							
			2.0	0.4	—	0.8	—	1.2	0.06
10	0.56	14.0							
			0.7	0.1	—	0.3	—	0.4	—
15	0.53	13.3							
			0.3	—	—	0.1	—	0.2	—
20	≈ 0.52	13.0							
			0.0						
30	≈ 0.52	13.0							
			0.0						
40	≈ 0.52	13.0							
			0.0						
50	≈ 0.52	13.0							
			0.0						
60	≈ 0.52	13.0							
$\Sigma \Delta h$, m					0.24		1.21		3.33

*$\Delta C = (C_{x-1} - C_x)$

where

$$v = 80 \text{ L/m}^2 \cdot \text{min} = 0.0080 \text{ L/cm}^2 \cdot \text{min}$$

$$\Delta C = 6.2 \text{ mg/L}$$

$$\Delta x = (0 \text{ cm} - 2 \text{ cm}) = -2 \text{ cm}$$

$$\Delta t = (2 \text{ hr} \times 60 \text{ min/hr} - 0 \text{ min}) = 120 \text{ min}$$

$$\Delta q = (-0.0080 \text{ L/cm}^2 \cdot \text{min})\left(\frac{6.2 \text{ mg/L}}{-2 \text{ cm}}\right) 120 \text{ min}$$

$$= 3.0 \text{ mg/cm}^3$$

e. The incremental head-loss values in columns (6), (8), and (10) are found by entering Fig. 12.33 with the incremental Δq values from columns (5), (7), and (9). Because the head loss in Fig. 12.33 is expressed in units of meters per centimeter of depth, the value

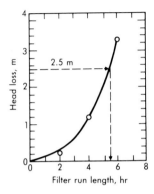

FIGURE 12.34

Head-loss buildup as a function of run length for Example 12.7.

obtained from the figure must be multiplied by the depth of the layer in question.

3. Estimate the run length to reach a head-loss value of 2.5 m.
   a. The run length is found by plotting the computed head loss versus the corresponding run length. The required plot is shown in Fig. 12.34.
   b. The run length from Fig. 12.34 is 5.5 hr.

Design of Granular-Medium Filters

The design of a granular-medium filter system involves the consideration and specification of the following items [12.5]:

1. Type of medium, size, and depth
2. Filtration rate
3. Pressure available for driving force
4. Method of filter operation (including cleaning)

At the present time there is no theoretical basis for the design of granular-medium filters. Design is based on the experience of others, the conduct of pilot-plant studies, and the designer's past experience [12.4]. Design data and information for the filters used for water and wastewater treatment are considered below.

Granular-Medium Filters for Water

Three types of granular-medium filters commonly used for water treatment include (1) single-medium shallow-bed (less than 1 m), (2) single-medium deep-bed (greater than 1 m), and (3) dual-medium. Representative design data for water filters are reported in Table 12.6.

TABLE 12.6

Typical Design Data for Granular-Medium Filters Used for the Treatment of Water

PARAMETER*	SINGLE MEDIUM†		DUAL MEDIUM		MULTIMEDIUM	
	Range	Typical	Range	Typical	Range	Typical
Garnet or ilmenite						
Depth, mm					75–200	100
Effective size, mm					0.2–0.35	0.25
Uniformity coefficient					1.3–1.7	1.6
Sand						
Depth, mm	500–900	600	150–500	300	150–400	300
Effective size, mm	0.35–0.70	0.45	0.45–0.6	0.5	0.45–0.6	0.5
Uniformity coefficient	1.3–1.7	1.5	1.4–1.7	1.6	1.4–1.7	1.6
Anthracite						
Depth, mm	900–1800	1500	400–600	500	400–600	500
Effective size, mm	0.7–1.0	0.75	0.8–1.4	1.0	0.8–1.4	1.1
Uniformity coefficient	1.4–1.8	1.6	1.4–1.8	1.6	1.4–1.8	1.6
Filtration rate, $L/m^2 \cdot min$	80–400	160	80–400	160	80–400	160
Backwashing	Air/water, surface wash		Air/water, surface wash		Air/water, surface wash	
Backwash rate, $L/m^2 \cdot min$	360–1000‡	500‡	500–1600	800	500–1600	800

*The effective size is defined as the 10 percent size by mass, d_{10}. The uniformity coefficient is defined as the ratio of the 60 to the 10 percent size by mass ($UC = d_{60}/d_{10}$).

†Separate sand and anthracite single-medium filters.

‡For single medium sand filter only

Granular-Medium Filters for Wastewater

In addition to the types of filters cited above, the shallow pulsed-bed filter is used extensively for effluent filtration. As noted earlier, the pulsed-bed filter is also used for the filtration of primary effluent [12.34]. Representative design data for wastewater filters are reported in Table 12.7.

Filter Backwashing and Cleaning

Perhaps the most critical problem in filter operation is the proper cleaning of the filter bed. Unless the filter bed is cleaned properly, material can accumulate within the bed. In water filtration plants, the large accumulations of fine material can grow into mud balls. The same problem occurs in wastewater filtration, but is aggravated by the presence of grease and bacterial slimes. Methods that have been used to achieve effective cleaning of granular-medium filters include backwashing, surface

TABLE 12.7

Typical Design Data for Granular-Medium Filters Used for the Treatment of Wastewater

PARAMETER*	SINGLE MEDIUM[†]		SINGLE MEDIUM[‡]		DUAL MEDIUM	
	Range	Typical	Range	Typical	Range	Typical
Sand						
Depth, mm	200–300	250	500–900	600	150–300	300
Effective size, mm	0.4–0.6	0.45	0.45–0.7	0.5	0.4–0.7	0.55
Uniformity coefficient	1.3–1.7	1.5	1.3–1.7	1.5	1.4–1.7	1.6
Anthracite						
Depth, mm			900–1800	1500	300–600	500
Effective size, mm			0.8–1.8	1.4	0.8–1.8	1.2
Uniformity coefficient			1.4–1.8	1.6	1.4–1.8	1.6
Filtration rate, $L/m^2 \cdot min$	80–320	160	80–400	160	80–400	160
Backwashing	Air pulse followed by water, chemical cleaning		Air/water, surface wash		Air/water, surface wash	
Backwash rate, $L/m^2 \cdot m$	360–800	600	360–1000[§]	500[§]	500–1600	800

*The effective size is defined as the 10 percent size by mass, d_{10}. The uniformity coefficient is defined as the ratio of the 60 to the 10 percent size by mass ($UC = d_{60}/d_{10}$).

[†] Pulsed-bed filter.

[‡] Separate sand and anthracite single-medium filters.

[§] For single medium sand filter only

washing, air scouring, chemical cleaning, and various combinations of these procedures.

Water Backwashing

The most common method used to clean dirty filters is to backwash them with filtered water. Based on experimental studies Amirtharajah [12.3] has found that optimum cleaning of a conventional filter bed occurs when the expanded porosity of the bed is in the range of 0.65 to 0.70. At this degree of expansion it appears that the shearing action of the rising backwash water is most effective in removing the accumulated material from the filtering medium.

In contrast to the conventional filter-backwash operation, where the backwash water is applied uniformly over the entire bed, is the pulsed-bed filter. In this filter the backwash water is applied to the bed through small-diameter orifices spaced about 100 mm apart in a rectangular grid. Sufficient water is applied to form a quick condition, and the sand is cleaned by the high shearing forces in the rising jet (Fig. 12.35) and by

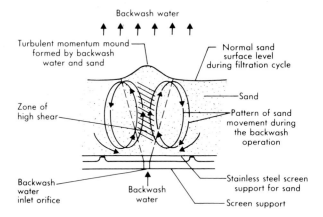

FIGURE 12.35

Definition sketch for the backwashing operation for a pulsed-bed filter.

abrasion as the sand moves in a circular motion. During backwashing the pulsed-bed filter acts as a spouted bed reactor used in chemical engineering applications.

Surface Washing

In surface washing, a surface wash is introduced through orifices on a fixed piping grid or on a rotating arm located 25 to 50 mm above the fixed bed (Fig. 12.36) [12.4]. Operationally, water in the filter is drained to the wash-water trough or below, and the surface wash is initiated for about 1 to 2 min before backwashing begins. The surface wash is continued while the filter is being backwashed until the backwash water begins to clear.

Air Scour

Air scour involves the application of air at the bottom of the filter. As the air flows upward through the filter it serves to break up accumulated deposits. Air may be applied prior to or simultaneously with the backwash water. When air is used alone, the water level in the filter is dropped below the wash-water trough to avoid the loss of filtering medium.

Chemical Cleaning (Pulsed-Bed Filter)

To overcome the effects of grease and slime buildup and to clean the filter bed effectively, a chemical clean cycle is incorporated in the operation of the pulsed-bed filter. As part of a routine maintenance program, the filter bed is periodically flooded with a mild bleach and detergent solution. After 30 min, the filter is backwashed. Grease and biological slime accumulated within the filter bed are eliminated, and the sand is restored to its original condition [12.34].

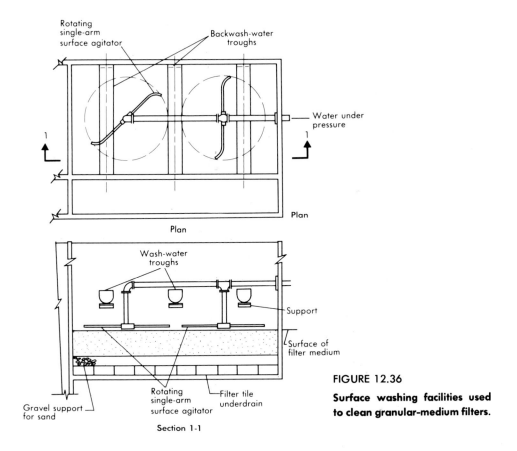

FIGURE 12.36

Surface washing facilities used to clean granular-medium filters.

Backwash Hydraulics

To expand the filter bed hydraulically the head loss must equal the buoyant mass of the granular medium in the fluid. Mathematically this relationship can be expressed as

$$h = l_e(1 - \alpha_e)\frac{\rho_s - \rho_w}{\rho_w}$$
(12.44)

where

h = head loss required to expand the bed, m

l_e = the depth of the expanded bed, m

α_e = the expanded porosity

ρ_s = density of the medium, kg/m³

ρ_w = density of water, kg/m³

Because the individual particles are kept in suspension by the drag force exerted by the rising fluid it can be shown from settling theory (see

Section 12.7) that

$$C_D A_p \rho_w \frac{v^2}{2} \phi(\alpha_e) = (\rho_s - \rho_w) g V_p \tag{12.45}$$

where

$v =$ face velocity of backwash water, m/s

$\phi(\alpha_e) =$ correction factor to account for the fact that v is the velocity of the backwash water and not the particle-settling velocity v_s

and other terms are as defined previously. From experimental studies [12.18, 12.19] it has been found that

$$\phi(\alpha_e) = \left(\frac{v_s}{v}\right)^2 = \left(\frac{1}{\alpha_e}\right)^9 \tag{12.46}$$

Thus

$$\alpha_e = \left(\frac{v}{v_s}\right)^{0.22} \tag{12.47}$$

or

$$v = v_s \alpha_e^{4.5} \tag{12.48}$$

However, because the volume of the filtering medium per unit area remains constant, $(1 - \alpha)l$ must be equal to $(1 - \alpha_e)l_e$ so that

$$\frac{l_e}{l} = \frac{1 - \alpha}{1 - \alpha_e} = \frac{1 - \alpha}{1 - \left(\dfrac{v}{v_s}\right)^{0.22}} \tag{12.49}$$

Thus the required backwash velocity and expanded depth can be computed using Eqs. (12.48) and (12.49), respectively, as illustrated in Example 12.8.

<hr>

EXAMPLE 12.8

DETERMINATION OF REQUIRED BACKWASH VELOCITIES FOR FILTER CLEANING

Determine the required backwash velocity to expand a granular-medium bed to a porosity of 0.70. Also, determine the depth of the expanded filter bed. (The expanded depth needs to be known to establish the minimum height of the wash-water troughs above the surface of the filter bed.) Assume the following data are applicable:

Granular medium = sand

Size of sand = 0.5 mm

Density of sand = 2600 kg/m^3

Depth of filter bed = 0.6 m

Temperature = 20°C

SOLUTION:

1. In Example 12.4 the settling velocity for a sand particle with a diameter of 5×10^{-4} m was found to be 0.083 m/s.

2. Determine the backwash velocity using Eq. (12.48).

$$v = v_s \alpha_e^{4.5}$$
$$= 0.083 \text{ m/s } (0.7)^{4.5}$$
$$= 0.0167 \text{ m/s}$$

3. Determine the expanded bed depth using Eq. (12.49).

$$\frac{l_e}{l} = \frac{1 - \alpha}{1 - (v/v_s)^{0.22}}$$

$$l_e = \frac{0.6 \text{ m } (1 - 0.4)}{1 - (0.0167/0.083)^{0.22}}$$

$$= 1.21 \text{ m}$$

COMMENT

In practice, the bottom of the backwash-water troughs are set from 25 to 100 mm above the expanded filter bed. The width and depth of the troughs should be sufficient to handle the volume of backwash water used to clean the bed with a minimum freeboard of 50 mm at the upper end of the trough [12.12].

12.9
ADSORPTION

Adsorption is the accumulation of a material at the interface between two phases. Most solutes in water accumulate at interfaces, and adsorption processes take advantage of this phenomenon to remove material from the liquid phase. A number of adsorbents are used in industry, but only one, activated carbon, is both an inexpensive and a nonpolar adsorber. Polar adsorbents attract water and therefore would not be very useful. Activated carbon is made through a two-step process. First an organic material (coal, bone, hardwood, or nut shells) is pyrolyzed (heated under oxygen-limited conditions). A residue results that is made up of carbon and hydrocarbons. Further oxidation of the hydrocarbons with steam and/or air produces a highly porous material with an activated surface. Pores may be as small as 10 Å, total surface area is usually between 500 and 1500 m^2/g, and dry density is about 500 kg/m^3.

Commercially available activated carbons are made from different materials and use variations of the basic activation process. For these reasons adsorption characteristics vary considerably with brand and even between manufacturing lots. Activated carbon can be purchased as a powder or in granular form. Virtually any desired sieve-size distribution can be obtained. Comparisons between carbons are usually made through use of such indexes as the iodine or molasses numbers. Iodine penetrates pores down to 10 Å, and molasses, a mixture of sugars and other organic compounds, penetrates pores down to about 30 Å. The respective numbers are estimates of the available surface area in pores greater than 10 and 30 Å.

Adsorption Models

Adsorption models used in water and wastewater treatment are based on equilibrium conditions, even though equilibrium is rarely achieved in practice [12.58]. As discussed in Chapter 7, the most commonly encountered models are the Freundlich [12.20] and Langmuir [12.31] adsorption isotherms. Discrimination between adsorption models is made by fitting the data to the expressions, as illustrated in Example 12.9.

EXAMPLE 12.9

FITTING ADSORPTION MODELS

The organic carbon adsorption data given below were obtained using two activated carbon samples, A and B. In each case 1.0 g of carbon and a liquid volume of 1.0 L were used. Determine the appropriate model to be used in each case.

INITIAL CONC., g/m^3	EQUILIBRIUM CONC., g/m^3	
	A	B
1.0	0.010	0.053
2.0	0.020	0.124
4.0	0.040	0.290
8.0	0.081	0.677
16.0	0.165	1.58
32.0	0.344	3.67
64.0	0.751	8.52
128.0	1.843	19.66

SOLUTION:

1. From Chapter 7, the linearized forms of the Freundlich and Langmuir adsorption isotherms are given below.
   a. Freundlich:

$$\log\left(\frac{x}{m}\right) = \log K_F + \frac{1}{n}\log C_e$$

b. Langmuir:

$$\frac{C_e}{(x/m)} = \frac{1}{ab} + \frac{C_e}{a}$$

2. Determine x/m values for each experiment and plot in linearized forms.
   a. For carbon A:

TOC, g/m³		x/m,	$\dfrac{C_e}{x/m}$,		
C_i	C_e	g/g	g/m³	$\ln C_e$	$\ln(x/m)$
1.0	0.010	0.001	10.0	− 6.92	− 6.91
2.0	0.020	0.002	10.0	− 3.91	− 6.22
4.0	0.040	0.004	10.0	− 3.22	− 5.53
8.0	0.081	0.008	10.1	− 2.51	− 4.84
16.0	0.165	0.016	10.3	− 1.80	− 4.15
32.0	0.344	0.032	10.8	− 1.07	− 3.45
64.0	0.751	0.063	11.9	− 0.29	− 2.76
128.0	1.843	0.127	14.5	− 0.61	− 2.07

b. For carbon B:

TOC, g/m³		x/m,	$\dfrac{C_e}{x/m}$,		
C_i	C_e	g/g	g/m³	$\ln C_e$	$\ln(x/m)$
1.0	0.053	0.0009	56.0	− 2.94	− 6.96
2.0	0.124	0.0019	65.2	− 2.09	− 6.28
4.0	0.290	0.0037	78.4	− 1.24	− 5.60
8.0	0.677	0.0073	92.7	− 0.39	− 4.92
16.0	1.58	0.0144	109.7	0.46	− 4.24
32.0	3.67	0.0283	129.7	1.30	− 3.56
64.0	8.52	0.0555	153.5	2.14	− 2.89
128.0	19.66	0.1080	182.0	2.98	− 2.22

3. The linearized isotherm plots are shown in Fig. 12.37. Carbon A is best described by a Langmuir isotherm and carbon B by a Freundlich isotherm.

4. The empirical coefficients are evaluated as follows:
   a. Langmuir coefficients for carbon A:

   $$\frac{1}{a} = \frac{5}{2}$$

   $$a = 0.4$$

   $$\frac{1}{ab} = 10$$

   $$b = 0.25 \text{ g/m}^3$$

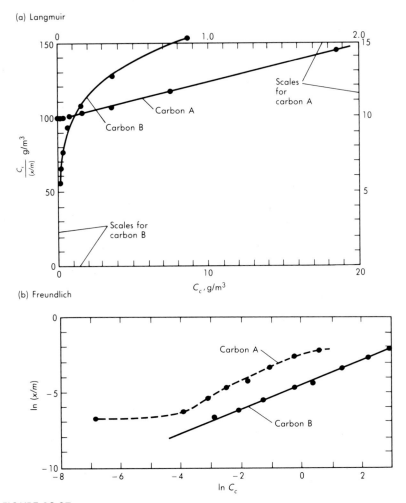

FIGURE 12.37

Linearized isotherms for Example 12.9: (a) Langmuir and (b) Freundlich.

b. Freundlich coefficients for carbon B:

$$\frac{1}{n} = \frac{4}{5}$$

$$n = 1.25$$

$$\ln K_F = -4.6, \; K_F = 0.01$$

Regeneration of activated carbon can be accomplished when the equilibrium concentration reaches a predetermined limit. Chemical or thermal methods can be used, although thermal regeneration is the most common.

Process Applications

Currently both granular and powdered activated-carbon adsorption operations are used for water and wastewater treatment [12.12, 12.28, 12.58]. Powdered carbon is often used for removing tastes and odors from drinking water [12.28]. In this case, the carbon is added to the rapid-mix unit, removed on the granular filter, and usually discarded with the backwash water rather than recovered and regenerated.

In some locations, the spent carbon is regenerated on-site. Most regeneration systems incorporate a thermal step, where water is driven off and adsorbed materials are pyrolized, and an activation step using steam. Regeneration costs are about one-tenth the cost of new carbon, but 5 to 10 percent of the carbon is lost during the process.

12.10
GAS STRIPPING

The stripping of dissolved gases from water has been discussed briefly in dealing with gas transfer. Until recently the use of gas stripping was almost entirely limited to wastewater treatment because noxious gases are rarely associated with usable domestic water supplies. Ammonia, hydrogen sulfide, sulfur dioxide, and phenol are the most commonly stripped gases, with all four being of industrial interest. Stripping is now being used to remove small concentrations of volatile organic materials such as TCE and DBCP from groundwaters prior to domestic use and to clean waters contaminated by leakage from chemical storage and landfills [12.39].

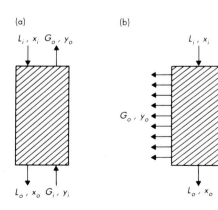

(a)   (b)

L_i, x_i G_o, y_o L_i, x_i

G_o, y_o G_i, y_i

L_o, x_o G_i, y_i L_o, x_o

FIGURE 12.38

Definition sketch for a gas stripping operation: (a) counter-current and (b) cross-current.

FIGURE 12.39

Gas stripping towers used to treat groundwater contaminated with organic compounds.

Source: Courtesy Aerojet General Corp.

The stripping operation is carried out in units similar to standard industrial cooling towers. Most systems allow the water to trickle downward over slats, rings, spheres, or corrugated surfaces. Clean air or other carrier gas is blown counter- or cross-current (Fig. 12.38). Stripping towers have porosities in the 95 percent range and surface-area-to-volume ratios of 80 to 100 m^2/m^3. In analysis of the stripping operation gas-liquid equilibrium is usually assumed as a reasonable approximation. Mass transfer rates are less often used than experience factors in choosing gas velocities and unit sizing. Typical superficial gas velocities are 1 to 2 m/s. Most stripping units are 5 to 10 m high. Units used to treat contaminated groundwater are shown in Fig. 12.39.

Gas Requirements

Gas requirements can be estimated from a simple overall material balance using Henry's law [Eq. (2.39)]. For the counter-current system of Fig. 12.38, the required materials balance is

$$L_i x_i - L_o x_o = G_o y_o - G_i y_i \qquad (12.50)$$

where

L_i = molar rate of liquid entering, mol/s

L_o = molar rate of liquid leaving, mol/s

x_i = mole fraction of dissolved gas entering in the liquid phase

x_o = mole fraction of dissolved gas leaving in the liquid phase

G_i = molar rate of clean gas entering, mol/s

G_o = molar rate of gas leaving, mol/s

y_i = mole fraction of dissolved gas entering in the gas phase

y_o = mole fraction of dissolved gas leaving in the gas phase

TABLE 12.8

Henry's Law Constants for NH_3 and H_2S in Water

T, °C	K_H, atm^{-1}	
	NH_3	H_2S
0	2.08	0.003731
10	1.28	0.002725
20	0.80	0.002070
25	0.64	0.001830
30	0.51	0.001642
40	0.34	0.001342

Source: From Ref. [12.40].

TABLE 12.9

Equilibrium Constants for NH_3 and H_2S

T, °C	$K_{NH_3} \times 10^5$, mol/L	$K_{H_2S} \times 10^7$, mol/L
0	1.374	0.262
5	1.479	
10	1.570	0.485
15	1.652	
20	1.710	0.862
25	1.774	
30	1.820	1.48
35	1.849	
40	1.862	2.44

The entering and leaving molar flow rates differ because of entrainment, evaporation, and condensation. Aerosols generated can be a significant fraction of the total flow and should be accounted for in a complete analysis. In most cases $y_i = 0$, and x_o is determined by the discharge requirements. Since the partial pressure of a gas is equal to the mole fraction at 1 atm, the value of y_o is determined using Henry's law and the x_i value. For example, Henry's law constants for NH_3 and H_2S are given in Table 12.8; values for other gases may be found in Table 2.15.

Because dissolved gases are in equilibrium with the ionized forms, pH is an important factor in treatment system performance. Equilibrium constants for NH_3 and H_2S are given in Table 12.9. Using the following relationships, the fraction of un-ionized NH_3 and H_2S can be calculated as a function of pH and temperature using the K_{NH_3} and K_{H_2S} values given in Table 12.9.

For ammonia:

$$[NH_3] = \frac{[NH_4^+][OH^-]}{K_{NH_3}} \tag{12.51}$$

$$\frac{[NH_3]}{[NH_3] + [NH_4^+]} = \frac{1}{1 + K_{NH_3}[H^+]/K_w} \tag{12.52}$$

For hydrogen sulfide:

$$[H_2S] = \frac{[HS^-][H^+]}{K_{H_2S}} \tag{12.53}$$

$$\frac{[H_2S]}{[H_2S] + [HS^-]} = \frac{1}{1 + K_{H_2S}/[H^+]} \tag{12.54}$$

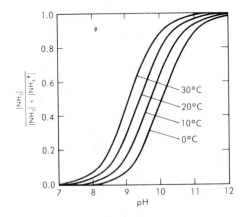

FIGURE 12.40

Fraction ammonia in NH$_3$ form as a function of temperature and pH.

Temperature Effects

Temperature effects on ionization for NH$_3$ and H$_2$S are illustrated in Figs. 12.40 and 12.41, respectively. Note that at a given pH value, the effect of temperature is not the same for the two systems. Because low temperatures result in less ionization of H$_2$S, while K_H increases with temperature, the stripping of H$_2$S is complex. Steam is usually used to enhance mass transfer and to "trap" the transferred H$_2$S in the ionized form. Stripping of H$_2$S from water requires considerable energy input but results in a concentrated condensate stream that is more desirable than an atmospheric discharge.

Process Application

To illustrate the application of the gas stripping process, the removal of ammonia and of volatile organic compounds is considered.

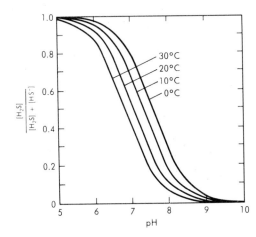

FIGURE 12.41

Fraction hydrogen sulfide in H$_2$S form as a function of temperature and pH.

Removal of Ammonia

Because of the equilibrium relationship that exists between the ammonium ion (NH_4^+) and gaseous ammonia (NH_3), ammonia stripping must be done at a relatively high pH (a value 10.5 is commonly used). Few wastewaters are naturally above pH 9, and pH adjustment is therefore necessary. Lime (CaO) is the least expensive base available for raising the pH, but a secondary problem of sludge disposal results from its use. Calcium carbonate (and phosphate) sludge also accumulates on stripping surfaces and must be cleaned off periodically. Accumulation of this scale on redwood slats at the South Lake Tahoe advanced wastewater treatment plant was so severe that use of the process was discontinued. The gas flow required for the removal of ammonia is estimated in Example 12.10.

EXAMPLE 12.10

AIR REQUIREMENTS FOR AMMONIA STRIPPING

A municipal wastewater contains 30 g/m^3 ammonia nitrogen, and a 5 g/m^3 discharge requirement must be met. Determine the gas flow rate required for a wastewater flow rate of 0.5 m^3/s and temperatures varying from 0 to 30°C.

SOLUTION:

1. Determine the operating pH.
   a. Required fraction of $NH_3 - N$ to be removed $= \frac{25}{30} = 0.83$.
   b. From Fig. 12.40, to obtain the required removal at 0°C the pH will need to be approximately 11.
2. Estimate removal as a function of temperature (at pH $= 11$).

T, °C	$\dfrac{[NH_3]}{[NH_3]+[NH_4^+]}$	$(NH_3 - N)_i$		$(NH_3 - N)_o$	
		g/m^3	mol/m^3	g/m^3	mol/m^3
0	0.89	26.70	1.91	4.45	0.32
10	0.95	28.50	2.04	4.75	0.34
20	0.98	29.40	2.10	4.90	0.35
30	0.99	29.70	2.10	4.95	0.35

3. Determine gas flow rate as function of temperature using Eq. (12.50).
   a. If it is assumed that $L_i = L_o$ and $G_i = G_o$, then Eq. (12.50) can be written as follows:

$$G = \frac{L(x_i - x_o)}{(y_o - y_i)}$$

where

L = molar rate of fluid flow, mol/s

$= (0.5 \text{ m}^3/\text{s})(1/18 \text{ g/mol})(10^6 \text{ g/m}^3)$

$= 27{,}778 \text{ mol/s}$

$$x_i \simeq \frac{[NH_3 - N]_i}{55.6 \text{ mol } H_2O/L} \text{ (liquid phase)}$$

$$x_o \simeq \frac{[NH_3 - N]}{55.6} \text{ (liquid phase)}$$

$$y_o = P_{NH_3} = \frac{x_i}{K_H}$$

Note: The partial pressure of gas is equal to the mole fraction of the gas at 1 atm.

$y_i = 0$ (clean air assumed)

b. Substitute the above values and solve for the gas flow rate G.

$$G = L\left(\frac{x_i - x_o}{y_o}\right) = L\left(\frac{[NH_3 - N_i]_i - [NH_3 - N]_o}{[NH_3 - N]_i}\right) K_H$$

$Q_g = (G \text{ mol})(\text{gas molar volume, m}^3/\text{mol})$

T, °C	G, mol/s	Q_g, m³/s
0	48,098	1077*
10	26,629	664
20	18,518	415
30	11,805	264

*$Q_g = 48{,}098 \text{ mol/s } (22.4 \times 10^{-3} \text{ m}^3/\text{mol})$

Removal of Volatile Organic Compounds

In Chapter 10 it was noted that gas stripping is used to remove volatile organic compounds from contaminated groundwaters. The removal of trace organic compounds by gas stripping depends on their tendency to volatilize as defined by Henry's law. In general, trace organic compounds are removed in packed towers (see Fig. 12.39). The off gas from the packed bed must also be treated to remove the contaminants for effective disposal, usually by some form of incineration or pyrolysis or by disposal in a hazardous-waste disposal site [12.39]. The air requirements for the removal of a volatile organic compound, TCE, found in a number of groundwaters is illustrated in Example 12.11.

EXAMPLE 12.11

AIR REQUIREMENTS FOR TRICHLOROETHYLENE (TCE) STRIPPING

A groundwater contaminanted with 10 g/m^3 TCE (molecular mass = 132) is to be treated by gas stripping to produce a product having no more than 0.5 g/m^3 TCE. Determine the minimum air flow needed to treat a water flow rate of 0.5 m^3/s. The Henry's law coefficient K_H is 1.65×10^{-3} atm^{-1} under the expected conditions of operation.

SOLUTION:

1. Use the modified form of Eq. (12.50) by assuming that $L_i = L_o$ and $G_i = G_o$. Thus

$$G = \frac{L(x_i - x_o)}{(y_o - y_i)}$$

2. Determine the entrance and exit liquid and gas phase TCE mole fractions using Eq. (2.11).

$$x_i \approx \frac{(10 \text{ g/m}^3)/(132 \text{ g/mol})}{55{,}600 \text{ mol/m}^3}$$

$$\approx 1.36 \times 10^{-6}$$

$$x_o \approx \frac{(0.5 \text{ g/m}^3)/(132 \text{ g/mol})}{55{,}600 \text{ mol/m}^3}$$

$$\approx 6.81 \times 10^{-8}$$

$y_i = 0$ (clean air is assumed)

$$y_o = \frac{x_i}{K_H} = \frac{1.36 \times 10^{-6}}{1.65 \times 10^{-3}}$$

$$= 8.24 \times 10^{-4}$$

3. Determine the minimum gas flow rate.

$$L = (0.5 \text{ m}^3/\text{s})(55{,}600 \text{ mol/m}^3)$$

$$= 27{,}800 \text{ mol/s}$$

$$G = \frac{L(x_i - x_o)}{y_o - y_i}$$

$$= \frac{(27{,}800)(1.36 \times 10^{-6} - 0.07 \times 10^{-6})}{8.24 \times 10^{-4}}$$

$$= 43.9 \text{ mol/s}$$

At standard temperature and pressure 1 mol of gas occupies 22.4 L or 0.0224 m^3. The volume varies linearly with temperature. Assuming that standard conditions exist, the minimum gas flow rate is

$$Q_g = (43.9 \text{ mols/s})(0.0224 \text{ m}^3/\text{mol})$$
$$= 0.98 \text{ m}^3/\text{s}$$

COMMENT

The minimum gas flow rate determined in this example is based on equilibrium conditions being attained. This is an impossibility and therefore a higher value ($Q_g > 0.98$ m^3/s) must be used. The higher the value of Q_g, the smaller the column required. However, limitations exist because of aerosol entrainment.

12.11
MEMBRANE PROCESSES

Dissolved solids can be removed from water and wastewater through the use of semipermeable membranes having pore diameters as small as 3 Å (one angstrom $= 10^{-10}$ m). When the separation is effected by water passing through the membrane, the process is termed *osmosis* or *hyperfiltration*. The reverse process, passage of solute molecules and ions through the membrane, is termed *dialysis*. Driving forces can be physical (pressure), chemical (concentration), thermal (temperature), or electrical (charge). The basic concepts of all membrane processes are similar, and only the case of a physical driving force, reverse osmosis, will be discussed here. Applications of membrane processes include desalinization of water for domestic and industrial use; treatment of industrial wastes; and recovery of economically important materials, such as precious metals, from waste streams [12.13, 12.30, 12.37]. An increasingly important application, medically and economically, is blood dialysis of individuals suffering temporary or permanent kidney failure.

Membrane Structure

Early membranes were made of cellulose acetate, but current materials are usually aromatic polyamides. In either case the membrane consists of a thin skin having a thickness of about 0.25 μm, supported by a more porous substructure about 100 μm thick. Separation of large molecules (such as organics) is accomplished at the skin, and the principal removal mechanism is straining, as shown in Fig. 12.42. Smaller ions, such as Na$^+$ and Cl$^-$, are excluded by adsorption of water on the skin surface (Fig. 12.43). The adsorbed layer is about one molecule in thickness (6 to 7 Å), and if the pore size is less than two water molecules in diameter

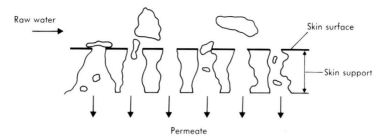

FIGURE 12.42

Removal of large molecules by sieve mechanism.

($\simeq 13$ Å), salt ions will be almost entirely excluded. Cellulose acetate membranes are particularly effective in removing Na^+ and Cl^-, with efficiencies above 95 percent possible.

Reverse Osmosis

When two solutions having different solute concentrations are separated by a semipermeable membrane, a difference in chemical potential will exist across the membrane (Fig. 12.44). Water will tend to diffuse through the membrane from the lower-concentration (higher-potential) side to the higher-concentration (lower-potential) side. In a system having a finite volume, flow continues until the pressure difference balances the chemical potential difference. This balancing pressure difference is termed the *osmotic pressure* and is a function of the solute characteristics and concentration and temperature.

If a pressure gradient opposite in direction and greater than the osmotic pressure is imposed across the membrane, flow from the more concentrated to the less concentrated region will occur and is termed *reverse osmosis* (Fig. 12.44c).

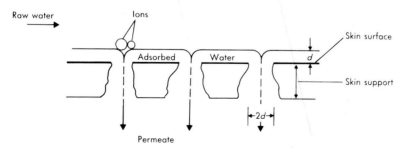

FIGURE 12.43

Rejection of small-diameter ions by adsorbed water layer.

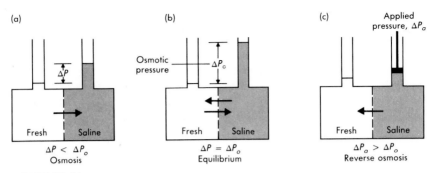

FIGURE 12.44

Conceptual model of osmotic flow: (a) osmotic flow, (b) osmotic equilibrium, and (c) reverse osmosis.

Flux of water F_w through the membrane is a function of the pressure gradient:

$$F_w = w(\Delta P_a - \Delta P_o) \tag{12.55}$$

where

$\quad F_w$ = flux of water, $\text{kg/m}^2 \cdot \text{s}$

$\quad w$ = flux rate coefficient involving temperature, membrane characteristics, and solute characteristics, s/m

$\quad \Delta P_a$ = imposed pressure gradient, kPa

$\quad \Delta P_o$ = osmotic pressure gradient, kPa

Some solute passes through the membrane in all cases. Solute flux is adequately described by an expression of the form

$$N_i = K_i \Delta C_i \tag{12.56}$$

where

$\quad N_i$ = flux of species i, $\text{kg/m}^2 \cdot \text{s}$

$\quad K_i$ = overall mass transport coefficient, m/s

$\quad \Delta C_i$ = salt concentration difference between the feed C_f and the permeate C_p, kg/m^3

Solute rejection ζ_i is defined as the efficiency of rejection from the permeate of a specific solute i:

$$\zeta_i = \frac{C_{i_f} - C_{i_p}}{C_{i_f}} \tag{12.57}$$

where C_{i_f} and C_{i_p} are solute concentrations in the feed and permeate, respectively.

Hollow-fiber polyamide membranes (Fig. 12.45) are used in most industrial waste treatment applications, while cellulose acetate remains the principal membrane for desalinization (Fig. 12.46). Applied pressures

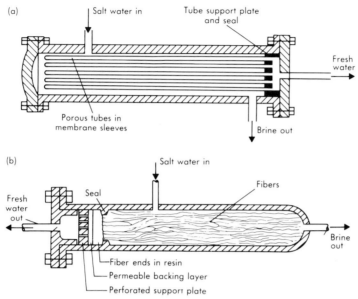

FIGURE 12.45

Membranes used for desalination: (a) membranes mounted on porous rods and (b) polyamide fibers mounted in resin.

required are dependent on the osmotic pressure and therefore on the solute concentration, but typical values are in the 2500- to 8000-kPa range. Water fluxes are typically 5 to 10×10^{-3} $m^3/m^2 \cdot d$ for hollow fibers and 0.5 to 1 $m^3/m^2 \cdot d$ for cellulose acetate sheets. Hollow fibers are considerably more space-efficient, and the output difference on a space-required basis is only one order of magnitude.

FIGURE 12.46

Reverse osmosis tubes being installed in a treatment facility for the purpose of desalinating treated wastewater effluent.

EXAMPLE 12.12

DETERMINATION OF MEMBRANE AREA REQUIRED FOR DESALINIZATION

A brackish water having a TDS concentration of 2500 g/m^3 is to be desalinized using cellulose acetate membranes having a flux rate coefficient w of 1.8×10^{-6} s/m and a mass transfer rate coefficient K_i of 1.2×10^{-6} m/s. The product water is to have a TDS of no more than 500 g/m^3. The flow rate is to be 0.010 m^3/s. The net pressure $(\Delta P_a - \Delta P_o)$ will be 4000 kPa.

SOLUTION:

1. The problem involves determination of the membrane area required to produce 0.010 m^3/s of water and the TDS concentration of the permeate. If the permeate TDS concentration is well below 500 g/m^3, blending of feed and permeate will reduce the membrane area required.

2. Estimate membrane area using Eq. (12.55).

   $$F_w = w(\Delta P_a - \Delta P_o)$$
   $$= (1.8 \times 10^{-6} \text{ s/m})(4000 \text{ kg/m} \cdot \text{s}^2) = 7.2 \times 10^{-3} \text{ kg/m}^2 \cdot \text{s}$$

   $$Q = F_w \left(\frac{1 \text{ m}^3}{1000 \text{ kg}} \right)(A)$$

   $$A = \frac{(0.01)(1000)}{7.2 \times 10^{-3}} = 1389 \text{ m}^2$$

3. Estimate permeate TDS concentration using Eq. (12.56).

   $$N_i = K_i \Delta C_i$$
   $$QC_p = N_i A = K_i(C_f - C_p)A$$
   $$C_p = \frac{K_i A C_f}{Q + K_i A}$$
   $$= \frac{(1.2 \times 10^{-6})(1389)(2.5)}{0.01 + 1.2 \times 10^{-6}(1389)}$$
   $$= 0.357 \text{ kg/m}^3$$

   The permeate concentration is lower than necessary. Reducing the area and blending is possible.

4. Estimate the decreased area and the blended flows using $C_p = 0.357$ kg/m^3 as an estimate of the permeate TDS.

   $$(Q_p + Q_f)(0.500 \text{ kg/m}^3) = Q_p(0.357 \text{ kg/m}^3) + Q_f(2.500 \text{ kg/m}^3)$$

   also $(Q_p + Q_f) = 0.010$ m^3/s

   Solving for Q_p and Q_f yields:

   $$Q_p = 0.0093 \text{ m}^3/\text{s and } Q_f = 0.0007 \text{ m}^3/\text{s}$$

a. The required area is

$$A = \frac{1000\, Q_p}{F_w}$$

$$= \frac{(1000\,\text{kg/m}^3)(0.0093\,\text{m}^3/\text{s})}{7.2 \times 10^{-3}\,\text{kg/m}^2 \cdot \text{s}}$$

$$= 1292\,\text{m}^2$$

b. Blend 0.0093 m³/s of permeate with 0.0007 m³/s of feed.

12.12
SOLIDS PROCESSING

In Section 12.1 of this chapter, it was noted that a product (sludge) is formed from the application of most physical treatment methods. Because of the importance of sludge disposal it is the purpose of this section to introduce the reader to the methods used for the processing of sludge.

Types and Sources of Solids and Sludge

The principal types and sources of solids and sludge from conventional water and wastewater treatment facilities are summarized in Table 12.10. As noted there, the end products from both water and wastewater treatment plants are dewatered solids, sludge, and other assorted residues. The ultimate disposal of these materials is typically by land spreading or landfilling [12.8, 12.35, 12.54].

Quantities of Sludge

The quantities of solids and sludge that must be processed and ultimately disposed of will depend on the characteristics of the water or wastewater to be treated, the required effluent quality, and the treatment methods that are used. For example, 60 to 70 percent of the incoming suspended solids in wastewater will be removed in a primary settling tank, and the solids content will vary from 4 to 8 percent. Representative data on the sludge volumes that can be expected are reported in Table 12.11.

If the solids content and the specific gravity are known, the volume of the sludge can be estimated as follows:

$$V_s = \frac{M_s}{S_s(S/100)\rho_w} \tag{12.58}$$

TABLE 12.10

**Types and Sources of Solids and Sludge from
Water and Wastewater Treatment Plants**

TYPE	SOURCE(s)
Water treatment	
Coarse solids, including wood and twigs	Screening facilities
Sand and silt	Presettling facilities, if used
Chemical sludge	Sedimentation tanks following coagulation/flocculation facilities
Backwash solids	Granular-medium filters
Dewatered solids, sludge, other residues	Dewatering and other sludge processing facilities
Wastewater treatment	
Coarse solids, including rags and wood	Screening facilities
Grit	Grit-removal facilities
Scum	Preaeration and primary sedimentation tanks (In some plants, grease-skimming tanks are used.)
Primary sludge	Primary settling tanks
Biological sludge	Secondary settling tanks
Backwash solids	Effluent granular-medium filters
Digested sludge	Aerobic and anaerobic digesters
Dewatered solids, sludge, other residues	Dewatering and other sludge processing facilities

where

V_s = volume, m³

M_s = mass of dry sludge solids, kg

S_s = specific gravity of sludge solids

S = solids content of sludge, %

ρ_w = density of water, 10^3 kg/m³ at 5°C

The specific gravity of the sludge solids composed of fixed and volatile solids can be determined using Eq. (12.59):

$$\frac{M_s}{S_s} = \frac{M_f}{S_f} + \frac{M_v}{S_v} \qquad (12.59)$$

where

M_f = mass of fixed solids, kg

S_f = specific gravity of fixed solids

M_v = mass of volatile solids, kg

S_v = specific gravity of volatile solids

TABLE 12.11

Sludge Volumes Resulting from Water and Wastewater Treatment

TREATMENT PROCESS	CONCENTRATION, kg/m³ dry solids		UNIT VOLUME, m³/10³ m³ flow	
	Range	Typical	Range	Typical
Water treatment				
Sedimentation	1–15	10	0.02–0.07	0.05
Filter backwash	0.001–0.003	0.002	20–50	30
Wastewater treatment				
Primary sludge	40–120	50	0.9–4.5	3
Primary and waste activated sludge	30–100	40	1–5.5	3.8
Primary and trickling filter humus	40–100	50	1–4.5	3
Primary with lime for phosphorus control	20–80	40	1.4–6	3.8
Secondary sludge and waste activated sludge with primary settling	5–15	7.5	4.5–20	5.7
Waste activated sludge without primary settling	7.5–25	12.5	2.8–13	6.8
Trickling filter humus	10–30	15	1.8–9	4.7
Flotation thickening (activated sludge)	30–60	40	1–3.3	1
Anaerobic digestion*				
Primary only	50–100	70	1–3.6	2.1
Primary and activated sludge	25–70	35	1.4–6.6	4.3
Primary and trickling filter	30–80	40	1.3–5.6	3.8

*Total in systems having digesters.

and other terms are as defined previously. Equation (12.59) can also be used to estimate the specific gravity of wet sludge by rewriting it as

$$\frac{M_{ws}}{S_{ws}} = \frac{M_s}{S_s} + \frac{M_w}{S_w} \qquad (12.60)$$

where

M_{ws} = mass of wet sludge, kg

S_{ws} = specific gravity of wet sludge

M_w = mass of water, kg

S_w = specific gravity of water

and other terms are as defined previously. The use of Eqs. (12.58), (12.59), and (12.60) is illustrated in Example 12.13.

EXAMPLE 12.13

ESTIMATION OF SLUDGE VOLUME

Wastewater entering a treatment plant contains 250 mg/L suspended solids. If 60 percent of these solids is removed during primary sedimentation, find the volume of primary sludge produced per thousand cubic meters of wastewater. Assume 40 percent of the solid matter in the sludge containing 92 percent water is composed of fixed mineral solids with a specific gravity of 2.5 and 60 percent is composed of volatile solids with a specific gravity of 1.0.

SOLUTION:

1. Determine the specific gravity of the primary sludge solids using Eq. (12.59). Assume M_s is equal to 1 kg for the purpose of computation.

$$\frac{1}{S_s} = \frac{0.4}{2.5} + \frac{0.6}{1.0}$$

$$S_s = \frac{1}{0.76} = 1.32$$

2. Determine the specific gravity of the wet sludge using Eq. (12.60). Assume M_{ws} is equal to 1 kg for the purpose of computation.

$$\frac{1}{S_{ws}} = \frac{0.08}{1.32} + \frac{0.92}{1}$$

$$S_{ws} = \frac{1}{0.98} = 1.02$$

3. Determine the volume of sludge removed per thousand cubic meters of wastewater treated noting that 60 percent of the influent solids are removed.

   a. The mass of dry solids removed per thousand cubic meters is

   $$M_s = 0.6 \times (250 \text{ g/m}^3) \times 10^3 \text{ m}^3 / (10^3 \text{ g/kg})$$
   $$= 150 \text{ kg}$$

   b. The volume of sludge per thousand cubic meters, determined using Eq. (12.58), is

   $$V_s = \frac{150 \text{ kg}}{1.02(8/100)10^3 \text{ kg/m}^3}$$
   $$= 1.8 \text{ m}^3$$

Sludge Processing Methods

The processing and disposal of solids and sludge resulting from the treatment of water and wastewater are some of the most difficult aspects of wastewater management because (1) the sludge contains much of the

TABLE 12.12

Methods used for Processing Water and Wastewater Treatment Plant Sludges

METHOD	APPLICATION	REPRESENTATIVE UNIT OPERATION, UNIT PROCESS, OR TREATMENT METHOD
Water treatment		
Thickening	Used to reduce the volume of water that must be removed in subsequent processing steps	Gravity thickening, centrifugation, heating, freezing, pressing, or vacuum filtration
Dewatering	Used to thicken sludge; lagoons used to store sludge	Drying beds, lagoons
Coagulant recovery	Used where the recovery of chemicals may be feasible	
Discharge to sewers	Used, where possible, as a simple and satisfactory method of sludge disposal; may upset operation of wastewater treatment facilities	
Wastewater treatment		
Preliminary operations	Used to improve treatability of the sludge, to reduce wear of pumps, or to reduce required capacity of other processing facilities	Sludge grinding, degritting, blending, and storage
Thickening	Used to reduce volume of sludge	Gravity thickening, flotation thickening, centrifugation
Stabilization	Used to reduce organic content of sludge (digestion) or to alter its characteristics so it will not cause nuisance conditions (oxidation and lime treatment)	Chlorine oxidation, lime stabilization, heat treatment, anaerobic digestion, aerobic digestion, composting
Conditioning	Used to make sludge more manageable in subsequent processing steps or to reduce chemical requirements	Chemical conditioning, elutriation, heat treatment
Dewatering	Used to reduce volume of sludge	Vacuum filtration, filter pressing, horizontal-belt filtration, centrifugation, drying (bed), lagooning
Drying	Used to further reduce volume of sludge by evaporating water	Flash drying, spray drying, rotary drying, multiple hearth drying
Disinfection	Used to control bacterial activity of microorganisms in sludge	Disinfection
Thermal reduction	Used to reduce volume of pre-thickened sludge by means of thermal reduction	Multiple hearth incineration, fluidized-bed incineration, flash combustion, coincineration with solid wastes, copyrolysis with solid wastes, wet-air oxidation

material that was offensive in the incoming liquid; (2) in wastewater treatment, waste biological sludge is organic and will decay; and (3) only a small part of the sludge is solid matter.

The principal methods now used for the processing of sludge are summarized in Table 12.12. The unit operations and chemical and biological processes used to accomplish each of these methods are also identified. Clearly an endless number of combinations is possible. Descriptions of each of these operations and processes is beyond the scope of this chapter; however, details of sludge processing may be found in Refs. [12.8], [12.15], [12.35], [12.47], and [12.54].

KEY IDEAS, CONCEPTS, AND ISSUES

- Commonly used physical treatment operations include screening, comminution, aeration, mixing, flocculation, sedimentation, filtration, adsorption, and air stripping.
- Flocculation efficiency is a function of the mean velocity gradient.
- Discrete-particle sedimentation involves particles falling freely without interaction or flocculation.
- Flocculant settling is more efficient than discrete-particle settling because larger particles have less drag and settle faster.
- Particle size distribution changes in sedimentation.
- The design of granular-medium filters is based on experience and/or on pilot-plant data.
- Adsorption on activated carbon is an effective method of removing volatile organics responsible for tastes and odors in water.
- Adsorption of dissolved organic materials from treated wastewater is an effective method of removing toxic materials and meeting stringent discharge requirements.
- Gas stripping is an effective and economical method of removing volatile trace organic materials from water.
- Disposal of the by-products produced by physical treatment operations is usually difficult and expensive.

DISCUSSION TOPICS AND PROBLEMS

12.1. The following data were obtained when a surface aerator was tested in tap water at 18°C. Using these data determine the $K_L a$ value at 10°C. Assume $\theta = 1.02$.

TIME, min	CONCENTRATION, g/m^3
0	0
15	2.45
30	4.55
45	5.95
60	6.85

12.2. If the α and β values for a surface aerator in a given application are 0.9 and 0.95, respectively, determine the field transfer rate at $23°C$ if the standard oxygen transfer rate (SOTR) is equal to 2.2 kg $O_2/kW \cdot h$. Use C_{O_2} values of 1, 2, and 4 g/m^3.

12.3. If it is assumed that the energy input needed to maintain completely mixed conditions in a tank with a diameter of 20 m and a depth of 5 m is 30 $kW/10^3$ m^3, estimate the rotational speed of a typical six-flat-blade turbine mixer with a diameter of 2 m.

12.4. Design a rapid mixer for a water treatment plant with a peak flow rate of 20,000 m^3/d. Assume the velocity gradient G in the mixer is to be 1,000 s^{-1} and the detention time is 30 s. Determine the size and configuration of the mixing tank, the power input, and the rotational speed of the mixer if the value of the power function ϕ is 6.0 in the turbulent range.

12.5. A water treatment plant for a community of 10,000 is designed for a constant flow rate of 4000 m^3/d. The flocculation system will consist of three tanks, each having an average hydraulic residence time of no greater than 45 min. It is desired to be able to take one tank out of service for maintenance without exceeding an average velocity gradient $\overline{dv/dy}$ of 50 s^{-1} in the remaining two units. Determine the motor output power, assuming that losses are 50 percent.

12.6. The data below were obtained in pilot-scale batch flocculation and sedimentation experiments. In each experiment the flocculation device was run for 15 min at a selected speed and then turned off. After 30 min, the turbidity of the clarified supernatant was measured. Plot

p, W	TURBIDITY, NTU
1	40
2	30
4	25
8	20
16	28
32	68
64	87

the mean velocity gradient versus turbidity and explain the shape of the curve. The pilot-plant volume was 1.0 m³, and the experimental temperature was 20°C. Turbidity of the original sample was 100 NTU.

12.7. Laboratory studies on flocculation have been conducted by subjecting destabilized colloid suspensions to controlled shear stresses for selected periods of time, then measuring the average particle size using photographic techniques. Data from a set of experiments made use one time period are below. Are the results reasonable? Explain your answer.

G, s^{-1}	MEAN PARTICLE DIAMETER, mm
10	0.02
20	0.04
30	0.09
40	0.18
50	0.32
60	0.30
70	0.20
80	0.08
90	0.04
100	0.04

12.8. A solids removal system is to be designed to remove particles from a water stream. It has been determined that all particles larger than 0.7 mm in diameter must be removed. If the particle density is 999 kg/m³, determine (a) an appropriate overflow rate and (b) an appropriate process configuration.

12.9. For quiescent conditions determine the settling velocity of spheres in water as a function of particle diameter. Use Eq. (12.27) or reasonable approximations. Assume the particle density is 1050 kg/m³ and use a particle diameter range of 0.01 to 10 mm.

12.10. A sediment-laden water is to be treated in a sedimentation tank with the dimensions given below:

Length = 50 m
 Width = 10 m
 Depth = 5 m

Flow rate into the tank is controlled at a constant value of 2.50 m³/s. A particle size analysis has been made and the results are listed on the following page. Given that $\rho_p = 2000$ kg/m³ and $\mu = 0.001$ N·s/m², determine (a) the effluent sediment concentration and (b) the effluent particle size distribution.

$\bar{d}_p$, mm	MASS CONCENTRATION, g/m^3
0.001	100
0.005	200
0.010	300
0.015	200
0.020	100

12.11. Sedimentation tanks are designed in two general configurations, rectangular (longitudinal flow) and circular (radial flow). An expression for ideal sedimentation in rectangular tanks was derived in Section 12.7 that related the overflow rate to the critical settling velocity. Derive the analogous relationship for circular tanks, beginning with consideration of the horizontal velocity as a function of radial distance.

12.12. Discuss the effect of the angle of tilt θ on the effectiveness of a tube settler (see Fig. 12.21). Would an optimum angle exist?

12.13. Rapid sand filters are normally operated on a constant head-loss basis that results in a large variation in flow rate. How is this variation matched to variation in community water demand?

12.14. Determine the filter area and clean-water storage volume required to meet the demands of the large community of Fig. 1.10. Assume that 80 percent of the treated water becomes wastewater and that at any time 10 percent of the filter area is out of service. Assume dual-media filters with a design flow rate of 0.3 $m^3/m^2 \cdot min$ will be used.

12.15. Discuss problems that would be expected in the operation of a filter treating sewage or primary effluent.

12.16. A secondary wastewater effluent having a TOC concentration of 12 g/m^3 is to be treated further by activated carbon adsorption. The objective is to produce a product having no more than 1.0 g/m^3 TOC. Determine the amount of a carbon that would be necessary per cubic meter of secondary effluent if the carbon characteristics were those of (a) carbon A in Example 12.9 and (b) carbon B in the same example. Assume the system is a batch process.

12.17. Solve Problem 12.16 using two batch processes operated in series (a two-stage process). Assume equal masses of carbon in the two units and that each stage is allowed to come to equilibrium. Explain why or why not this process is different from a single system. Would a PFR be different from a batch process? Why or why not?

12.18. A wastewater containing 50 g/m^3 $NH_3 - N$ is to be treated by stripping to produce a product with 1 g/m^3 $NH_3 - N$. For a flow rate of 1.0 m³/s, determine the appropriate operating pH and the gas flow requirements in cubic meters per second.

12.19. Blowdown water from an industrial reaction process contains 300 g/m³ SO_2. The Henry's law coefficient for SO_2 is approximately 0.021 atm^{-1} at 30°C. Determine the air flow required to produce a treated water having a SO_2 concentration of (a) 10 g/m³ and (b) 1 g/m³. Assume the water temperature is 30°C.

12.20. A saline water having a TDS value of 1000 g/m³ has an osmotic pressure of 50 kPa. A hollow-fiber reverse-osmosis unit is to be designed for a flow of 0.05 m³/s. If the operating pressure difference is to be 3000 kPa and the rate coefficient w is 3×10^{-8} s/m, determine the fiber area required.

12.21. The reverse-osmosis unit of Problem 12.20 is expected to operate with a solute rejection of 0.95. Determine the TDS concentration in the product water and brine.

12.22. Estimate the amount of primary sludge produced per day by a city of 200,000 people. Assume a settled sludge concentration of 50 kg/m³.

12.23. Determine the liquid volume before and after anaerobic digestion and the percent reduction for 10^3 kg of primary sludge with the following characteristics. Assume 60 percent of the volatile solids in the primary sludge are destroyed during the digestion process.

ITEM	PRIMARY	DIGESTED
Solids, %	5	10
Volatile matter, %	60	—
Specific gravity of fixed solids	2.5	2.5
Specific gravity of volatile solids	1.0	1.06

REFERENCES

12.1. American Society of Civil Engineers, (1983), *Development of Standard Procedures for Evaluating Oxygen Transfer Devices*, U.S. Environmental Protection Agency, EPA-600/2-83-102, Cincinnati, Ohio.

12.2. Amirtharajah, A., (1978), "Design of Flocculation Systems" in *Water Treatment Plant Design for the Practicing Engineer*, edited by R. L. Sanks, Ann Arbor Science Publishers, Ann Arbor, Mich.

12.3. ———, (1978), "Optimum Backwashing of Sand Filters," *J. Environmental Engineering Division*, ASCE, vol. 104, no. EE5, p. 917.

12.4. Barnes, D., P. J. Bliss, B. W. Gould, and H. R. Vallentine, (1981), *Water and Wastewater Engineering Systems*, Pitman Publishing, Marshfield, Mass.

12.5. Baumann, E. R., (1978), "Granular-Media Deep-Bed Filtration" in *Water Treatment Plant Design for the Practicing Engineer*, edited by R. L. Sanks, Ann Arbor Science Publishers, Ann Arbor, Mich.

12.6. _____, (1978), "Precoat Filtration" in *Water Treatment Plant Design for the Practicing Engineer*, edited by R. L. Sanks, Ann Arbor Science Publishers, Ann Arbor, Mich.

12.7. Bernado, L. D., and J. L. Cleasby, (1980), "Declining Rate Versus Constant Rate Filtration," *J. Environmental Engineering Division*, ASCE, vol. 106, no. EE6, p. 1023.

12.8. Borchardt, J. A., W. J. Redman, G. E. Jones, and R. T. Sprague (eds.), (1981), *Sludge and Its Ultimate Disposal*, Ann Arbor Science Publishers, Ann Arbor, Mich.

12.9. Camp, T. R., and P. C. Stein, (1943), "Velocity Gradients and Internal Work in Fluid Motion," *Boston Society of Civil Engineers*, vol. 30, no. 2, p. 219.

12.10. Carman, P. C., (1937), "Fluid Flow Through Granular Beds," *Transactions of Institute of Chemical Engineers*, London, vol. 15, p. 150.

12.11. Cleasby, J. L., (1972), "Filtration" in *Physicochemical Process for Water Quality Control*, edited by W. J. Weber, Jr., Wiley Interscience, New York.

12.12. Cox, C. R., (1964), *Operation and Control of Water Treatment Processes*, World Health Organization, Geneva.

12.13. Cruver, J. E., (1972), "Membrane Processes" in *Physicochemical Process for Water Quality Control*, edited by W. J. Weber, Jr., Wiley Interscience, New York.

12.14. Culp, G. L., and R. L. Culp, (1974), *New Concepts in Water Purification*, Van Nostrand Reinhold Company, New York.

12.15. Culp, R. L., G. M. Wesner, and G. L. Culp, (1978), *Handbook of Advanced Wastewater Treatment*, Van Nostrand Reinhold, New York.

12.16. Eckenfelder, W. W., Jr., (1980), *Principles of Water Quality Management*, CBI Publishing Company, Boston.

12.17. _____ and D. L. Ford, (1970), *Water Pollution Control: Experimental Procedures for Process Design*, Pemberton Press, Austin, Tex.

12.18. Fair, G. M., J. C. Geyer, and D. A. Okun, (1968), *Water and Wastewater Engineering*, John Wiley & Sons, New York.

12.19. _____ and L. P. Hatch, (1933), "Fundamental Factors Governing the Streamline Flow of Water Through Sand," *J. American Water Works Association*, vol. 25, no. 11, p. 1551.

12.20. Freundlich, H., (1926), *Colloid and Capillary Chemistry*, Methuen and Company, London.

12.21. Harris, H. S., W. E. Kaufman, and R. B. Krone, (1966), "Ortho-kinetic Flocculation in Water Purification," *J. Sanitary Engineering Division*, ASCE, vol. 92, SA6, p. 95.

12.22. Hazen, A., (1925), *The Filtration of Public Water Supplies*, 2d ed., John Wiley & Sons, New York.

12.23. Holland, F. A., and F. S. Chapman, (1966), *Liquid Mixing and Processing in Stirred Tanks*, Reinhold Publishing Corporation, New York.

12.24. Hudson, H. E., Jr., (1981), *Water Clarification Processes: Practical Design and Evaluation*, Van Nostrand Reinhold Company, New York.

12.25. Ives, K. J., (ed.), (1978), *The Scientific Basis of Flocculation*, Sijhoff and Noordhoff, 1978.

12.26. _____, (1969), "Theory of Filtration," Special Subject No. 7, International Water Supply Congress and Exhibition, London.

12.27. Iwasaki, T., (1937), "Some Notes on Sand Filtration," *J. American Water Works Association*, vol. 29, no. 10, p. 1596.

12.28. Kornegay, B. H., (1978), "Activated Carbon for Taste and Odor Control" in *Water Treatment Plant Design for the Practicing Engineer*, edited by R. L. Sanks, Ann Arbor Science Publishers, Ann Arbor, Mich.

12.29. Kozeny, J., (1927), "Uber Grundwasserbewegung," *Wasserkraft and Wasserwirtschaft*, vol. 22, nos. 4, 6, 7, 8, 10.

12.30. Lacey, R. E., (1972), "Membrane Separation Processes," *Chemical Engineering*, vol. 79, no. 19, p. 56.

12.31. Langmuir, I., (1918), "The Absorption of Gases on Plane Surfaces of Glass, Mica, and Platinum," *J. American Chemical Society*, vol. 40, no. 9, p. 1361.

12.32. Larsen, P., (1977), *On the Hydraulics of Rectangular Sedimentation Basins*, Experimental Studies Report No. 1001, Department of Water Resources, Lund Institute of Technology, University of Lund, Lund, Sweden.

12.33. McCabe, W. L., and J. C. Smith, (1976), *Unit Operations of Chemical Engineering*, 3d ed., McGraw-Hill Book Company, New York.

12.34. Matsumoto, M. R., T. M. Galeziewski, and G. Tchobanoglous, (1982), "Filtration of Primary Effluent," *J. Water Pollution Control Federation*, vol. 54, no. 12, p. 1581.

12.35. Metcalf & Eddy, Inc., (1977), *Wastewater Engineering*, 2d ed., revised by G. Tchobanoglous, McGraw-Hill Book Company, New York.

12.36. *Municipal Wastewater Treatment Plant Design*, (1977), Manual of Practice No. 8, Water Pollution Control Federation, Washington, D.C.

12.37. Nusbaum, I., and A. B. Reidinger, (1978), "Water Quality Improvement by Reverse Osmosis" in *Water Treatment Plant Design for the Practicing Engineer*, edited by R. L. Sanks, Ann Arbor Science Publishers, Ann Arbor, Mich.

12.38. Oldshue, J. Y., (1983) *Fluid Mixing Technology*, McGraw-Hill Book Company, New York.

12.39. *Organic Chemical Contaminants in Groundwater: Transport and Removal*, (1981), AWWA Seminar Proceedings Report No. 20156, AWWA, Denver.

12.40. Perry, R. H., and D. W. Green (eds), (1984), *Chemical Engineers Handbook*, 6th ed., McGraw-Hill Book Company, New York.

12.41. Reynolds, T. D., (1982), *Unit Operations and Processes in Environmental Engineering*, Brooks/Cole Engineering Division, Monterey, Calif.

12.42. Rose, H. E., (1945), "An Investigation of the Laws of Flow of Fluids Through Beds of Granular Materials," *Proc. Institute of Mechanical Engineers*, vol. 153, p. 141.

12.43. Rushton, J. H., (1952), "Mixing of Liquids in Chemical Processing," *Industrial Engineering Chemistry*, vol. 44, no. 12, p. 2931.

12.44. Schamber, D. R., and B. E. Larock, (1983), "Particle Concentration Predictions in Setting Basins," *J. Environmental Engineering Division*, ASCE, vol. 109, no. 3, p. 753.

12.45. Schemidtke, N. W., and D. W. Smith (eds), (1983), *Scale-Up of Water and Wastewater Treatment Processes*, Butterworth Publishers, Stoneham, Mass.

12.46. Schroeder, E. D., (1977), *Water and Wastewater Treatment*, McGraw-Hill Book Company, New York.

12.47. *Sludge Dewatering*, (1983), Manual of Practice 20, Water Pollution Control Federation, Washington, D.C.

12.48. Smoluchowski, M., (1918), "Versuol einer mathematischen Theorie der Koagulations kinetic Killoider Lasungen," *Z. Phys. Chem.*, vol. 92, p. 155.

12.49. Stenstrom, M. K., and R. G. Gilbert, (1981), "Affects of Alpha, Beta and Theta Factor upon the Design, Specification and Operation of Aeration Systems," *Water Research*, vol. 15, no. 6, p. 643.

12.50. Sundstrom, D. W., and H. E. Klei, (1979), *Wastewater Treatment*, Prentice-Hall, Englewood Cliffs, N.J.

12.51. Swift, D. L., and S. K. Friedlander, (1964), "The Coagulation of Hydrosols by Brownian Motion and Laminar Shear Flow," *Colloids Science*, vol. 15, p. 621.

12.52. Tchobanoglous, G., and R. Eliassen, (1970), "Filtration of Treated Sewage Effluent," *J. Sanitary Engineering Division*, ASCE, vol. 96, no. SA2 p. 243.

12.53. Tebbutt, T. H. Y., (1977), *Principles of Water Quality Control*, 2d ed., Pergamon Press, Oxford, England.

12.54. Vesilind, P. A., (1974), *Treatment and Disposal of Wastewater Sludges*, Ann Arbor Science Publishers, Ann Arbor, Mich.

12.55. Walker, D. J., (1978), "Sedimentation" in *Water Treatment Plant Design for the Practicing Engineer*, edited by R. L. Sanks, Ann Arbor Science Publishers, Ann Arbor, Mich.

12.56. *Water Quality and Treatment: A Handbook of Public Water Supplies*, (1971), 3d ed., American Public Work Association, McGraw-Hill Book Company, New York.

12.57. *Water Treatment Plant Design*, (1969), American Water Works Association, New York.

12.58. Weber, W. J., Jr., (1972), *Physicochemical Processes for Water Quality Control*, Wiley Interscience, New York.

13

Chemical Treatment Methods

With chemical treatment methods, changes in water quality are brought about through chemical reactions. Typically, chemicals must be added to the water or wastewater being treated to effect a change. Although a large number of chemical treatment processes are available, only the most commonly used in water and wastewater treatment are discussed here. The order of presentation in this chapter is based on the relative importance of the treatment method rather than on its probable placement in a process flow sheet.

13.1
APPLICATIONS OF CHEMICAL TREATMENT

The most important chemical treatment methods are those used for disinfection, precipitation of dissolved materials, coagulation (destabilization) of colloids, oxidation, and ion exchange. Disinfection is used in the treatment of both domestic water and wastewater. Industrially, disinfection is used to control biological slime buildup in piping and bacterial counts in food processing. Precipitation is used domestically and industrially for water softening and iron removal and for the removal of soluble ions such as PO_4^{-3} from wastewaters. Coagulation is used almost entirely for the destabilization of colloids found in surface waters. Chemical oxidation, which actually includes chlorination and ozonation, is used in the removal or breakdown of ions such as Fe^{+2}, Mn^{+2}, and CN^- (cyanide). Iron and manganese are much more soluble in the $+2$ state than in the $+3$ state and are easily oxidized with oxygen. Thus their removal is a combination oxidation-precipitation process. Cyanide removal involves oxidation to the innocuous end products CO_2 and N_2. The most common application of ion exchange is for the removal of hardness from residential water supplies.

13.2
DISINFECTION

Disinfection is the term applied to the selective destruction of disease-causing organisms. The complete destruction of all organisms is termed *sterilization*. Disinfection in water and wastewater treatment involves the exposure of the disease-causing organisms in water to some destructive agent. Methods and means that have been used for disinfection include (1) the addition of chemicals; (2) the application of physical agents, such as heat and light; (3) mechanical means; and (4) exposure to electromagnetic, acoustic, and particle radiation. The addition of chemical agents is the most common method used throughout the world.

The most widely used method in the United States is the application of chlorine gas (Cl_2). Unfortunately, the use of chlorine can result in production of carcinogenic compounds such as trihalomethanes and chloroform, which is a major disadvantage. In many European countries, chlorination of wastewater is not practiced because of the possible formation of chlorinated hydrocarbons. Two other halogens, bromine and iodine, are good disinfectants, but cost and difficulty of application have prevented their widespread use. Ozone is used as the primary disinfectant in some countries, but the necessity for on-site generation and the greater safety and health problems associated with ozone have limited its general use in the United States. Ozone has the disadvantage of not leaving a residual concentration to prevent reinfection of disinfected water. Thus the use of either chlorine or ozone involves trade-offs that must be considered on a case-by-case basis.

Chlorination

Chlorination is required for virtually all domestic waters supplied from surface waters, most groundwaters used as domestic supplies, recycled cooling and process waters, and many treated wastewaters. Disinfection in water and wastewater treatment typically involves the destruction of viruses, bacteria, and protozoans. Because chlorine is an oxidant, the amount required for disinfection is a function of the organic and NH_3-N concentration. As would be expected, considerably more chlorine is required for disinfection of wastewaters (40 to 60 g/m^3) than for domestic water supplies (2 to 5 g/m^3).

Chlorine is usually added in the form of Cl_2 gas, chlorine dioxide (ClO_2), sodium hypochlorite (NaOCl), and calcium hypochlorite [$Ca(OCl)_2$]. Although the use of sodium or calcium hypochlorite is more expensive than the use of chlorine gas, many cities have switched to the use of these chemicals because of the greater safety involved in their handling as compared with chlorine gas [13.7]. Sodium and calcium hypochlorite are commonly used in small treatment plants.

Chlorine Chemistry

When chlorine in the form of the gas Cl_2 is added to water, two reactions occur—hydrolysis and ionization. In the hydrolysis reaction, hypochlorite (HOCl) is formed, and the chlorite ion (OCl^-) is formed in the ionization reaction. The two reactions and the corresponding equilibrium expressions are given below.

Hydrolysis reaction:

$$Cl_2 + H_2O \rightleftarrows HOCl + H^+ + Cl^- \qquad (13.1)$$

$$K_1 = \frac{[HOCl][H^+][Cl^-]}{[Cl_2]} = 4.5 \times 10^{-4} \qquad \text{at } 25°C \qquad (13.2)$$

Ionization reaction:

$$HOCl \rightleftarrows H^+ + OCl^- \qquad (13.3)$$

$$K_2 = \frac{[H^+][OCl^-]}{[HOCl]} = 3.7 \times 10^{-8} \qquad \text{at } 25°C \qquad (13.4)$$

As noted in Eqs. (13.1) and (13.3), chlorine adds acidity to the water but rarely enough to greatly alter the pH.

Free Residual Chlorine

Free residual chlorine, defined as the sum of [HOCl] and [OCl^-], is often used as the measure of chlorination effectiveness [13.7, 13.9]. Maintaining a specified free residual chlorine value (0.5 to 1.0 g/m^3) is a standard objective in water supply systems. From Eqs. (13.1) through (13.4) it can be shown that most of the applied chlorine is in either the [HOCl] or [OCl^-] form. Thus free residual chlorine is also a measure of the available chlorine in solution.

Because [HOCl] is by far the most effective disinfectant of the three forms, the species distribution is very important (Fig. 13.1). As noted, dissolved gaseous chlorine is insignificant. The percentage distribution of HOCl as a function of pH can be computed using Eq. (13.5):

$$\frac{[HOCl]}{[HOCl] + [OCl^-]} = \frac{1}{1 + [OCl^-]/[HOCl]} = \frac{1}{1 + K_2/[H^+]} \qquad (13.5)$$

The distribution of [HOCl] and [OCl^-] as a function of pH for a temperature of 25°C is shown in Fig. 13.2. The distribution of the chlorine species at other temperatures can be computed using the values of the ionization constant given in Table 13.1.

Reactions with Ammonia

Reactions of hypochlorite with ammonia result in the formation of chloramines and the gases molecular nitrogen (N_2) and nitrous oxide (N_2O). The formation of chloramines due to the addition of chlorine can

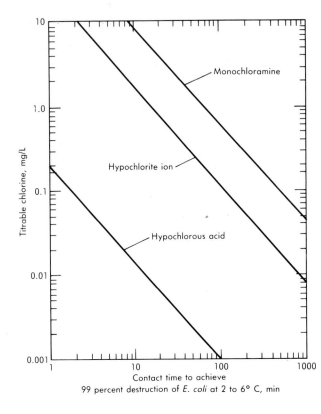

FIGURE 13.1

Comparison of germicidal efficiency of hypochlorous acid, hypochlorite ion, and monochloramine for 99 percent destruction of *E. coli* at 2 to 6°C.

Source: From Ref. [13.5].

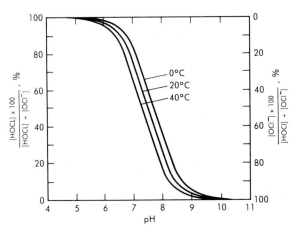

FIGURE 13.2

Distribution of [HOCL] and [OCl⁻] as a function of pH and temperature.

TABLE 13.1

Values of the Ionization Constant for Hypochlorous Acid as a Function of Temperature

TEMPERATURE, °C	$K_2 \times 10^8$, mol/L
0	2.0
5	2.3
10	2.6
15	3.0
20	3.3
25	3.7

Source: From Ref. [13.5].

be defined as follows:

$$HOCl + NH_3 \rightleftarrows NH_2Cl \text{ (monochloramine)} + H_2O \qquad (13.6)$$

$$HOCl + NH_2Cl \rightleftarrows NHCl_2 \text{ (dichloramine)} + H_2O \qquad (13.7)$$

$$HOCl + NHCl \rightleftarrows NCl_3 \text{ (nitrogen trichloride)} + H_2O \qquad (13.8)$$

The above reactions are very dependent on the pH, temperature, contact time, and initial ratio of chlorine to ammonia. In most cases, the predominant species are monochloramine (NH_2Cl) and dichloramine ($NHCl_2$). The chlorine present in these compounds is called the *combined available chlorine*. Chloramines are also disinfectants, but their action is much slower than that of hypochlorite.

Breakpoint Chlorination

Breakpoint chlorination involves the addition of sufficient chlorine to result in a free chlorine residual. Conceptually, four steps are involved, as shown in Fig. 13.3. In the first step, easily oxidized substances such as Fe^{+2}, H_2S, and organic matter react with chlorine and reduce it to chloride. In the second step, chloramines and chloroorganic compounds are produced. Continued addition results in the oxidation of these compounds, with the production of N_2O, chloride, and N_2, as illustrated in Eqs. (13.9) and (13.10).

$$NH_2Cl + NHCl_2 + HOCL \rightarrow N_2O + 4HCl \qquad (13.9)$$

$$2NH_2Cl + HOCl \rightarrow N_2 + H_2O + 3HCl \qquad (13.10)$$

At the breakpoint, virtually all of the chloramines and a significant

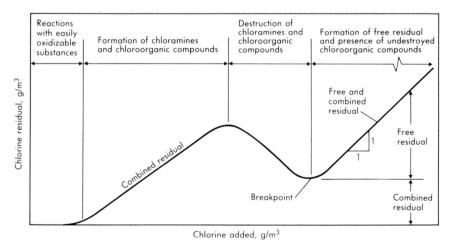

FIGURE 13.3

Schematic representation of the breakpoint chlorination curve.

fraction of the chloroorganic compounds have been oxidized. Further addition of chlorine results in a free residual of $[HOCl]$ and $[OCl^-]$.

The reactions involved in the formation of N_2O and N_2, as cited above, are complex and to some extent undefined, but the resultant removal of nitrogen from solution is widely used in wastewater treatment. Two notable examples of chlorination for nitrogen removal are the treatment processes at South Lake Tahoe and Truckee, California. In both cases, wastewater discharge requirements include strict total nitrogen limitations, and, because the process is effective and easily controlled, chlorination is used to remove any ammonia nitrogen remaining after treatment.

Factors Affecting Disinfection (With Chlorine)

Important factors that affect the disinfection process in water and wastewater, particularly with chlorine, include (1) initial contact; (2) contact time; (3) form and concentration of disinfectant; (4) type and number of organisms; and (5) environmental variables, including pH and temperature. These factors are considered further in the following discussion. Additional details on these and other factors may be found in Refs. [13.5], [13.8], and [13.20].

Method of Initial Contact

Where chlorine is used as the disinfectant, it has been demonstrated that the initial contact between the chlorine compounds and the organisms to be destroyed is of major importance. Unfortunately, this aspect of disinfection is still not appreciated fully in practice. Some effective methods that have been used to achieve effective initial contact are shown in Fig. 13.4.

Contact Time

The contact time required for a selected germicidal action is a function of residual chlorine concentration, temperature, presence of interfering materials, and the organism of concern [13.20]. At the same HOCl concentration, deactivation of Coxsackie virus requires about 500 times the contact time necessary for deactivation of the adenovirus [13.4]. Bacteria are, in general, much easier to kill than viruses, but encysted bacteria are very resistant to chlorination. Many of the protozoa of concern in water quality management also have encysted states that are more resistant to disinfection (e.g., *Giardia lamblia*; see discussion in Chapter 3). Fortunately, protozoa are relatively large, and most are removed by filtration.

A number of disinfection-rate models have been proposed. Most available data can be fitted satisfactorily with a first-order function or a

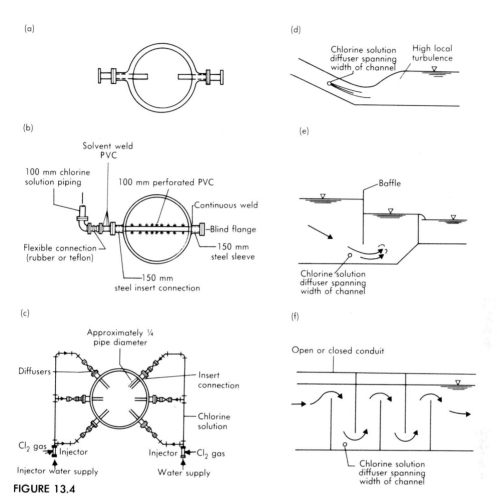

FIGURE 13.4

Methods used to achieve effective initial contact of chlorine with water and wastewater.

first-order power function of the form given by Eq. 13.11 [13.9]:

$$\frac{N_t}{N_0} = e^{-kt^m} \tag{13.11}$$

where

N_t = number of organisms present at time t

N_0 = number of organisms present at time t_0

k = bacterial decay constant, t^{-1}

t = time, t

m = empirical constant, usually equal to 1

(a)   (b)

FIGURE 13.5

Typical plug flow chlorine contact basins used in wastewater treatment plants.

The value of the decay constant k is obtained by plotting $(-\ln N_t/N_0)$ versus t on log-log paper.

In practice, chlorine contact basins are designed to perform as plug-flow reactors to ensure that most of the fluid to be disinfected is in contact with the chlorine for an adequate period of time. Typical chlorine contact basins are shown in Fig. 13.5. The application of Eq. (13.11) is illustrated in Example 13.1.

Concentration of Disinfectant

The effect of the concentration of disinfectant (chlorine) has been modeled using with the following empirical equation [13.8]:

$$C^n t = \text{constant} \tag{13.12}$$

where

$C =$ concentration of disinfectant, g/m^3

$n =$ constant

$t =$ time, min

When the value of n is greater than 1, the efficiency of the disinfectant decreases rapidly with dilution. When n is less than 1, contact time is more important than dosage. When n equals 1, both concentration and dosage are of equal importance.

Concentration of Organisms

The effect of the concentration of organisms has been modeled using Eq. (13.13):

$$\frac{C^m}{N_r} = \text{constant} \tag{13.13}$$

where

C = concentration of disinfectant, g/m³

m = empirical constant

N_r = concentration of organisms reduced by a constant
    percentage in a given time

Environmental Variables

The major environmental variables of importance in the disinfection process are temperature and pH. The effects of these variables were studied in a classic paper by Butterfield et al. [13.5]. Based on their findings, it can be shown that the ratio of the times required to achieve equal percentage kills is given by Eq. (13.14).

$$\ln \frac{t_1}{t_2} = \frac{E(T_2 - T_1)}{R T_1 T_2} \qquad (13.14)$$

where

t_1, t_2 = times for equal percentage kills at temperatures T_1 and T_2

E = activation energy, J/mol

T_2, T_1 = temperature, °K

R = gas constant, 8.314 J/mol · °K

Values of E for various pH values and chlorine compounds are given in Table 13.2.

TABLE 13.2

Typical Activation Energy Values for the Effectiveness of Chlorine in the Destruction of E. Coli

CHLORINE FORM	pH	E, J/mol
Aqueous chlorine	7.0	34,330
	8.5	26,800
	9.8	50,240
	10.7	52,800
Chloramines	7.0	50,240
	8.5	58,600
	9.5	83,740

Source: From Ref. [13.5].

EXAMPLE 13.1

ANALYSIS OF CHLORINATION DATA

Using the following data, obtained from batch chlorination studies, determine the coefficients in Eq. (13.11). Also determine the coefficients in Eq. (13.12) for 99 percent kill.

FREE AVAILABLE CHLORINE, g/m³	PERCENT SURVIVAL				
	Contact time, min				
	1	3	5	10	20
0.05	97	82	63	21	0.3
0.07	93	60	22	0.5	—
0.14	75	11	0.7	—	—

SOLUTION:

1. Determine the coefficients in the Eq. (13.11) for each concentration.

$$\frac{N_t}{N_0} = e^{-kt^m}$$

a. Taking the negative logarithm of the above equation yields

$$-\ln \frac{N_t}{N_0} = kt^m$$

b. Taking the log of the above expression yields

$$\log\left(-\ln \frac{N_t}{N_0}\right) = \log k + m \log t$$

c. The constants k and m can be determined by plotting the above expression on log-log paper.

d. Arrange the data for plotting.

C, g/m³	T, min	$\dfrac{N_t}{N_0}$	$-\ln \dfrac{N_t}{N_0}$
0.05	1	0.97	0.030
	3	0.82	0.198
	5	0.63	0.462
	10	0.21	1.561
	20	0.003	5.809
0.07	1	0.93	0.073
	3	0.60	0.511
	5	0.22	1.514
	10	0.005	5.298
0.14	1	0.75	0.288
	3	0.11	2.207
	5	0.007	4.962

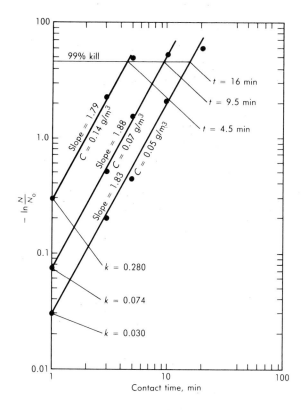

FIGURE 13.6

Plot of In(N/N_0) versus contact time for Example 13.1.

e. The above data are plotted in Fig. 13.6.

f. The resulting coefficients determined from Fig. 13.6 for each concentration are

$$C = 0.05 \text{ g/m}^3 \qquad \ln\frac{N_t}{N_0} = -0.280t^{1.79}$$

$$C = 0.07 \text{ g/m}^3 \qquad \ln\frac{N_t}{N_0} = -0.074t^{1.88}$$

$$C = 0.14 \text{ g/m}^3 \qquad \ln\frac{N_t}{N_0} = -0.030t^{1.83}$$

2. Determine the coefficients in Eq. (13.12).

$c^n t = \text{constant}$

a. Taking the log of the above equation yields

$n \log C + \log t = \log (\text{constant})$

b. To obtain the coefficients for a 99 percent kill, the required times for each concentration are determined from Fig. 13.6. The values

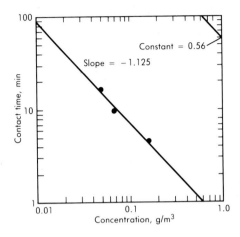

FIGURE 13.7

**Plot of contact time for 99 per-
cent kill versus concentration for
Example 13.1.**

are

C, g/m^3	TIME FOR 99 PERCENT KILL, min
0.05	16.0
0.07	9.5
0.14	4.5

c. Plot the above data to determine the coefficients for Eq. (13.12). The required plot is shown in Fig. 13.7.

d. The coefficients determined from the plot are

$$n = -1.125$$
$$\text{Constant} = 0.56$$

e. The final form of Eq. (13.12) for 99 percent kill is

$$C^{-1.125}t = 0.56$$

COMMENT

Because the value of n in Eq. (13.12) is less than 1 in this example, the contact time is more important than the chlorine dosage. In general, contact time is one of the most important variables in the design of chlorination facilities for both water and wastewater.

Dechlorination

Dechlorination is important in cases where a chlorinated wastewater comes in contact with fish and other aquatic animals [13.6, 13.16, 13.20]. Sulfur dioxide, sodium sulfite, sodium metabisulfite, and activated carbon have been used for this purpose, but sulfur dioxide is the preferred compound. Dechlorination reactions with sulfur dioxide, sodium metabisulfite, and carbon are given below. Similar reactions occur with chloramines.

Dechlorination with sulfur dioxide:

$$SO_2 + Cl_2 + 2H_2O \rightarrow H_2SO_4 + 2HCl \tag{13.15}$$

Dechlorination with sodium metabisulfite:

$$NaHSO_3 + Cl_2 + H_2O \rightarrow NaHSO_4 + 2HCl \tag{13.16}$$

Dechlorination with carbon:

$$C + Cl_2 + 2H_2O \rightarrow 4HCl + CO_2 \tag{13.17}$$

13.3
PRECIPITATION

Precipitation is used to remove metal ions such as Ca^{+2} and Mg^{+2} and ecologically important anions such as PO_4^{-3}. As was noted in Chapter 2, Ca^{+2} and Mg^{+2} are the principal sources of hardness in water, and hardness is important because of the associated precipitation with soap and scale formation at elevated temperatures. Phosphate is the growth-limiting nutrient in many lakes, and therefore increased discharges may result in rapid eutrophication.

Precipitation Reactions

Common precipitation reactions and their solubility products are listed in Table 13.3. All of the cations listed are important in water quality

TABLE 13.3

Common Precipitation Reactions and their Solubility Products

REACTION	pK_{sp} at 25°C
$Al(OH)_3 = Al^{+3} + 3OH^-$	31.2
$AlPO_4 = Al^{+3} + PO_4^{-3}$	22.0
$CaCO_3 = Ca^{+2} + CO_3^{-2}$	8.4
$Ca(OH)_2 = Ca^{+2} + 2OH^-$	5.4
$Ca_3(PO_4)_2 = 3Ca^{+2} + 2PO_4^{-3}$	26.0
$CaSO_4 = Ca^{+2} + SO_4^{-2}$	4.6
$FeCO_3 = Fe^{+2} + CO_3^{-2}$	10.4
$Fe(OH)_2 = Fe^{+2} + 2OH^-$	14.5
$Fe(OH)_3 = Fe^{+3} + 3OH^-$	38.0
$FePO_4 = Fe^{+3} + PO_4^{-3}$	21.9
$MgCO_3 = Mg^{+2} + CO_3^{-2}$	4.9
$Mg(OH)_2 = Mg^{+2} + 2OH^-$	9.2

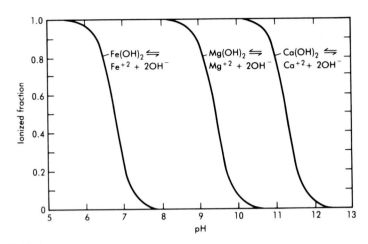

FIGURE 13.8

Solubility of Fe(OH)$_2$, Mg(OH)$_2$, and Ca(OH)$_2$ in pure water.

management, and all are relatively easy to remove, although factors such as pH (Fig. 13.8) may set constraints. Equilibrium concentrations are important also. For example, lime [Ca(OH)$_2$] is important because it is inexpensive and can be used to remove hardness and to precipitate anions such as PO$_4^{-3}$. However, Ca(OH)$_2$ is very soluble at pH values below 12, as shown below:

$$Ca(OH_2) \rightleftarrows Ca^{+2} + 2OH^- \tag{13.18}$$

From Table 13.3

$$[Ca^{+2}][OH^-]^2 \le K_{sp} = 3.98 \times 10^{-6} \tag{13.19}$$

If no other species are present,

$$[H^+][OH^-] = 10^{-14}$$

$$[Ca^{+2}] \le \frac{3.98 \times 10^{-6}}{(10^{-14}/[H^+])^2} = 3.98 \times 10^{22}[H^+]^2$$

At a pH of 12,

$$[Ca^{+2}] \le 3.98 \times 10^{-2} \text{ mol/L}$$
$$\le 1.59 \text{ g/L} = 1590 \text{ g/m}^3$$

This is still an excessive amount of Ca^{+2} in solution in terms of water quality. For example, the hardness would be 3.18 eq/L or 3180 meq/L, and waters in the 4 to 10 meq/L range are considered very hard.

Hardness Removal

The removal of hardness is accomplished in two ways: through precipitation with lime and soda ash (Na_2CO_3) and through ion exchange. The lime–soda-ash process used for the removal of calcium and magnesium is described in Eqs. (13.20) through (13.24) below.

Conversion of carbon dioxide in untreated water:

$$CO_2 + Ca(OH_2) \rightarrow \underline{CaCO_3} + H_2O \qquad (13.20)$$

Before any softening can be accomplished, the carbon dioxide in the untreated water must be converted. The addition of lime is used to convert the CO_2 to $CaCO_3$, a precipitate that can be removed by settling.

Removal of calcium and magnesium carbonate hardness:

$$Ca(HCO_3)_2 + Ca(OH)_2 \rightarrow \underline{2CaCO_3} + 2H_2O \qquad (13.21)$$

$$Mg(HCO_3)_2 + 2Ca(OH_2) \rightarrow \underline{2CaCO_3} + \underline{Mg(OH)_2} + 2H_2O \qquad (13.22)$$

Removal of calcium and magnesium noncarbonate hardness:

$$Ca^{+2} + Na_2CO_3 \rightarrow \underline{CaCO_3} + 2Na^+ \qquad (13.23)$$

$$Mg^{+2} + Ca(OH)_2 \rightarrow \underline{Mg(OH)_2} + Ca^{+2} \qquad (13.24)$$

Recarbonation for the removal of excess lime and pH control (pH $\approx$ 9.5):

$$Ca(OH_2) + CO_2 \rightarrow \underline{CaCO_3} + H_2O \qquad (13.25)$$

$$Mg(OH)_2 + CO_2 \rightarrow MgCO_3 + H_2O \qquad (13.26)$$

Recarbonation for pH control (pH $\approx$ 8.5):

$$CO_3{}^{-2} + CO_2 + H_2O \rightarrow 2HCO_3{}^- \qquad (13.27)$$

Stoichiometric Chemical Requirements

Based on a consideration of the softening reactions as defined by Eqs. (13.20) through (13.24) it can be shown that the stoichiometric requirements for lime and soda expressed in equivalents per cubic meter are as follows:

$$\text{Lime required, eq/m}^3 = CO_2 + HCO_3{}^- + Mg^{+2} + \text{excess} \qquad (13.28)$$

$$\text{Soda required, eq/m}^3 = Ca^{+2} + Mg^{+2} - \text{alkalinity} \qquad (13.29)$$

Approximately 1 eq/m^3 of lime in excess of the stoichiometric requirement must be added to bring the pH to a value above 11 and to ensure the complete precipitation of $Mg(OH)_2$. When the precipitates have been removed by sedimentation and filtration, recarbonation is carried out to remove the excess lime and to bring the pH down to a value in the 9.2 to 9.7 range. The purpose is to leave the water in a slightly encrustive

(a)

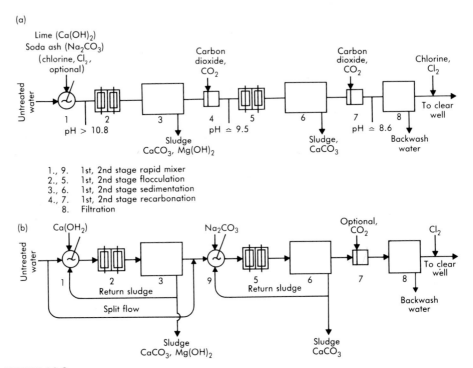

1., 9.   1st, 2nd stage rapid mixer
2., 5.   1st, 2nd stage flocculation
3., 6.   1st, 2nd stage sedimentation
4., 7.   1st, 2nd stage recarbonation
   8.   Filtration

FIGURE 13.9

Typical flow sheets for the lime-soda softening of water: (a) conventional and two-stage and (b) split treatment.

Source: Adapted from Ref. [13.14].

condition [see Eqs. (3.4) and (3.5)]. A thin layer of $CaCO_3$ that offers some deterrence to corrosion will gradually form on pipes, as noted in Chapter 3. The practical limits of the lime–soda-ash process are approximately 1 eq/m^3. Flow sheets for the lime–soda-ash process are shown in Fig. 13.9. Additional details on water softening may be found in Refs. [13.11] and [13.14]. Determination of the chemical requirements for softening is illustrated in Example 13.2.

EXAMPLE 13.2

CHEMICAL REQUIREMENTS FOR WATER SOFTENING WITH LIME AND SODA ASH

A water having the analysis given below is to be softened by the excess-lime–soda-ash method. Determine the amount of each chemical required per cubic meter treated.

CONSTITUENT	CONCENTRATION, eq/m³
Ca^{+2}	4.50
Mg^{+2}	1.50
HCO_3^-	3.00
CO_3^{-2}	0.02
Cl^-	2.00
SO_4^{-2}	1.0
$T, °C$	25

SOLUTION:

1. Estimate the solution pH and alkalinity from the carbonate equilibrium relationships developed in Chapter 2.

   a. Estimate the pH using Eq. (2.46).

   $$[H^+] = \frac{K_2[HCO_3^-]}{[CO_3^{-2}]}$$

   $$[CO_3^{-2}] = (\tfrac{1}{2}\text{mol/eq } CO_3^{-2})(0.02 \text{ eq/m}^3)(10^{-3} \text{ m}^3/\text{L})$$

   $$= 10^{-5} \text{ mol/m}^3$$

   $$[H^+] = \frac{4.68 \times 10^{-11}(3.00 \times 10^{-3} \text{ mol/L})}{(10^{-5} \text{ mol/m}^3)^2}$$

   $$= 1.41 \times 10^{-8} \text{ mol/L}$$

   $$pH = 7.85$$

   b. Determine the alkalinity in equivalents per cubic meter using Eq. (2.59).

   $$\text{Alkalinity} = (HCO_3^-) + (CO_3^{-2}) + (OH^-) - (H^+)$$

   $$= 3.00 + 0.02 + \left(\frac{10^{-14}}{1.4 \times 10^{-8}} \times 10^3\right) - 1.41 \times 10^{-8}$$

   $$= 3.00 + 0.02 + 0.0007 - 1.4 \times 10^{-8}$$

   $$= 3.02 \text{ eq/m}^3$$

2. Determine the lime required using Eq. (13.28).

   $$\text{Lime required} = CO_2 + HCO_3^- + Mg^{+2} + \text{excess}$$

   $$= [0.0 + 3.0 + 1.5 + 1(\text{excess})] \text{ eq/m}^3$$

   $$= 5.5 \text{ eq/m}^3$$

   $$= 203.5 \text{ g/m}^3$$

   $$= 275 \text{ g/m}^3 \text{ as } CaCO_3$$

3. Determine the soda ash required using Eq. (13.29).

$$\text{Soda required} = Ca^{+2} + Mg^{+2} - \text{alkalinity}$$
$$= (4.5 + 1.5) \text{ eq/m}^3 - 3.02 \text{ eq/m}^3$$
$$= 2.98 \text{ eq/m}^3$$
$$= 157.9 \text{ g/m}^3$$
$$= 149 \text{ g/m}^3 \text{ as } CaCO_3$$

COMMENT

An excess-lime–soda-ash softening system consists of two reactors, two flocculators, two sedimentation tanks, and a filter operated in series (see Fig. 13.9). Flocculation is required because the precipitate formed consists of colloidal-sized particles. Sedimentation is effective, but a small amount of fine particles will remain, making filtration necessary.

Phosphate Removal

Concern over phosphate has been relatively recent and resulted from concern with another environmental problem, foaming. Detergents became widely available during the late 1940s. The major advantage of the early detergents was that the cleaning agent, alkyl benzene sulfonate (ABS), does not form a precipitate with divalent cations, as is the case with the triglycerides of soap. Bacterial communities develop the enzyme systems necessary for ABS metabolism only when stressed. When other organic sources are available, ABS is rarely attacked. For this reason, ABS passes through conventional wastewater treatment plants unscathed.

By the early 1960s, foaming at treatment plants, outfalls, and riffles was an aesthetic problem. Legislation enacted in 1965 required that a biodegradable surfactant linear alkyl sulfonate (LAS) be substituted for ABS, and virtually all detergents sold today (including those marked biodegradable and sold at extraordinary prices) are biodegradable. Unfortunately, the hydrophyllic portion of LAS has a phosphate group, and biodegradation releases phosphate into solution. Domestic wastewater in the United States typically has phosphorus concentrations of about 10 g/m³. Adequate substitutes for LAS have not been found. A promising alternative, the strong chelating agent nitrilotriacetatic acid (NTA) was found to cause brain damage in infants. The current solution to foaming and phosphate-caused eutrophication is to use LAS detergents and precipitate the phosphate with Fe^{+3}, Al^{+3}, or Ca^{+2} where eutrophication is thought to be a problem.

Phosphate removal is carried out in a manner similar to softening. The choice of a precipitant ion is dependent upon discharge requirements, wastewater pH, and chemical costs. The pertinent reactions for the precipitation of phosphate with alum and lime follow.

Precipitation with alum (simplified):

$$Al_2(SO_4)_3 + 2PO_4^{-3} \rightarrow 2AlPO_4 + 3SO_4^{-2} \tag{13.30}$$

Precipitation with lime (simplified):

$$5Ca^{+2} + 4OH^- + 3HPO_4^{-2} \rightarrow Ca_5(OH)(PO_4)_3 + 3H_2O \tag{13.31}$$

Typically lime is the only chemical added in large quantities, although small amounts of alum may be used to enhance flocculation. The operating pH for PO_4^{-3} removal with lime is normally above 11, because flocculation is best in this range. Thus, the amount of lime needed is controlled by the alkalinity of the wastewater and not by the stoichiometric requirement for precipitation.

13.4
COAGULATION

Coagulation is the process of destabilizing colloidal particles so that particle growth can occur during flocculation. Colloidal particles typically have a net negative surface charge. The size of colloids (0.001 to 1 μm) is such that the attractive body forces between particles are considerably less than the repelling forces of the electrical charge. Under these stable conditions, particle growth does not occur, and Brownian motion keeps the particles in suspension. As particles grow, gravity sedimentation, filtration, and other inexpensive particle-separation procedures become feasible.

Coagulation Mechanisms

Four major coagulation mechanisms exist: (1) double-layer compression, (2) charge neutralization, (3) interparticle bridging, and (4) precipitate enmeshment [13.13, 13.17, 13.18]. *Double-layer compression* is brought about by increasing the total ion concentration (Fig. 13.10). By compressing the electrical double layer, the effect of the surface charge is limited to a thin layer around the particles. As a result, the attractive body forces become larger than the repelling electrical forces, so particle growth can occur if the particles collide during flocculation. *Charge neutralization* can be effected through pH control or by providing cations that adsorb on particle surfaces. *Interparticle bridging* occurs when polymers form and are adsorbed on several particles as the floc sweeps through the water (Fig. 13.11). *Precipitate enmeshment* results when a precipitate forms and traps colloidal particles (Fig. 13.12). With the

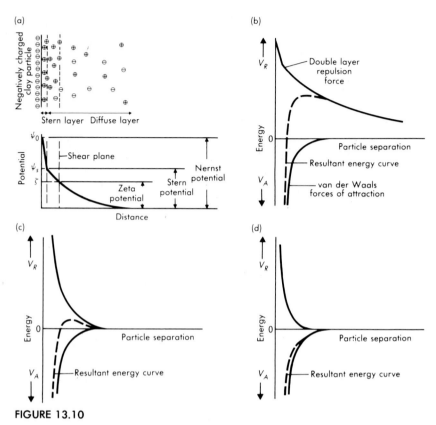

FIGURE 13.10

Compression of the electrical double layer by the addition of counter-ions.

Source: From Ref. [13.18].

exception of double-layer compression, all of the mechanisms are important in coagulation as applied to water and wastewater treatment. Which mechanism predominates depends on the coagulant chosen and the operating conditions.

Jar Testing

Jar testing (Fig. 13.13) provides a method of determining the best coagulant, operating pH, and dosage for a given water or wastewater. The concept is simple. The pH of several water samples is adjusted to preselected values. An equal amount of the coagulant being tested is added to each water sample under conditions of rapid mixing. After a short period during which the coagulation reactions and the initial particle aggregation occur, the mixing is slowed and particle growth through flocculation begins. After mixing is stopped, the particles settle and the turbidity of the supernatant liquor is measured. A minimum

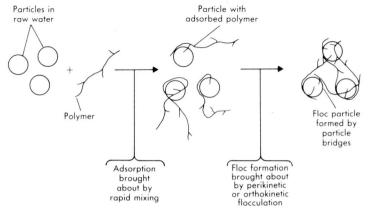

FIGURE 13.11

Coagulation by interparticle bridging with polymers.

Source: Adapted from Ref. [13.13].

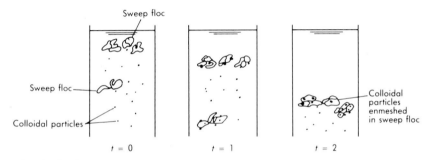

FIGURE 13.12

Enmeshment of particles in sweep floc during settling.

FIGURE 13.13

Jar test apparatus used in determining coagulant dosages. To avoid vortexing and to achieve effective mixing and flocculation, beakers equipped with internal baffles, as shown in the photograph, or square (battery) jars should be used.

turbidity value, associated with the optimal operating pH, is usually found, and a similar set of tests is then run using a constant pH and varying coagulant dosages.

Coagulant Chemicals

A number of coagulants are usually tested. The most common are alum $[Al_2(SO_4)_3 \cdot 18H_2O]$, ferric chloride ($FeCl_3$), lime [CaO or Ca(OH)$_2$], and various polyelectrolytes. Each coagulant has a pH range within which it is effective. The effective concentrations and pH ranges for alum are shown in Fig. 13.14. As shown there, the effective zone for coagulation is bound by the various equilibrium expressions for alum [13.1]. The pertinent reaction for alum is

$$Al_2(SO_4)_3 \cdot 18H_2O + 3Ca(HCO_3)_2 \rightleftarrows 3CaSO_4 + 2Al(OH)_3 + 6CO_2 + 18H_2O$$

| Aluminum sulfate (666.7) | Calcium bicarbonate (3×162) | Calcium sulfate (3×136) | Aluminum hydroxide (2×78) | Carbon dioxide (6×44) | Water (18×18) |

$$(13.32)$$

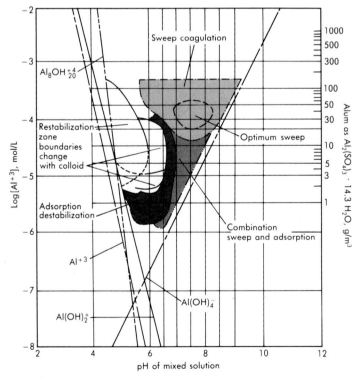

FIGURE 13.14

Operating ranges for alum coagulation.

Source: From Ref. [13.1], reprinted with permission, copyright © 1982, The American Water Works Association.

The insoluble aluminum hydroxide forms a gelatinous floc that settles slowly, sweeping out suspended material as it settles. The quantity of sludge produced is estimated in Example 13.3.

EXAMPLE 13.3

ESTIMATION OF CHEMICAL AND SLUDGE QUANTITIES IN ALUM COAGULATION

Determine the amount of natural alkalinity needed for the treatment of a surface water with alum at a dosage of 30 g/m^3. If the flow rate is 4000 m^3/d, estimate the amount of sludge produced, assuming that the specific gravity of the dry solids is 2.25 and the aluminum hydroxide sludge as collected contains 4 percent solids.

SOLUTION:

1. Determine the amount of alkalinity needed.
   a. The reaction of alum with natural alkalinity is given by Eq. (13.32).

$$Al_2(SO_4)_3 \cdot 18H_2O + 3Ca(HCO_3)_2 \rightleftarrows 3CaSO_4 + 2Al(OH)_3 + 6CO_2 + 18H_2O$$
$$(666.7) \qquad (3 \times 162) \qquad (3 \times 136) \quad (2 \times 78) \quad (6 \times 44) \quad (18 \times 18)$$

   b. Determine the alkalinity needed on a mass basis. From the above equation, 1 mol of alum reacts with 3 mol of $Ca(HCO_3)_2$; thus

$$\frac{g/m^3 \ Ca(HCO_3)_2}{g/m^3 \ Al_2(SO_4) \cdot 18H_2O} = \frac{3(162)}{666.7} = 0.73$$

   If it is assumed that $Ca(HCO_3)_2$ represents the alkalinity of the natural water, then 0.73 g/m^3 of alkalinity is needed per gram per cubic meter of alum.
   c. For an alum dosage of 30 g/m^3, the natural alkalinity needed is

   Alkalinity needed expressed as $g/m^3 \ HCO_3^-$
$$= \frac{122}{162}(0.73) \times 30 \ g/m^3$$
$$= 16.5 \ g/m^3$$

   Alkalinity needed expressed as $g/m^3 \ CaCO_3$ $= 13.5 \ g/m^3$

2. Determine the volume of alum hydroxide sludge produced when treating a flow of 4000 m^3/d.

a. Determine the specific gravity of the wet sludge using Eq. (12.60). Assume M_{ws} is equal to 1 kg for the purpose of computation.

$$\frac{M_{ws}}{S_{ws}} = \frac{M_s}{S_s} + \frac{M_w}{S_w}$$

$$\frac{1}{S_{ws}} = \frac{0.04}{2.25} + \frac{0.96}{1}$$

$$S_{ws} = 1.02$$

b. Determine the volume of wet sludge using Eq. 12.58.

$$V_s = \frac{M_s}{S_s(S/100)\rho_w}$$

(1) The mass of dry solids removed per day is

$$M_s = \frac{(4000\ \text{m}^3/\text{d})(30\ \text{g/m}^3)}{10^3\ \text{g/kg}} \times \frac{2 \times 78}{666.7}$$

$$= 28.1\ \text{kg/d}$$

(2) The volume of wet sludge is

$$V_s = \frac{28.1\ \text{kg/d}}{1.02(4/100)10^3\ \text{kg/m}^3}$$

$$= 0.69\ \text{m}^3/\text{d}$$

Polyelectrolytes are synthetic polymers having many ionizable groups and can be manufactured as anionic, cationic, or nonionic. In the ionized form the polymers act as very large ions. Thus, when a cationic polyelectrolyte is added to a solution containing negatively charged colloids, electrochemical coagulation can occur. Each commercial polyelectrolyte has specific properties, and utilization involves experimental selection rather than theoretical analysis. Because of the complexity of water and wastewater treatment anionic or nonionic polymers often prove best suited, even though cationic polymers are the intuitive choice. Polyelectrolytes are often used in combination with other coagulants. Quite often, improved coagulation results from considerably lower coagulant dosage.

13.5
CHEMICAL OXIDATION

Chemical oxidants such as oxygen, chlorine, permanganate, ozone, and hydrogen peroxide are used to some extent in both water and wastewater treatment. An example—breakpoint chlorination, where both organic matter and ammonia nitrogen are oxidized—has already been discussed. Three other important oxidation reactions are used for the removal of iron, manganese, and cyanide from water. Both iron and manganese are slightly soluble in the +2 valence state, and both cause water quality problems. Iron is particularly bad because of the red water and stains resulting from oxidation. Waters containing significant iron concentrations usually contain manganese, a potentially toxic material. Fortunately both iron and manganese are easily oxidized to nonsoluble products.

Iron and Manganese Removal

Iron and manganese removal usually is accomplished through oxidation with atmospheric oxygen (see Fig. 11.1). Air is bubbled through the water and the reactions shown below take place:

$$2Fe^{+2} + \tfrac{1}{2}O_2 + 5H_2O \rightarrow 2Fe(OH)_3 + 4H^+ \tag{13.33}$$

$$Mn^{+2} + \tfrac{1}{2}O_2 + H_2O \rightarrow MnO_2 + 2H^+ \tag{13.34}$$

In most cases, production of H^+ in the reactions is not significant. For example, a concentration of 0.9 g/m^3 Fe^{+2} (see Table 3.2, Biloxi, Miss.) is equivalent to 1.6×10^{-5} mol/L, and oxidation to $Fe(OH)_3$ would result in the production of 3.2×10^{-5} mol/L of H^+. Alkalinity of waters containing iron and manganese is usually sufficient to neutralize the acid.

Oxidation of Fe^{+2} is considerably faster than oxidation of Mn^{+2}. At near-neutral pH values, about 15 min is necessary for 90 to 95 percent conversion of Fe^{+2}. Raising the pH increases the rate of reaction, or catalysts can be added if necessary.

Cyanide Removal

Cyanide salt is used in extraction of gold and silver and in the surface hardening of metals and metal plating. The cyanide ion CN^- is a constituent of the wastewaters from these processes; because of its high toxicity CN^- must be removed before discharge to municipal sewers or receiving waters. Cyanide toxicity results from an irreversible reaction with the iron in hemoglobin and the resulting loss of the ability to carry oxygen in blood.

Oxidation of cyanide (CN^-) to cyanate (CNO^-) with Cl_2 or NaOCl is the most common method of treatment [13.3, 13.9, 13.10, 13.12]. If Cl_2 is used, NaOH must be added also, as shown in Eq. (13.35):

$$CN^- + 2NaOH + Cl_2 \rightarrow CNO^- + 2NaCl + H_2O \tag{13.35}$$

The reaction for the oxidation of CN^- with NaOCl is shown in Eq. (13.36):

$$CN^- + NaOCl \rightarrow CNO^- + NaCl \tag{13.36}$$

The above reactions are carried out at alkaline pH values, usually between 8.5 and 11. If the pH is then lowered to less than 7, the cyanate hydrolyzes:

$$CNO^- + 2H^+ + H_2O \rightarrow NH_4^+ + CO_2 \tag{13.37}$$

Adding excess Cl_2 at moderately basic pH levels results in oxidation of CNO^- to N_2 and CO_2, as given by Eq. (13.38):

$$2CNO^- + 3Cl_2 + 4NaOH \rightarrow N_2 + 2Cl^- + 4NaCl + 2H_2O + 2CO_2 \tag{13.38}$$

Ozone can also be used for the oxidation of cyanide to cyanate. For small systems, the use of ozone is more expensive than chlorine [13.7].

13.6
ION EXCHANGE

Ion exchange processes are used to remove undesirable ions such as Ca^{+2}, Mg^{+2}, Fe^{+2}, and NH_4^+ from water and wastewater. The exchange medium consists of a solid phase of naturally occurring minerals or a synthetic resin having a mobile ion attached to an immobile functional acid or base group. In the exchange process the mobile ions are exchanged with solute ions having a stronger affinity for the functional group (e.g., Ca^{+2} replaces Na^+ or SO_4^{-2} replaces Cl^-).

Naturally occurring ion exchange materials, known as *zeolites*, are used for water softening and ammonium ion removal [13.2, 13.15]. Zeolites used for water softening are complex aluminosilicates with sodium as the mobile ion. Ammonium exchange is accomplished using a naturally occurring zeolite clinoptilolite. Synthetic aluminosilicates are manufactured, but most synthetic ion exchange materials are resins or phenolic polymers. Four types of ion exchange resins are in use: (1) strong-acid cation exchangers having a strong-acid functional group; (2) weak-acid cation exchangers having a weak-acid functional group; (3) strong-base anion exchangers having strong-base functional groups; and (4) weak-base anion exchangers having weak-base functional groups [13.2, 13.15].

Typical Exchange Reactions

Typical ion exchange reactions for natural and synthetic ion exchange materials are given below.

For natural zeolites (z):

$$Na_2Z + \begin{Bmatrix} Ca^{+2} \\ Mg^{+2} \\ Fe^{+2} \end{Bmatrix} \rightleftarrows \begin{Bmatrix} Ca^{+2} \\ Mg^{+2} \\ Fe^{+2} \end{Bmatrix} Z + 2Na^+ \tag{13.39}$$

For synthetic resins (R):
Strong-acid cation exchange:

$$RSO_3H + Na^+ \rightleftarrows RSO_3Na + H^+ \tag{13.40}$$

$$2RSO_3Na + Ca^{+2} \rightleftarrows (RSO_3)_2Ca + 2Na^+ \tag{13.41}$$

Weak-acid cation exchange:

$$RCOOOH + Na^+ \rightleftarrows RCOONa + H^+ \tag{13.42}$$

$$2RCOONa + Ca^{+2} \rightleftarrows (RCOO)_2Ca + 2Na^+ \tag{13.43}$$

Strong-base anion exchange:

$$RR_3'NOH + Cl \rightleftarrows RR_3'NCl + OH^- \tag{13.44}$$

$$2RR_3'NCl + SO_4^{-2} \rightleftarrows (RR_3'N)_2SO_4 + 2Cl^- \tag{13.45}$$

Weak-base anion exchange:

$$RNH_3OH + Cl^- \rightleftarrows RNH_3Cl + OH^- \tag{13.46}$$

$$2RNH_3Cl + SO_4^{-2} \rightleftarrows (RNH_3)_2SO_4 + 2Cl^- \tag{13.47}$$

Application of Exchange Resins

The application of exchange resins for the production of softened and demineralized water is shown in Fig. 13.15. Ion exchange units used to produce demineralized water are shown in Fig. 13.16.

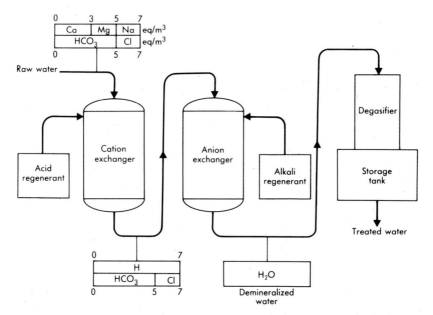

FIGURE 13.15

Flow sheet for the removal of hardness and for the complete demineralization of water using ion exchange resins.

FIGURE 13.16

Typical ion exchange reactor units used to produce demineralized water.

Regeneration of Ion Exchangers

After a period of operation, ion exchange systems become saturated; that is, they approach equilibrium with the feed solution. Regeneration is quite simple. A brine of the original mobile ion is brought in contact with the resin, and the equilibrium is shifted to one that favors the original condition. In water softening and most other cation exchange applications, a NaCl brine is used. If all cations are to be removed, a strong acid such as H_2SO_4 is used for regeneration. The used brine may present a difficult disposal problem. Communities with large numbers of home water softeners usually allow brine discharge to the sanitary sewers. Increase in water TDS through such a community can be expected to be approximately twice that of communities without home water softeners.

Selectivity

Affinity of ions for the functional groups is a function of charge and size. Ions with a larger ionic charge (Ca^{+2} versus Na^+) have greater affinity, and therefore, the equilibrium will favor the higher-charged ion on the solid phase. Affinity varies inversely with effective size. Therefore ions that are highly hydrated, and consequently have a larger effective size, have relatively less affinity for functional groups. Relative affinities of common ions are given below:

$$Ag^+ > Cs^+ > K^+ > Na^+ > Li^+$$
$$Ba^{+2} > Sr^{+2} > Ca^{+2} > Mg^{+2}$$
$$I^- > NO_3^- > CN^- > HSO_4^- > NO_2^- > CL^- > HCO_3^-$$

Interactions between functional groups on resins and solute ions affect affinity also. For example, if the functional group is similar to a precipitating ion (e.g., a phosphoric resin and Ca^{+2}) affinity increases. Effectiveness of weak-acid and weak-base ion exchange resins is related to solution pH because ionization of the functional groups is essentially complete only in certain pH ranges. For example, a weak-acid resin ($RCOOH$) is fully ionized only at pH values above 4, and therefore such a resin should not be used if the pH is near or below 7.

Affinity can be quantitatively described by the selectivity coefficient. The ion exchange reaction between the mobile ion H^+ and the solute ion Na^+ results in the following expression:

$$RH + Na^+ \rightleftarrows RNa + H^+ \qquad\qquad (13.48)$$

The equilibrium expression is

$$\frac{x_{RNa}[H^+]}{x_{RH}[Na^+]} = K_{H \to Na} \qquad\qquad (13.49)$$

where

x_{RNa} = mole fraction of Na^+ on exchange resin expressed in terms of sorbed species

x_{RH} = mole fraction of H^+ on exchange resin expressed in terms of sorbed species

$[H^+]$ = concentration of H^+ in solution

$[Na^+]$ = concentration of Na^+ in solution

$K_{H \to Na}$ = selectivity coefficient

The selectivity coefficient depends on the nature and valence of the ion and the concentration of the ion in solution. In practice, selectivity coefficients are determined by laboratory measurements and are usually only valid for the conditions measured.

Exchange Capacity

Zeolites and resins are rated on the basis of exchange capacity, the equivalents of cations or anions that can be exchanged per unit mass. Typical resin exchange capacities are in the 2- to 10-eq/kg range. Zeolite cation exchangers have exchange capacities of 0.05 to 0.1 eq/kg. Exchange capacity is measured by placing the resin in a known form. A cationic resin might be washed with a strong acid to place all exchange sites in the H^+ form or washed with a strong NaCl brine to place all sites in the Na^+ form. A solution of known concentration of an exchangeable ion (e.g., Ca^{+2}) can then be added until exchange is complete and the amount of exchange capacity can be measured or, in the acid case, the resin is titrated with a strong base.

EXAMPLE 13.4

ION EXCHANGE TREATMENT FOR REMOVAL OF CALCIUM

A cation exchange resin, initially in the sodium form, is to be used for removal of calcium from water. The resin exchange capacity is 3.5 eq/kg and the value of the sodium-calcium selectivity coefficient is 70. If the initial and final calcium concentrations are 100 g/m^3 and 1 g/m^3, respectively, determine the amount of resin required to treat 1 m^3 of water.

SOLUTION:

1. Modify Eq. (13.49) so that the sorbed mole fraction values of Ca^{+2} and Na^+ can be expressed in terms of molar quantities.

   a. Equation (13.49) is

   $$\frac{x_{RCa}[Na^+]}{x_{RNa}[Ca^{+2}]} = K_{Na \to Ca}$$

   b. The mole fractions of Ca^{+2} and Na^+ are given by

   $$x_{RCa} = \frac{n_{Ca}}{n_{Ca} + n_{Na}}$$

   $$x_{RNa} = \frac{n_{Na}}{n_{Ca} + n_{Na}}$$

   where

   n_{Ca} = number of mole of Ca^{+2} sorbed on resin

   n_{Na} = number of moles of Na^+ sorbed on resin

   c. Substituting the above expressions for x_{RCa} and x_{RNa} in Eq. (13.49) and simplifying yields

   $$\frac{n_{Ca}[Na^+]}{n_{Na}[Ca^{+2}]} = K_{Na \to Ca}$$

2. Determine the values of the terms in the modified selectivity equation developed in step 1.

   a. Determine n_{Na} as a function of n_{Ca} for 1 kg of resin by noting that 2 mol of Na^+ are exchanged for each mole of Ca^{+2}.

   $$n_{Na} = 3.5 - 2n_{Ca}$$

   b. Determine the moles of calcium removed, n_{Ca}, as a function of the liquid volume expressed in liters.

   $$\Delta n_{Ca} = V\left(\frac{100 \times 10^{-3} \text{ g/L} - 1 \times 10^{-3} \text{ g/L}}{40 \text{ g/mol}}\right) = 2.475 \times 10^{-3} V \text{ mol}$$

c. Determine the final $[Ca^{+2}]$ concentration in solution.

$$[Ca^+] = \frac{1 \text{ g/m}^3}{40 \text{ g/mol}} 10^{-3} \text{ m}^3/\text{L} = 2.5 \times 10^{-5} \text{ mol/L}$$

d. Determine the final $[Na^+]$ concentration in solution.

$$[Na^+] = \frac{2 \Delta n_{Ca}}{V} = \frac{2(2.475 \times 10^{-3} V)}{V}$$

$$= 4.95 \times 10^{-3} \text{ mol/L}$$

3. Determine liquid volume using the modified form of Eq. (13.49) developed in step 1.

$$\frac{n_{Ca}[Na^+]}{n_{Na}[Ca^{+2}]} = K_{Na \to Ca}$$

a. Substitute known values and solve for the liquid volume.

$$\frac{2.475 \times 10^{-3} V (4.95 \times 10^{-3})}{[3.5 - 2(2.475 \times 10^{-3}) V] (2.5 \times 10^{-5})} = 70$$

$$\frac{0.49 V}{3.5 - 4.95 \times 10^{-3} V} = 70$$

$$V = 293 \text{ L} = 0.29 \text{ m}^3$$

b. The liquid volume processed per kilogram of resin is

$$V/\text{kg} = \frac{0.29 \text{ m}^3}{1.0 \text{ kg}} = 0.29 \text{ m}^3/\text{kg}$$

4. Determine the mass of resin required to treat 1.0 m^3 of water.

$$R_m = \frac{1}{0.29} = 3.45 \text{ kg/m}^3$$

COMMENT

The low volume treated per unit mass of resin results from the high treatment efficiency (99 percent) required. Note that the balance between n_{Ca} and n_{Na} controls the volume requirement. A larger value of $K_{Na \to Ca}$ would have only a moderate effect on the volume required.

KEY IDEAS, CONCEPTS, AND ISSUES

- Disinfection is the term used to describe the selective destruction of pathogenic organisms. The complete destruction of all organisms is termed sterilization.

- Factors affecting the effectiveness of the disinfection process with chlorine include (1) initial contact, (2) contact time, (3) concentration of chlorine, (4) type and number of organisms, and (5) pH and temperature.
- Disinfection with chlorine is an effective method of eliminating pathogenic bacteria from water.
- Reactions of chlorine with organic compounds may result in the production of toxic or carcinogenic products.
- Precipitation of multivalent cations and anions is an economical method for their removal.
- The lime–soda-ash softening process can effectively reduce the total hardness in water to approximately 1 eq/m^3.
- Removal of precipitates by sedimentation and filtration is usually necessary.
- The four major coagulation mechanisms are (1) double-layer compression, (2) charge neutralization, (3) interparticle bridging, and (4) particle enmeshment.
- Iron and manganese removal involves a combination of chemical oxidation and precipitation.
- Ion exchange is an effective method of removing undesirable ions from solution. The ion exchange process is used extensively for water softening at individual residences and for the production of demineralized water at industrial facilities.

DISCUSSION TOPICS AND PROBLEMS

13.1. Determine and plot the distribution of Cl_2, HOCl, and OCl^- as a function of pH for a closed system.

13.2. Chlorine disinfection studies have been run on a treated wastewater and the results are given below. Determine the bacterial decay coefficient k of Eq. (13.11).

FREE AVAILABLE CHLORINE, g/m^3	PERCENT SURVIVAL				
	Contact time, min				
	1	2	4	8	16
50	94	80	43	4	9×10^{-4}
80	90	72	35	3	1×10^{-3}
100	82	57	20	1	3×10^{-4}

13.3. Determine the contact time necessary to attain a 99 percent kill at the three chlorine concentrations of Problem 13.2.

13.4. A wastewater has been chlorinated for disinfection purposes but must be dechlorinated prior to discharge into a sensitive river. If the flow rate is $0.5 \ m^3/s$ and the free residual chlorine concentration is 4 g/m^3, estimate the sulfur dioxide requirement.

13.5. Estimate the pH change brought about by dissolving 3 g/m^3 of Cl_2 in water having an initial alkalinity and pH of 6 eq/m^3 and 7.0, respectively.

13.6. For the water in Problem 13.5 determine the effect of dechlorination with SO_2 on the pH.

13.7. Determine and plot the solubility of $Fe(OH)_2$ as a function of pH. Would pH modification be necessary to precipitate $Fe(OH)_2$ from most natural waters?

13.8. Estimate the mass and volume of sludge produced per day from a water containing 5 g/m^3 Fe^{+2} if complete removal were possible by chemical oxidation and precipitation. Express the answer in terms of kilograms per cubic meter of water treated. Assume a sludge concentration of 5 percent solids.

13.9. A water analysis has resulted in the data listed below. Estimate the quantity of lime and soda ash necessary to soften the water.

CONSTITUENT	CONCENTRATION, g/m^3
Ca^{+2}	90
Mg^{+2}	45
Fe^{+2}	0.1
Na^+	15
HCO_3^-	366
CO_3^{-2}	0.009
SO_4^{-2}	192
Cl^-	31

13.10. A metal plating wastewater contains 140 g/m^3 of CN^-. Approximately 200 m^3 of the wastewater is produced per day, 5 days per week. If chlorine can be purchased for \$3.50/kg, determine the cost of the chemical for 1 year.

13.11. A well water used as a source of industrial boiler water has the chemical characteristics given below. Determine the amount of lime and soda ash needed to soften the water.

CONSTITUENT	CONCENTRATION, g/m^3
Ca^{+2}	60
Mg^{+2}	73
HCO_3^-	122
Cl^-	185
pH	7.0

13.12. Estimate the sludge produced per day for the water of Problem 13.11 if the treated flow is 4 m^3/s. Water content of the sludge can be expected to be 90 percent by mass.

13.13. Chemical costs and jar test data for a municipal water source are given below. Determine the least-cost method of coagulation such that a treated water turbidity of 3 NTU can be attained.

CHEMICAL	COST, $/kg
Alum	0.14
$FeCl_3$	0.20
Polyelectrolyte	1.28

ALUM, g/m^3	TURBIDITY, NTU
10	12
15	6
20	4
25	2
30	5
35	9

$FeCl_3$, g/m^3	TURBIDITY, NTU
10	6
15	5
20	3
25	4
30	7
25	10

ALUM + 1 g/m^3 POLYELECTROLYTE, g/m^3	TURBIDITY, NTU
10	10
15	4
20	3
25	3
30	5
35	7

$FeCl_3$ + 1 g/m^3 POLYELECTROLYTE, g/m^3	TURBIDITY, NTU
10	5
15	3
20	3
25	5
30	7
35	9

13.14. A column containing 2 kg of a cation exchange resin is placed in a regenerated state by washing with NaCl brine until all cations except Na^+ have been removed. At that time, 200 g of Na^+ has been adsorbed on the resin. A 100-g sample of the resin is then placed in a 100-mL volume of water initially containing 350 g/m^3 of $CaCl_2$.

After 30 min the Ca^{+2} and Na^+ concentrations are determined to be 15 g/m^3 and 193 g/m^3, respectively. Determine the value of the selectivity coefficient.

13.15. The resin of Problem 13.14 is removed from the solution, allowed to drain and placed in a second 100-mL volume of water initially containing 350 g/m^3 $CaCl_2$. Estimate (a) the equilibrium liquid phase concentrations of Ca^{+2} and Na^+ and (b) the cation exchange capacity of the resin.

13.16. A cation ion exchange resin originally in the Na^+ form and having an exchange capacity of 4.0 eq/kg is to be used to soften the water of Problem 13.11. The selectivity coefficient for the divalent cations versus sodium is 80. Determine the amount of resin required per cubic meter of water to reduce the hardness to 1 eq/m^3.

REFERENCES

13.1.   Amirtharajah, A., and K. M. Mills, (1982), "Rapid-Mix Design for Mechanisms of Alum Coagulation," *J. American Water Works Association*, vol. 74, no. 4, p. 210.

13.2.   Applebaum, S. B., (1968), *Demineralization by Ion Exchange*, Academic Press, New York.

13.3.   Benefield, L. D., J. F. Judkins, and B. L. Weand, (1982), *Process Chemistry for Water and Wastewater Treatment*, Prentice-Hall, Englewood Cliffs, N.J.

13.4.   Berg, G., (1964), "The Virus Hazard in Water Supplies," *J. New England Water Works Assoc.*, vol. 78, p. 79.

13.5.   Butterfield, C. T., E. Wattie, S. Megregian, and C. W. Chambers, (1943), "Influence of pH and Temperature on the Survival of Coliforms and Enteric Pathogens when Exposed to Free Chlorine," U.S. Public Health Report No. 58.

13.6.   *Chlorination in Sewage Treatment*, (1933), Report of Committee on Sewage Disposal, American Public Health Association.

13.7.   *Chlorination of Wastewater*, (1976), Manual of Practice No. 4, Water Pollution Control Federation, Washington, D.C.

13.8.   Esvelt, L. A., W. J. Kaufman, and R. E. Selleck, (1971), "Toxicity Removal from Municipal Wastewater," University of California Sanitary Engineering Research Laboratory Report 71-7, Richmond.

13.9.   Fair, G. M., and J. C. Geyer, (1954), *Water Supply and Waste-Water Disposal*, John Wiley & Sons, New York.

13.10. Gurnham, C. F. (ed.), (1965), *Industrial Wastewater Control*, Academic Press, New York.

13.11. Merrill, D. T., (1978), "Chemical Conditioning for Water Softening and Corrosion Control," in *Water Treatment Plant Design for the Practicing Engineer*, edited by R. L. Sanks, Ann Arbor Science Publishers, Ann Arbor, Mich.

13.12. Nemerow, N. L., (1971), *Liquid Waste of Industry: Theories, Practices, and Treatment*, Addison-Wesley Publishing Company, Reading, Mass.

13.13. O'Melia, C. R., (1978), "Coagulation" in *Water Treatment Plant Design for the Practicing Engineer*, edited by R. L. Sanks, Ann Arbor Science Publishers, Ann Arbor, Mich.

13.14. Reh, C. W., (1978), "Lime Soda Softening Processes" in *Water Treatment Plant Design for the Practicing Engineer*, edited by R. L. Sanks, Ann Arbor Science Publishers, Ann Arbor, Mich.

13.15. Sanks, R. L. (ed.), (1978), *Water Treatment Plant Design for the Practicing Engineer*, Ann Arbor Science Publishers, Ann Arbor, Mich.

13.16. Stone, R. W., W. J. Kaufman, and A. J. Horne, (1973), "Long Term Effects of Toxicants and Biostimulants on the Waters of Central San Francisco Bay," University of California Sanitary Engineering Research Laboratory Report 73-1.

13.17. Stumm, W., and O'Melia, C. R., (1968), "Stoichiometry of Coagulation," *J. American Water Works Association*, vol. 60, no. 5, p. 514.

13.18. Van Olphen, H., (1977), *An Introduction to Clay Colloid Chemistry*, 2d. ed., Wiley Interscience, New York.

13.19. Weber, W. J., Jr., (1972), *Physiochemical Processes for Water Quality Control*, Wiley Interscience, New York.

13.20. White, G. C., (1972), *Handbook of Chlorination*, Van Nostrand Reinhold Company, New York.

14

Biological Treatment Methods

Microbial cultures are the active agents for removing impurities from water in biological treatment processes. Basic concepts of biological reactions were discussed in Part I as part of biochemical oxygen demand and in Part III in the development of stream and estuary modeling. Using microbial cultures to modify water quality under controlled circumstances requires a somewhat different emphasis, however. First, the reactor system is a matter of choice. Second, the engineer has a number of decisions to make with respect to process configuration—the most important involving volume, reaction rate, and stoichiometry trade-offs. Third, effluents produced by biological treatment are not suitable for potable use, and thus economic utilization of biological treatment is limited to wastewaters.

14.1
APPLICATIONS OF BIOLOGICAL TREATMENT

The principal use of biological treatment is for removal of organic material from wastewater (see Chapter 11). The characteristics of typical domestic wastewaters in industrialized countries are given in Table 14.1. Four other common uses of biological treatment are in the oxidation of ammonia nitrogen (nitrification), the reduction of oxidized nitrogen (denitrification) to gaseous nitrogen (N_2 and N_2O), the removal of phosphorus, and the oxidation/stabilization of organic sludges. The sludges include solids removed from wastewater during primary sedimentation and biological solids produced in biological treatment. It is important to note that differences exist in primary and biological, or secondary, sludges. Primary sludges are denser, easier to dewater and easier to stabilize biologically than secondary sludges. A possible reason is that the composition of the organic material in primary sludges is less complex in organization as compared with cell tissue. Another reason is that natural decomposition processes have begun before the material enters the wastewater stream.

TABLE 14.1

Typical Composition of Dry-Weather Domestic Wastewater in Industrialized Countries

CHARACTERISTIC	UNITED STATES	JAPAN	UNITED KINGDOM	FEDERAL REPUBLIC OF GERMANY
Per-capita flow, m^3/d	0.35	0.30	0.18	0.16
BOD_5, g/m^3	200	140	350	400
COD, g/m^3	500	200	500	570
Suspended solids, g/m^3	200	70	350	400
TKN,* g/m^3	40	21	50	57
P, g/m^3	10	3.5	15	17

*TKN = organic nitrogen + ammonia.

TABLE 14.2

Major Biological Treatment Processes Used for Wastewater Treatment

TYPE	COMMON NAME	USE*
Aerobic processes		
Suspended growth	Activated sludge process	
	Conventional (plug flow)	
	Continuous-flow stirred-tank	
	Sequencing batch reactor	
	Step aeration	
	Pure oxygen	Carbonaceous BOD removal (nitrification)
	Modified aeration	
	Contact stabilization	
	Extended aeration	
	Oxidation ditch	
	Suspended-growth nitrification	Nitrification
	Aerated lagoons	Carbonaceous BOD removal (nitrification)
	Aerobic digestion	
	Conventional air	Stabilization, carbonaceous BOD removal
	Pure oxygen	Stabilization, carbonaceous BOD removal
	High-rate aerobic algal ponds	Carbonaceous BOD removal
Attached growth	Trickling filters	
	Low-rate	Carbonaceous BOD removal (nitrification)
	High-rate	Carbonaceous BOD removal
	Roughing filters	Carbonaceous BOD removal
	Rotating biological contactors	Carbonaceous BOD removal (nitrification)
	Packed-bed reactors	Nitrification
Combined processes	Trickling filter-activated sludge	Carbonaceous BOD removal (nitrification)
	Activated sludge-trickling filter	Carbonaceous BOD removal (nitrification)
Anoxic processes		
Suspended growth	Suspended-growth denitrification	Denitrification
Attached growth	Fixed-film denitrification	Denitrification

596

14.2

TYPES OF BIOLOGICAL TREATMENT PROCESSES

Biological treatment processes can be classified in a number of ways. Common divisions are aerobic and anaerobic, i.e., according to metabolic activity, and suspended and attached growth, i.e., according to the location of the microorganisms. In some systems, processes from these two classifications are combined. The principal biological treatment processes represented by these classifications are reported in Table 14.2.

Each type of biological treatment process has advantages and disadvantages. Suspended-growth processes, as typified by the activated sludge

TABLE 14.2 *(Cont.)*

TYPE	COMMON NAME	USE*
Anaerobic processes		
Suspended growth	Anaerobic digestion	
	Standard rate, single stage	
	High rate, single stage	Stabilization, carbonaceous BOD removal
	Two stage	
	Anaerobic contact process	Carbonaceous BOD removal
Attached growth	Anaerobic filter	Carbonaceous BOD removal, stabilization (denitrification)
	Anaerobic lagoons (ponds)	Carbonaceous BOD removal (stabilization)
Aerobic/anoxic or anaerobic processes		
Suspended growth	Single stage nitrification-denitrification	Carbonaceous BOD removal, nitrification, denitrification, phosphate removal
Attached growth	Nitrification-denitrification	Nitrification, denitrification
	Land treatment	
	Slow rate	Carbonaceous BOD removal (nitrification, denitrification)
	Rapid infiltration	
	Overland flow	
Combined processes	Facultative lagoons (ponds)	Carbonaceous BOD removal
	Maturation or tertiary ponds	Carbonaceous BOD removal (bacterial decay, nitrification)
	Anaerobic-facultative lagoons	Carbonaceous BOD removal
	Anaerobic-facultative-aerobic lagoons	
On-site systems	Septic tank-leach fields	Treatment and disposal of wastewater from individual residences and other buildings in areas not served with sewers
	Septic tank-mounds	
	Septic tank-evapotranspiration	

*Major use is presented first; other uses are identified in parentheses.
Source: Adapted from Ref. [14.31].

process, require the least space and are the most flexible in terms of operation. Performance of suspended-growth processes is more variable [14.34, 14.35], and operation is more complex than with attached-growth systems. Attached-growth systems are relatively simple to operate and are usually quite stable in terms of performance. Effluent quality from attached-growth processes is not as good as from suspended-growth processes, and oxygen transfer limitations serve to constrain the acceptable influent waste strength [14.44].

Systems utilizing photosynthetic production of oxygen by algae, such as oxidation ponds, are simple to construct and operate, but the algae produced are difficult to remove, and numerous examples exist where the effluent BOD and suspended-solids concentrations exceed those of the influent [14.32]. Land treatment systems provide high performance and reliability, but their space requirements are the greatest of any biological treatment system. On-site systems are effective methods of treatment of small wastewater flows.

14.3
FUNDAMENTALS OF BIOLOGICAL TREATMENT

The fundamental mechanisms involved in biological treatment are the same for all processes. Microorganisms, principally bacteria, utilize organic material and inorganic ions present in wastewater to support growth. A portion of the material is oxidized, and the energy released is used to convert the remaining material into new cell tissue. The oxidation reactions are collectively referred to as *catabolism*, and the production or synthesis of new cell tissue is called *anabolism*. Catabolism and anabolism, while conceptually different, are coupled together in many ways, and the combined processes are known as *metabolism*.

Organic and other materials are present in wastewater as suspended and colloidal particles and dissolved molecules. Particles and macromolecules cannot be ingested by bacteria directly but must first be hydrolyzed into small subunits before they can be transported across the cytoplasmic membrane (see Fig. 2.23). For example, starch (a branched polymer of glucose molecules) is broken into glucose and maltose molecules (di-glucose) by extracellular enzymes, and these subunits are transported into the cell for further metabolism. The process of contaminant removal by bacteria can be visualized as a series of steps involving:

Liquid phase transport of contaminant to cell surface → Sorption of contaminant on and in cell capsule →

Hydrolysis of contaminant into subunits → Transport of contaminant subunits to cytoplasmic membrane →

Transport of hydrolyzed submits into cell across cytoplasmic membrane → Metabolism

(14.1)

Any one of the steps may be rate-limiting, varying with the type of process, limiting nutrient, hydraulic regime, and environmental conditions.

It is important to remember that cells are always a product of metabolism, and hence always a product of biological wastewater treatment. If some method of wasting the excess cells produced each day is not provided, the mass of the microbial culture will increase until system failure occurs. The most common types of failure are related to the inability to remove the high microbial-mass concentrations in a gravity sedimentation tank and the inability to meet the high oxygen transfer rates required by the high microbial-mass concentration. In either case, the concentration of organic matter in the effluent will increase. It is also important to remember that the wasted microbial mass presents an auxiliary treatment problem. Satisfactory disposal of microbial sludges is a difficult and expensive problem to solve.

14.4
ACTIVATED SLUDGE PROCESSES

In activated sludge processes, a suspended aerobic microbial culture is used to treat the incoming wastewater. At the end of the reaction period the microbial culture is separated from the liquid being treated. Most of the culture, as shown in Fig. 14.1, is returned and mixed with incoming wastewater. Under conditions existing in the activated sludge process, the microbial culture grows in clumps or *flocs* that contain large numbers of bacteria held together by the secreted polymers that accumulate on their capsules. Typical flocs are shown in the scanning electron micrograph of Fig. 14.2(a). The bacterial cells are scattered though the floc interior and actually make up only about 10 to 25 percent of the floc volume, as can be seen in the transmission electron micrograph of Fig. 14.2(b). Maximum floc size is controlled by shear stress in the reactor, and minimum floc size is controlled by the gravity sedimentation process used to separate the culture from the treated effluent. Organisms other than bacteria (fungi, protozoa, etc.) live in or on the flocs but do not ordinarily occur in large numbers. Some free-swimming organisms such as nematode worms and rotifers are found in activated sludge also. Protozoa (see Fig. 2.28) and rotifers (see Fig. 2.30) feed on free-swimming bacteria and are therefore aids in producing an effluent having low turbidity.

Because the actual bacterial population is very difficult to measure, the concentration of suspended solids or volatile suspended solids in the aeration tank is used as an estimate of the cell concentration. The mixture of wastewater and suspended culture is referred to as the *mixed liquor*, and the respective suspended solids concentrations as *mixed-liquor suspended solids* (*MLSS*) and *mixed-liquor volatile suspended solids* (*MLVSS*).

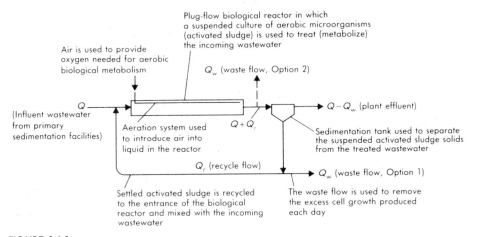

Plug-flow biological reactor in which
a suspended culture of aerobic microorganisms
(activated sludge) is used to treat (metabolize)
the incoming wastewater

Air is used to provide
oxygen needed for aerobic
biological metabolism

Q_w (waste flow, Option 2)

Q
(Influent wastewater
from primary
sedimentation facilities)

Aeration system used
to introduce air into
liquid in the reactor

$Q + Q_r$

$Q - Q_w$ (plant effluent)

Sedimentation tank used to separate
the suspended activated sludge solids
from the treated wastewater

Q_r (recycle flow)

Q_w (waste flow, Option 1)

Settled activated sludge is recycled
to the entrance of the biological
reactor and mixed with the incoming
wastewater

The waste flow is used to remove
the excess cell growth produced
each day

FIGURE 14.1

Definition sketch for the activated sludge process.

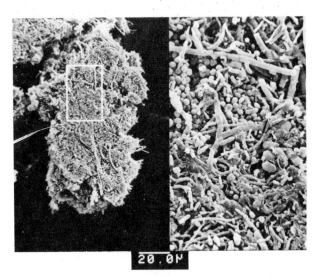

20.0μ

FIGURE 14.2

**Typical example of an activated
sludge floc. The bar shown below the micrograph on the left
represents a length of 20 microns. The micrograph on the
right is a 5 × enlargement of
the area within the rectangle
shown in the micrograph on the
left.**

Source: Courtesy of Audrey Levine,
University of California, Davis, Calif.

Process Configurations

The three basic activated sludge process configurations in use are nominal plug-flow (PF); continuous-flow, stirred-tank (CFST); and batch
(Fig. 14.3). Nominal PF is the most common. Mixing within the reactor
is usually provided by the aeration system. The original activated sludge
configuration was a single-unit batch reactor [14.3]. Hydraulic problems
with the decant portion of the operating cycle resulted in the develop-

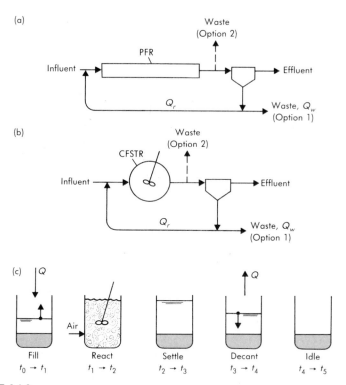

FIGURE 14.3

Representative configurations of the activated sludge process: (a) plug-flow, (b) continuous-flow stirred-tank, and (c) batch.

ment of continuous-flow systems utilizing a separate tank for culture/liquid separation. Until the 1950s, virtually all activated sludge processes were nominal PF with a separate sedimentation tank. Settled cells were recycled or wasted as necessary. In the past 30 years, CFST activated sludge systems have become quite common, although most new processes are still nominal PF. Batch processes have reappeared as methods of decanting improved [14.5, 14.22, 14.24].

Method of Aeration

Oxygen is transferred into the liquid phase of activated-sludge reaction vessels, or aeration tanks, by a number of means. The most common method is the introduction of compressed air through diffusers at the bottom of the aeration tank (Fig. 14.4a). Mechanical turbines, either surface or submerged, are also in widespread use (Fig. 14.4b). Submerged turbines are often coupled to compressed air systems to provide intensive contact between the compressed air bubbles and the liquid. Surface

(a)   (b)

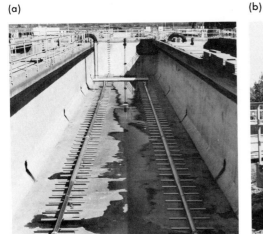

FIGURE 14.4

Common methods of aeration in the activated sludge process: (a) submerged coarse-bubble diffusers (see also Fig. 12.4d) and (b) mechanical turbine aerators.

turbines act as pumps, pulling the liquid upward and throwing it radially in small droplets that make contact with the air. For additional details on aeration see Section 12.4.

Process Modeling

The overall biological reaction in the activated sludge process is the same as that in the BOD process discussed in Chapter 2.

$$\text{Organic material} + O_2 + \text{Nutrients} \xrightarrow{\text{Bacteria}} \text{New cells} + CO_2 + H_2O \quad (2.68)$$

The rate of disappearance of the organic material and oxygen and the rate of formation of new cells are of particular interest in process design. In most cases, it is assumed that the mixing induced by the aeration system eliminates liquid-phase mass transport limitations. In the rate expressions in common use all of the steps from sorption through metabolism [see Eq. (14.1)] are lumped together.

Rate of Removal of Organic Material

The rate of organic removal, r_o, has been found to be a saturation-type function of the form

$$r_o = -\frac{kCX}{K+C} \quad (14.2)$$

where

r_o = rate of organic removal, $g/m^3 \cdot d$

k = maximum specific growth rate, d^{-1}

C = organic concentration (usually as BOD_u), g/m^3

X = mixed-liquor volatile suspended solids concentration, g/m^3

K = half-saturation constant, g/m^3

Reported values of the half-saturation constant K vary with the organic species under consideration. Values of K for pure compounds such as glutamic acid and glucose are usually in the range of 1 to 10 g/m^3. When organic concentration is measured as BOD_u or COD, the K values are most often in the 20- to 30-g/m^3 range. Municipal primary effluents rarely have BOD_u values in excess of 350 g/m^3. Consequently, the rate of organic removal (r_o) cannot be assumed to be saturated.

Rate of Cell Growth

The rate of growth of new cells is estimated as the rate of production of volatile suspended solids, r_g. As would be expected from stoichiometric considerations, the growth rate function is similar to the removal rate function, but an additional term—the maintenance rate $-k_dX$—accounting for basal metabolism must be included:

$$r_g = Y\frac{kCX}{K + C} - k_dX \tag{14.3}$$

where

r_g = cell growth rate, $g/m^3 \cdot d$

Y = cell yield coefficient, g cells produced/g organic matter removed

k_d = maintenance rate coefficient, d^{-1}

and other terms are as defined previously.

The existence of the maintenance rate effectively establishes a minimum organic concentration necessary for a culture's survival. When the organic concentration decreases below the minimum value, the growth rate becomes negative, and the culture begins to deteriorate. The effluent from wastewater treatment processes operated at extremely low growth rates will often contain small or pinpoint flocs made up of organic debris that are difficult to remove by sedimentation.

Rate of Oxygen Uptake

The oxygen uptake rate is related stoichiometrically to the organic removal rate and the growth rate. If the organic concentration is given as BOD_u, the oxygen required to carry out the conversion shown in Eq. (2.68) is the difference between the initial BOD_u value and the oxygen equivalents of the cells produced. The approximate conversion factor between the volatile suspended solids concentration and its oxygen equivalents is 1.42 g of oxygen per gram of volatile solids. The value of 1.42 is obtained by considering the amount of oxygen required to oxidize cell tissue ($C_5H_7NO_2$) to carbon dioxide, ammonia, and water (see Section 2.6). Thus the oxygen uptake rate is

$$r_{O_2} = r_o^* + 1.42r_g \tag{14.4}$$

where

r_{O_2} = rate of oxygen uptake, $g/m^3 \cdot h$

r_o^* = rate of organic removal as BOD_u, $g/m^3 \cdot h$

Although the rate of oxygen uptake in Eq. (14.4) is shown as the sum of the organic removal and growth rates, it is important to remember that the sign of the term r_o^* is negative. Because r_o^* is always greater than r_g [see Eqs. (14.2) and (14.3)], the rate of oxygen uptake is always negative.

Typical Process Coefficients

Typical coefficient and parameter values for the activated sludge process are given in Table 14.3. Most activated sludge processes are operated in the MLSS range of 1500 to 3000 g/m^3. The MLVSS value is usually between 80 and 90 percent of the MLSS concentration, depending upon operating conditions and the characteristics of the wastewater.

TABLE 14.3

Typical Coefficient Values (Based on BOD_u) and Design Parameters for Activated Sludge Processes for Average Wastewater Temperatures in the Range from 13–18°C

| | TYPE OF REACTOR | | |
PARAMETER	CFST	PF	Batch
θ_H, hr	3–8	3–8	2, 4, 1*
θ_c, d	4–10	4–10	> 10†
k, g BOD_u/g cells · d	2–4	2–4	2–4
K, g BOD_u/m^3	10–40	10–40	10–40
k_d, d^{-1}	0.04–0.06	0.04–0.06	0.04–0.06
MLSS, g/m^3	1500–3500	1500–3500	1500–3500
MLVSS, g/m^3	1100–2500	1100–2500	1100–2500
X_r, g/m^3	5000–10,000	5000–10,000	N/A
$X_e^{‡}$, g/m^3	10–35	5–35	5–35
Y, g cells/g BOD_u removed	0.3–0.65	0.3–0.65	0.3–0.65

*Fill, react, settle-decant times used for design based on average flow.

† Full-scale processes currently in operation. Current research is on systems with θ_c < 10 d.

‡ Typical values, not a controllable parameter.

Sludge Age

Recycle of the MLSS in the activated sludge process allows separation of the average retention times of the wastewater and the microbial culture (see Fig. 14.1). The sludge age θ_c, often called the *SRT* (*solids retention time*) or *MCRT* (*mean cell residence time*), is defined as the average time the microbial culture stays in the system and is controlled primarily by the wasting rate Q_w. Wasting rates are usually set to give a sludge age between 3 and 10 d. The sludge age θ_c is calculated from a mass balance on the entire system (Fig. 14.5). In making the mass balance when actual field data are not available, the average suspended-solids concentration in the sedimentation tank is assumed to be equal to the MLSS in the aeration tank:

$$\theta_c = \frac{(V_A + V_s)\,\overline{X}}{Q_w X_r + (Q - Q_w)\,X_e} \tag{14.5}$$

where

θ_c = sludge age, d

V_A = aeration tank volume, m³

V_s = sedimentation tank volume, m³

$\overline{X}$ = average aeration tank suspended-solids concentration, g/m³

Q_w = wastage flow rate, m³/d

X_r = recycle sludge concentration, g/m³

Q = wastewater flow rate, m³/d

X_e = effluent suspended-solids concentration, g/m³

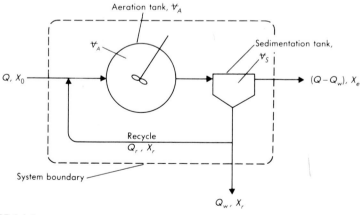

FIGURE 14.5

Definition sketch for determination of sludge age for the activated sludge process.

Calculation of the sludge age is illustrated in Example 14.1. The relationship between sludge age and cell growth is illustrated in Example 14.2.

EXAMPLE 14.1

CALCULATION OF SLUDGE AGE

Determine the sludge age for an activated sludge process with the following operating parameters:

$$Q = 2 \ \text{m}^3/\text{s}$$
$$Q_w = 0.015 \ \text{m}^3/\text{s}$$
$$Q_r = 0.50 \ \text{m}^3/\text{s}$$
$$V_A = 43,000 \ \text{m}^3$$
$$V_s = 7000 \ \text{m}^3$$
$$X_e = 20 \ \text{g/m}^3$$
$$X = 2000 \ \text{g/m}^3$$

SOLUTION:

1. Determine the recycled/wasted sludge concentration X_r by preparing a solids mass balance around the sedimentation tank.
   a. Assuming steady-state conditions and no generation, the appropriate mass balance is

   $$0 \quad = \underbrace{(Q + Q_r) X}_{} - \underbrace{(Q_r + Q_w) X_r - (Q - Q_w) X_e}_{} + 0$$

   Accumulation =   Inflow   −   Outflow   + Generation

   b. Solving for X_r yields

   $$X_r = \frac{(Q + Q_r) X - (Q - Q_w) X_e}{Q_r + Q_w}$$

   c. Substitute the given values and solve for X_r:

   $$X_r = \frac{(2.5 \ \text{m}^3/\text{g})(2000 \ \text{g/m}^3) - (1.985 \ \text{m}^3/\text{s})(20 \ \text{g/m}^3)}{0.515 \ \text{m}^3/\text{s}}$$

   $$= 9632 \ \text{g/m}^3$$

2. Determine the sludge age using Eq. (14.5). Assume the equivalent solids concentration in the secondary settling tank is equal to 2000 g/m, the concentration of suspended solids in the aeration tank.

   $$\theta_c = \frac{(43,000 \ \text{m}^3 + 7000 \ \text{m}^3)(2000 \ \text{g/m}^3)}{(0.015 \ \text{m}^3/\text{s})(9632 \ \text{g/m}^3) + (1.985 \ \text{m}^3/\text{s})(20 \ \text{g/m}^3)}$$

   $$= 5.43 \times 10^5 \ \text{s} = 6.28 \ \text{d}$$

COMMENT

Usually the solids in the sedimentation tank and effluent solids losses are dropped from the calculation for convenience. In this example, the result of this approximation is about 10 percent higher, which is a satisfactory estimate of the true value.

$$\theta_c \approx \frac{(43{,}000 \text{ m}^3)(2000 \text{ g/m}^3)}{(0.015 \text{ m}^3/\text{s})(9632 \text{ g/m}^3)} = 5.95 \times 10^5 \text{ s}$$

$$\approx 6.89 \text{ d}$$

Additional details on the parameters for the activated sludge process may be found in Ref. [14.49].

EXAMPLE 14.2

RELATIONSHIP BETWEEN SLUDGE AGE AND CELL GROWTH RATE IN A STEADY-STATE CFST ACTIVATED SLUDGE PROCESS

Determine the steady-state growth rate in a CFST activated sludge process (Fig. 14.5) as a function of sludge age.

SOLUTION:

1. Prepare a steady-state mass balance on suspended solids around the reactor.

$$0 \quad = \underbrace{QX_i + Q_r X_r}_{} - \underbrace{(1+\alpha)QX}_{} + \underbrace{r_g V_A}_{}$$

Accumulation = Inflow $-$ Outflow $+$ Generation

where $\alpha = Q_r/Q$ and other terms are as defined previously.

2. Prepare a steady-state mass balance on suspended solids around the sedimentation tank.

$$0 \quad = \underbrace{(1+\alpha)QX}_{} - \underbrace{Q_r X_r - Q_w X_r - (Q - Q_w)X_e}_{}$$

Accumulation = Inflow $-$ Outflow

3. Assume QX_i is negligible and solve for r_g using the expressions developed in steps 1 and 2.

$$r_g = \frac{1}{V_A}\left[Q_w X_r + (Q - Q_w)X_e\right]$$

which is equal to

$$r_g = \frac{1}{V_A}\frac{(V_A + V_S)X}{\theta_c}$$

Neglecting V_S, the approximate value of r_g is

$$r_g \simeq \frac{X}{\theta_c}$$

COMMENT

The term r_g/X is commonly called the *specific growth rate* μ. For situations where solids storage in the sedimentation tank and solids losses in the effluent are negligible,

$$\mu = \theta_c^{-1}$$

Note that this relationship is only true for a CFST activated sludge process operating under steady-state conditions. Specific growth rates in batch and PF activated sludge processes vary with time and position, respectively. However, the concept of the average specific growth rate remains useful in comparing loading rates and performance among processes of different configurations.

Application of Rate Expressions

Application of the rate expressions to process design differs for each system. In most cases, continuous-flow systems are assumed to be operating at steady state. Batch processes in which the removal steps are separated in time are inherently non-steady state. Because Eqs. (14.2), (14.3), and (14.4) have been developed from experimental work with continuous-flow systems, they are not directly applicable to batch processes. It is quite possible that the form of the batch expressions for the reaction stage will be the same as for continuous-flow systems, but the coefficient values will undoubtedly be different.

Process Design and Operating Parameters

As can be seen from Table 14.3, the design parameter values of the three basic configurations of activated sludge differ very little. The two continuous-flow variations have been studied extensively, and considerable information is available on their performance, although complete agreement is still lacking [14.6, 14.23, 14.31, 14.44].

Batch processes are receiving increased interest from designers. Although the configuration was the first used for activated sludge [14.3], it was quickly dropped because of operational difficulties. In the 1950s, Pasveer [14.38] suggested a quasi-batch operation of a lightly loaded activated-sludge-process modification termed the *oxidation ditch*, and Goronzy [14.22] developed this idea extensively. Irvine and Busch [14.24], working from the direction of optimizing process performance, introduced the concept of anoxic fill, microprocessor control, and sequencing reactors in parallel, and they are responsible for making sequencing batch reactors (SBRs) a viable alternative to continuous-flow processes.

Secondary Sedimentation

Separation of the microbial mass from the treated wastewater is a required step in the activated sludge process. As noted in the previous sections, recycling the settled microbial sludge allows separation of the hydraulic and solids retention times. Equally important is the fact that the microbial culture would be a major pollution load on the receiving waters, as 30 to 50 percent of the influent BOD is converted to cells. Microbial material in the effluent decays with time and has a BOD_u of approximately 1.42 g O_2/g cells.

Discrete particle sedimentation was discussed in Chapter 12. Sedimentation of thick suspensions, as is typical of activated sludge, is more complex than discrete particle settling because of fluidic and mechanical interaction between particles. However, the design of secondary clarifiers is based on overflow rate, as in the case of discrete particle settling. Typical design values are in the range from 15 to 32 $m^3/m^2 \cdot d$, based on average dry-weather flow (see Table 12.3). Where wide variations in flow occur (e.g., small communities or communities with high infiltration rates in the sewers) conservative overflow-rate values should be used. Overflow rates based on peak flow should not exceed 60 $m^3/m^2 \cdot d$. Solids loading rates are also used in secondary clarifier design. Typical values are in the range from 70 to 140 $kg/m^2 \cdot d$, based on the average flow rate plus recycle, with a peak loading of 200 $kg/m^2 \cdot d$.

Activated sludge remains biologically active in the sedimentation tank. If the settled solids remain in the tank bottom for an extended period of time, gases will be generated and will become entrained in the floc. The result is rising sludge and loss of solids in the effluent. In

(a)   (b)

FIGURE 14.6

Secondary clarifier equipped with a suction manifold to remove settled activated sludge.

activated sludge systems where nitrification is occurring, denitrification can also occur, adding to the rising sludge problem. Prevention of rising sludge is accomplished by pumping settled solids from the bottom using a suction manifold attached to the sludge-scraping system (Fig. 14.6). Hydraulic residence times in secondary sedimentation tanks based on average flow will vary from 2 to 5 hr depending on the depth of the tank.

EXAMPLE 14.3

PROCESS DESIGN OF A CFST ACTIVATED SLUDGE PROCESS

A domestic wastewater having an average BOD_5 of 200 g/m³ after primary sedimentation is to be treated using a CFST activated sludge process. Average and peak design flow rates are 0.5 m³/s and 0.75 m³/s, respectively. At peak flow the wastewater BOD_5 is 150 g/m³ after primary sedimentation. Nitrification is not desired, and the design sludge age is to be equal to or less than 6 d. Effluent BOD_5 and suspended-solids concentrations are to be below 30 g/m³. Determine the MLSS concentration, the oxygen uptake rate in the aeration tank, the aeration and sedimentation tank volumes, and the waste sludge produced per day.

SOLUTION:

1. Calculate influent BOD_u values using the first-order model (see Section 2.6) and a first-order rate constant of 0.23 d^{-1}.

$$\text{Average } BOD_u = \frac{BOD_5}{0.68} = \frac{200}{0.68} = 294 \text{ g/m}^3$$

$$\text{Peak } BOD_u = \frac{BOD_5}{0.68} = \frac{150}{0.68} = 221 \text{ g/m}^3$$

2. Assume the effluent suspended-solids concentration $X_e = 20$ g/m³, and determine the approximate effluent BOD_5 contributed by effluent suspended solids.

$$BOD_u X_e = (1.42 \text{ g BOD/g SS})(X_e)$$
$$= (1.42 \text{ g BOD/g SS})(20 \text{ g/m}^3)$$
$$= 28.4 \text{ g/m}^3$$
$$BOD_5 \approx 0.68 \, X_e \quad \text{(see Section 2.6)}$$
$$= 19.3 \text{ g/m}^3$$

3. Calculate filtrable BOD allowable in effluent.

$$FBOD_5 = 30 \text{ g/m}^3 - 19.3 \text{ g/m}^3 = 10.7 \text{ g/m}^3$$
$$FBOD_u \approx (10.7 \text{ g/m}^3)/0.68 = 15.7 \text{ g/m}^3$$

4. Determine the effluent $FBOD_u$ using Eq. (14.3) and a steady-state

materials balance on MLSS. Apply the result of Example 14.2 and Eq. (14.3).

$$r_g = Y\frac{kCX}{K+C} - k_d X \simeq \frac{X}{\theta_c}$$

a. Solving for C, the effluent $FBOD_u$ yields

$$C = \frac{K(1+k_d\theta_c)}{Yk\theta_c - (1+k_d\theta_c)}$$

b. Choose conservative coefficient and parameter values from Table 14.3 and solve for C.

$k = 2$ g BOD_u/g cells · d

$k_d = 0.06$ d^{-1}

$K = 30$ g BOD_u/m^3

$Y = 0.4$ g cells/g BOD_u removed

$\theta_c = 5$ d

$$C = \frac{(30 \text{ g/m}^3)[1+(0.06 \text{ d}^{-1})(5 \text{ d})]}{0.4(2 \text{ d}^{-1})(5 \text{ d}) - [1+(0.06 \text{ d}^{-1})(5 \text{ d})]}$$

$$= 14.4 \text{ g/m}^3$$

5. Determine the MLSS concentration using a steady-state materials balance on $FBOD_u$ around the aeration tank.

$$0 = \underbrace{QC_i + \alpha QC_r}_{\text{Inflow}} - \underbrace{(1+\alpha)QC}_{\text{Outflow}} + \underbrace{r_o V}_{\text{Generation}}$$

The most conservative solution is to assume no biological reaction in the sedimentation tank or recycle line, which results in $C_r = C$. The materials balance then reduces to

$$0 = QC_i - QC + r_o V$$

Solving for r_o yields

$$r_o = \frac{Q(C-C_i)}{V_A} = \frac{C-C_i}{\theta_H}$$

From Eq. (14.2)

$$r_o = -\frac{kCX}{K+C}$$

Combining the above equations and solving for X yields

$$X = \frac{(K+C)(C_i-C)}{kC\theta_H}$$

Because activated sludge processes do not respond sharply to changes in loading, average influent BOD and average θ_H values can be used. Choose $\theta_H = 4$ hr (0.167 d) and determine the MLSS in the aeration tank.

$$X = \frac{(30 \text{ g/m}^3 + 14.4 \text{ g/m}^3)(294 \text{ g/m}^3 - 14.4 \text{ g/m}^3)}{(2 \text{ d}^{-1})(14.4 \text{ g/m}^3)(0.167 \text{ d})}$$

$$= 2581 \text{ g/m}^3$$

6. Determine the oxygen uptake rate in the aeration tank. From Eq. (14.4)

$$r_{O_2} = r_o^* + 1.42 r_g$$

$$= \frac{C - C_i}{\theta_H} + 1.42 \left(Y \frac{kCX}{K + C} - k_d X \right)$$

$$= \frac{14.4 \text{ g/m}^3 - 294 \text{ g/m}^3}{0.167 \text{ d}} + 1.42$$

$$\times \left[0.4 \frac{(2 \text{ d}^{-1})(14.4 \text{ g/m}^3)(2581 \text{ g/m}^3)}{30 \text{ g/m}^3 + 14.4 \text{ g/m}^3} \right.$$

$$\left. - (0.06 \text{ d}^{-1})(2581 \text{ g/m}^3) \right]$$

$$= -1674 \text{ g/m}^3 \cdot \text{d} + 731 \text{ g/m}^3 \cdot \text{d} = -943 \text{ g/m}^3 \cdot \text{d}$$

7. Check oxygen uptake rate at peak flow rate, assuming MLSS and effluent BOD concentrations do not change.

$$\theta_H = \frac{0.5 \text{ m}^3/\text{s}}{0.75 \text{ m}^3/\text{s}} (4 \text{ hr}) = 2.67 \text{ hr} = 0.11 \text{ d}$$

$$r_{O_2} = \frac{14.4 \text{ g/m}^3 - 221 \text{ g/m}^3}{0.11 \text{ d}} + 1.42$$

$$\times \left[0.4 \frac{(2 \text{ d}^{-1})(14.4 \text{ g/m}^3)(2581 \text{ g/m}^3)}{30 \text{ g/m}^3 + 14.4 \text{ g/m}^3} \right.$$

$$\left. - (0.06 \text{ d}^{-1})(2581 \text{ g/m}^3) \right]$$

$$= -1878 \text{ g/m}^3 \cdot \text{d} + 731 \text{ g/m}^3 \cdot \text{d} = -1147 \text{ g/m}^3 \cdot \text{d}$$

8. Determine the aeration tank volume based on the average flow rate.

$$V_A = Q\theta_H$$

$$= (0.5 \text{ m}^3/\text{s})(0.167 \text{ d})(86,400 \text{ s/d})$$

$$= 7214 \text{ m}^3$$

Note: At least two and preferably three aeration tanks should be used in all cases. Using three tanks, each 5 m deep, the volume and area of each tank would be 2485 m³ and 481 m². These are conveniently sized units.

9. Determine the surface area and volume of the sedimentation tanks using a hydraulic loading rate of 24 m³/m² · d based on average flow (see Table 12.3). The detention time is determined in step 10. The solids loading rate, based on average flow plus recycle, is checked in step 11.

    Using a design hydraulic loading rate of 24 m³/m² · d

$$A_s = \frac{Q}{\text{HLR}}$$

$$= \frac{(0.5 \text{ m}^3/\text{s})(86{,}400 \text{ s/d})}{24 \text{ m}^3/\text{m}^2 \cdot \text{d}}$$

$$= 1800 \text{ m}^2$$

Secondary sedimentation tanks should have an average depth of approximately 4.0 m, giving a total sedimentation tank volume of

$$V_s = (1800 \text{ m}^2)(4.0 \text{ m}) = 7200 \text{ m}^3$$

Note: As in the case of aeration tanks, a number of sedimentation tanks are desirable to provide maximum flexibility and redundancy. Sedimentation-tank diameters are limited by equipment availability to approximately 60 m. If six tanks are used, the approximate diameter will be 20 m.

10. Determine hydraulic detention time in sedimentation tanks at average flow.

$$\theta_H = \frac{V_s}{Q_{\text{avg}}}$$

$$= \frac{7200 \text{ m}^3}{(0.5 \text{ m}^3/\text{s})(3600 \text{ s/hr})} = 4.0 \text{ hr}$$

11. Check solids loading rate to secondary sedimentation tank. Conservative design practice is to allow for 100 percent recycle.

$$\text{Solids loading rate} = \frac{(Q_{\text{avg}} + Q_r)X}{A_s}$$

$$= \frac{(0.5 \text{ m}^3/\text{s} + 0.5 \text{ m}^3/\text{s})(86{,}400 \text{ s/d})(2581 \text{ g/m}^3)}{1800 \text{ m}^2 (1000 \text{ g/kg})}$$

$$= 124 \text{ kg/m}^2 \cdot \text{d} \left(5.2 \text{ kg/m}^2 \cdot \text{h}\right)$$

12. Calculate waste sludge production rate. The waste sludge production rate is equal to $r_g V_A$.

$$r_g V_A = \left(Y \frac{kCX}{K+C} - k_d X \right) V_A$$

$$= \left\{ 0.4 \left[\frac{2 \, \mathrm{d}^{-1} (14.4 \, \mathrm{g/m^3})(2581 \, \mathrm{g/m^3})}{30 \, \mathrm{g/m^3} + 14.4 \, \mathrm{g/m^3}} \right] \right.$$

$$\left. - 0.06 \, \mathrm{d}^{-1} (2581 \, \mathrm{g/m^3}) \right\} 7214 \, \mathrm{m^3}$$

$$= 3.7 \times 10^6 \, \mathrm{g/d} = 3700 \, \mathrm{kg/d}$$

COMMENT

The complete design of an activated sludge process requires consideration of the peak and minimum flow conditions and varying recycle rates. Values calculated in this example are approximately correct, but actual θ_c values need to be determined based on wastage rates and losses in the effluent. Design of PF and SB activated sludge systems is more complex because of the spatial and time variation of the rates, but design values will be similar to those calculated in this example.

Other Activated Sludge Configurations

A number of modifications of the three basic activated-sludge-process flow sheets exist. The most important are the contact stabilization, the step aeration, the oxidation ditch, the pure oxygen, and the minimal aeration systems.

Contact Stabilization Process

The contact-stabilization activated sludge system was developed by Ulrich and Smith [14.56] at the Austin, Texas, wastewater treatment plant (Fig. 14.7a). In the contact stabilization process, the organic adsorption step is physically separated from the metabolism step. A small aerated contact tank ($\theta_H = 0.5$ hr) is followed by a sedimentation tank. The settled sludge is then aerated for several hours in a reaeration basin. Because the settled sludge is three to six times as concentrated as the sludge in the contact tank the overall system is smaller than conventional processes. Niku and Schroeder [14.34] found that contact stabilization systems are somewhat less stable than conventional activated sludge processes.

(a)

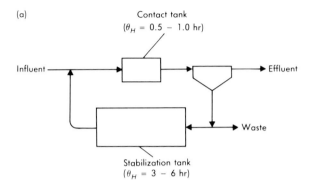

Contact tank
$(\theta_H = 0.5 - 1.0 \text{ hr})$

Influent

Effluent

Waste

Stabilization tank
$(\theta_H = 3 - 6 \text{ hr})$

(b)

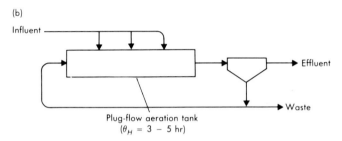

Influent

Effluent

Waste

Plug-flow aeration tank
$(\theta_H = 3 - 5 \text{ hr})$

(c)

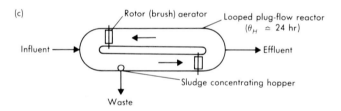

Rotor (brush) aerator   Looped plug-flow reactor
$(\theta_H \simeq 24 \text{ hr})$

Influent

Effluent

Sludge concentrating hopper

Waste

(d)

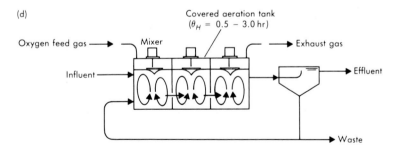

Covered aeration tank
$(\theta_H = 0.5 - 3.0 \text{ hr})$

Oxygen feed gas   Mixer   Exhaust gas

Influent   Effluent

Waste

FIGURE 14.7

Definition sketch for alternative activated sludge processes: (a) contact stabilization, (b) step aeration, (c) oxidation ditch, and (d) pure oxygen.

Step Aeration Process

In a step-aeration activated sludge system, incoming wastewater is introduced into a nominally plug-flow aeration tank at several points along its length (Fig. 14.7b). The result is that the load is distributed over the system more uniformly, and the process behaves more like a CFST activated sludge process.

Oxidation Ditch Process

The oxidation ditch system, developed by Pasveer in Holland [14.38], is a variation of the conventional PF activated sludge process in which a continuous-loop PF reactor is used (Fig. 14.7c). The contents of the reactor are aerated, mixed, and recirculated continuously using a brush aerator (Fig. 14.8). The original oxidation ditch was developed to provide low-cost treatment for small communities. The oxidation ditch is operated in the extended aeration mode (i.e., with long hydraulic detention times, high solids retention times, and low organic loading rates). The oxidation ditch process has found wide application in the United States and throughout the world because of its performance, simplicity, and low cost. Additional details on the oxidation ditch may be found in Ref. [14.29].

Pure Oxygen Process

The pure-oxygen activated sludge process was first studied by Okun in 1947 [14.36]. Twenty years later, Union Carbide Corporation developed a commercial version of the pure oxygen process (Fig. 14.7d). Since

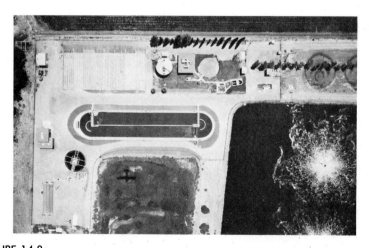

FIGURE 14.8

The oxidation-ditch activated sludge process.

Source: Courtesy of George S. Nolte and Associates.

1970, a large number of pure-oxygen activated sludge processes have been put into operation, including several plants treating over 400,000 m^3/d. In large plants (over 40,000 m^3/d) oxygen is generated cryogenically. Smaller plants, partially because of scale and partially because of wider swings of inflow, use a selective sorption process—pressure swing adsorption (PSA)—for oxygen production.

Advantages of the pure oxygen system include higher oxygen transfer gradients and rates and smaller reaction tanks. These advantages are important in treating strong industrial wastewaters. Overall cost and energy benefits have not been established for either air or pure oxygen systems. Air systems have the advantage of stripping out excess CO_2, which makes pH control simpler.

Minimal Aeration Process

This modification is a unique method that minimizes the air required for the activated sludge process and is applicable where more than one tank is used [14.60]. Flow to each aeration tank is maintained at a level near the design value. Because of flow variation, maintenance of flow to all tanks on around-the-clock basis is impossible, and some tanks are taken off-line for periods of several hours per day. The contents of the off-line tanks are not aerated, and, consequently, considerable aeration energy is saved. To take advantage of this process the aeration system must be designed so that portions of the system can be isolated and taken out of service.

Growth of Filamentous Microorganisms

The growth of filamentous microorganisms (Fig. 14.9) is the most common operational problem in all activated sludge systems. There are several genera of microorganisms that cause the problem, most notably *Microthrix*, *Sphaerotilus*, *Beggiatoa*, *Thiothrix*, *Nocardia*, *Lecicothrix*, and *Geotrichum* [14.11, 14.18, 14.53]. These organisms are present in normal activated sludge in small numbers but, given appropriate conditions, grow to such an extent that the floc becomes fluffy, less dense, and difficult to compact. Secondary clarification and thickening become difficult and often fail when cultures become filamentous (or bulky). Observed and reported causes of filamentous bulking include low organic concentration, low nutrient (N, P, Fe) concentration [14.63], and low dissolved-oxygen concentration. More than one cause is probable, but none is well understood.

Control of filamentous bulking can be achieved kinetically by increasing the average growth rate through increased wasting, by providing high growth rate conditions in a region near the aeration tank inlet, or chemically by applying chlorine (1 to 10 g Cl_2 per kg sludge) to the recycled sludge. Chemical control is the most common approach, but

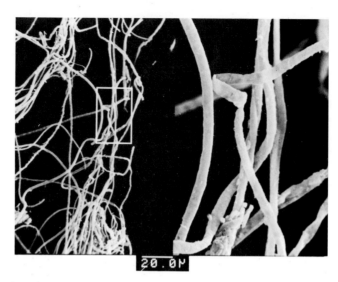

FIGURE 14.9

Typical example of a filamentous activated sludge culture. Note the fluffy character of the floc as compared to the dense floc shown in Fig. 14.2. The bar shown below the micrograph on the left represents a length of 20 microns. The micrograph on the right is a 5 × enlargement of the area within the rectangle shown in the micrograph on the left.

Source: Courtesy of Audrey Levine, University of California, Davis, Calif.

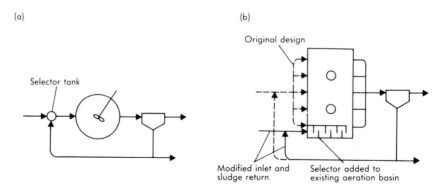

FIGURE 14.10

Use of a selector or contact tank to control the growth of filamentous organisms; (a) selector tank added to complete-mix activated sludge process and (b) selector constructed into existing activated sludge reactor.

Source: Adapted from Refs. [14.11] and [14.13].

kinetic control is becoming better understood and appears to have considerable promise. It is now well established that installation of a short-residence-time mixing volume at the aeration tank inlet (Fig. 14.10) will give floc-forming bacteria a kinetic advantage and retard filamentous growth [14.11, 14.12, 14.13, 14.45].

Nocardia are gram-positive obligate aerobic bacteria that grow in a mycelial form that has few cross-walls to separate nucleoids. Although *Nocardia* are sometimes associated with filamentous bulking, organisms of this genus can also cause severe foaming problems in the activated sludge process. The foam is difficult to break down and is often described as having the consistency of styrofoam. Elimination of the organisms using one of the methods cited previously is the best method of controlling foaming caused by *Nocardia*.

14.5
INORGANIC NUTRIENT REMOVAL IN ACTIVATED SLUDGE PROCESSES

Nitrogen and phosphorus removal in activated sludge processes has received increasing interest since 1960. Some nitrogen and phosphorus are removed by assimilation into new cell tissue and subsequent sludge wasting, but in most municipal and many industrial wastewaters an excess exists that is not removed by conventional treatment. Nitrogen can be removed by nitrification followed by denitrification, and phosphorus can be removed by modifying the process to encourage *luxury uptake*—that is, polyphosphate storage within the cells in bodies called volutin granules.

Nitrogen Removal

Nitrification has been discussed in Section 2.5, and the principles described there are applied in biological wastewater treatment. A complication occurs in that the mixotrophic nitrifying organisms are not competitive for dissolved oxygen with heterotrophic organisms unless organic concentrations are very low [14.50]. Thus SRT values used are normally greater than 7 d. Both nitrification and denitrification rates appear to follow Monod-type functions. Saturation coefficients for all of the reactions are in the 0.05- to 0.5-g/m^3 range, and consequently, the effective reaction rates are of nearly zero order.

Because the aerobic nitrification reactions must precede the anaerobic denitrification reactions, a second operating constraint occurs. As noted above, organic concentrations must be low for nitrification to proceed, but an organic carbon and energy source is needed to drive the denitrification reactions. Wastewater can be used by bypassing a portion of the flow to the denitrification reactor (Fig. 14.11a), but the organic and

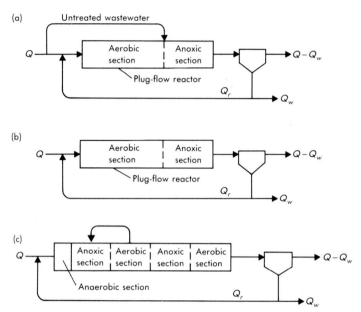

FIGURE 14.11

Alternative flow sheets for denitrification.

ammonia nitrogen present in the bypassed wastewater are not removed. The net effect is a maximum nitrogen removal of approximately 60 percent. Endogenous respiration can be used to drive denitrification, but the rates are an order of magnitude less than where an organic source is provided. Organics such as methanol [14.33] or high COD/N industrial wastes can be used (Fig. 14.11b). Methanol is relatively expensive, and suitable industrial wastes are often unavailable. Abufayed [14.1] has investigated using primary wastewater sludge as an organic source for denitrification. In his laboratory-scale studies, organic nitrogen release was minor, and effluent total nitrogen concentrations below 1.0 g/m³ were attainable. Further work needs to be done at pilot- and full-scale to determine the economic feasibility of this approach.

Barnard [14.4] developed a unique process configuration in which denitrification is carried out effectively using wastewater as the organic source (Fig. 14.11c). There is a high rate of recirculation between the aerobic second cell where nitrification occurs and the anaerobic first cell where most of the denitrification takes place. The third cell is anaerobic and allows endogenous denitrification of the remaining nitrate. A final aerobic cell is used to stop denitrification and prevent rising sludge in the clarifier. The process is proprietary and goes under the trade name of Bardenpho. This process has been used widely in South Africa and is increasingly used in the United States and Canada. Bardenpho processes are loaded very lightly in terms of COD and BOD and are classified as extended aeration systems.

Phosphorus Removal

Biological phosphorus removal can be accomplished by operating an activated sludge process in an anaerobic-aerobic sequence [14.4, 14.9]. A number of bacterial species respond to this sequence by accumulating large excesses of polyphosphate within the cell in volutin granules [14.9]. During the anaerobic period, a release of phosphate takes place. In the aerobic phase, the released phosphate and an additional increment are taken up and stored as polyphosphate, giving a net removal, coincident with organic removal and metabolism. Phosphate can be removed as waste sludge or through use of a second anaerobic phase. During the second anaerobic phase the stored phosphate is released; the cells can be separated and recycled and the released soluble phosphate removed by precipitation. The proprietary Bardenpho process described above follows the anaerobic-aerobic schematic and is used for phosphate as well as nitrogen removal.

Two other proprietary phosphate removal processes are marketed —the Phostrip process, which was the first of the three proposed and uses a second anaerobic phase and precipitation, and the A/O process. In the Phostrip process the anaerobic phosphate release is in the sludge recycle and is followed by a precipitation of the released phosphate. Conceptually, the A/O process is similar to the Bardenpho process, except without nitrification-denitrification as a necessary step. Phosphate in the A/O process is removed as excess sludge.

14.6
TRICKLING FILTER PROCESSES

Trickling filters are the oldest form of engineered biological treatment (Fig. 14.12). Sometimes called biological filters, these systems developed somewhat naturally from attempts to filter municipal and other wastewaters. Because microbial growth on the filter medium quickly clogged the filters, larger and larger media were used until a rock size of 50 to 100 mm was reached, and clogging was minimal. Performance under the hydraulic and organic loading conditions imposed on these empirically designed units is excellent, but the unit size requirements are so great that the so-called low-rate (standard) trickling filter is uneconomical for use with large flows. (The breakpoint is approximately 4000 m³/d.) Two major modifications of the basic process have been developed—high-rate and plastic-medium trickling filters.

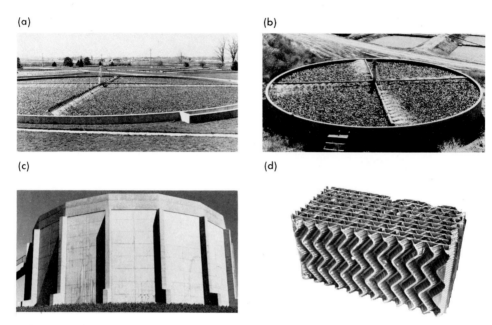

FIGURE 14.12

Typical trickling filters and filter mediums: (a) and (b) conventional high-rate rock filters, (c) covered high-rate tower trickling filter packed with a plastic filter medium, and (d) typical plastic filter medium (0.6 × 0.6 × 1.2 m).

FIGURE 14.13

Distributors used to apply wastewater to trickling fiters: (a) rotary and (b) fixed.

Source: Part (b) courtesy of Neptune Microfloc, Inc.

High-rate trickling filters always include effluent recycle. The recycle increases the shearing rate and controls the buildup of attached microbial slime on the filter medium, as well as damping out the effects of flow variation and improving the distribution of organic material through the filter depth. Plastic media were first used in 1954 [14.7, 14.43]; because of their higher porosity and light mass relative to rock, they can be used in a deep, space-saving configuration. Plastic media are more expensive than rock, and for this reason they are used almost exclusively for highly loaded systems where the high porosity can be used to greatest advantage.

Performance of high-rate rock- and plastic-medium trickling filters is generally lower than low-rate systems in terms of effluent BOD and SS concentrations. Low-rate processes typically produce effluents having about 20 g/m^3 of these two constituents, while high-rate rock- and plastic-medium trickling filters usually have effluent concentrations equal to or greater than 30 g/m^3 [14.35].

Physical Characteristics

Most trickling filters are designed using a rotating distributor that evenly sprinkles wastewater over the upper surface (Fig. 14.12). Rock trickling filters are usually between 1 and 3 m deep, and the diameter, which is determined by the loading parameters, is rarely more than 50 m. Trickling filters with plastic media are usually 3 to 12 m in depth. The distributor, if of the rotating type, must have geometrically spaced ports so that the wastewater will be evenly spread over the surface. Both fixed-nozzle (Fig. 14.13) and reciprocating rectangular distributors exist, but most current designs incorporate rotating units. Slotted vitrified-clay tile underdrains are used with conventional rock trickling filters because of the required compressive strength, durability, and corrosion resistance. Other materials can be used for plastic-medium units because the strength required is much less. Where high hydraulic loading rates are used, care must be taken to ensure that airflows through the underdrains are satisfactory [14.31, 14.44].

Secondary Sedimentation

Most of the suspended solids in trickling-filter effluents are sloughed biomass. The mass of solids produced is roughly equivalent to an activated sludge process operated at an extremely long SRT. Sloughing is caused by a number of factors, including shear stress and anoxic conditions near the slime-media interface. In low-rate systems, slime thickness usually builds up to depths of several millimeters and solids slough periodically, while in high-rate units sloughing is nearly continuous. In either case sedimentation tanks must be provided, however.

Sloughed biological solids are denser than activated sludge floc and settle well under quiescent conditions. Solids from high-rate systems tend to be small in size, and removal by sedimentation is not generally as satisfactory as with low-rate systems or activated sludge processes [14.6, 14.22, 14.31].

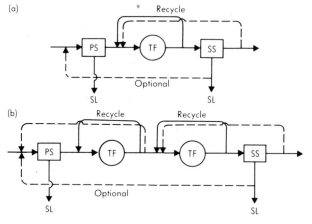

FIGURE 14.14

Typical high-rate trickling-filter flow configurations: (a) single-stage filters and (b) two-stage filters.

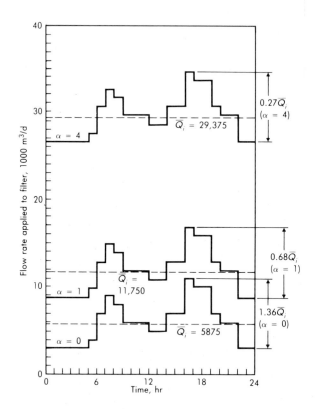

FIGURE 14.15

Damping of the influent flow variation with internal recycle in Example 14.4.

Recycle

Effluent recycle in high-rate systems, including those with plastic filter media, may be from a point ahead of, or following, the sedimentation tank (Fig. 14.14). Presedimentation recycle provides an advantage in that sloughed cells are mixing with incoming wastewater, thus enhancing the reaction rate; but it also has the disadvantage of increasing the possibility of plugging the unit. Postsedimentation recycle increases the sedimentation tank loading and tends to dilute the wastewater without adding a reactant, but does not have the plugging potential of the presedimentation recycle format. Both configurations provide flow equalization.

EXAMPLE 14.4

DETERMINATION OF EQUALIZATION EFFECT OF RECYCLE

The influent flow to a trickling-filter plant varies over a typical day, as shown in Fig. 14.15. Determine the applied flow for recycle ratios of 1 and 4.

SOLUTION:

1. Determine the total applied flow Q_T. The total flow applied to the filter is equal to the incoming flow rate plus the recycle flow rate:

$$Q_T = Q_i + \alpha \overline{Q}_i$$

where

Q_T = total applied flow rate, m^3/d
Q_i = inflow flow rate, m^3/d
α = recycle ratio, dimensionless
$\alpha = Q_r / \overline{Q}_i$
Q_r = recycle flow rate, m^3/d
$\overline{Q}_i$ = average daily inflow flow rate, m^3/d

2. Determine the average daily flow rate $\overline{Q}_i$.
   a. The average flow rate is equal to the area under the Q_i curve divided by the elapsed time:

$$\overline{Q}_i = \frac{\text{area under } Q_i \text{ curve}}{\text{elapsed time}}$$

   b. The area under the curve is 141,000 $m^3 \cdot hr/d$.
   c. The average flow rate is

$$\overline{Q}_i = \frac{141,000 \ m^3 \cdot hr/d}{24 \ hr}$$
$$= 5875 \ m^3/d$$

3. Plot the total applied flow rate using the equation given in step 1 for recycle ratios of 1.0 and 4.0. The solutions are plotted in Fig. 14.15. The difference between the high and low flow rates for the

three flow conditions is as follows:

α	DIFFERENCE BETWEEN HIGH AND LOW FLOWS
0	$1.36\overline{Q}_i$
1	$0.68\overline{Q}_i$
4	$0.27\overline{Q}_i$

Process Design and Operating Parameters

Important process design and operating parameters for trickling filters include the hydraulic loading rate (HRL) and the organic loading rate (ORL). Investigation of removal organic mechanisms in trickling filters has included study of the formation of the biofilm [14.8, 14.43, 14.55] and the formulation of mathematical models based on heterogeneous and pseudohomogeneous reaction expressions [14.17, 14.41, 14.54, 14.58]. Commonly used trickling-filter design formulas are given in Table 14.4. Application of the Eckenfelder equation [Eq. (14.7)] is illustrated in Example 14.5.

TABLE 14.4

Commonly Used Trickling Filter Design Formulas

FORMULA		DEFINITION OF TERMS
Velz [14.58] $\dfrac{C_e}{C_i^*} = e^{-kz}$	(14.6)	C_e = effluent BOD concentration, g/m^3 C_i^* = influent BOD concentration (including recycle), g/m^3 k = removal rate constant, m^{-1} z = depth, m
Eckenfelder [14.17] $\dfrac{C_e}{C_i^*} = \exp\left\{ -k\left[\dfrac{A_s}{(1+\alpha)Q}\right]^n za_v^m \right\}$	(14.7)	C_e = effluent BOD concentration, g/m^3 C_i^* = influent BOD concentration (including recycle), g/m^3 k = removal rate constant, variable units A_s = cross-sectional area of filter, m^2 α = recycle ratio Q = influent flow rate, m^3/d z = depth, m a_v = specific surface area of filtering medium, m^2/m^3 m, n = empirical constants

Hydraulic Loading Rate

The hydraulic loading rate is an empirically derived design and operating parameter that relates to ponding, surface shearing rate, and hydraulic detention time. Usually hydraulic loading rate is reported in units of volume of wastewater, including recycle, per unit cross-sectional area per day. Because most rock-medium trickling filters are between 1 and 2 m in depth, the volumetric loading rate used in some countries is easily translated into flow per unit area-time. Hydraulic loading rates in common use are from 0.5 to 3.0 $m^3/m^2 \cdot d$ for low-rate units, 8 to 30 $m^3/m^2 \cdot d$ for high-rate rock trickling filters, and up to 50 $m^3/m^2 \cdot d$ for high-rate plastic-medium trickling filters.

Organic Loading Rate

Waste-material loading on trickling filters is characterized by the organic loading rate in terms of kilograms of BOD_5 per cubic meter-day (kg $BOD_5/m^3 \cdot d$). There is no parameter for solids loading, and solids removal in trickling filters has not been characterized in any predictive manner. It is known that standard-rate systems usually have effluent BOD_5 and suspended-solids concentrations in the 20-g/m^3 range and that high-rate trickling-filter effluent BOD_5 and SS concentrations are typically 30 g/m^3 or higher. Typical organic loading rate ranges are 0.1 to 0.4, 0.4 to 1.8, and 0.5 to 3.0 kg $BOD_5/m^3 \cdot d$ for low-rate and high-rate rock- and plastic-medium trickling filters, respectively.

Standard Design Parameters

A number of parameters have been discussed above: depth, area, porosity, specific surface, loading rate, and recirculation ratio. A summary of standard values and typical performance characteristics is given in Table 14.5. The ranges result from differences in waste characteristics

TABLE 14.5
Typical Design Parameters and Operating Characteristics of Trickling Filters for Average Wastewater Temperatures in the Range from 13–20°C

CHARACTERISTICS	LOW-RATE ROCK	HIGH RATE	
		Rock	Plastic
Depth, m	1–3	1–2.5	3–12
Specific surface, m^2/m^3	40–70	40–70	80–100
Porosity	0.45–0.55	0.45–0.55	0.90–0.97
Hydraulic loading rate, $m^3/m^2 \cdot d$	0.5–3.0	8–30	10–50
Organic loading rate, kg $BOD_5/m^3 \cdot d$	0.1–0.4	0.4–1.8	0.5–3.0
Recirculation ratio	0	1–4	1–4
Sloughing	Intermittent	Continuous	Continuous
Nitrification	Yes	At lower organic loading rates	Not in economic operating range
Effluent BOD_5, g/m^3	< 25	≥ 30	≥ 30
Effluent SS, g/m^3	< 25	≥ 30	≥ 30

and flow variation. For example, a system designed to treat a strong waste might have a high organic loading and a low hydraulic loading. Average annual flow is used to determine loading rates and system sizing. Peak flows are used to determine hydraulic capacity, except where large seasonal variations occur.

EXAMPLE 14.5

SIZING OF A TRICKLING FILTER

A wastewater has an average annual flow rate of 15,000 m^3/d and a BOD_5 concentration of 400 g/m^3. Using the data in Table 14.5, determine the appropriate size of (a) a high-rate rock and (b) a plastic-medium trickling filter having a recycle ratio of 2. Use the Eckenfelder formula given in Table 14.4, with $k = 0.1$, $m = 1.0$, and $n = 0.7$, to check the design arrived at using the data in Table 14.5. The effluent BOD_5 must be less than or equal to 30 g/m^3.

SOLUTION:

1. Determine the total average applied flow rate.

   $Q_i = Q + Q_r = 3Q = 45,000$ m^3/d

2. Determine necessary cross-sectional areas using average values from Table 14.5.
   a. Rock medium:

   Average hydraulic loading rate $(HLR) = 19$ $m^3/m^2 \cdot d$

   $$A = \frac{45,000 \text{ m}^3/\text{d}}{\text{HLR}}$$

   $A = 2368$ m^2

   b. Plastic medium:

   Average $HLR = 30$ $m^3/m^2 \cdot d$

   $A = 1500$ m^2

3. Determine necessary volumes.
   a. Rock medium:

   Average organic loading rate $(OLR) = 1.1$ kg $BOD_5/m^3 \cdot d$

   $$OLR = \frac{QC_i + \alpha QC_e}{V}$$

   Assume $C_e = 30$ g/m^3.

   $$V = \frac{(15,000 \text{ m}^3/\text{d})(0.400 \text{ kg/m}^3) + 2(15,000 \text{ m}^3/\text{d})(0.030 \text{ kg/m}^3)}{OLR}$$

   $V = 6273$ m^3

b. Plastic medium:

Average OLR = 1.75 kg BOD_5 /$m^3 \cdot d$

$V = 3943$ m^3

4. Determine the size of the rock-medium filter using a typical depth of 2.0 m.

$$A = \frac{6273 \text{ m}^3}{2} = 3137 \text{ m}^2$$

This area corresponds to a HLR of 14.3 m^3/$m^2 \cdot d$, which is within the recommended range.

5. Determine the size of the plastic-medium filter using a typical depth of 6.0 m.

$$A_s = \frac{3943 \text{ m}^3}{6 \text{ m}} = 657 \text{ m}^2$$

This area corresponds to a HLR of 68 m^3/$m^2 \cdot d$, which exceeds the maximum hydraulic loading rate, thus the maximum HLR will control the design.

Let HLR = 50 m^3/$m^2 \cdot d$

$$A = 900 \text{ m}^2$$

The increased surface area will allow for a reduction in the depth to 4.4 m.

6. Calculate effluent BOD_5 using the Eckenfelder formula [Eq. (14.7) in Table 14.4].

$$\frac{C_e}{C_i^*} = \exp\left\{-k\left[\frac{A_s}{(1+\alpha)Q}\right]^n za_v^m\right\}$$

a. For the rock medium ($a_v = 55$ m^2/m^3):

$$\frac{C_e}{C_i^*} = \exp\left\{-0.1\left[\frac{3137 \text{ m}^2}{(1+2)(15,000 \text{ m}^3/\text{d})}\right]^{0.7}(2.0 \text{ m})(55 \text{ m}^2/\text{m}^3)\right\}$$

$$= 0.18$$

where

$$C_i^* = \frac{C_i + \alpha C_e}{1 + \alpha}$$

Solving for C_e,

$$C_e = \frac{0.18 C_i}{1 + 0.82\alpha}$$

$$= \frac{0.18(400 \text{ g/m}^3)}{1 + 0.82(2)}$$

$$= 27.3 \text{ g/m}^3$$

b. For the plastic medium ($a_v = 90 \text{ m}^2/\text{m}^3$):

$$\frac{C_e}{C_i^*} = \exp\left\{-0.1\left[\frac{900 \text{ m}^2}{(1+2)(15{,}000 \text{ m}^3/\text{d})}\right]^{0.7} (4.4 \text{ m})(90 \text{ m}^2/\text{m}^3)\right\}$$

$$= 0.08$$

$$C_e = \frac{0.08 C_i}{1 + 0.92\alpha}$$

$$= \frac{0.08(400 \text{ g/m}^3)}{1 + 0.92(2)}$$

$$= 11.3 \text{ g/m}^3$$

7. Summary:

CHARACTERISTIC	ROCK	PLASTIC
HLR, $\text{m}^3/\text{m}^2 \cdot \text{d}$	14.3	50
OLR, $\text{kg/m}^3 \cdot \text{d}$	1.1	1.75
Area, m^2	3137	900
Depth, m	2.0	4.4

COMMENT

The average values of the hydraulic and organic loading rates used in this example worked well as first estimates. Note that the OLR controlled the high-rate rock-medium filter design, while the HLR controlled the high-rate plastic-medium filter design. The values used for k (0.1) and the exponents m (1.0) and n (0.7) in the Eckenfelder equation are typical, but considerable variation is found in the literature. Reported values for the removal rate constant, k, and the exponent, n, vary from 0.06 to 0.12 and 0.5 to 1.0 respectively. Values of k, usually given for 20°C, can be adjusted for other temperatures using Eq. (2.79) with a temperature coefficient value of 1.04. The most commonly used value for the exponent m is 1.0. Where industrial wastes are to be treated, pilot-plant studies are recommended.

Oxygen Transfer

Air is usually supplied to trickling filters through natural drafts resulting from temperature differences between the ambient and the internal air. Deep plastic-medium filters often require the use of compressed air to supply a forced draft. Maximum oxygen transfer rates in natural-draft trickling filters are about 28 $g/m^2 \cdot d$, and this corresponds to uptake rates expected in the biofilm for applied BOD_u concentrations of about 400 g/m^3. The applied BOD_u concentration of Example 14.5 would be approximately

$$BOD_u \approx \frac{1}{0.67} \frac{(400 \text{ g/m}^3)(1) + (30 \text{ g/m}^3)(2)}{3} = 229 \text{ g/m}^3$$

where 0.67 is the factor used to convert BOD_5 to BOD_u.

Many industrial wastes are considerably stronger than 400 g BOD_5/m^3, and anoxic conditions within trickling filters can occur. The result can include rather nasty odor problems. A case in point is a situation that developed when a new 6-m-deep plastic-medium trickling-filter system was brought on line at the same time the local canning season began. A combination of high-strength wastewater, poor wastewater distribution, and marginal air-flow capacity resulted in poor process performance and severe odor problems for a distance of more than 2 km. Control of the odor problem required modifying the distributors, capping the system, using forced air, and passing the exhaust air through activated carbon.

Temperature Effects

Temperature effects can be applied quantitatively only if a reaction model is used in process design. Such models are in common use but are beyond the scope of this textbook. For further information see Refs. [14.6], [14.23], and [14.31].

14.7

ROTATING BIOLOGICAL CONTACTOR PROCESSES

Rotating biological contactors (RBCs) are used in a process in which honeycombed or corrugated disks act as a support medium for attached microbial cultures (Fig. 14.16). A number of disks are placed on a shaft and rotated through a tank containing wastewater (Fig. 14.17). Most RBC processes are designed to operate in a series of stages, with each shaft-disk unit constituting a stage. Disk submergence is usually 40 percent of the diameter, and disk size is limited to 3.6 m by the height of interstate highway overpasses. Shaft lengths are typically 7 to 8 m, having a total surface area of approximately 8400 m^2, with over 95 percent of the area supporting biofilm growth. Rotational speeds used are usually 1.5 to 2 r/min and result in peripheral tangential velocities of 0.09 to 0.13 m/s.

In modeling RBC processes, the organic concentration in each stage is assumed to be uniform, and the organic removal rate is set up as a heterogeneous reaction model. A number of design expressions have been

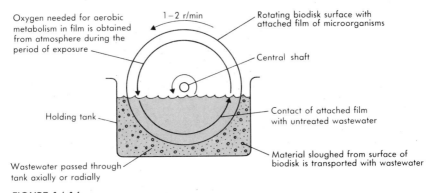

Oxygen needed for aerobic metabolism in film is obtained from atmosphere during the period of exposure

1 – 2 r/min

Rotating biodisk surface with attached film of microorganisms

Central shaft

Holding tank

Contact of attached film with untreated wastewater

Wastewater passed through tank axially or radially

Material sloughed from surface of biodisk is transported with wastewater

FIGURE 14.16

Definition sketch for an RBC treatment system.

(a)

(b)

FIGURE 14.17

RBC treatment facility and equipment: (a) typical covered bio-disk installation and (b) bio-disk unit being assembled.

Source: Part (b), courtesy of Crane Co.—Cochrane Environmental Systems.

proposed, but none has proved to describe process performance adequately. Recently, Opatken has proposed a second-order model based on filtered BOD [14.37].

Performance

Effluent quality from RBC systems similar to that from activated sludge processes can be obtained if enough stages are used, although economic constraints may become important. Failures of both the plastic (polyethylene, polypropylene, and polyurathane have been used) disks and the shafts were major problems in early units. Plastic failure resulted from exposure to sunlight (UV radiation) and heavy biomass growth that caused structural overloading. The latter problem was also the cause of shaft failure. These problems have been solved by covering units (which may result in a forced-air ventilation requirement) and restricting organic loading rates.

Design Loadings

RBC systems are designed on the basis of organic (in units of kilograms of BOD per square meter of surface area-day) and hydraulic loading rates [14.20]. Because design values are based on experience, they will differ somewhat with the type of disk. Typical organic loading rate values vary from 0.03 to 0.15 kg $BOD_u/m^2 \cdot d$, based on the total disk area. Hydraulic loadings are usually less than 0.2 $m^3/m^2 \cdot d$, and hydraulic detention times of 0.5 hr per stage are typical. An obvious problem results in that the initial stages may be quite overloaded, with both oxygen transfer limitations and overgrowth of biomass resulting. This

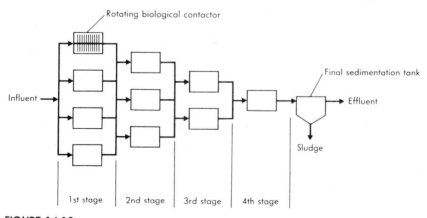

FIGURE 14.18

A tapered RBC system.

problem can be mitigated by tapering the number of units per stage, as shown in Fig. 14.18. As indicated there, a sedimentation unit must follow the RBCs, and recycle is not normally used.

14.8
OXIDATION POND PROCESSES

Oxidation ponds are essentially earthen basins in which wastewater is treated by natural processes. Four main types of pond processes are used. They are facultative ponds, maturation ponds, anaerobic pretreatment ponds, and high-rate aerobic ponds. Although all four types of ponds are described below, the remainder of this section is devoted to a discussion of facultative ponds.

Facultative ponds, or *wastewater lagoons*, as they are often called, are simple biological treatment processes that incorporate a symbiotic relationship between algae (see Fig. 2.26) and heterotrophic bacteria. Conventional oxidation ponds are earthen basins filled with screened and, in some cases, comminuted raw wastewater or primary effluent. Large solids settle out to form an anaerobic sludge layer. Soluble and colloidal organic materials are oxidized by aerobic and facultative bacteria using oxygen produced by algae growing abundantly near the surface. Carbon dioxide produced in the organic oxidation serves as a carbon source for the algae. Anaerobic breakdown of the solids in the sludge layer results in the production of dissolved organics and gases such as CO_2, H_2S, and CH_4, which are either oxidized by the aerobic bacteria or vented to the atmosphere. A schematic diagram of an oxidation pond system is shown in Fig. 14.19.

Maturation ponds are used as a second stage following facultative ponds. The principal use of maturation ponds, especially in less developed countries, is for the destruction of pathogens [14.30]. Maturation ponds are aerobic throughout their depth (usually 1 to 1.5 m). Detention times vary from 4 to 12 d. The loading rate will vary with the degree of pretreatment achieved in the facultative ponds.

Anaerobic ponds are used to treat high-strength wastes that have a high solids content [14.30]. Because of the high organic loading these ponds are anaerobic throughout their depth. To eliminate odors, a surface cap of fresh water is often maintained over the anaerobic material in the pond. In some cases, small, specially designed aerators are used to maintain aerobic conditions in the surface layers of the pond. The partly clarified effluent is usually discharged to facultative ponds for further treatment.

High-rate aerobic ponds are designed to maximize algal growth so as to achieve a high protein yield. High-rate ponds are shallow (0.15 to 0.5 m) and require effective and continuous management for proper

(a)

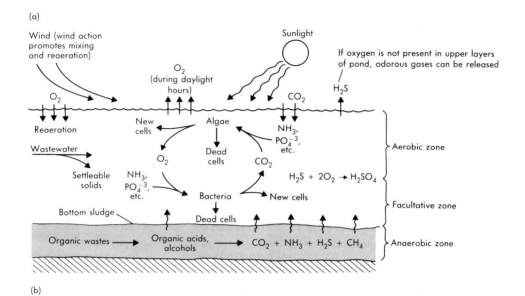

(b)

FIGURE 14.19

Facultative oxidation pond: (a) definition sketch for the operation and (b) typical ponds, Kingston, Jamaica.

operation. They are not routinely used unless the production of algal protein is the prime objective.

Operation of Oxidation Ponds

Process operation is uncontrolled. Problems develop when oxygen production by the algae does not meet the oxygen demand of the bacterial population. Under these "overloaded" or "unbalanced" conditions, organic removals will be unsatisfactory and odor problems will result. Thus algal growth and photosynthesis are included in the design process. In most cases winter conditions control the design process because sunlight and photosynthesis are less then (see Table 1.15), while the mass rate of organic input $(\overline{QC_i})$ remains relatively constant. In some cases, short-term anaerobic conditions can occur in the summer months during the early morning hours. Algae respire during the night but cannot produce oxygen without sunlight. During the summer months, algae populations become very great, and algal respiration combined with that of bacteria can deplete oxygen stored during daylight hours. Conversely, photosynthesis by the large algae populations may result in supersaturation with oxygen during the daylight hours.

The principal advantages of oxidation ponds relate to their ease of construction and management. Earth walls are riprapped to prevent damage from waves, and a clay liner is needed to prevent leakage to the groundwater. Most ponds are rectangular, but other shapes are used at particular sites. Residence times used range from 6 to 30 d, and organic loadings used range from 15 to 80 kg BOD$_5$/ha · d, depending on temperature and available sunlight. Pond depth is usually 1.5 to 2 m.

Management of Oxidation Ponds

Oxidation ponds are most often used in small communities and consequently should be designed with the objective of minimizing management requirements. Unless regular maintenance is performed, both structural and performance failures can occur, however. The principal maintenance problem is controlling plant growth in shallow areas and on dikes. Plant roots break up dikes, which results in gradual structural breakdown. Aquatic plants provide a protected area for mosquito larvae and a place where debris accumulates. Regular removal of growth is required in all installations [14.32].

Most oxidation ponds are neither preceded nor followed by any form of treatment other than disinfection by chlorination. Wastewater solids settle to the bottom and a buildup occurs over a period of time that requires occasional solids removal. If a pond has only one inlet point, there will be a local solids buildup that requires removal much sooner than if several inlet points are used. Multiple outlets are also desirable to minimize dead volume and short circuiting.

Performance of Oxidation Ponds

Effluent quality from oxidation ponds is highly variable. Wastewater organic removal efficiency is very high, but a large fraction of the algae do not settle and remain in the effluent. Summer effluent suspended-solids and BOD concentrations often each exceed 100 g/m³. Winter effluent quality is usually in the 30- to 50-g/m³ range. Removal of pathogenic organisms in facultative and maturation ponds is quite good because of the long residence times, and this fact, as noted previously, is the principal reason for utilization of these systems in developing nations [14.30].

Process Design

Process design is usually based on organic loading rates and hydraulic residence times [14.21]. As noted in Table 14.6, the ranges in common use are 15 to 80 kg $BOD_5/ha \cdot d$ and 6 to 30 d, respectively. Large systems are often designed as CFSTRs, and the analysis is similar to that used for simple lake models (see Example 9.1), although two or three CFSTRs in series would probably be more appropriate. A second approach is to use the Wehner and Wilhelm equation [see Eq. (6.21) and Fig. 6.13]. Considerable dead volume should be expected in oxidation ponds, particularly in single inlet systems. In applying CFSTR or PFR-with-dispersion models to oxidation pond process design, first-order kinetics and typical BOD-rate coefficient values (e.g., 0.2 to 0.4 d^{-1}) are generally used. The design of an oxidation pond is illustrated in Example 14.6.

TABLE 14.6

Typical Design Parameters and Operating Characteristics of Oxidation Ponds

CHARACTERISTICS	OXIDATION PROCESS			
	Facultative	Maturation	Anaerobic	Aerated
Depth, m	1–2.5	1–1.5	3–6	2–4
Pond size (in multiples), ha	1–4	1–4	0.1–1	1–4
Detention time, d	6–30	4–12	10–50	2–10
Organic loading rate,* kg BOD_5/ha · d	15–80	5–20	200–1000	–*
BOD_5 conversion	80–95	40–80	50–85	80–95
Effluent SS, g/m³	30–100	10–40	80–200	60–200
Application	Municipal and industrial wastewater treatment	Polishing step following facultative pond	High-strength wastes, pretreatment for facultative pond	Municipal and industrial wastewater treatment

*Organic loading depends on the oxygen-transfer capacity of the aeration system.

EXAMPLE 14.6

OXIDATION POND PROCESS DESIGN

A facultative oxidation pond is to be designed for a community of 5000 people. Summer wastewater flow is 2000 m^3/d, and the BOD_5 is 180 g/m^3. Winter flow and BOD_5 values are 6000 m^3/d and 90 g/m^3, respectively. Average pond temperatures are 20°C in the summer and 10°C in the winter. Assuming that a CFSTR model is appropriate and that oxygen transfer will not be a problem, determine the pond volume required to reduce the influent BOD_5 to 20 mg/L.

SOLUTION:

1. Determine the summer hydraulic residence time and volume requirement assuming $k = 0.30$ d^{-1} and steady-state conditions. Note that because the solution involves a ratio of BOD values it is possible to use BOD_5 rather than BOD_u.

$$C_e = \frac{C_i}{1 + k\theta_H}$$

$$\theta_{H_s} = \frac{1}{k}\left(\frac{C_i}{C_e} - 1\right)$$

$$= \frac{1}{0.3 \text{ d}^{-1}}\left(\frac{180}{20} - 1\right)$$

$$= 27 \text{ d}$$

$$V_s = Q\theta_{H_s}$$

$$= (2000 \text{ m}^3/\text{d})(27 \text{ d}) = 54{,}000 \text{ m}^3$$

2. Determine the winter BOD-reaction-rate constant. Assume the value of the temperature coefficient θ is equal to 1.135.

$$k_{10°C} = k_{20°C}\theta^{T-20}$$

$$= (0.30 \text{ d}^{-1})(1.135^{-10})$$

$$= 0.085 \text{ d}^{-1}$$

3. Determine the winter residence time and volume requirements.

$$\theta_{H_w} = \frac{1}{0.085 \text{ d}^{-1}}\left(\frac{90}{20} - 1\right)$$

$$= 41 \text{ d}$$

$$V_w = Q\theta_{H_w}$$

$$V_w = (6000 \text{ m}^3/\text{d})(41 \text{ d}) = 246{,}000 \text{ m}^3$$

COMMENT

Use of the large volume required under winter conditions will result in very high conversion efficiencies during the summer months. As stated above, effluent quality will depend on algae concentrations.

Removal of Algae from Oxidation Pond Effluents

Because oxidation ponds are simple to construct and management does not require sophisticated equipment or highly trained staff, methods of upgrading effluent quality by removing algae have received considerable interest. Methods that have been used include upflow rock filtration, slow-rate intermittent sand filtration, land treatment (see Section 14.9), coagulation, flocculation and sedimentation, microstraining, and precoat filtration. Detailed descriptions of these processes can be found in Ref. [14.32]. Precoat filtration of algae was developed by Dodd for use in harvesting algae from high-rate oxidation ponds. Harvested algae have a high protein content and can be used as feed for ruminant animals [14.16].

Algae removal has proved to be a difficult and expensive process. In choosing to upgrade an oxidation-pond effluent, the engineer must keep in mind the original objective: to provide a simple, effective, and inexpensive system for wastewater treatment. Algae removal has not proved to be cost-effective in many situations, but the possibility should always be investigated. Harvesting algae as a protein source is limited to high-rate pond applications in tropical countries. The system appears promising and will undoubtedly be increasingly used in future years.

14.9
LAND TREATMENT PROCESSES

Five types of land treatment systems are currently in use: slow-rate treatment or irrigation, rapid infiltration, overland flow, wetland application, and septic tank-leach field systems. Characteristics of these systems are summarized in Table 14.7. In all land treatment systems, soil microorganisms utilize nutrients in the wastewater. Percolation though the soil is an important aspect of the slow-rate process, rapid infiltration, and leach fields. Although land disposal has always been practiced, engineered treatment systems incorporating land as part of the process are quite recent innovations [14.27, 14.39, 14.40, 14.42]. Wetland systems and septic tank-leach field systems are considered separately in Sections 14.10 and 14.12, respectively.

TABLE 14.7

Design Parameters and Operating Characteristics of Land Treatment Systems

CHARACTERISTICS	SLOW RATE	RAPID INFILTRATION	OVERLAND FLOW	WETLANDS	SEPTIC TANK-LEACH FIELD
Hydraulic loading rate, $m^3/m^2 \cdot wk$	0.01–0.1	0.1–2.0	0.1–0.3		
Organic loading rate, g $BOD_5/m^2 \cdot wk$	2–20	50–400	20–60	60–80	
Application method	Sprinkler or surface	Surface	Sprinkler or surface	Surface	Subsurface
Minimum preapplication treatment	Sedimentation	Sedimentation	Grit removal and comminution	Screening and/or sedimentation	Sedimentation
Need for vegetation	Required	Optional	Required	Required	
Disposition of wastewater	Evapotranspiration and percolation	Percolation	Surface runoff and evapotranspiration	Surface runoff and evapotranspiration	Percolation
Effluent BOD_5, g/m^3	2–5	5–10	10–15	5–20	
Effluent SS, g/m^3	1–5	2–5	10–20	5–20	

Slow-Rate Treatment

Slow-rate systems involve use of wastewater to irrigate a water-tolerant crop such as feed corn, pasture, or forests (Figure 14.20a). Application is made throughout the growing season, and wastewater is stored during the remainder of the year. Application rates used depend on soil permeability and the nature of the crop. Where the crop is highly profitable, application rates are based on optimal production. Application on less desirable crops is based on maximizing evapotranspiration (see Chapter 1) and percolation.

Sprinkler or ridge-and-furrow systems are used for distributing the wastewater. Preapplication treatment required is screening and grit removal, but primary sedimentation is desirable. Because some storage is

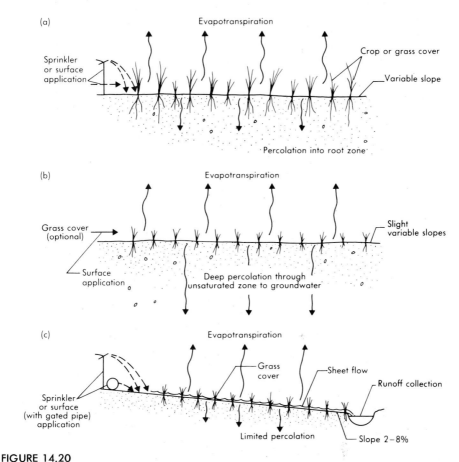

FIGURE 14.20

Land treatment processes: (a) slow rate, (b) rapid infiltration, and (c) overland flow.

usually included as part of a slow-rate system, sedimentation is nearly always provided.

Rapid Infiltration

Rapid-infiltration systems are used where highly permeable sandy soils are available (Fig. 14-20b). Relatively small areas (0.1 to 0.2 ha) are separated by low dikes and kept flooded to depths of 100 to 400 mm for periods of up to several days. Loading is followed by drainage periods of one to two weeks before a particular area is loaded again. There is no cover crop and the surface is occasionally scarified to break up material plugging the surface layer.

Application rates to rapid-infiltration systems are typically in the 0.1- to 2-m/wk range, not including drainage time. Thus the total annual application to a particular unit might range from 1.7 to 35 m. This wide range results from variation in soils, weather, and wastewater characteristics, as well as from a tendency toward conservative design of systems for which there is little operational control.

Preapplication primary treatment and scum removal are highly desirable where rapid-infiltration systems are used. Debris should not be allowed to clutter up the beds or serve as an attraction for rodents, cockroaches, and other vermin [14.40].

Overland Flow

Where impermeable soils are available, overland flow is often an inexpensive method of treatment that produces an effluent superior to most other biological systems (Fig. 14-20c). In overland flow, wastewater is applied to grass-covered slopes or terraces and allowed to flow as a sheet to a collection ditch (Fig. 14.21). Water-tolerant grasses such as bermuda, alta-tall fescue, and reed canary are used to minimize channeling and erosion. The bacterial culture that grows on the thatch that accumulates at the soil surface carries out organic biooxidation, nitrification, and denitrification. Some income can be derived from hay sales, but harvesting requires an extended period of drying before ordinary farm equipment can be brought onto the slopes, in addition to the time necessary to cut, dry, and bale the hay. Water is a product also, and in arid areas the treated wastewater can be used for further crop irrigation. Some salt concentration occurs because of evapotranspiration, but in most cases the runoff is suitable for a wide variety of crops.

Hydraulic loading rates used vary from 15 to 40 mm/d, based on the total treatment area. Because system geometry is not standardized, wastewater application is intermittent, and application is near or at the top of the slope. More appropriate loading-rate parameters are volume of flow per unit slope width per hour and kilograms of BOD per unit slope width per hour [14.46, 14.47], with appropriate values in the range of 0.05

(a)   (b)

(c)   (d)

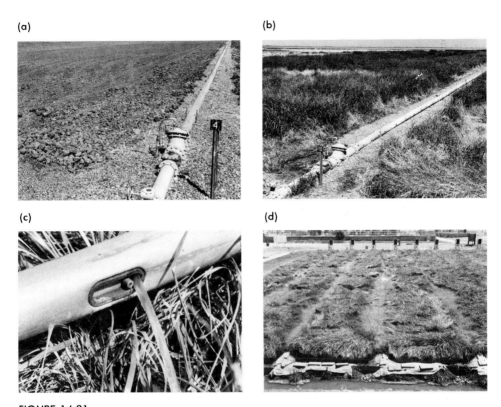

FIGURE 14.21

Views of overland flow systems: (a) unplanted slopes, (b) planted slopes, (c) gated distribution pipe, and (d) view of experimental slopes planted with sod.

to 0.4 m^3/m · hr and 20 to 40 g BOD/m · hr, respectively. Application periods in common use vary between 6 and 16 hr/d, although there is reason to believe that shorter periods (e.g., 2-hr application, 4-hr drain) may result in improved effluent quality [14.26]. Slopes between 2 and 8 percent have proved satisfactory with respect to process performance, but 3 percent is a practical minimum because of construction and maintenance problems.

Overland flow systems are one of the few methods of wastewater treatment that can be affected by precipitation. Figueiredo [14.19] reported that the mass rate of discharge of suspended solids increases at the beginning of a storm, peaks, and decreases to lower than normal values when precipitation stops. Solids discharged during storms are more highly mineralized than wastewater solids and probably include some soil as well as oxidized organic material applied as part of the wastewater.

14.10

WETLAND TREATMENT SYSTEMS

An aquatic treatment system consists of a natural or prepared wetlands to which wastewater is applied for the purpose of treatment. Within the context of aquatic treatment systems, a wetlands is a comparatively shallow (less than 0.6 m typically) body of slow-moving water in which dense stands of water-tolerant plants such as cattails, bulrushes, or water hyacinths are grown (Fig. 14.22). Types of engineered wetlands used as aquatic treatment systems include artificial marshes, ponds, and trenches. Wetlands are effective as wastewater treatment processes for a number of reasons. Bacteria attached to the submersed roots and stems of aquatic plants growing in the wetlands are of particular importance in the removal of soluble and colloidal BOD from a wastewater. The quiescent water conditions found in a wetlands are conducive to the sedimentation of wastewater solids. The adsorption/filtration potential of the roots and stems of the aquatic plants, the ion exchange/adsorption capacity of the wetlands sediments, and the emersed parts of the aquatic plants that reduce the perturbating effects of climatic variables on the wetlands aquatic environment contribute to the effectiveness of wetland treatment systems. Wetland treatment systems are discussed in detail in Ref. [14.52].

Performance of Wetland Treatment Systems

Based on the results from numerous research and demonstration projects (Fig. 14.23) in which domestic wastewater was applied to wetlands, it

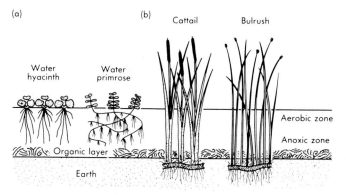

FIGURE 14.22

Definition sketch for plants used in aquatic treatment systems: (a) floating (water hyacinth and water primrose) and (b) emergent (cattail and bulrush).

Source: From Ref. [14.51].

FIGURE 14.23

Views of experimental aquatic wastewater treatment systems. Water hyacinths are shown in the foreground and cattails in the background.

appears that this type of process is capable of producing effluents with BOD_5, SS, and total N concentrations as low as 5, 5, and 1 g/m^3, respectively. The removal of other wastewater contaminants such as P, heavy metals, refractory organics, and pathogens are very much dependent on site- and wastewater-specific factors. Clearly, wetlands wastewater treatment systems can be used as secondary (effluent < 30 g BOD_5/m^3 and < 30 g SS/m^3) or advanced secondary wastewater treatment processes [14.15, 14.51, 14.52].

Removal of BOD

In terms of actual performance, wetlands systems appear to be able to remove BOD_5 at rates as high as and higher than 110 kg/ha · d during the warmer seasons and when the amount of bacterial support structure provided by aquatic plants in the form of roots and stems is greatest. During the cooler seasons, the BOD removal performance of wetlands systems decreases because of the slowing of the metabolic activity of both the bacteria and the plants that support and in part sustain the bacteria. If temperatures cool to the point where plants die back and/or initiate autolysis, the BOD removal performance of the system is reduced further. This reduction in performance occurs not only because the metabolic activity of the bacteria is reduced but because the total mass of bacteria in immediate contact with the wastewater is reduced because of the loss of the plant structure.

Removal of Suspended Solids

The removal of SS in a wetlands system is usually a concomitant aspect of system design in the typical wastewater treatment situation in which a significant reduction in BOD is to occur. Thus SS removal in and of itself is not a design variable in the normal sense, though solids deposition and accumulation in wetlands systems must be considered during system design.

Removal of Nitrogen

It appears that the total N removal of wetlands systems can be as high as 45 kg/ha · d. The principal nitrogen-removal mechanism operative in wetlands systems is bacterial nitrification/denitrification (not plant uptake); consequently, nitrogen removal in wetlands systems is a function of climatic conditions, as is the case for BOD removal. However, nitrogen removal is even more sensitive to climatic factors, as nitrifying bacteria in wastewater environments are ineffective below approximately 10°C.

Design Considerations for Wetland Systems

The three critical considerations in the design of wetland systems are mosquito control, wastewater pretreatment, and wastewater distribution within the wetlands systems. Of these three, mosquito control is the most critical and has profound effects on the other two.

Mosquito Control

The wastewater marsh and other wetlands systems combine shallow, polluted water with submerged, floating, and emergent vegetation, creating habitats ideal for mosquito propagation. In most locations, control of the mosquito population is a necessary condition for use of wetlands wastewater treatment systems. At this time the best approach to controlling mosquito production is to design the system so that natural predators of mosquito larvae (e.g., mosquito fish, dragonfly and damselfly nymphs, and a variety of water beetles) will thrive. Because most natural predators of mosquito larvae are strict aerobes, wetlands systems should be designed so that anaerobic conditions are avoided, particularly in the immediate vicinity of influent points, where mosquito production is most likely to occur. To avoid anaerobic conditions the wastewater BOD load to the wetlands system must be kept low (e.g., equal to less than 110 kg/ha · d) and well distributed over the wetlands surface area [14.52].

Pretreatment

The necessary function of pretreatment in the design of wetlands systems is to remove from the wastewater materials, such as grit, plastics, and grease, that are not readily biodegradable. Grit settles out in the manifold influent piping of wetlands systems and in concentrated areas in the wetlands itself. Grit deposition, particularly in conjunction with grease and plastic, in manifold pipes and wetlands can cause organic overloading and shallow water depths in some areas of the wetlands and hydraulically static conditions in other areas. All of these conditions promote the production of mosquitoes.

Distribution

In wetlands systems, wind and thermally induced mixing are much reduced in comparison with related systems, such as oxidation ponds, because of the physical presence of the aquatic plants. Unfortunately, the consequences of "overloaded" areas in wetlands systems are much more severe than in stabilization ponds. For this reason, many stabilization ponds that have been converted to wetlands systems without extensive modification of the existing influent distribution system have become mosquito and odor nuisances. If a wastewater influent to a wetlands has a relatively high BOD and SS, it must be spread mechanically over much of the surface of the wetlands to prevent the production of mosquitoes and odors.

Design of Wetland Systems

Several systems incorporating a wetlands for wastewater treatment are shown in Fig. 14.24. As shown, different pretreatment options are used. Wetlands for wastewater treatment should be designed as plug-flow reactors with aspect ratios greater than 15 (length to width). Recom-

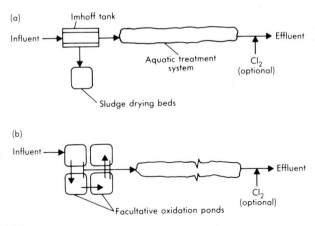

FIGURE 14.24

Low-cost aquatic treatment systems for small rural communities: (a) Imhoff tank followed by wetland treatment system and (b) facultative oxidation pond followed by wetland treatment system.

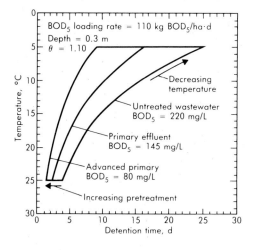

FIGURE 14.25

Design curves for aquatic treatment systems.

Source: From Ref. [14.52].

mended design criteria include (1) a BOD_5 loading rate of 110 kg/ha · d, (2) a depth of about 0.3 m, and (3) a temperature coefficient of 1.10. Depending on the geographic location, either cattails or water hyacinths could be used. Cattails, the plant of choice in all but the southern United States, like water hyacinths can build up excessive biomass. Under these conditions, harvesting or plant management is necessary to maintain a well-functioning biosystem. It should be noted that wetlands systems will function effectively even when covered with a foot or more of ice, as long as detention time is maintained under the ice cap. Depending on the temperature and the degree of pretreatment, it is anticipated that a design envelope such as the one shown in Fig. 14.25 will encompass the actual operating conditions found in the field.

14.11

ANAEROBIC BIOLOGICAL TREATMENT

Anaerobic treatment processes are used almost exclusively for high-strength wastewaters such as organic sludges (which are approximately 95 percent water) and high-BOD-concentration industrial wastes. One important exception is denitrification, an anaerobic process ordinarily associated with the treatment of municipal wastewater. However, as noted in Section 14.5, nitrate and nitrite serve analogously to oxygen as a terminal electron acceptor, and denitrification functionally operates as an aerobic process. In conventional anaerobic treatment processes, specifically those used for high-strength organic wastewaters, internally generated organic compounds and CO_2 are used as terminal electron acceptors in a complex series of reactions. Many of the bacterial species involved are obligate anaerobes, and the most unique group, the methanogenic bacteria, produce methane gas, which has a heat value of approximately 36,500 kJ/m³, in economically recoverable quantities.

Anaerobic Fermentation

A schematic diagram of the anaerobic fermentation process reaction sequence is shown in Fig. 14.26. The process can be entered at any step, but the mixture or organics contained in most wastewaters is largely confined to components of the first stage [14.28, 14.48]. Complex organics include particulate material that must be hydrolyzed into soluble components. Soluble materials are oxidized to low-molecular-mass organic acids, including acetic acid, the principal precursor of methane. As shown in Fig. 14.26, hydrogen gas is produced, and this compound is extremely important in the system ecology. The thermodynamic balance for the reactions converting alcohols and volatile acids to acetic acid is extremely sensitive to the hydrogen partial pressure. If the hydrogen partial pressure is maintained at a sufficiently low level, the free energy of conversion to acetic acid is negative, and the reactions proceed. Thus a symbiotic relationship exists between the bacteria using H_2 and CO_2 to produce CH_4 and those splitting acetic acid. The principal methane fermenting reactions are listed below:

$$H_2 + \tfrac{1}{4}CO_2 \rightarrow \tfrac{1}{4}CH_4 + \tfrac{1}{2}H_2O \tag{14.8}$$

$$CH_3COOH \rightarrow CH_4 + CO_2 \tag{14.9}$$

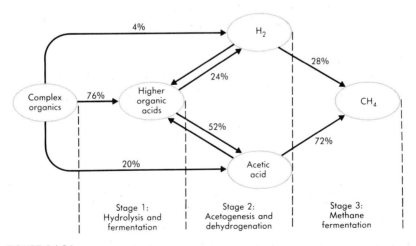

FIGURE 14.26

Steps in the anaerobic fermentation process, with energy flow represented as percent COD.

Source: From Ref. [14.48]. Reprinted with permission from the American Chemical Society.

Because methanogenesis is the controlling step in anaerobic treatment, knowledge of the characteristics and requirements of methanogenic bacteria is critical to effective process design and operation. Accumulation of this knowledge has been slow because these bacteria (now classified as members of the group *Archaebacteria*) are extremely slow-growing and have very unusual nutrient requirements. Trace elements of particular importance are iron, cobalt, nickel, and sulfide. Nickel is a required component of a cofactor unique to methanogens. The sulfide requirement is interesting because the presence of sulfides in anaerobic systems results in precipitation of trace metal nutrients. Excess sulfide concentrations can effectively shut down methane fermentation.

Quantities of Methane Produced

Methane production stoichiometry and rate are major factors in determining if anaerobic treatment is feasible. Buswell and Mueller [14.10] suggested the stoichiometric relationship given in Eq. (14.10):

$$C_nH_aO_b + \left(n - \frac{a}{4} - \frac{b}{2}\right)H_2O \rightarrow$$

$$\left(\frac{n}{2} - \frac{a}{8} + \frac{b}{4}\right)CO_2 + \left(\frac{n}{2} + \frac{a}{8} - \frac{b}{4}\right)CH_4 \qquad (14.10)$$

For carbohydrates 0.35 m³ of CH_4 will be produced per kilogram of COD removed according to Eq. (14.10). It should be noted that cell production is not considered in that equation. For CFSTRs, the following expression can be used to estimate the gas production, taking into account cell yield:

$$M_{CH_4} = 0.35 \left(nQC_i - 1.42r_gV\right) \qquad (14.11)$$

where

n = fraction of biodegradable COD converted (usually about 0.85)
Q = volumetric flow rate, m³/s or m³/d
C_i = concentration of COD applied, kg/m³
r_g = growth rate, g/m³ · s or g/m³ · d
V = reactor volume, m³
M_{CH_4} = methane production rate at 1 atm pressure, m³/s or m³/d

Because the growth rate is generally low, it can often be neglected. In typical methane fermentation systems, 60 to 70 percent of the gas produced (by volume) will be methane, with most of the remainder being CO_2. The application of Eq. (14.11) is illustrated in Example 14.7.

EXAMPLE 14.7

ESTIMATION OF METHANE PRODUCTION POTENTIAL OF A CHICKEN FARM

A chicken farm maintains a population of 200,000 birds of laying age. Each hen produces an average of 1.9×10^{-4} m³/d of manure having a COD of 150,000 g/m³. Pilot-plant studies have been run using a CFSTR digester operated at a 12-d hydraulic residence time. The growth rate in these studies was 600 g/m³ · d, the fraction of COD converted was 0.79, and 0.37 m³ CH_4 was produced per kilogram COD destroyed. Estimate the methane production rate, assuming that a CFSTR digester operated at a 12-d hydraulic residence time will be used.

SOLUTION:

1. Determine the waste flow rate.

$$Q = (1.9 \times 10^{-4} \text{ m}^3/\text{hen} \cdot \text{d})(200{,}000 \text{ hens})$$
$$= 38 \text{ m}^3/\text{d}$$

2. Determine the digester volume.

$$V = Q \cdot \theta_H$$
$$= (38 \text{ m}^3/\text{d})(12 \text{ d})$$
$$= 456 \text{ m}^3$$

3. Determine the methane production rate.
   a. The modified form of Eq. (14.11) is

$$M_{CH_4} = 0.37 \left(nQC_i - 1.42 r_g V \right)$$

   b. Substitute the given values and solve for M_{CH_4}.

$$M_{CH_4} = 0.37 \text{ m}^3/\text{kg} \left[0.79(38 \text{ m}^3/\text{d})(150 \text{ kg/m}^3) \right.$$
$$\left. - 1.42(0.6 \text{ kg/m}^3 \cdot \text{d})(456 \text{ m}^3) \right]$$
$$= 1522 \text{ m}^3/\text{d}$$

COMMENT

Methane has a heat value of approximately 3.65×10^4 kJ/m³ at 20°C and 1 atm pressure. The total potential energy production is thus approximately 5.6×10^7 kJ/d, or 648 kW. Assuming a 30 percent conversion efficiency the usable energy production rate would be 194 kW.

Process Configurations

The earliest anaerobic processes were carried out in unmixed, unheated tanks. Solids settled to the bottom, and a large fraction of the soluble organics were contained in the supernatant liquor (Fig. 14.27b). Methane fermentation was slow and little use was made of the gas. Mixing and heating of the tanks resulted in greatly increased rates of fermentation. Mixing brought the bacterial culture into contact with a larger fraction of the organic material, and operating at temperatures between 30 and 35°C greatly increased the overall conversion rate. The completely mixed, heated reactor is now considered the standard process configuration for the digestion of municipal wastewater sludges (Fig. 14.27c and d). Because growth rates are low, methods of separating the solids retention time from the hydraulic retention time have been of continuing interest.

A number of alternative process configurations in current use are shown in Fig. 14.28. An early modification was the anaerobic contact process which is a straightforward adaptation of the activated sludge process. A major operating problem has been degasification in the separator. Upflow packed beds (often called anaerobic trickling filters) were introduced in the early 1960s. Upflow packed beds were best suited to treating soluble and colloidal organics because of a tendency to plug. One solution to the plugging problem was to use a downflow packed bed; two other solutions involve the use of fluidized and expanded beds [14.25]. Sludge-blanket systems are really heated, high-rate modifications of the original stratified anaerobic digesters. The higher flow and organic loading rates make these processes much more efficient than their predecessors, however. Recent proposals [14.48, 14.57] for separating hydraulic and solids retention times include the baffled reactor, the two-stage leaching bed and leachate filter, and membrane solids separation. In the latter case supernatant is filtered though a continuously cleaned membrane. Photographs of a bag digester used for the processing of dairy wastes are shown in Fig. 14.29.

Operating Parameters

Principal operating parameters for anaerobic treatment systems are temperature, pH, gas-production rate, CH_4/CO_2 ratio, alkalinity, and volatile acids concentration [14.28, 14.48]. Heated and mixed anaerobic digesters are usually operated at hydraulic residence times between 10 and 15 d. Packed and fluidized bed systems typically have hydraulic residence times of several hours. Alkalinity and volatile acid concentrations are typically in the 200- to 500-g/m³ range, but effective operation can be obtained at much higher values. Methane fermentation rates decrease sharply at temperatures below 30°C and at pH values below 6.5 and above 8.5.

Process Performance

Performance of anaerobic treatment processes must be considered in terms of conversion rates, effluent quality, and system reliability. Conventional processes are about as reliable as other biological processes, but

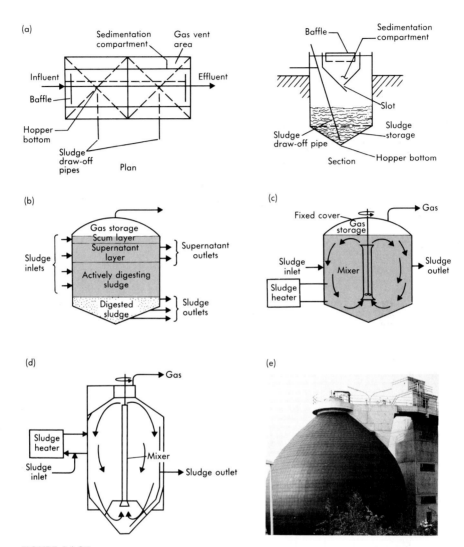

FIGURE 14.27

Anaerobic digesters used for the treatment of wastewater sludges: (a) Imhoff tank (early unheated, open-top digester combined with sedimentation tank); (b) early unheated, unmixed covered digester; (c) circular-heated, completely mixed digester; and (d) and (e) egg-shaped heated, completely mixed digester.

because of the odors produced, failures are very noticeable. Packed beds have been very reliable in laboratory- and pilot-scale, but prototype experience is lacking. A number of anaerobic contact process installations are currently in use, with the greatest number associated with food processing wastewaters, but information on long-term reliability is not available.

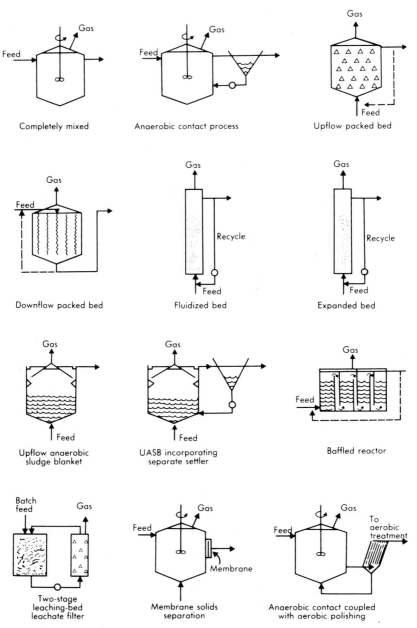

FIGURE 14.28

Typical reactor configurations used in anaerobic wastewater treatment.

Source: Adapted from Ref. [14.48]. Reprinted with permission from the American Chemical Society.

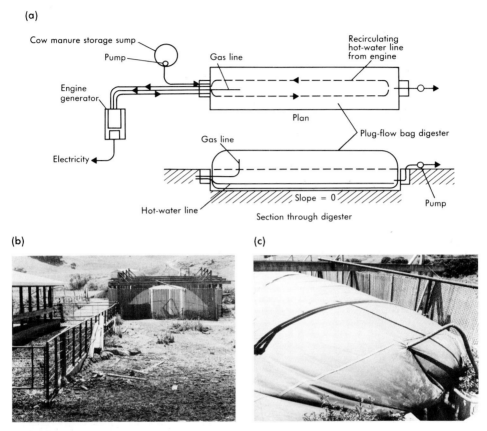

FIGURE 14.29

Plug-flow anaerobic bag digester used for dairy wastes: (a) schematic and (b) and (c) photographic views.

Comparative data on selected anaerobic treatment processes are reported in Table 14.8. Corresponding operating parameters are given in Table 14.9. In considering the COD per day rates given in Table 14.9, it is interesting to note that air-activated sludge processes are constrained to COD loading rates below $2 \ kg/m^3 \cdot d$ because of oxygen transfer limitations. This seeming advantage for anaerobic treatment is offset by the requirement that the wastewater must be of high strength and the fact that 90 percent COD removal from a strong waste leaves a high concentration in the effluent. Neither aerobic nor anaerobic processes offer panaceas for biological wastewater treatment.

TABLE 14.8

Comparison of Selected Anaerobic Treatment Processes

TYPE OF PROCESS	SUITABLE WASTES*	SOLIDS CONTENT, %	SOLIDS RETENTION TIME, d	DEGREE OF MIXING	OPERATING TEMPERATURE, °C	GAS PRODUCTION	DEGREE OF CONTROL/AUTOMATION
Anaerobic contact	Industrial (agricultural)	≤ 5	0.5–5	Not required	30–35	Continuous	High/automatic
Anaerobic filter	Industrial	≤ 5	0.5–5	Continuous	30–35	Continuous	High/automatic
Batch	Agricultural	≤ 25	30–60 +	Limited	30–35	Irregular	Low/usually manual
Expanded bed	Industrial (agricultural, municipal)	4–10	0.5–5	Not required	30–35	Continuous	High/automatic
High rate	Agricultural, industrial	4–15	5–20	Continuous	30–35	Continuous	High/automatic
High rate municipal Primary	Wastewater sludges	4–10	10–25	Continuous	30–35	Continuous	High/automatic
Secondary	Waste from primary digester	4–10	20–60	None	Unheated	Not collected	Low/semiautomatic
Plug flow (horizontal)	Agricultural (low volume)	6–15	25–60	None	30–35	Continuous	Low/usually manual

*Most common applications are listed first; other applications appear in parentheses.

TABLE 14.9
Operating Parameters for Selected Anaerobic Treatment Processes

TYPE OF PROCESS	OPERATING TEMPERATURE, °C	SOLIDS RETENTION TIME, d	LOADING RATE, kg/m³·d		GAS PRODUCTION, m³/kg VS DESTROYED	COD REMOVAL*, %
			Volatile solids	COD		
Anaerobic contact	30–35	0.5–5		1–6		80–95
Anaerobic filter	30–35	0.5–5		1–6		80–95
Batch	30–35	30–60 +		1–10		80–95
Expanded bed	30–35	0.5–5		1–20		80–95
High rate	30–35	5–20	2–8	1–6	0.38–0.39	75–95
High rate municipal						
Primary	30–35	10–25	1.5–6	1–6	0.75–1.10	80–95
Secondary	30–35	20–60	0.5–4			
Plug flow (horizontal)	Unheated	25–60	1–2	1–3		

*Reported values are based on the biodegradable fraction. For municipal wastewater sludges the biodegradable fraction of the total COD is approximately 0.6. For most industrial wastes the biodegradable fraction is greater than 0.8.

14.12

ON-SITE WASTEWATER MANAGEMENT SYSTEMS

Wastewater from individual dwellings or small commercial units in unsewered locations is usually disposed of using on-site treatment and disposal systems. Although a variety of on-site systems have been used, the most common involves the use of a septic tank for the partial treatment of the wastewater and a subsurface-soil absorption system for the disposal of the effluent from the septic tank.

On-Site System Components

The principal components of the most common type of on-site wastewater management system, as cited above, are a septic tank and a leach field (Fig. 14.30). Each of these components is considered separately in the following discussion.

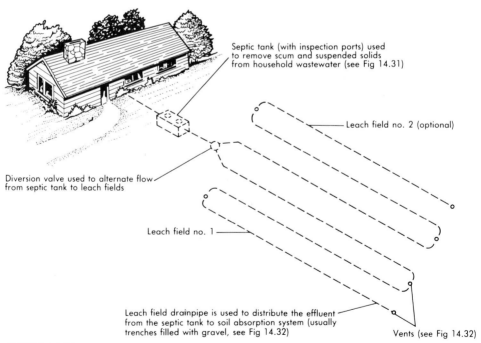

Septic tank (with inspection ports) used to remove scum and suspended solids from household wastewater (see Fig 14.31)

Leach field no. 2 (optional)

Diversion valve used to alternate flow from septic tank to leach fields

Leach field no. 1

Leach field drainpipe is used to distribute the effluent from the septic tank to soil absorption system (usually trenches filled with gravel, see Fig 14.32)

Vents (see Fig 14.32)

FIGURE 14.30

Definition sketch for an on-site wastewater management system.

Septic Tank

A septic tank, as shown in Fig. 14.31, is essentially a watertight tank that serves as a combined settling and skimming tank and as an unheated, unmixed anaerobic digester. Typically one interior baffle is used to divide the tank, and manholes are provided to permit inspection and cleaning. Most septic tanks are made of concrete or fiberglass, although other materials such as redwood have been used. The capacity of septic tanks used at individual residences typically varies between 2.8 and 5.6 m^3.

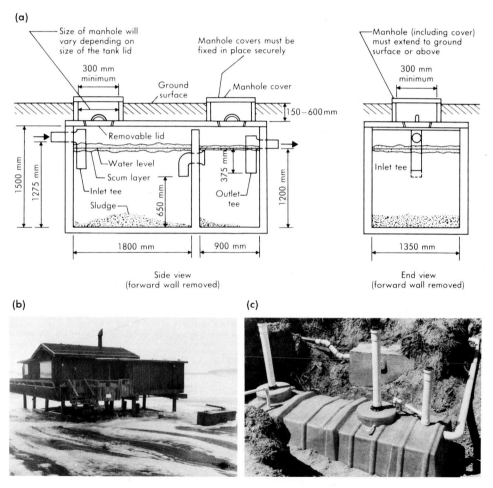

FIGURE 14.31

Septic tanks used in on-site systems: (a) typical concrete tank; (b) concrete tank exposed by ocean wave action in Stinson Beach, Calif.; and (c) plastic tank.

Settleable solids in the incoming wastewater settle and form a sludge layer at the bottom of the tank. Greases and other light materials float to the surface where a scum layer is formed as floating materials accumulate. Partially treated wastewater flows from the clear space between the scum and sludge layers to the soil absorption system. The organic material retained in the bottom of the tank undergoes facultative and anaerobic decomposition and is converted to more stable compounds and gases such as carbon dioxide (CO_2), hydrogen sulfide (H_2S), and methane (CH_4). Because of this conversion, the volume of the material being deposited is being reduced continually, although there is always a net accumulation in the tank. Because the accumulation of scum and solids reduces the effective settling capacity of the tank, the contents of the tanks should be pumped periodically (e.g., every 2 to 4 years). Additional details on the operation of septic tanks may be found in Refs. [14.14], [14.59], [14.61], and [14.62].

Soil Absorption System

The subsurface-soil absorption system is used to dispose of the septic tank effluent in the soil mantle. Typically the soil absorption system, commonly known as a *leach field*, consists of a series of narrow, deep trenches filled with gravel (Fig. 14.32). Perforated pipe is used to distribute the septic tank effluent to the leach field. Depending on the liquid level in the trench, the septic tank effluent must infiltrate into the soil mantle through the bottom and side walls of the trench. Additional treatment of the septic tank effluent occurs as it percolates through the soil.

In a study conducted in 1955 [14.2], settled wastewater was spread on five California soils of varying initial permeabilities (about 17-fold). The objective was to determine the factors governing the infiltration and the percolation of wastewater into soil formations and the steady-state infiltration rates that could be expected under continual inundation. The important results of this study are presented graphically in Figs. 14.33 and 14.34. Without doubt, the most significant finding from this study, as shown in Fig. 14.33, was that, regardless of the initial soil permeability and after varying periods of time, all of the soils reached a steady-state or an equilibrium infiltration rate under continual inundation. Based on these results, it was concluded that the infiltration capacity of a soil absorption system is essentially controlled by the nature of the solids accumulations at or near the soil surface—not by the soil structure. Another significant finding of this study, as shown in Fig. 14.34, was the beneficial effect of resting the soil percolation system. When the soil-wastewater interface was exposed to the atmosphere between dosings, the quasi-equilibrium or steady-state infiltration rate could be increased by a factor of 2 or more.

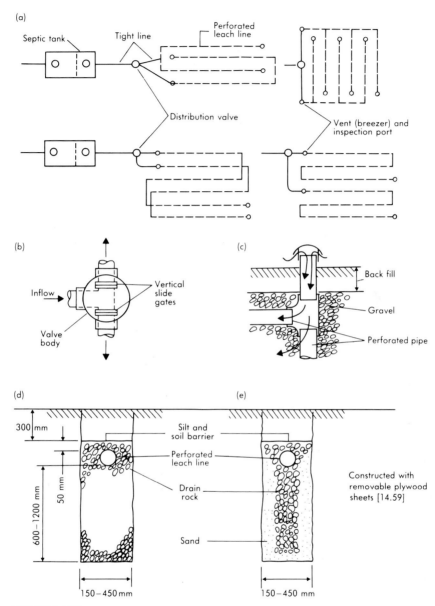

FIGURE 14.32

Typical soil absorption systems and details used for disposal of septic tank effluent:
(a) leach field configurations, (b) distribution valve, (c) vent and inspection port,
(d) conventional leach field trench design, and (e) special leach field trench design.

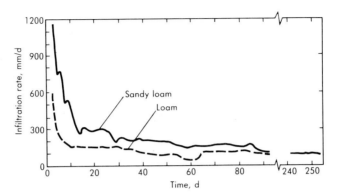

FIGURE 14.33

Infiltration rate as a function of time for two soils continuously inundated with wastewater.

Source: Adapted from Ref. [14.2].

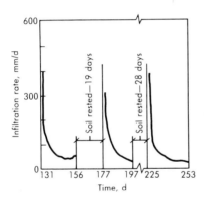

FIGURE 14.34

Beneficial effect on infiltration rate of intermittent application of settled wastewater.

Source: Adapted from Ref. [14.2].

Based on the results of the study [14.2] cited above and other more recent studies, the operation of a subsurface-soil absorption system that is continually inundated may be described as follows. Suspended solids are filtered out by the soil at the interface or are deposited in the soil pore spaces in the immediate vicinity of the interface. Under anaerobic conditions, decomposition of both suspended and dissolved organic material results in bacterial growth. The bacterial decomposition of organic matter also results in the production of polysaccharides (slimes) and ferrous sulfide, a black insoluble precipitate, both of which tend to fill the soil pore space. The net result of the accumulation of effluent solids, bacterial growths, and bacterial decomposition products on and

(a)

(b)

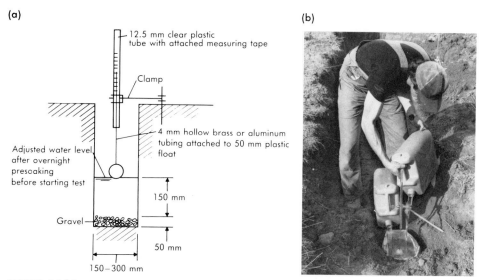

FIGURE 14.35

Definition sketch for conduct of percolation test: (a) test and equipment details and (b) field test. Water is being poured into a perforated paper bag to avoid splashing and unnecessary clogging of sidewall surface area. This technique was developed by J. T. Winneberger [14-61].

near the liquid-soil interface is a reduction of the infiltration capacity at the soil interface. As shown in Fig. 14.34, where the leaching field is intermittently inundated (e.g., exposed to the atmosphere between liquid applications) the net production of polysaccharides and ferrous sulfide is minimized and higher infiltration rates can be maintained.

There are several practical implications in the above discussion:

1. The results of the standard percolation test (see Current Design Practice below) are of little value in establishing loading rates, as most soils will achieve the same equilibrium loading rate when inundated completely with partially treated effluent.

2. The design of soil absorption systems on the basis of equilibrium rates requires little or no information on the soil formation other than the fact that the soil will drain. The equilibrium freshwater rate should be about 3 to 5 times the proposed equilibrium effluent loading rate.

3. More effective treatment of the wastewater will prolong the useful life of a soil absorption system and will allow use of a higher equilibrium loading rate.

4. Periodic resting of a soil absorption system will restore the infiltration capacity and will allow for the maintenance of a higher equilibrium loading rate.

Current Design Practice

Current design practice for individual on-site systems involves the use of a percolation test to determine the required size of the soil absorption system. In the percolation test, holes varying in diameter from 150 to 300 mm and in depth from 0.6 to 1.0 m are dug in the location where the leach field is to be placed. After the formation around the test hole has been soaked for a period of 24 hr, the percolation rate is determined by measuring either (1) the time required for the water surface to drop a specified distance (reported as minutes per millimeter) or (2) the depth the water surface falls in a specified period of time (see Fig. 14.35). Exact details in performing the percolation test vary throughout the United States [14.14].

Once percolation tests have been performed, the allowable hydraulic loading rate, in cubic meters per square meter-day ($m^3/m^2 \cdot d$), for the soil absorption system is determined from a table or curve relating the percolation rate (in minutes per millimeter) to the allowable loading rate, again in cubic meters per square meter-day ($m^3/m^2 \cdot d$). Because of the findings reported above regarding equilibrium infiltration rates, many agencies are now starting to use an equilibrium loading rate. Where the equilibrium design is used, the results of the percolation test are only used to assess whether the soil is sufficiently permeable to accept any wastewater. (Five times the application rate is recommended.) Also, a number of agencies are requiring the installation of dual soil absorption systems (see Fig. 14.32). Where two systems are installed, higher equilibrium loading rates can be used. Additional details on soil absorption systems may be found in Refs. [14.2], [14.59], [14.61], and [14.62]. The design of an on-site wastewater management system is illustrated in Example 14.8.

EXAMPLE 14.8

DESIGN OF ON-SITE WASTEWATER MANAGEMENT SYSTEM

Design an on-site system using a trench-type soil absorption system for an individual residence in a nonsewered area. The maximum occupancy of the residence will be five persons. Use an equilibrium design for the soil absorption system based on the side-wall area of the trench. The long-term equilibrium design loading rate for the soil absorption system is to be 0.010 $m^3/m^2 \cdot d$. Use alternating fields in the final design and a dosing chamber to achieve a more uniform application of the septic tank effluent and to improve the operation of the soil absorption system.

SOLUTION:

1. Estimate the daily flow rate using the data given in Table 1.3.

   Flow rate $= (200 \text{ L/capita} \cdot \text{d}) \times 5$ persons

   $\qquad = 1000 \text{ L/d}$

   $\qquad = 1.0 \text{ m}^3/\text{d}$

2. Determine the average detention time in the septic tank. Based on the local building code, the minimum required septic tank volume is 4500 L. Assuming 30 percent of the volume is lost because of sludge and scum accumulations, the average detention time is

   $$\text{Dentention time} = \frac{0.7 \,(4500 \text{ L})}{1000 \text{ L/d}}$$

   $\qquad\qquad = 3.2 \text{ d}$

   The minimum acceptable detention time should be about 2 d, although some agencies use a value of 1.0 d.

3. Determine the required length of the trench soil absorption system. Assume the maximum trench depth is to be fixed at 1200 mm to conform to local code requirements and that the maximum liquid depth in the trenches is to be 800 mm.

   a. The surface area available per meter of trench length is

      $\text{Area/m} = 2(0.8 \text{ m/m} \times 1.0 \text{ m})$

      $\qquad\qquad = 1.6 \text{ m}^2/\text{m of trench}$

   b. The required soil absorption surface area is

      $$\text{Area required} = \frac{\text{flow rate}}{\text{surface loading rate}}$$

      $$= \frac{1.0 \text{ m}^3/\text{d}}{0.010 \text{ m}^3/\text{m}^2 \cdot \text{d}}$$

      $\qquad = 100 \text{ m}^2$

   c. The required trench length is

      $$\text{Trench length} = \frac{100 \text{ m}^2}{1.6 \text{ m}^2/\text{m}}$$

      $\qquad\qquad = 62.5 \text{ m}$

   d. Use two 32-m trenches arranged as shown in Fig. 14.30.

COMMENT

The loading rate for the soil absorption systems will usually be fixed by local codes. Also the soil absorption area to be considered for design purposes will vary. In this example, the side-wall area (the preferred method) was used. In many locations, only the trench bottom area is considered. In other locations both areas are considered. Although there is little standardization, the design procedures given in Ref. [14.14] can be used as a guide.

Alternative Systems

Because conventional soil absorption systems cannot be used in some locations, several alternative systems have been developed [14.11]. Two common examples are the mound system and the evapotranspiration (ET) bed. Mound systems are used when groundwater is near the surface, and ET beds are used where it is desired to prevent percolation to the groundwater. Additional details on these and other alternative systems may be found in Ref. [14.14].

KEY IDEAS, CONCEPTS, AND ISSUES

- The principal use of biological treatment is for the removal of organic materials from wastewaters. Biological treatment is also used for the removal of selected inorganic ions and sludge stabilization.
- Although many types and configurations of biological treatment processes exist, all are based on the same principles of microbial metabolism.
- Microbial growth occurs in all biological treatment processes, and the excess microbial mass is generally a secondary waste.
- The principal activated sludge process configurations are plug-flow, continuous-flow stirred-tank, and batch.
- Key activated sludge process design parameters include hydraulic residence time, sludge age, oxygen transfer rate, recycle rate, and sludge wasting rate.
- Filamentous bulking is a common cause of activated sludge process failure.
- Aerobic attached-growth biological treatment processes such as trickling filters and rotating biological contactors utilize microbial cultures growing on inert media to remove constituents from wastewaters.
- Hydraulic and organic loading rates are the principal design parameters used with attached growth systems.
- Oxidation ponds incorporate a symbiotic relationship between algae and heterotrophic bacteria. Algal solids in oxidation pond effluent often exceed 100 g/m^3 during the summer months and rarely drop below 30 g/m^3 in the winter.

- Land treatment systems are classified in three groups: slow rate, rapid infiltration, and overland flow.
- Wetland treatment systems incorporate dense stands of water-tolerant plants to support the microbial cultures responsible for the treatment of the contaminants in wastewaters.
- Anaerobic biological treatment systems are most suitable for treatment of high-strength wastewaters such as organic sludges and industrial wastes.
- Methane gas is a major product of anaerobic treatment.
- On-site treatment systems, usually consisting of a septic tank followed by a soil absorption system, are used for homes and other facilities in unsewered areas. Properly designed, these systems provide a high degree of treatment for household wastewaters.

DISCUSSION TOPICS AND PROBLEMS

14.1. Assuming that approximately 65 percent of the suspended solids in a typical domestic waste are settleable and that typical primary sludge concentrations are 50,000 g/m^3, determine the primary sludge volume produced per capita-day in each of the countries of Table 14.1.

14.2. Sedimentation tanks often are the limiting factor in the performance of biological wastewater treatment processes and usually their design is based on average daily flow. Determine the variation in overflow rate for sedimentation tanks, assuming that peak instantaneous flow rate is 3.5 times the average flow on a peak day, which in turn is 2.5 times the annual average flow rate, and minimum instantaneous flow is 0.2 times the annual average flow rate.

14.3. Given the flow rate and BOD concentration variation shown in Fig. 14.36 determine the mass loading rate in grams of BOD_u per hour as a function of time and the daily average.

14.4. Using the relationship $\mu = \theta_c^{-1}$ developed in Example 14.2, determine effluent BOD as a function of sludge age. Use the coefficient values given in Table 14.3 and plot C_e versus θ_c for a CFST activated sludge process.

14.5. A simple suspended-growth biological treatment process known as an aerated lagoon consists of an aeration tank without a sedimentation tank and without recycle. For the case of an ideally mixed aerated lagoon, determine the growth rate and effluent BOD as a function of sludge age.

14.6. The simplest type of activated sludge system to model is a CFSTR with wasting from the mixed liquor at steady state (see Fig. 14.3b,

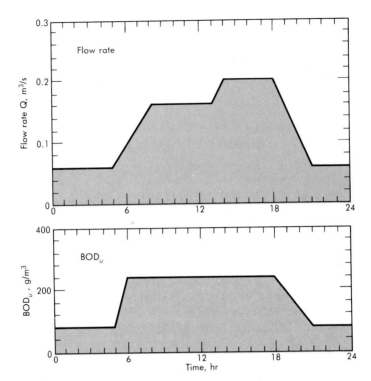

FIGURE 14.36

Definition sketch for Problem 14.3

sludge wasting option 2). Plot the effluent substrate and MLSS concentration as functions of θ_c for a wastewater having BOD_u of 250 g/m³. Use $Y = 0.4$ g cells/g BOD_u, $k = 4.0$ d^{-1}, $k_d = 0.05$ d^{-1}, $K = 30$ g/m³, and $X_e = 20$ g/m³, and ignore the effect of solids in the sedimentation tank. Use θ_H values of 4 and 8 h.

14.7. For the system of Problem 14.6 determine and plot the rate of oxygen uptake as a function of sludge age.

14.8. Conventional activated sludge aeration devices can transfer up to 1800 g/m³·d O_2. For a plant treating a wastewater having the characteristics given in Problem 14.6, determine the minimum reactor (aeration) time as a function of sludge age.

14.9. An activated sludge process with a hydraulic residence time of 4.5 hr is used to treat a wastewater containing 250 g/m³ BOD_u and 35 g/m³ NH_3—N. If the effluent concentrations of BOD_u, NH_3—N, and NO_3^-—N are 5, 0, and 23, respectively, determine the oxygen transfer rate.

14.10. Using the flow rate data in Fig. 14.15 and the SBR sequence times given in Table 14.3 determine the number of batch reactors required and the daily cycle of each unit, including the idle time when the reactor is empty. Assume that one-third of the reactor contents is retained after decanting.

14.11. A wastewater has an average BOD_u of 275 g/m³ and an average flow rate of 0.2 m³/s. Determine the total area of low- and high-rate rock-medium trickling filters necessary to treat the wastewater.

14.12. Using the Eckenfelder equation and the coefficient values below determine the effluent BOD from the three trickling filters of Problem 14.11.

$$k = 0.08$$

$$\alpha = 2.5$$

$$m = 0.8$$

$$n = 0.75$$

14.13. For the wastewaters of Table 14.1 determine if hydraulic or organic loading rate is more likely to control trickling filter design.

14.14. Effluent from a primary sedimentation tank is to be treated using a high-rate rock-medium trickling filter. Flow varies between 15,000 and 30,000 m³/d, with an average value of 22,000 m³/d, and the settled BOD_5 varies between 110 g/m³ at low flow and 220 g/m³ at peak flow, with an average value of 160 g/m³. Determine appropriate values for the trickling filter diameter and depth based on parameter values given in Table 14.5.

14.15. Repeat Problem 14.14 for a high-rate plastic-medium trickling filter.

14.16. Determine the total surface area of an RBC system that would be required to treat the wastewater of Problem 14.14.

14.17. Discuss the problems that can be expected in the operation of oxidation ponds in cold climates.

14.18. Determine the facultative oxidation pond volume required for 90 percent conversion of influent BOD for a flow of 1000 m³/d and a $k = 0.25$ d^{-1} if the system can be modeled as two CFSTRs in series. For an influent BOD_5 of 100 g/m³ and a depth of 2 m, determine the organic loading rate and compare the value to standard design values.

14.19. Assuming that the data given in Problem 14.18 are for 20°C conditions, determine the acceptable volume and organic loading rate at 10°C.

14.20. Solve Problem 14.18 using a single reactor, the Wehner and Wilhelm equation [Eq. (6.21)] and a D/uL value of 0.5.

14.21. Using the parameters in Table 14.6 compare the land requirements of the various land treatment systems for a city of 5000 people, (a) assuming that a separate sewer system exists and (b) assuming that combined sewers are in use and that the combined flow is two times the normal sanitary dry weather flow. Include 4-months winter storage for the slow-rate system.

14.22. A slow-rate land treatment system is to be used by a community of 5000 people located at 45° north latitude. If the cover crop is to be corn and the soil is a highly permeable clayey-sand, estimate the appropriate system size. If soil temperatures will be below 0°C for approximately 90 d/y determine the storage required.

14.23. A domestic wastewater is to be treated using a wetland system. Pretreatment being considered includes primary sedimentation and trickling filtration.
   a. Discuss the advantages and disadvantages of the two methods.
   b. Assuming that primary effluent BOD_5 is 150 g/m^3 and trickling filter effluent BOD_5 is 40 g/m^3, compare the wetland-system area requirements.
   c. Discuss the need for secondary sedimentation in the trickling filter system.

14.24. Discuss the advantages of wetland system development in (a) urban and (b) rural environments. Consider disease transmission and potential nuisance problems in your discussion.

14.25. Estimate the methane that can be produced per day from a primary sludge having a BOD_u of 20,000 g/m^3. Assume the digester is an ideal CFSTR, the residence time is 10 d, and the growth rate is negligible.

14.26. Determine the quantity of methane that will be produced from fermentation of a tricarboxylic acid having the empirical formula $C_3H_8O_6$.

14.27. An industrial wastewater has an average temperature of 15°C. If heat losses in an anaerobic treatment system are 10 percent, determine the wastewater BOD required to operate a methane fermentation system at 35°C. Use Eq. (14.11) to estimate CH_4 production per unit volume of wastewater. Assume the growth rate is negligible.

REFERENCES

14.1. Abufayed, A., (1984), "Use of Primary Sewage Sludge as an Energy Source for Denitrification in Sequencing Batch Reactors," Ph.D. thesis, Department of Civil Engineering, University of California, Davis.

14.2. *An Investigation of Sewage Spreading on Five California Soils*, (1955), I.E.R. Series 37, Technical Bulletin 12, Sanitation Engineering Research Laboratory, University of California, Berkeley.

14.3. Ardern, E., and W. T. Lockett, (1914), "The Oxidation of Sewage Without Aid of Filters," *J. Chemical Industry*, vol. 33, pp. 523, 1122.

14.4. Barnard, J. L., (1975), "Biological Nutrient Removal Without Chemicals," *Water Research*, vol. 9, no. 516, p. 485.

14.5. Barth, E. F., (1981), "Sequencing Batch Reactors for Municipal Wastewater Treatment," presented at the 9th United States-Japan Conference on Sewage Treatment Technology, Cincinnati, Ohio.

14.6. Benefield, L. D., and C. W. Randall, (1980), *Biological Process Design for Wastewater Treatment*, Prentice-Hall, Englewood Cliffs, N.J.

14.7. Bryan, E. H., (1982), "Development of Synthetic Media for Biological Treatment of Municipal and Industrial Wastewaters," *Proc. First International Conference on Fixed Film Biological Processes*, edited by Y. C. Wu, E. D. Smith, R. D. Miller, and E. J. Opatken, Kings Island, Ohio, p. 89.

14.8. Bryers, J. D., (1982), "Processes Involved in Early Biofilm Development," *Proc. First International Conference on Fixed Film Biological Processes*, edited by Y. C. Wu, E. D. Smith, R. D. Miller, and E. J. Opatken, Kings Island, Ohio, p. 155.

14.9. Buchan, L., (1983), "Possible Biological Mechanism of Phosphorus Removal," *Water Science and Technology*, vol. 15, p. 87.

14.10. Buswell, A. M., and H. F. Mueller, (1952), "Mechanics of Methane Fermentation," *J. Industrial Engineering Chemistry*, vol. 44, no. 3, p. 550.

14.11. Chambers, B., and E. J. Tomlinson (eds.), (1982), *Bulking of Activated Sludge: Preventative and Remedial Methods*, Ellis Horwood, Chichester, England.

14.12. Chiesa, S. C., (1982), "Growth and Control of Filamentous Microbes in Activated Sludge," Ph.D. thesis, Department of Civil Engineering, University of Notre Dame, Notre Dame, Ind.

14.13. Chudoba, J., V. Ottova, and V. Madera, (1973), "Control of Activated Sludge Filamentous Bulking—I. Effect of Hydraulic Regime or Degree of Mixing in Aeration Tank," *Water Research*, vol. 7, no. 8, p. 1163.

14.14. *Design Manual: Onsite Wastewater Treatment and Disposal Systems*, (1980), U.S. Environmental Protection Agency, Municipal Environmental Research Laboratory, Cincinnati, Ohio.

14.15. Dinges, R., (1982), *Natural Systems for Water Pollution Control*, Van Nostrand Reinhold Company, New York.

14.16. Dodd, J. C., and J. L. Anderson, (1977), "An Integrated High Rate Pond-Algae Harvesting System," *Progress in Water Technology*, vol. 9, no. 3, p. 713.

14.17. Eckenfelder, W. W., Jr., (1967), "Trickling Filter Design and Performance," *J. Sanitary Engineering Division*, ASCE, vol. 87, no. SA4, p. 87.

14.18. Eikelboon, D. H., (1977), "Identification of Filamentous Organisms in Bulking Activated Sludge," *Progress in Water Technology*, vol. 8, no. 6, p. 153.

14.19. Figueiredo, R. F., (1982), "Effects of Rainfall on the Performance of the Overland Flow Wastewater Treatment System," Ph.D. thesis, Department of Civil Engineering, University of California, Davis.

14.20. Friedman, A. A., R. C. Woods, and R. C. Wilkey, (1976), "Kinetic Response of Rotating Biological Contactors," *Proc. 31st Annual Purdue Industrial Waste Conference*, Purdue, Ind., p. 420.

14.21. Gloyna, E. F., J. F. Malina, and E. M. Davis (eds.), (1976), *Ponds as A Wastewater Treatment Alternative*, University of Texas Press, Austin.

14.22. Goronzy, M. D., (1979), "Intermittent Operation of Extended Aeration Processes for Small Systems," *J. Water Pollution Control Federation*, vol. 51, no. 2, p. 278.

14.23. Grady, C. P. L., Jr., and H. L. Lim, (1980), *Biological Wastewater Treatment*, Marcel Dekker, New York.

14.24. Irvine, R. L., and A. W. Busch, (1979), "Sequencing Batch Reactors, an Overview," *J. Water Pollution Control Federation*, vol. 51, no. 2, p. 238.

14.25. Jewell, W. J., (1982), "Anaerobic Attached Film Expanded Bed Fundamentals," *Proc. First International Conference on Fixed-Film Biological Processes*, edited by Y. C. Wu, E. D. Smith, R. D Miller, and E. J. Opatken, Kings Island, Ohio, p. 17.

14.26. Kruzic, A., (1984), "Control of Nitrogen Removal in the Overland Flow Wastewater Treatment System," Ph.D. thesis, Department of Civil Engineering, University of California, Davis.

14.27. Loehr, R. C., W. J. Jewell, J. D. Novak, W. W. Clarkson, and G. S. Friedman, (1979), *Land Application of Wastes*, vols. I and II, Van Nostrand Reinhold Company, New York.

14.28. McCarty, P. L., (1964), "Anaerobic Waste Treatment Fundamentals," *Public Works*, vol. 95, nos. 9–12.

14.29. Mandt, M. G., and B. A. Bell, (1982), *Oxidation Ditches in Wastewater Treatment*, Ann Arbor Science Publishers, Ann Arbor, Mich.

14.30. Mara, D. D., (1976), *Sewage Treatment in Hot Climates*, John Wiley & Sons, Chichester, England.

14.31. Metcalf & Eddy, Inc., (1979), *Wastewater Engineering*, 2d ed., revised by G. Tchobanoglous, McGraw-Hill Book Company, New York.

14.32. Middlebrooks, E. J., D. H. Falkenberg, R. F. Lewis, and D. J. Enreth (eds.), (1974), *Upgrading Wastewater Stabilization Ponds to Meet New Discharge Standards*, Utah Water Research Laboratory Publication PRW G1 59-1, Utah State University, Logan.

14.33. Moore, S. F., and E. D. Schroeder, (1970), "An Investigation of the Effect of Residence Time on Denitrification," *Water Research*, vol. 4, no. 10, p. 685.

14.34. Niku, S., and E. D. Schroeder, (1981), "Stability of Activated Sludge Processes Based on Statistical Measures," *J. Water Pollution Control Federation*, vol. 53, no. 4, p. 457.

14.35. _____, _____, and R. S. Haugh, (1982), "Reliability and Stability of the Trickling Filter Process," *J. Water Pollution Control Federation*, vol. 54, no. 2, p. 189.

14.36. Okun, D. A., (1949), "A System of Bioprecipitation of Organic Matter from Activated Sludge," *Sewage Works Journal*, vol. 21, no. 5, p. 763.

14.37. Opatken, E. J., (1982), "Rotating Biological Contactors-Second Order Kinetics," *Proc. First International Conference on Fixed Film Biological Processes*, edited by Y. C. Wu, E. D. Smith, R. D. Miller, and E. J. Opatken, Kings Island, Ohio, p. 210.

14.38. Pasveer, A., (1960), "New Developments in the Application of Kessener Brushes (Aeration Rotors) in the Activated Sludge Treatment of Trade Waste Waters" in *Waste Treatment*, edited by P. C. G. Isaac, Proc. Second Symposium on the Treatment of Waste Waters, Pergamon Press, New York.

14.39. *Process Design Manual: Land Treatment of Municipal Wastewater*, (1981), U.S. Environmental Protection Agency, Cincinnati, Ohio.

14.40. Reed, S. C., and R. W. Crites, (1984), *Handbook of Land Treatment Systems for Industrial and Municipal Wastes*, Noyes Data Corporation, Park Ridge, N.J.

14.41. Rittman, B. G., and P. L. McCarty, (1980), "Evaluation of Steady-State-Biofilm Kinetics," *Biotechnology and Bioengineering*, vol. 22, p. 2359.

14.42. Sanks, R. L., and T. Asano (eds.), (1975), *Land Treatment and Disposal of Municipal and Industrial Wastewater*, Ann Arbor Science Publishers, Ann Arbor, Mich.

14.43. Sarner, E., (1980), *Plastic-Packed Trickling Filters*, Ann Arbor Science Publishers, Ann Arbor, Mich.

14.44. Schroeder, E. D., (1977), *Water and Wastewater Treatment*, McGraw-Hill Book Company, New York.

14.45. Silverstein, J. A., (1982), "Control of Activated Sludge Characteristics in a Sequencing Batch Reactor Wastewater Treatment Process," Ph.D. thesis, Department of Civil Engineering, University of California, Davis.

14.46. Smith, R. G., (1980), "Development of a Predictive Model to Describe Organic Removal by the Overland Flow Treatment Process,"

Ph.D. thesis, Department of Civil Engineering, University of California, Davis.

14.47. _____ and E. D. Schroeder, (1983), "Physical Design of Overland Flow Systems," *J. Water Pollution Control Federation*, vol. 55, no. 3, p. 255.

14.48. Speece, R. E., (1983), "Anaerobic Biotechnology for Industrial Wastewater Treatment," *Environmental Science and Technology*, vol. 17, no. 9, p. 416a.

14.49. Stall, T. R., and J. H. Sherrard, (1978), "Evaluation of Control Parameters for the Activated Sludge Process," *J. Water Pollution Control Federation*, vol. 50, no. 3, p. 450.

14.50. Steinmuller, W., and E. Bock, (1976), "Growth of *Nitrobacter* in the Presence of Organic Matter," *Archives of Microbiology*, vol. 108, p. 299.

14.51. Stowell, R., R. Ludwig, J. Colt, and G. Tchobanoglous, (1981), "Concepts in Aquatic Treatment Systems Design," *J. Environmental Engineering Division*, ASCE, vol. 107, no. EE5, p. 919.

14.52. _____, A. S. Weber, G. Tchobanoglous, B. A. Wilson, and K. R. Townzen, (1983), "Mosquito Considerations in the Design of Wetland Systems for the Treatment of Wastewater," Department of Civil Engineering, University of California, Davis.

14.53. Strom, P. F., and D. Jenkins, (1984), "Identification and Significance of Filamentous Microorganisms In Activated Sludge," *J. Water Pollution Control Federation*, vol. 56, no. 5, p. 449.

14.54. Swilley, E. L., and B. Atkinson, (1963), "A Mathematical Model for the Trickling Filter," *Proc. 18th Industrial Waste Conference*, Purdue University, Purdue, Ind.

14.55. Trulear, M. G., and W. G. Characklis, (1982), "Dynamics of Biofilm Processes," *J. Water Pollution Control Federation*, vol. 54, no. 9, p. 1288.

14.56. Ulrich, A. H., and M. W. Smith, (1951), "The Biosorption Process for Sewage and Waste Treatment," *Sewage and Industrial Wastes*, vol. 23, no. 9, p. 1248.

14.57. van den Berg, L., and K. J. Kennedy, (1983), "Comparison of Advanced Anaerobic Reactors," Third International Symposium on Anaerobic Digestion, Boston.

14.58. Velz, C. J., (1948), "A Basic Law for the Performance of Biological Beds," *Sewage Works Journal*, vol. 20, no. 4, p. 607.

14.59. Warshall, P., (1979), *Septic Tank Practices*, Anchor Press, Doubleday, Garden City, N.J.

14.60. Wilderer, P., and R. Staud, (1980), "Optimierung und Betreibliche Stenerung des Belebtschlammver Fahrens Nach dem Prinzip des

Zu-und Abschaltens von Karelementen," *ATV-Bevichte*, vol. 32, p. 347.

14.61. Winneberger, J. H. T., (1984), *Septic Tank Systems: A Consultant's Tool Kit, Volume I: Subsurface Disposal of Septic Tank Effluents*, Butterworth Publishers, Boston.

14.62. _____, (1984), *Septic Tank Systems: A Consultant's Tool Kit, Volume II: The Septic Tank*, Butterworth Publishers, Boston.

14.63. Wood, D. K., and G. Tchobanoglous, (1975), "Trace Elements in Biological Waste Treatment," *J. Water Pollution Control Federation*, vol. 47, no. 7, p. 1933.

14.64. Young, J. C., and P. L. McCarty, (1969), "The Anaerobic Filter for Waste Treatment," *J. Water Pollution Control Federation*, vol. 41, no. 5, p. R160.

15

Synthesizing Water
and Wastewater
Treatment Systems

Water and wastewater treatment systems are composed of a number of unit operations and processes linked together in sequences (see Chapter 11). How these sequences are conceptualized, selected, designed, tested, and operated is the subject of this chapter. Treatment system synthesis involves making choices, as illustrated in Fig. 15.1, among a number of alternatives available in almost every step. An optimal solution to a given treatment problem may not exist in the mathematical sense because of the number of alternatives and the lack of knowledge about true performance characteristics of processes and equipment. For this reason, experience is a key factor in system design. However, innovations in equipment and processes are continually appearing, and there is a need to bring new approaches into use, a factor that tends to run counter to reliance on experience. The best method of incorporating both factors is to pilot-test systems, or portions of systems. Use of pilot testing will be discussed later in the chapter.

The purpose of this chapter is to introduce the methods of structuring the treatment system synthesis process. These methods are outlined in separate sections dealing with (1) the decision-making process, (2) process selection, (3) limits on process performance, (4) pilot-plant testing, and (5) implementation of treatment processes. Each situation is unique, but the hierarchical structure provided here allows a logical progression from the general level of problem identification to the highly specific level of unit design and equipment selection. The important issue of operation and maintenance, which affects the long-term performance of treatment plants, is also considered.

(a)   (b)

(c)   (d)

FIGURE 15.1

Typical choices available in water treatment: (a) reactor clarifier or (b) conventional flocculator followed by a sedimentation tank. Typical choices available in wastewater treatment: (c) activated sludge process or (d) tower trickling filters or combinations of the two processes.

Source: Part (c) courtesy of Consoer Townsend and Assoc., Inc.

15.1

THE DECISION-MAKING PROCESS

Choosing the sequence of steps to be used in the solution of a particular treatment problem involves two levels of decision making. The first level is related to constraints on the system that are imposed by the quantity and characteristics of the water or wastewater and the source or receiving water. When these constraints have been identified, the second level of decision making, which involves the actual selection and design of the units in the sequence, can be carried out. The more important constraints that must be considered are reviewed below.

Primary Constraints

The first three steps in treatment system synthesis are always the same: (1) characterizing the quantity and quality of the water or wastewater to be treated, (2) determining the requirements or standards that the treated water must meet, and (3) establishing the environmental impacts of the project. These three constraints affect all other aspects of treatment plant system synthesis. The primary constraints are closely related and cannot be considered independently of each other. For example, standards will be set by a regulatory agency on the basis of the water or wastewater characteristics and the possible environmental impacts associated with withdrawing water from a source or discharging treated wastewater to a receiving stream or lake. Conversely, a designer will select processes on the basis of the characteristics of the water or wastewater and the standards set by the regulatory agencies concerned.

Characterization of Water and Wastewater
The importance of characterization of a water or a wastewater in the synthesis of treatment plant systems is related to process selection and sizing. For example, a water supply characterized by low pH and low alkalinity will need to be neutralized, and alkalinity will have to be added during treatment to prevent corrosion of pipes and appliances. Wastewaters containing heavy metals in excess of the discharge requirements set by regulatory agencies will require treatment (usually precipitation) for the removal of the metals. Toxic materials such as THMs, pesticides, and herbicides are often found in groundwaters, rivers, and wastewaters and must be removed to acceptable levels before the treated water is delivered to the users. Characterization of the waters and wastewaters is required to determine the degree of treatment required before use of a potential water source or discharge of a wastewater to the environment.

Characterization is also important in process sizing. This fact is most easily understood by referring to the process loading-rate data given in Chapters 12, 13, and 14. Loading-rate constraints have been developed for virtually all treatment unit operations and processes. Examples include the hydraulic and solids loading rates for sedimentation tanks and the organic and hydraulic loading rates for trickling filters.

Requirements for Treated Water Quality
Both water and wastewater treatment facilities are subject to requirements set by regulatory agencies, but those set on wastewater treatment systems are far more situation-dependent. Examples have been given in

previous chapters of the need for certain types or levels of treatment in specific situations. Nutrient removal is required only where nutrients in the discharge would result in a serious deterioration of receiving water quality. Specialized treatment to remove toxic compounds is not needed in most locations. Cooling of heated discharges is rarely required, but may be necessary in critical locations with low receiving water flow rates and the presence of cold-water fish species (e.g., trout).

The fact that limitations on effluent nutrient content, toxicity, and temperature will affect treatment system design is clear, but the numerical values used in establishing the requirements are often as important as the type of requirements. For example, if a discharge must meet a 30 g/m^3 value for both BOD_5 and suspended solids, conventionally designed high-rate trickling filters would be an inappropriate choice for secondary biological treatment [15.13]. If the requirements are 10 g/m^3 for both BOD_5 and suspended solids, filtration of the secondary effluent will probably be necessary. A few activated sludge plants consistently meet a 10 g/m^3 standard [15.6, 15.14], but this situation is rare. Land and wetland treatment systems can often meet stringent discharge standards, but the land requirements are generally prohibitive for treatment of large flows.

Most wastewater discharge requirements are set in three components: a 30-d mean value, a 7-d mean value, and a maximum allowable value. For example, the effluent BOD_5 value might have a limitation of 30, 45, and 90 g/m^3 for the 30-d average, 7-d average, and maximum allowable values, respectively. These three values result from an approximately log-normal frequency distribution. A violation occurs if any of the three values is exceeded. In practice, one of the requirements is usually more stringent than the other two and, as a consequence, controls the design process [15.12].

Environmental Impacts

Environmental impacts resulting from development of water and wastewater treatment systems are extremely varied. Primary impacts include changes in flow rates of small streams and decreases in groundwater levels due to withdrawals for water supplies, increases in contaminant concentrations of receiving waters due to discharges, increases in river temperatures resulting from decreases in flows or discharges of warm effluents, development of odors due to the anaerobic degradation of organic wastes, and increases of eutrophication rates resulting from nutrient discharges and temperature increases. Secondary, and more subtle, impacts are population increases made possible by increased availability of water or increased wastewater treatment capacity, potential for local aesthetic problems resulting from temporary failure of wastewater treatment processes, restrictions on land or water use required to protect a water supply, and growth-induced impacts such as air pollution, transportation and high population density.

Although environmental impacts are usually perceived as being negative, it is equally important to recognize the positive environmental results of a project. Public health protection and enhancement is one of the basic responsibilities of government. Water and wastewater treatment systems are among the most important factors in the maintenance of public health and should be given a priority similar to fire and police protection. Wastewater treatment will usually result in improved aesthetic conditions in receiving waters and can often be coupled with the development of ecological preserves and recreational facilities.

Secondary Constraints

Secondary considerations include community size and population density, geographical location, climate, local energy costs, and similar parameters and variables.

Community Size and Population Density

Community size is important in system synthesis for a number of reasons. Water demand and wastewater flow variation decrease as population increases; certain types of treatment (e.g., septic tank-leach fields) are almost never cost-effective or manageable in larger communities; the financial base necessary for construction, operation, and management of mechanically sophisticated treatment systems increases with community size; and the impact of the discharges from large communities is usually greater than that from smaller communities.

Geographical Location

Geographical location has significance other than its relation to climate. Chemical costs are greater in remote areas, as is the cost of equipment maintenance and repair. In some areas, specific chemicals are by-products of industrial processes and can be obtained at prices that virtually dictate the choice of treatment process. Use of waste pickle liquor from steel mills for the precipitation of phosphorus is an example of the latter situation. In the Great Lakes region pickle liquor is available at very low cost because of the large number of steel plants in the area, making it very unlikely that the option of biological phosphorus removal will be competitive.

Climate

Climate enters into consideration because reaction rates are affected by temperature of the water or wastewater and because treatment systems operated in regions of low temperature must often be housed to prevent freezing. In some areas wind and dust are problems because of wind-induced mixing and increased maintenance costs for some types of equipment.

TABLE 15.1

**Energy Requirements for Alternative Treatment Systems
with a Plant Capacity of 3800 m³ / d**

SYSTEM OR SUBSYSTEM	ENERGY USAGE, kW · h/yr × 10^{-3}*					TOTAL kW · h/yr × 10^{-3}
	Primary Energy		Secondary Energy			
	Electricity	Fuel	Plant construction‡	Chemicals, etc.	Parts and supplies	
Primary	50	90	44	—	8	192
Activated sludge	237	209	156	60	34	694
Trickling filter	130	180	165	60	30	565
Facultative pond	30	40	90	30	16	206
Facultative pond plus aquatic wetland	52	70	140	30	74	316

*To convert to Btu/yr, multiply the given value by 10,800.

† With the exception of primary chlorination.

‡ Based on a return period of 20 years.

Source: Developed from Ref. [15.22].

Energy Cost

Energy consumption is an increasingly important factor in process selection and design [15.16]. Although energy costs are generally rising everywhere, considerable variation exists on a local and regional basis. Electrical energy costs are usually lowest in areas with large hydroelectric generation capacities, such as the Niagara Falls and the Pacific northwest regions. The fossil-fuel resources of the southwest and the Pennsylvania-West Virginia areas make these regions stable with respect to energy resources, if not inexpensive. In some cases, the choice of a land-intensive treatment system can be justified on the basis of the differential energy cost of a more sophisticated treatment system requiring much less space. Typical energy usage values for several types of treatment processes are summarized in Table 15.1.

15.2
PROCESS SELECTION

In this section, the selection of unit operations and processes and process flow sheets is considered. Important factors that must be considered in selecting and evaluating unit operations and processes are given in Table 15.2. For the most part, the factors given there are self-explanatory, and the table can be used as a checklist in the selection of unit operations and processes. In what follows, a number of important issues—including the role of the engineer; the client's needs; the importance of past experience; the impact of regulatory requirements; the availability of equipment; the availability of trained personnel; construction, operation, and maintenance costs; and energy usage—that relate to the factors in Table 15.2 are examined in greater detail. Process performance and pilot-plant testing are examined separately in Sections 15.3 and 15.4, respectively.

Role of the Engineer

The role of the engineer in process selection varies with the relationship to the client. Consulting engineers act as advisors during the process selection and design stages and as agents for the client during construction and start-up of a system. The consultant's selections of processes and equipment should be viewed as suggestions that will be reviewed by the engineering staff of the client, and also by the city council, the board of directors, or an equivalent group. Some large communities, utility districts, and industries have engineering staffs capable of carrying out the design of water and wastewater treatment systems. In this case, the advisory role and the decision-making process are, to a large extent, combined, since there is no outside technical evaluation. A situation similar to that of staff design occurs when a governmental body or a company has no technical staff capable of reviewing the work of a consultant. In such cases, the roles of consultant and reviewer are, for practical purposes, combined. Employing a technical reviewer (e.g., a second consultant) during the conceptualization phase of large projects is often an excellent idea. The role of the reviewer must be delineated carefully, but the incorporation of a second approach will often result in a greatly improved project.

Client's Needs

Conforming the treatment system to the client's needs includes consideration of future expansion, design life, actual versus desirable product requirements, and the client's financial situation. Few communities or industries stay the same in size or character for extended periods of time, and some consideration must be given to the need for larger or different facilities in the future. Future community changes will affect the type of system chosen as well as the sizing of certain parts of the system, such as pump stations and piping. All facilities have a useful life and will need to be replaced at some point in time. In some cases, the design life is relatively short, and this should be recognized in the design phase. Two examples where short-design-life systems are desirable are for areas of rapid community growth where regionalization is planned at a future time and for seasonal industries where it is generally advisable to

TABLE 15.2

**Important Factors that must Be Considered when Selecting
and Evaluating Unit Operations and Processes**

FACTOR	COMMENT
1. Process applicability	The applicability of a process is evaluated on the basis of past experience, data from full-scale plants, and data from pilot-plant studies. If new or unusual conditions are encountered, pilot-plant studies are necessary.
2. Applicable flow range	The process should be matched to the expected flow range. For example, stabilization ponds are not suitable for extremely large flows.
3. Applicable flow variation	Most unit operations and processes work best with a constant flow rate, although some variation can be tolerated. If the flow variation is too great, flow equalization may be necessary.
4. Influent wastewater characteristics	The characteristics of the influent affect the types of processes to be used (e.g., chemical or biological) and the requirements for their proper operation.
5. Inhibiting and unaffected constituents	What constituents are present that may be inhibitory, and under what conditions? What constituents are not affected during treatment?
6. Climatic constraints	Temperature affects the rate of reaction of most chemical and biological processes. Freezing conditions may affect the physical operation of the facilities.
7. Reaction kinetics and reactor selection	Reactor sizing is based on the governing reaction kinetics. Data for kinetic expressions usually are derived from experience, the literature, and the results of pilot-plant studies. The effect of reaction kinetics on reactor selection is considered in Chapter 6.
8. Performance	Performance is most often measured in terms of effluent quality, which must be consistent with the given effluent-discharge requirements.
9. Treatment residuals	The types and amounts of solid, liquid, and gaseous residuals produced must be known or estimated. Often pilot-plant studies are used to identify residuals properly.

minimize capital investment. It must be remembered that steps taken to minimize capital investment or operational costs cannot result in decreased system performance.

Meeting the actual needs of a client usually requires considerably less investment than providing a system that produces the best possible product. For example, water softening is often economically desirable

TABLE 15.2   *(Cont.)*

FACTOR	COMMENT
10. Sludge-handling constraints	Are there any constraints that would make sludge handling expensive or infeasible? In many cases, a treatment method should be selected only after the sludge processing and handling options have been explored.
11. Environmental constraints	Nutrient requirements must be considered for biological treatment processes. Environmental factors, such as the prevailing winds and wind directions, may restrict the use of certain processes, especially where odors may be produced.
12. Chemical requirements	What resources and what amounts must be committed for a long period of time for the successful operation of the unit operation or process?
13. Energy requirements	The energy requirements, as well as probable future energy costs, must be known if cost-effective treatment systems are to be designed.
14. Other resource requirements	What, if any, additional resources must be committed to the successful implementation of the proposed treatment system using the unit operation or process in question?
15. Reliability	What is the long-term record of the reliability of the unit operation or process under consideration? Is the operation or process easily upset? Can it stand periodic shock loadings? If so, how do such occurrences affect the quality of the effluent?
16. Complexity	How complex is the process to operate under routine conditions and under emergency conditions such as shock loadings? What level of training must the operator have to operate the process?
17. Ancillary processes required	What support processes are required? How do they affect the effluent quality, especially when they become inoperative?
18. Compatibility	Can the unit operation or process be used successfully with existing facilities? Can plant expansion be accomplished easily? Can the type of reactor be modified?

Source: Adapted from Ref. [15.9].

but is not a necessity in terms of meeting drinking water standards. The ability to pay for a treatment system includes far more than financing the construction costs. Many small communities can obtain construction funding through grants or bonds but do not have the income to operate and maintain a sophisticated treatment facility. Solutions to such problems include tying into the system of a larger, nearby community or developing a simpler treatment system with lower operating and maintenance costs (e.g., using slow sand filters or groundwater for a water supply and land treatment for wastewater treatment).

Past Experience

Experience with treatment systems is an important factor in the design process. The principal responsibility of the design engineer is to provide a system that meets the requirements established for the treated water or wastewater. Until this responsibility is met, questions of minimizing cost or maximizing quality cannot be raised. Included in the statement that a system "meets requirements" is the assumption that the various units and pieces of equipment will work and be reliable. The more experience available to the designer—through personal work, the work of colleagues within the same firm or agency, and/or reports in the technical literature —the greater the probability that the system being designed will represent a near-optimal solution to the problem. It is important to recognize the need to include the critical review of colleagues in the design process.

Equally important is the necessity of keeping up with innovations reported in the literature. Innovations in treatment technology are continually being presented at professional conferences, in professional journals, and at seminars sponsored by federal and state regulatory agencies and universities. Examples of recent innovations include methods of filtration for improved removal of *Giardia* cysts [15.8], application of granular activated carbon adsorption for water treatment in the United States [15.4], suggestions for improved clarifier performance [15.17], thermophillic, attached-film, anaerobic expanded beds for treatment of high-strength wastewaters [15.20], and use of genetic engineering techniques in biological wastewater treatment [15.7]. This brief list of innovations was selected from the literally thousands of recently published papers on water and wastewater treatment.

Regulatory Agency Requirements

The effect of regulatory agency requirements on the synthesis of water treatment systems is, as noted above, much less situation-dependent than requirements set on wastewater treatment systems. Most of the requirements are related to providing safe water for human consumption. Constraints are set on the quality of the source because the presence of some constituents (e.g., toxic or carcinogenic chemicals) or the concentration of other constituents is either a public health problem (e.g., high TDS is related to hypertension) or makes treatment difficult (e.g., high turbidity and high organic concentration reduce the effectiveness of disinfection). It should be noted that chemical sludges are a significant by-product of water and wastewater treatment in most cases, and disposal of these sludges is a difficult problem. Discharge of water treatment sludges to the source is unacceptable because the sludges are (1) relatively dense and the discharge will result in the formation of sludge deposits, and (2) the chemicals added in the treatment process may be undesirable in the source.

Equipment Availability

Availability of equipment refers as much to parts and servicing as to purchase of the original item. Virtually any desired piece of equipment can be supplied to any location (particularly if cost is no object), but the time necessary to obtain parts and service is often a critical factor in equipment selection. Some pieces of equipment (large electric motors are an example) require several months to obtain and consequently both repair services and adequate parts inventories are essential. Process selection must depend, to an extent, on the type of equipment redundancy and parts inventory required for the alternatives under consideration. These considerations are particularly important in remote areas, in regions where weather affects transportation, and in developing nations. An example is the selection of processes to be used in locations such as Yellowstone National Park. The park is remote from major transportation terminals, many of the facilities are essentially inaccessible during the winter, and seasonal reliability must be high.

Availability of Trained Personnel

Operation of water and wastewater treatment systems entails considerable responsibility. The cost of treatment facilities is usually one of the major expenditures in a community's budget. In addition, the direct effect of water treatment on community health must be stressed. For these reasons, personnel assigned to operate and maintain a treatment facility must be trained appropriately. The more sophisticated the facility, the greater must be the level of training. Qualified individuals are usually available in metropolitan areas, as is financial support for their employment. Small communities often have a problem in finding both the personnel and the money with which to pay them. Selection of processes for small communities, or communities in developing nations, should reflect the level of training of the operations and maintenance personnel.

Initial Construction and Future Operation and Maintenance Costs

Project costs must be calculated in a number of ways. The most obvious cost is the total construction cost of the facility. However, the total construction cost is only part of the total expense incurred by the owner

688 SYNTHESIZING WATER AND WASTEWATER TREATMENT SYSTEMS

of a treatment system. Operation and maintenance costs and interest on bonds or loans must also be included. Often the annual expense is more important to a community or industry than the total construction cost, and this factor must be considered in process selection. Finally, a project may be constructed in phases to minimize the initial costs. Certain expenses such as land acquisition must be included in the initial cost and will affect the choice of processes in many cases.

Two situations exist where operation and maintenance costs are often less important to the systems owner than capital cost: communities with very limited bonding capacity and communities with seasonal industries such as canneries. The opposite situation exists where governmental grants are available to pay for a large share of the construction cost but not for the operation and maintenance expense. Granting agencies (such as the U.S. Environmental Protection Agency) review project grant proposals to determine if the least total cost, including capital, operation, and maintenance expenses, is being used.

Energy Usage

Although energy costs can be included in the cost of operation, they are worthy of separate consideration. Since 1973, the cost of energy has risen dramatically, and economists predict that the price of energy will continue to increase for the foreseeable future. Energy conservation should be included in the design of all new systems, but care must be given to examining the entire system and not one component [15.5, 15.10]. For example, activated sludge is an energy-intensive wastewater treatment process, but the total system—including sludge disposal and the needed additional treatment, such as filtration, to meet discharge requirements—may require less energy than competitive systems.

15.3
LIMITS ON PROCESS PERFORMANCE

Water and wastewater treatment systems are expected to provide products and effluents that meet defined standards. The standards vary in certain details, depending on local conditions and needs, but are for the most part very similar. Water distributed for human consumption must be free of pathogenic organisms (as demonstrated by the MPN value) and have concentrations of specific chemical constituents such as cadmium and trihalomethanes below values determined to be toxic or carcinogenic by regulatory agencies (see Appendix D). Effluents from wastewater treatment plants are expected to meet requirements on BOD, suspended solids, grease, toxic substances, and MPN. In designing processes to meet these requirements the engineer must remember that actual limits on process performance exist and that process performance varies in both explainable and inexplainable ways [15.2, 15.12, 15.13]. Therefore, product and effluent characteristic values used in design will generally need to be lower than the defined standards.

Variations in Treatment System Performance

Treatment system performance varies for a number of reasons. Characteristics of water and wastewater change with time. Reaction rates vary with temperature and reactant concentration. Operations such as sedimentation are affected by flow rate and stratification induced by temperature/density differences. Mechanical and electrical equipment such as chemical feeders, pumps, sludge rakes, and meters break down. In some cases, breakdowns cause a unit to be taken out of service and other units to be overloaded during the repair period. Failures of meters and other devices used to collect, record, and transmit performance data often result in incorrect information being sent to the plant operator. These facts emphasize the need for preventive maintenance and regular performance testing of the equipment as well as of the overall process.

Reliability and Stability of Treatment Systems

System reliability is defined as the probability that a set of performance standards will be met. Determination of reliability requires statistical evaluation of a system's performance over a time period long enough to establish typical operating patterns [15.14, 15.23]. In most cases the minimum period required is 1 year. New facilities must be designed using experience with similar operating systems or with pilot-scale systems. Performance data collected over a suitable time period is used to establish a *probability density function* (*PDF*) that can be used to satisfactorily describe the distribution of values. For example, the variation in effluent suspended-solids concentrations from a treatment facility over the course of a year might be found to fit a log-normal PDF. The characteristics of the PDF can then be used to predict the probability that the process will produce an effluent meeting the discharge requirements, i.e., to predict process reliability.

Most PDFs used to describe variations in performance are based on the assumption that variation is random. For this reason, nonrandom factors must be "filtered out." Demand for water, as noted in Chapter 1, is strongly related to time of day, day of week, and week of year. Variation due to time of day and day of week are primarily functions of social behavior in a community and are largely independent of external influences such as temperature and weather. However, variation associated with the week of year is strongly related to weather and other seasonal characteristics. For example, a ski resort can be expected to have low water demand for much of the year, relatively high demand during the winter, and peak demands on weekends and holidays. Variation due to nonrandom events can be accounted for in process design. For example, the peak water demand at a ski resort can be included in the design of both the water and wastewater treatment systems. Inherent system variability results from uncontrolled factors and can be accounted for using statistical methods.

Prediction of Wastewater Treatment System Reliability

Relationships that can be used to predict wastewater treatment system reliability have been developed using data obtained from a large number of facilities in the United States [15.6, 15.12, 15.13, 15.14]. A summary of the relationships is given in Fig. 15.2, in which reliability is plotted as a function of two variables, the normalized mean m_x/X_s and the coefficient of variation V_x. The normalized mean is the arithmetic mean value of the chosen variable over a suitable time period divided by the effluent standard value. The coefficient of variation is the standard deviation of the data divided by the mean value. Typical values found for the coefficient of variation of activated sludge and trickling filter systems are

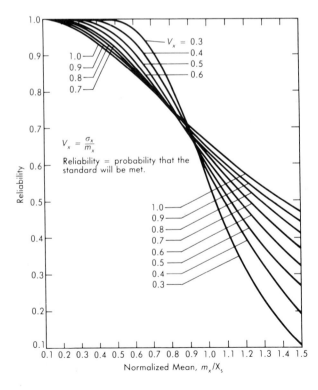

FIGURE 15.2

Reliability of a treatment system as a function of the normalized mean effluent concentration and the coefficient of variation.

Source: From Ref. [15.14].

0.6 and 0.5, respectively. Comparable information is not available for water treatment plants. The use of Fig. 15.2 is illustrated in Example 15.1.

EXAMPLE 15.1

PREDICTION OF RELIABILITY OF AN ACTIVATED SLUDGE SYSTEM

An activated sludge system under design will be required to have a 30-d average BOD_5 of less than 30 g/m³. The performance of a similar system designed by the same firm has been analyzed and found to have $V_x = 0.5$. The firm has decided to design the new system using the probability that the effluent requirement will be met at least 90 percent of the time. Use Fig. 15.2 to determine the design value for the effluent BOD_5 concentration.

SOLUTION:

1. Enter Fig. 15.2 on the ordinate at the selected reliability value, 0.9.
2. Move to the 0.5 V_x curve and then vertically to the abscissa to obtain a normalized mean value.
3. Determine the value of m_x/M_s from Fig. 15.2.

$$\frac{m_x}{X_s} = 0.62$$

4. Calculate the design effluent BOD_5 value.

$$m_x = 0.62(X_s)$$
$$BOD_5 = m_x = 0.62(30 \text{ g/m}^3)$$
$$= 19 \text{ g/m}^3$$

Process Stability

Process stability is defined as the degree of variability of performance. Quantitative measurement of stability can be made using several statistical methods. A detailed description of the statistical parameters used in these measurements is beyond the scope of this text. Further information is available in Refs. [15.1], [15.2], [15.3], and [15.14]. Examples of stability estimates include the relative variation as described by the ratio of the standard deviation to the mean value, the actual variation as described by the standard deviation or the variance, and the range of values. Variance, standard deviation, and range for the variable X are defined below.

$$\text{Variance of } X = \frac{1}{n-1} \sum (X - m_x)^2 \tag{15.1}$$

where

m_x = mean value of X

n = number of values of X

Standard deviation of $X = \sqrt{\text{variance}}$ (15.2)

Range of X = maximum value of X − minimum value of X (15.3)

In studies of the performance of wastewater treatment plants [15.12, 15.13], it was concluded that the standard deviation gave the best estimate of process stability.

Equipment Performance and Reliability

Equipment is selected on the basis of performance specifications. Failure to meet the performance specifications can result in process failure at the outset. More often the result is reduced life of the equipment or the need to expand or replace a system before the design capacity has been reached. The latter situation often occurs several years after a system has been in operation, and accountability is difficult to assign. Design engineers have the responsibility of setting clear and definitive performance specifications that can be used over the life of the equipment. It is important to test equipment upon installation to establish that the performance specifications can be met. Failure to meet the performance specifications may result in the need to replace equipment, as was the case with the aeration equipment shown in Fig. 15.3.

The concept of reliability can be applied to mechanical equipment in the same manner as it is applied to processes and systems. There is some difficulty in the application of equipment reliability estimates because variation occurs both with the type of equipment, with the manufacturer, the quality of installation and the quality of maintenance. In addition,

(a)

(b)

FIGURE 15.3

Aeration equipment replaced because of mechanical problems.

reliability of a particular brand may change with time as new models are developed or changes in manufacturing technique are instituted.

15.4
PILOT-PLANT TESTING

In many cases, the information available for the design of a water or wastewater treatment process is judged to be inadequate [15.9]. Examples of such conditions include cases where a new type of process is under consideration, where two or more processes are to be compared prior to final selection, where the operating conditions expected are outside the range of normal experience, where particular operating conditions need to be simulated, where optimal operating conditions need to be determined, and where it must be established that a proposed process will meet regulatory agency or legal requirements. A summary of considerations made in pilot-plant testing is given in Table 15.3.

Setting Up Pilot Testing Programs

Pilot-plant testing programs are expensive in terms of both time and money. It is necessary to plan the programs carefully to obtain the desired information. Factors to be considered include (1) the processes to be tested, (2) the physical scale required, (3) the operating conditions to be simulated, and (4) nonphysical factors. Examples of pilot-plant facilities are shown in Fig. 15.4.

Processes to Be Tested

Where the process to be pilot-tested is to be added to an existing system, on-site studies are possible. The pilot plant is brought to, or constructed on, the site so the actual water or wastewater that is to be treated can be used. Where the entire treatment facility is to be constructed but only a portion of the system is to be pilot-tested, it may be possible to test that process at another treatment plant that has similar water or wastewater. In some cases the entire treatment system must be tested because of interrelationships between units or the nature of the water or wastewater. For example, industrial wastewaters often are peculiar to a particular manufacturing facility, and pilot-plant studies need to be made on-site.

Physical Scale of Pilot-Plant Facilities

The scale of a pilot treatment system can vary from a small unit operated on a laboratory bench to a full-scale treatment facility processing thousands of cubic meters per day. Choice of scale is dependent on the critical parameters being considered and the type of answers required

TABLE 15.3
Considerations in Setting Up Pilot-Plant Test Programs

ITEM	CONSIDERATION
Reasons for conducting pilot testing	Test new process Simulate a process Predict process performance Document process performance Optimize system design Satisfy regulatory agency requirements Satisfy legal requirements

Water and wastewater treatment processes commonly pilot-tested

ITEM	CONSIDERATION
Physical-chemical processes	Sedimentation Coagulation Filtration Gas transfer-aeration Chlorination Adsorption Ion exchange Reactor hydraulics Sludge dewatering
Biological processes	Aerobic treatment Aquatic treatment Anaerobic treatment Sludge stabilization
Pilot-plant size	Laboratory bench Laboratory-scale model Pilot-scale tests Full- (prototype) scale tests
Nonphysical design factors	Available time, money, and labor Degree of innovation involved Quality of water or wastewater Location of facilities Complexity of process Similar testing experience Dependent and independent variables
Physical design factors	Scale-up factors Size of prototype Flow variations expected Facilities and equipment required Materials of construction
Design of pilot testing program	Dependent variables Independent variables Test ranges for variables Time required Test facilities Statistical design of data acquisition program

Source: Adapted from a presentation by M. Tolaney of J. M. Montgomery Engineers, Inc., given at the California Water Pollution Control Association annual meeting, 1980.

(a)   (b)

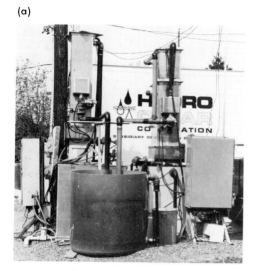

FIGURE 15.4

Examples of pilot-plant test facilities for (a) filtration of primary settled effluent and (b) anaerobic treatment of brewery wastewater.

[15.19]. Sedimentation processes cannot be described adequately using small units and, in fact, cannot be quantitatively scale-modeled at all because of the importance of secondary currents and density stratification (see Fig. 12.20). Certain factors such as potential solids removal under ideal conditions and sludge compactability can be predicted from relatively small-scale units, however. Coagulation efficiency is primarily dependent on the mixing and reaction environment (chemical characteristics of water or wastewater, reactant, concentration, pH, and temperature). Because mixing processes are extremely difficult to scale up, the size of the pilot system should be carefully considered. Nevertheless, most coagulation studies are performed at the laboratory bench.

Gas transfer rates depend on gas partial pressures and liquid-phase concentrations. Gas transfer operations utilizing bubbles (diffused aeration) or film flow (tower stripping) need to be pilot-tested using full-scale depth or height. Where mixers are to be used (submerged turbines with spargers or surface aerators), the transfer equipment must be full-scale and the tankage must be large enough to eliminate wall effects. Adsorption and ion exchange processes need to be studied at two scales. Laboratory studies are used to determine the potential treatment possible using a particular sorbent or ion exchange resin. When the material to be used has been chosen, performance studies are made using columns of prototype-scale depth but smaller diameter.

Pilot testing of biological wastewater treatment processes is done at both laboratory- and larger-scale. In most biological wastewater treatment studies, the major concerns are wastewater treatability (which can be determined in the laboratory), determination of stoichiometric and kinetic coefficients (which can be determined in the laboratory), and correction or prevention of a particular operating problem such as filamentous bulking. Operating problems are often quite different in the laboratory and the field because of differences in dissolved-oxygen concentration, temperature, mixing, and flow variation. Some conditions, such as spatial variation in oxygen concentration, are difficult to model in the laboratory. Thus the nature of the information desired will govern the scale of the pilot plant used.

Adequate time must be set aside to perform pilot studies. Most important is that the conditions studied in the pilot testing program represent those that will be experienced in the full-scale system [15.19]. A major objective in conducting a pilot testing program is to avoid mistakes in process selection and design. Problems in pilot system operation provide important information and should not be ignored because of tight schedules or limited funds.

Operating Conditions

Pilot-plant studies should be initiated only after specific objectives have been selected with respect to operating conditions. A purpose of pilot-plant testing is to ensure that discharge requirements can be met and to gain experience in process operation under the most stressful conditions. For example, pilot testing of an adsorption-chlorination system for the removal of taste and odors caused by algae should be operated during the time when these problems occur, and the system should be loaded in increasing increments to determine the upper and lower limits of operation.

Nonphysical Factors in Pilot-Plant Testing

The decision of whether to conduct pilot-plant testing and the extent of the program required is dependent on a number of factors. Time, money, and staff availability are always important considerations. However, because a well-designed and conducted pilot testing program will save money in the long run, other factors are more important. The degree of innovation being considered is an important factor in determining the type of pilot testing program necessary. If the engineer has had considerable experience with the processes being considered or if the literature contains adequate information, the testing program may be limited. However, if a new treatment concept is being tested, extensive studies would be appropriate.

Process complexity is an important factor because the more complex the process the greater the probability that problems will occur. The number of dependent and independent variables that need to be studied greatly affects the design of a pilot testing program. Sampling and analysis of the samples is time-consuming and expensive. If diurnal variation of variables is important, two or three shifts of personnel will be required and must be included in the study plan. In some cases the pilot facility is at a location that presents problems in staff housing and analysis of samples. Mobile laboratories and pilot-plant equipment can be used in situations where permanent facilities are not available.

Physical Design of Pilot-Plant Facilities

Design of pilot-plant facilities should include consideration of scale-up factors, size of the prototype process, flow variation to be expected in the prototype, and the construction requirements of the pilot facility. As noted above, some process characteristics can be scaled up to prototype size directly, and others are extremely difficult to model. In general, reaction processes scale up more easily than flow processes, and therefore reaction processes can be piloted in smaller units. In all cases the closer to full-scale the pilot plant is, the more precise can be the modeling effort. When large systems are being designed it is sometimes possible to construct full-scale modules and use them as pilot systems for the design of the rest of the system. Similarly, existing modules of large systems can be used to pilot-test structural, equipment, or operational modifications under consideration.

All treatment systems are subject to flow variations. The greatest flow variations are experienced in wastewater treatment facilities, and it is important that consideration of loading be included in the testing programs. Hydraulic loading variations affect flow patterns in sedimentation tanks and packed beds. Organic loading variations affect performance of biological treatment units and oxygen requirements.

Design of Pilot-Testing Programs

Pilot-testing programs require careful design to ensure that the necessary information will be obtained. Dependent and independent variables must be identified; ranges of independent variable values to be used must be selected. Because pilot testing is expensive, the experimental program must be carefully laid out with respect to the time schedule and the type and quantity of data required. Statistical design of the data acquisition program is necessary to ensure validity of the pilot-testing program results as well as to minimize program expense [15.1, 15.2, 15.3].

Construction of Pilot-Plant Facilities

Pilot-plant facilities should be constructed of durable materials that will not detract from the testing program by falling apart, corroding, or providing a source of contaminants to the water or wastewater. Depending on the size of the pilot plant a variety of materials from glassware, to shipping and storage containers such as metal or plastic drums, to specially constructed units can be used. Tanks and other facilities do not need to look like those of the prototype and, because of the scaling laws, generally should not be geometrically similar [15.19]. Where previously used containers are utilized it is important to clean them carefully to prevent introduction of unsuspected contaminants into the experimental systems.

15.5
IMPLEMENTATION OF TREATMENT SYSTEMS

When information on the performance requirements, water or wastewater characteristics, community, climate, weather, performance of candidate processes, and costs have been accumulated and preliminary decisions made with respect to appropriate process choices, the treatment train to be used can be selected. The steps are (1) selection of processes and operations, (2) selection of design criteria, (3) process sizing, (4) design of plant layout, including future expansion, (5) preparation of hydraulic profiles, (6) preparation of a solids balance, and (7) preparation of construction drawings and specifications.

Selection of Processes

Selection of the processes to be used in a particular situation is accomplished in two steps: elimination of processes that cannot be expected to meet performance criteria and evaluation of processes capable of meeting performance criteria on the basis of cost, future requirements, client needs and environmental impacts. In choosing between two processes, both capable of meeting the performance criteria, cost will normally be the governing factor in making the selection. However, if, because of community size or location, operation and maintenance will be a severe problem, or if significant changes in performance requirements are, in the engineer's judgment, likely during the design life of the facility, a higher-cost process might be the appropriate choice. An example of a complex flow sheet involving the selection of hundreds of process components and items of equipment is shown in Fig. 15.5.

Selection of Design Criteria

Final design criteria are based on the designer's experience, the performance of similar processes reported in the literature, the results of pilot testing, and regulatory agency requirements. All other aspects of design depend on the design criteria chosen, making this an extremely important step.

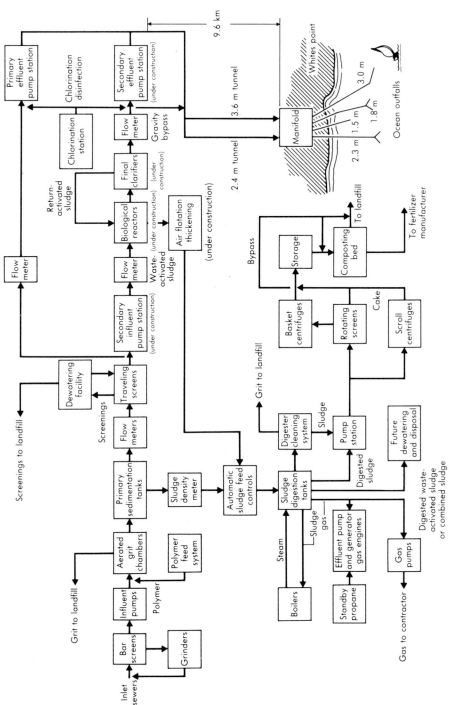

FIGURE 15.5   **Flow diagram for the County Sanitation Districts of Los Angeles County Joint Water Pollution Control Plant as of October 1982.**   Source: Courtesy of County Sanitation Districts of Los Angeles County.

Process Sizing

Process sizing is based on the design criteria, the need for redundancy in the system, and the plant layout. Commonly used design criteria have been summarized in Chapters 12, 13, and 14. More detailed information can be found in Refs. [15.11], [15.18], [15.24], and [15.25]. Modifications to the plant layout, and in some cases to the design criteria, may result from difficulty in process sizing. Three factors should be remembered:

1. It is desirable to have at least two of each type of unit to provide redundancy when one unit is down for repair.
2. Mechanical equipment can generally be purchased in a limited range of sizes and in limited increments of size.
3. Many units, such as sedimentation tanks, have practical size limitations and related operational constraints such as current generation by wind action.

For these reasons process sizing using design criteria is only the first step. Subsequent steps include determination of the sizes of equipment that are available and selection of the final unit size based on this information. This equipment selection is a parallel step to process sizing.

Plant Layout

Design of plant layout involves two steps. The first is a schematic layout that is used to delineate the proper relationship of the various steps in the treatment train. The second involves placing the individual units in the available area, including required access for maintenance and repair of equipment, and minimizing the total space requirements. Typical layouts for water and wastewater treatment plants are shown in Figs. 15.6 and 15.7. Plant layout is an iterative process because the later steps, particularly final process sizing and equipment selection, will result in necessary changes.

Preparation of Hydraulic Profiles

Water surface elevations through treatment units and conduits must be calculated for minimum, average, and maximum flow conditions so that the necessary head can be provided, tank elevations can be set, needed freeboards can be established, conduits can be sized, and the range and type of pumps to be used can be chosen. It is generally desirable to design the treatment system to operate on the hydraulic grade line so that pumping can be minimized. This type of design usually saves energy and, more importantly, minimizes hydraulic problems related to pump or power failures. Because operation of wastewater treatment plants at the hydraulic gradeline is not possible for most cities located in flat areas, sufficient standby power must be available so that the incoming waste-

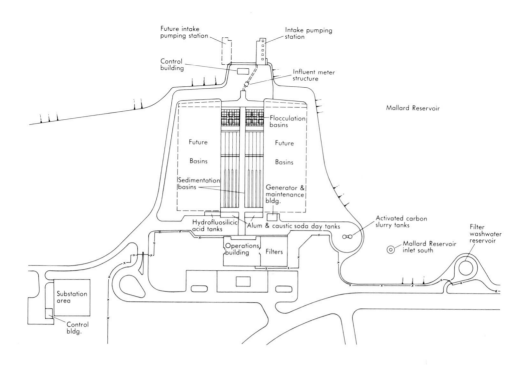

FIGURE 15.6

Typical water treatment plant incorporating flocculation, sedimentation, and filtration: (a) plant layout and (b) aerial view. Source: Courtesy of Contra Costa Water District and James M. Montgomery, Consulting Engineers, Inc.

(a)

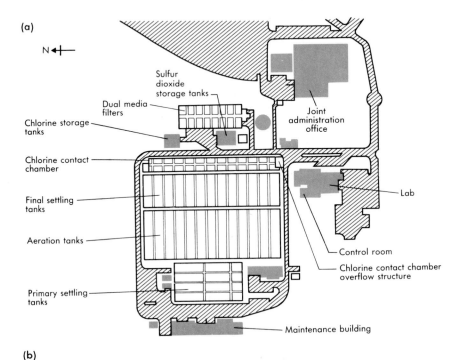

(b)

FIGURE 15.7

Typical wastewater treatment plant incorporating the activated sludge process: (a) plant layout and (b) aerial view.

Source: Courtesy of County Sanitation Districts of Los Angeles County.

water can be passed through the treatment facilities with at least partial treatment.

Preparation of a Solids Balance

Solids balances were discussed briefly in Chapter 11 and will be discussed in detail in Volume 2 of this series. They are important because waste solids management and disposal are extremely expensive, often accounting for over 60 percent of the cost of wastewater treatment. Performing a solids balance is usually an iterative process that is most easily done using a computer.

Preparation of Construction Drawings and Specifications

Construction drawings and specifications, known as the *contract documents*, are used by contractors to make cost estimates and to guide the construction of the facilities. Contract documents often run to hundreds of pages and are extremely costly to produce. Precision and accuracy of these documents is essential because they form the basis for the contractual agreement for the construction of a facility. Contractors are required to construct the facilities as described in the drawings and specifications unless changes are negotiated. Changes are almost always made during construction, and a second set of drawings, the "as-built drawings" must also be prepared. Care in making the as-built drawings is essential because maintenance and repair services require knowledge of where all valves, connections, piping, and wiring are located. Pumps are easily seen because of their size, but the purpose of a pump within the treatment system is difficult to decipher without proper as-built drawings and schematics.

15.6
OPERATION AND MAINTENANCE OF TREATMENT SYSTEMS

Once a treatment system has been constructed, operation and maintenance become the keys to meeting performance standards. The designer should provide an operation and maintenance (O&M) manual and should direct the facility's start-up. Trained operators are essential, as has been pointed out previously. Most states have operator certification programs for both water and wastewater treatment plant operators. The certification level necessary in a particular situation depends on the type and size of plant. Continued training through short courses and professional society programs is usually available and is advisable to keep the quality of operation high.

Maintenance programs should be highly structured. Preventive maintenance schedules must be set up and strictly followed. Performance testing of equipment at appropriate intervals and regular checks for structural deterioration will prevent breakdowns and allow for repairs before major damage occurs. Maintenance program development guides are available from manufacturers, state and federal agencies, and professional societies such as the American Water Works Association, the Water Pollution Control Federation, and the American Society of Civil Engineers.

KEY IDEAS, CONCEPTS, AND ISSUES

- The three major constraints that affect the design and implementation of treatment facilities are the characteristics of the water or wastewater to be treated, the regulatory requirements that must be met, and the environmental impacts of the project.

- The selection of treatment processes must include the consideration of a large number of factors (summarized in Table 15.2), some of which require qualitative judgments based on the past experience of the designer, the owner's preferences, environmental concerns, and consideration of site-specific conditions.

- The factors listed in Table 15.2 should be used as a checklist for the selection of unit operations and processes.

- All treatment processes have limitations with respect to reliability and stability.

- Equipment reliability must be considered in process selection and design and in writing equipment specifications.

- Pilot-plant testing programs should be conducted whenever doubt exists as to the potential process performance in a given situation.

- Pilot-plant testing is particularly important in the combined treatment of municipal and industrial wastewaters and where new types of processes are being considered.

- Pilot plants should be operated as closely as possible to the manner in which the prototype is to be operated. Flow variation is an important factor in pilot-plant testing.

- Process implementation involves (1) process selection, (2) selection of design criteria, (3) process sizing, (4) design of plant layout, (5) preparation of hydraulic profiles, (6) preparation of solids balances, and (7) preparation of construction drawings and specifications.

- A tightly structured operation and maintenance program is a requirement for the successful and economical long-term operation of water and wastewater treatment systems.

DISCUSSION TOPICS AND PROBLEMS

15.1. Using the material presented in Chapters 11 through 15 develop a possible flow sheet for treatment of the wastewater having the characteristics given below:

$$Q = 0.01 \text{ m}^3/\text{s} \ (8 \text{ hr}/\text{d}, 5 \text{ d}/\text{wk})$$
$$BOD_u = 6{,}000 \text{ g}/\text{m}^3$$
$$COD = 15{,}000 \text{ g}/\text{m}^3$$
$$NH_3 - N = 600 \text{ g}/\text{m}^3$$
$$pH = 3.7$$
$$T = 67°C$$

15.2. A water having the characteristics given below is to be used as a domestic supply and as a boiler-feed water. Discuss possible treatment flow sheets that would result in a product suitable for both uses.

CONSTITUENT	CONCENTRATION, g/m^3
Na^+	69
Ca^{+2}	73
Mg^{+2}	85
HCO_3^-	488
CO_3^{-2}	15
Cl^-	160

15.3. A rural community of 1100 people located in northern Minnesota must construct a new wastewater treatment facility. Discuss factors to be considered in choosing between the processes listed below:
a. Oxidation ponds
b. Extended aeration activated sludge
c. Trickling filters
d. Slow-rate land treatment

15.4. Explain why community size strongly affects process selection from an economic standpoint.

15.5. Discuss the reasons for variation in peak-to-mean flow ratio in sewers and peak-to-mean demand rate in domestic water supply systems.

15.6. Many communities with hard water utilize individually owned ion exchange units instead of a central softening system. Give possible reasons why communities would choose this method of water softening, assuming that there is a factor of 3 cost difference per unit volume of water softened. Estimate the cost difference for a household, assuming outside taps (used for lawn watering and irrigation) are untreated.

15.7. An activated sludge plant under design must meet a 30-g/m³-BOD$_5$ effluent standard with a 95 percent reliability. Treatment plants designed by the same consulting firm have had coefficients of varia-tion V_x values of 0.6 during the 7-year period preceding this design. Determine the design effluent BOD_5 value.

15.8.   Discuss the use of the ratio of the standard deviation to the mean value as a measure of stability.

15.9.   Large communities (e.g., Houston and Chicago) have followed differ-ent strategies in developing wastewater treatment facilities. Some have a number of relatively small plants (< 10 m^3/s), while others have a few very large plants (> 100 m^3/s). Discuss the factors that would need to be considered in developing one of these strategies.

15.10.   Disinfection using chlorine has proved to be a simple and effective method of virtually eliminating pathogenic bacteria in domestic water supplies. What steps could be taken to minimize the danger of producing chlorinated hydrocarbons, many of which are known or suspected carcinogens, during the chlorination process? Include con-sideration of concentration of these compounds and the risk factor in your discussion.

15.11.   Why has ozone not been widely used, or accepted, as a disinfectant in the United States?

REFERENCES

15.1.   Benjamin, J. R., and A. Cornell, (1970), *Probability, Statistics and Decision Analysis for Civil Engineers*, McGraw-Hill Book Company, New York.

15.2.   Bertheoux, P. M., (1974), "Some Historical Statistics Related to Future Standards," *J. Environmental Engineering Division*, ASCE, vol. 100, EE2, p. 423.

15.3.   Box, G. E. P., and G. M. Jenkins, (1976), *Time Series Analysis: Forecasting and Control*, 2d ed., Holden-Day, San Francisco.

15.4.   Culp, R. L., and R. M. Clark, (1983), "Granular Activated Carbon Installations," *J. American Water Works Association*, vol. 75, no. 8, p. 398.

15.5.   *Energy Conservation at Wastewater Treatment Plants*, (1980), Manual of Operation No-38, Water Pollution Control Federation, Washing-ton, D.C.

15.6.   Hovey, W. H., E. D. Schroeder, and G. Tchobanoglous, (1979), "Activated Sludge Effluent Quality Distributions," *J. Environmental Engineering Division*, ASCE, vol. 105, no. EE5, p. 819.

15.7.   Kulpa, C. F., R. I. Irvine, and S. A. Sojka (eds.), (1982), *Impact of Applied Genetics in Pollution Control*, symposium held at the Univer-sity of Notre Dame, Notre Dame, Ind.

15.8. Logsdon, G. S., J. M. Symons, R. L. Hoye, Jr., and M. M. Arozarena, (1981), "Alternative Filtration Methods for Removal of *Giardia* Cysts and Cyst Models," *J. American Water Works Association*, vol. 73, no. 1, p. 111.

15.9. Metcalf & Eddy, Inc., (1977), *Wastewater Engineering*, 2d ed, revised by G. Tchobanoglous, McGraw-Hill Book Company, New York.

15.10. Mills, R. A., and G. Tchobanoglous, (1976), "Energy Consumption in Wastewater Treatment," in *Proc. Seventh National Agricultural Waste Management Conference*, edited by W. S. Jewell, Ann Arbor Science Publishers, Ann Arbor, Mich.

15.11. *Municipal Wastewater Treatment Plant Design*, (1977), Manual of Practice No. 8, Water Pollution Control Federation, Washington, D.C.

15.12. Niku, S., and E. D. Schroeder, (1981), "Factors Affecting Effluent Variability from the Activated Sludge Process," *J. Water Pollution Control Federation*, vol. 53, no. 5, p. 546.

15.13. _____, _____, and R. S. Haugh, (1982), "Reliability and Stability of Trickling Filter Processes," *J. Water Pollution Control Federation*, 54, no. 2, p. 189.

15.14. _____, _____, and F. J. Samaniego, (1979), "Performance of Activated Sludge Processes and Reliability Based Design," *J. Water Pollution Control Federation*, vol. 51, no. 12, p. 2841.

15.15. *Nutrient Control*, (1983), Manual of Practice No. FD-7, Water Pollution Control Federation, Washington, D.C.

15.16. Owen, W. F., (1982), *Energy in Wastewater Treatment*, Prentice-Hall, Englewood Cliffs, N.J.

15.17. Parker, D. S., (1983), "Assessment of Secondary Clarification Design Concepts," *J. Water Pollution Control Federation*, vol. 56, no. 4, p. 349.

15.18. Sanks, R. L. (ed.), (1978), *Water Treatment Plant Design for the Practicing Engineer*, Ann Arbor Science Publishers, Ann Arbor, Mich.

15.19. Schemidtke, N. W., and D. W. Smith (eds.), (1983), *Scale-up of Water and Wastewater Treatment Processes*, Butterworth Publisher, Stoneham, Mass.

15.20. Schraa, G., and W. J. Jewell, (1984), "High Rate Conversions of Soluble Organics with a Thermophillic Anaerobic Attached Film Expanded Bed," *J. Water Pollution Control Federation*, vol. 56, no. 3, p. 226.

15.21. Schroeder, E. D., (1977), *Water and Wastewater Treatment*, McGraw-Hill Book Company, New York.

15.22. Tchobanoglous, G., J. Colt, and R. W. Crites, (1979), "Energy

Resource Consumption in Land and Aquatic Treatment Systems," *Proc. U.S. Department of Energy Optimization of Water and Wastewater Management for Municipal and Industrial Applications Conference*, vol. 2, Argonne National Laboratory, Energy and Environmental System Division, ANL/EES-TM-96, New Orleans, La.

15.23. Thomann, R. V., (1970), "Variability of Waste Treatment Plant Performance," *J. Sanitary Engineering Division*, ASCE, 96, SA3, p. 819.

15.24. *Water Reuse*, (1983), Manual of Practice No. SM-3, Water Pollution Control Federation, Washington, D.C.

15.25. *Water Treatment Plant Design*, (1969), American Water Works Association, New York.

Appendixes

A

SYMBOLS AND NOTATION

Symbols and notation used in this book are, to the extent possible, those in common use in the field of environmental engineering. Because of the breadth of topics covered, a number of symbols are used for more than one purpose. For this reason, definitions of terms are given where presented in the text as well as in this appendix. Symbols and notation used in a limited number of chapters are noted by listing the chapters in parentheses at the end of the definition.

SYMBOL	DEFINITION
$a, b, c, \ldots$	Stoichiometric number in chemical reaction, mol (2, 5)
$A, B, C, \ldots$	Atomic mass of chemical species $A, B, C, \ldots$, g/mol (2, 5)
A	Alkalinity, eq/m^3 (2)
A	van't Hoff-Arrhenius coefficient, variable units (2, 5)
A	Area, m^2
A_s	Surface area, m^2
b	Adsorption saturation coefficient, m^3/g (7)
b	Effective length of ocean outfall diffuser, m (8)
B_1, B_2	Blank BOD dissolved-oxygen concentration at t_1 and t_2, g/m^3 (2)
BOD	Biochemical oxygen demand concentration, g/m^3
BOD$_5$	Five-day biochemical oxygen demand, g/m^3
BOD$_r$	BOD remaining at time t, g/m^3
BOD$_u$	Ultimate BOD concentration, g/m^3
C	Fire-flow coefficient related to type of construction (1)
C	Concentration of solute in liquid phase, g/m^3
C_D	Coefficient of drag (12)
C_i	Concentration of constituent i, mol/m^3
C_p	Heat capacity, J/g $\cdot$ K

C_s	Saturation concentration of dissolved gas, g/m^3
CBOD	Carbonaceous BOD, g/m^3
$CBOD_u$	Ultimate carbonaceous BOD, g/m^3
CFSTR	Continuous-flow stirred-tank reactor
d	Grain size of porous medium, m (10)
d_p	Diameter of diffuser discharge port, m (8)
d_i	Impeller diameter, m (12)
D	Degree of mixing (12)
D	Coefficient of dispersion, m^2/s
D_x	Coefficient of dispersion in x direction, m^2/s
D_T	Transverse dispersion coefficient, m^2/s (8)
D_T	Diameter of mixing tank, m (12)
D_m	Coefficient of molecular diffusion, m^2/s
D_1, D_2	Sample dissolved-oxygen concentration at t_1 and t_2, g/m^3 (2)
D_{O_2}	Oxygen deficit, g/m^3 (8)
D_c	Critical oxygen deficit, g/m^3
E	Activation energy, J/mol
E	Evaporation rate, m^3/d (9)
E_f	Efficiency (12)
f	Ratio of seed to sample in BOD test (2)
f	Friction factor (12)
F	Fire flow required, m^3/s (1)
F_A	Mass flux of species A, $g/m^2 \cdot s$
F_D	Buoyant force, N (12)
F_D	Drag force, N (12)
F_G	Gravitational force, N (12)
FBOD	Filtrable BOD, g/m^3
FCOD	Filtrable COD, g/m^3
FTOC	Filtrable TOC, g/m^3
g	Gravitational constant, m/s^2
G	Mean velocity gradient, s^{-1} (12)
G	Molar flow rate of gas, mol/s (12)
h	Head loss, m (12)
h	Depth of liquid, m
h_0	Depth of diffuser centerline, m (8)
h_{d_i}	Heat dispersion, J/s (9)
h_{i_i}	Heat inflow, J/s (9)
h_{o_i}	Heat outflow, J/s (9)
h_{sz_i}	Direct insulation, J/s (9)
h_{w_i}	Advected heat, J/s (9)
H	Average depth of flow, m (8)
H	Depth of idealized sedimentation tank, m (12)
H	Depth of tube settler, m (12)
H_d	Heat dissipation rate, J/s

H_i	Heat content of ith lake section, J (9)
H_i	Height of impeller mixer from tank bottom, m (12)
H_L	Height of liquid in mixing tank, m (12)
HLR	Hydraulic loading rate, $m^3/m^2 \cdot d$
k	Reaction rate constant, variable units
k_T	Reaction rate constant at temperature T
k_d	Maintenance rate coefficient, d^{-1} (14)
k_2	Stream reaeration coefficient, d^{-1} (8, 9)
k	Characteristic property of porous system, m (12)
K	Coefficient of mass transfer, m/hr (7)
K	Reaction rate saturation coefficient, g/m^3 (5, 14)
K	Equilibrium coefficient, variable units (2, 13)
K	Hydraulic conductivity, m/s (10)
K_{BC}	Blaney-Criddle consumptive use coefficient (1)
K_D	Distribution (partition) coefficient (7)
K_F	Freundlich adsorption coefficient (7, 12)
K_H	Henry's Law coefficient, atm^{-1} (2, 12)
$K_L a$	Overall gas-liquid mass transfer coefficient, s^{-1} (2, 6, 12)
K_{SD}	Soil distribution coefficient, m^3/g (7)
K_{SP}	Solubility product, variable units (2)
K_W	Ionization constant of water, $(mol/L)^2$
l	Depth of unexpanded filter, m (12)
l	Length of finite section (8)
l_e	Depth of expanded filter bed, m (12)
L	Characteristic length, m (6)
L	Length of tube settler, m (12)
L	Length of idealized sedimentation tank, m (12)
L	Ultimate BOD, g/m^3 (8)
L_d	Length of diffuser, m (8)
L_H	Flux of water through soil, $m^3/m^2 \cdot s$ (10)
L_i, L_o	Molar flow rate in liquid phase, mol/s (12)
LI	Langlier stability index (3)
L_m	Distance to point of complete mixing, m (8)
m	Molality, mol/kg (2)
m	Mass of material adsorbed, g (7, 12)
m_x	Mean design variable value (15)
M	Molarity, mol/L
M_{O_2}	Oxygen mass transfer rate, $g/m^3 \cdot s$
M_S	Mass of dry sludge, kg (12)
MDTOC	Minimum detectable threshold odor concentration, g/m^3 (3)
MLSS	Mixed-liquor suspended solids, g/m^3 (14)
MLVSS	Mixed-liquor volatile suspended solids, g/m^3 (14)
MPN	Most probable number of coliform organisms, number/100 mL (2)
n	Reaction order (5)
n	Manning's roughness coefficient (8)

n	Rotations per minute (12)
n_i	Number of moles of species i
$n_1, n_2, n_3, \ldots$	Sample size in MPN test, mL (2)
N	Normality, eq/L
N_A	Molar flux of species A, $mol/m^2 \cdot s$ (10)
N_{DF}	Densimetric Froude number (9)
N_F	Froude number (12)
N_P	Power number (12)
N_R	Reynolds number
N_t	Number of organisms present at time t (13)
NTU	Nephelometric turbidity unit (2)
OLR	Organic loading rate, $kg/m^3 \cdot d$
OTR_f	Oxygen transfer rate (field), $kg\ O_2/kW \cdot h$ (12)
p	Percent of annual daytime hours in time increment (1)
p	Percent of annual average demand in time increment (1)
p	Power input, W (12)
p_v	Vapor pressure, atm (7)
pH	Negative log of hydrogen ion concentration
pK	Negative log of equilibrium coefficient
$p_1, p_2, p_3, \ldots$	Number of positive tubes in MPN test (2)
pfu	Plaque-forming units, pfu/100 mL (3)
ppm	Parts per million (2)
P	Precipitation rate, m^3/d (9)
P	Power input per unit volume, W/m^3 (12)
P	Decimal fraction of seed in BOD bottle (2)
P_i	Partial pressure of species i, atm (2)
PFR	Plug-flow reactor
q	Blade width on impeller mixer, m (12)
$q_1, q_2, q_3, \ldots$	Number of negative tubes in MPN test (2)
q_w	Wastewater flow rate, m^3/s (8)
Q	Volumetric flow rate, m^3/s
Q_r	Recycle flow rate, m^3/s (6, 14)
Q_w	Wastage flow rate, m^3/s (14)
r	Reaction rate, $mol/m^3 \cdot d$ or $g/m^3 \cdot d$
r	Blade length on impeller mixer, m (12)
r_g	Rate of microbial growth, $g/m^3 \cdot d$
r_i	Reaction rate of species i, $g/m^3 \cdot d$
r_o	Rate of organic removal, $g/m^3 \cdot d$
r_o^*	Rate of organic removal in terms of BOD_u, $g/m^3 \cdot d$
r_{AP}	Rate of algal photosynthesis, $g/m^3 \cdot d$ (8)
r_{AR}	Rate of algal respiration, $g/m^3 \cdot d$ (8)
r_B	Rate of benthal oxygen demand, $g/m^3 \cdot d$ (8)
r_B^*	Rate of benthal oxygen demand, $g/m^2 \cdot d$ (8)
r_{O_2}	Rate of oxygen uptake, $g/m^3 \cdot d$ (8, 12, 14)
R	Universal gas constant, $J/mol \cdot K$

R_H	Hydraulic radius, m (8)
RI	Ryznar stability index (3)
s	Diameter of central disk on impeller mixer, m (12)
S	Hydraulic gradient (8)
S	Mass of solute sorbed per unit mass of dry soil, g/g (7)
S	Shape factor (12)
S	Solids content of sludge, percent (12)
S_s	Specific gravity of sludge solids (12)
S_1, S_2, S_3	Initial, dispersion, and decay dilution factors for ocean outfalls (8)
S_B	Distributed oxygen demand loading, $g/m^3 \cdot d$ (8)
SAR	Sodium adsorption ratio (3)
SBR	Sequencing batch reactor
SOTR	Standardized oxygen transfer rate, kg O_2/kW $\cdot$ h (12)
SS	Suspended solids concentration, g/m^3
t	Time, s, min, d, yr
T	Temperature, °C, K
TDS	Total dissolved solids concentration, g/m^3
ThOD	Theoretical oxygen demand concentration, g/m^3
TOC	Total organic carbon concentration, g/m^3
TOD	Total oxygen demand concentration, g/m^3
TS	Total solids concentration, g/m^3
TVS	Total volatile solids concentration, g/m^3
u	Velocity, m/s
u	Average advective velocity, m/s (8)
U	Consumptive use of water in irrigated system, mm (1)
U	Maximum tidal velocity, m/s (8)
U^*	Shear velocity, m/s (8)
U_c	Current velocity over ocean outfall, m/s
U_p	Velocity of wastewater leaving ocean outfall discharge port, m/s (8)
v	Velocity, m/s
v	Superficial velocity in porous medium, m/s (10, 12)
v_{sc}	Critical discrete particle settling velocity, m/s (12)
v_s	Discrete particle settling velocity, m/s (12)
V	Volume, m^3
VSS	Volatile suspended solids concentration, g/m^3
w	Stream width, m (8)
w	Water flux rate coefficient, s/m (12)
W	Mass rate of reactant input, g/d (8)
W	Width of idealized sedimentation tank, m (12)
W_b	Width of baffle in mixing tank, m (12)
x	Mass of sorbate, g (7, 12)
X_A	Mole fraction of species A

X_{R_i} Sorbed mole fraction of species i on ion exchange resin (13)
X MLSS or MLVSS concentration, g/m^3 (14)
X_e Effluent suspended-solids concentration, g/m^3 (14)
X_r Recycle suspended-solids concentration, g/m^3 (14)
X_s Effluent standard value (15)

y Mole fraction in gas phase (7, 12)
y Width of aquifer, m (10)
Y Microbial growth yield coefficient, g/g (14)
Y_{obs} Observed microbial yield, g/g (14)

z_x Depth to groundwater at position x, m (10)

α Recycle ratio (6, 14)
α Porosity of porous medium (7, 12)
α Correction factor for oxygen (12)
β Constant in calculation of dilution from ocean outfall (8)
β Ratio of absorbed to incoming radiation (9)
β Correction factor for oxygen saturation concentration (12)
γ_i Activity coefficient for species i (2)
γ Specific weight of water, kN/m^3 (10)
ΔP_o Osmotic pressure differential, kPa (12)
ΔP_a Imposed pressure differential across membrane, kPa (12)
ΔH_v Molecular latent heat of vaporization, J/mol (7)
ζ Solute rejection efficiency (12)
ζ Bulk extinction coefficient, m^{-1} (9)
θ Temperature coefficient for reaction rates
θ_c Sludge age, d (14)
θ_H Hydraulic residence time, hr
λ Coliform density, number/mL (2)
μ Ionic strength (2, 3)
μ Dynamic viscosity, $kg/m \cdot s$
μ Specific growth rate, d^{-1} (14)
ν Kinematic viscosity, m^2/s
ρ Density, kg/m^3
ρ_{sw} Density of seawater, kg/m^3 (8)
ρ_{ww} Density of wastewater, kg/m^3 (8)
τ Shear stress, N/m^2 (12)
ϕ Power function (12)
ϕ Shape factor for sand grains (12)

[ ]   Molar concentration, mol/L

B

METRIC CONVERSION FACTORS (SI UNITS TO U.S. CUSTOMARY)

MULTIPLY THE SI UNIT		BY	TO OBTAIN THE U.S. CUSTOMARY UNIT	
Name	Symbol		Symbol	Name
Acceleration				
meters per second squared	m/s^2	3.2808	ft/s^2	feet per second squared
meters per second squared	m/s^2	39.3701	in/s^2	inches per second squared
Area				
hectare (10,000 m^2)	ha	2.4711	acre	acre
square centimeter	cm^2	0.1550	in^2	square inch
square kilometer	km^2	0.3861	mi^2	square mile
square kilometer	km^2	247.1054	acre	acre
square meter	m^2	10.7639	ft^2	square foot
square meter	m^2	1.1960	yd^2	square yard
Energy				
kilojoule	kJ	0.9478	Btu	British thermal unit
joule	J	2.7778×10^{-7}	$kW \cdot h$	kilowatt-hour
joule	J	0.7376	$ft \cdot lb_f$	foot-pound (force)
joule	J	1.0000	$W \cdot s$	watt-second
joule	J	0.2388	cal	calorie
kilojoule	kJ	2.7778×10^{-4}	$kW \cdot h$	kilowatt-hour
kilojoule	kJ	0.2778	$W \cdot h$	watt-hour
megajoule	MJ	0.3725	$hp \cdot h$	horsepower-hour
Flow rate				
cubic meters per day	m^3/d	264.1720	gal/d	gallons per day
cubic meters per day	m^3/d	2.6417×10^{-4}	Mgal/d	million gallons per day
cubic meters per second	m^3/s	35.3147	ft^3/s	cubic feet per second
cubic meters per second	m^3/s	22.8245	Mgal/d	million gallons per day

MULTIPLY THE SI UNIT		BY	TO OBTAIN THE U.S. CUSTOMARY UNIT	
Name	Symbol		Symbol	Name
cubic meters per second	m^3/s	15,850.3	gal/min	gallons per minute
liters per second	L/s	22,824.5	gal/d	gallons per day
liters per second	L/s	0.0228	Mgal/d	million gallons per day
liters per second	L/s	15.8508	gal/min	gallons per minute
Force				
newton	N	0.2248	lb_f	pound force
Length				
centimeter	cm	0.3937	in	inch
kilometer	km	0.6214	mi	mile
meter	m	39.3701	in	inch
meter	m	3.2808	ft	foot
meter	m	1.0936	yd	yard
millimeter	mm	0.03937	in	inch
Mass				
gram	g	0.0353	oz	ounce
gram	g	0.0022	lb	pound
kilogram	kg	2.2046	lb	pound
megagram (10^3 kg)	Mg	1.1023	ton	ton (short: 2000 lb)
megagram (10^3 kg)	Mg	0.9842	ton	ton (long: 2240 lb)
Power				
kilowatt	kW	0.9478	Btu/s	British thermal units per second
kilowatt	kW	1.3410	hp	horsepower
watt	W	0.7376	$ft \cdot lb_f/s$	foot-pounds (force) per second

MULTIPLY THE SI UNIT		BY	TO OBTAIN THE U.S. CUSTOMARY UNIT	
Name	Symbol		Symbol	Name
Pressure (force/area)				
pascal (newtons per square meter)	Pa (N/m^2)	1.4504×10^{-4}	lb$_f$/ft^2	pounds (force) per square inch
pascal (newtons per square meter)	Pa (N/m^2)	2.0885×10^{-2}	lb$_f$/ft^2	pounds (force) per square foot
pascal (newtons per square meter)	Pa (N/m^2)	2.9613×10^{-4}	inHg	inches of mercury (60°F)
pascal (newtons per square meter)	Pa (N/m^2)	4.0187×10^{-3}	inH$_2$O	inches of water (60°F)
kilopascal (kilonewtons per square meter)	kPa (kN/m^2)	0.1450	lb$_f$/in^2	pounds (force) per square inch
kilopascal (kilonewtons per square meter)	kPa (kN/m^2)	0.0099	atm	atmosphere (standard)
Temperature				
degree Celsius (centigrade)	°C	$1.8(°C) + 32$	°F	degree Fahrenheit
degree kelvin	K	$1.8(K) - 459.67$	°F	degree Fahrenheit
Velocity				
kilometers per second	km/s	2.2369	mi/hr	miles per hour
meters per second	m/s	3.2808	ft/s	feet per second
Volume				
cubic centimeter	cm^3	0.0610	in^3	cubic inch
cubic meter	m^3	35.3147	ft^3	cubic foot
cubic meter	m^3	1.3079	yd^3	cubic yard
cubic meter	m^3	264.1720	gal	gallon
cubic meter	m^3	8.1071×10^{-4}	acre · ft	acre · foot
liter	L	0.2642	gal	gallon
liter	L	0.0353	ft^3	cubic foot
liter	L	33.8150	oz	ounce (U.S. fluid)

C

PROPERTIES OF WATER

The principal physical properties of water are summarized in Table C.1. They are described briefly below.

Specific Weight

The specific weight of a fluid, γ, is the gravitational attractive force acting on a unit volume of a fluid. In SI units, the specific weight is expressed as kilonewtons per cubic meter (kN/m^3). At normal temperatures γ is 9.81 kN/m^3.

Density

The density of a fluid, ρ, is its mass per volume. For water, ρ is 1000 kg/m^3 at 4°C. There is a slight decrease in density with increasing temperature. The relationship between γ, ρ, and the acceleration due to gravity g is $\gamma = \rho g$.

Modulus of Elasticity

For most practical purposes, liquids may be regarded as incompressible. The bulk modulus of elasticity E is given by

$$E = \frac{\Delta p}{\Delta V / V}$$

where Δp is the increase in pressure, which when applied to a volume V, results in a decrease in volume ΔV. In SI units the modulus of elasticity is expressed as kilonewtons per square-meter (kN/m^2). For water, E is approximately 2.150 kN/m^2 at normal temperatures and pressures.

Dynamic Viscosity

The viscosity of a fluid, μ, is a measure of its resistance to tangential or shear stress. In SI units the viscosity is expressed as newton-seconds per square meter ($N \cdot s/m^2$).

Kinematic Viscosity

In many problems concerning fluid motion, the viscosity appears with the density in the form μ/ρ, and it is convenient to use a single term ν, known as the kinematic viscosity. The kinematic viscosity of a liquid, expressed as square meters per second (m^2/s) in SI units, diminishes with increasing temperature.

TABLE C.1
Physical Properties of Water

TEMPERATURE, °C	SPECIFIC WEIGHT γ, kN/m^3	DENSITY ρ, kg/m^3	MODULUS OF ELASTICITY* $E/10^6$, kN/m^2	DYNAMIC VISCOSITY $\mu \times 10^3$, N·s/m^2	KINEMATIC VISCOSITY $\nu \times 10^6$, m^2/s	SURFACE TENSION[†] σ, N/m	VAPOR PRESSURE p_v, kN/m^2
0	9.805	999.8	1.98	1.781	1.785	0.0765	0.61
5	9.807	1000.0	2.05	1.518	1.519	0.0749	0.87
10	9.804	999.7	2.10	1.307	1.306	0.0742	1.23
15	9.798	999.1	2.15	1.139	1.139	0.0735	1.70
20	9.789	998.2	2.17	1.002	1.003	0.0728	2.34
25	9.777	997.0	2.22	0.890	0.893	0.0720	3.17
30	9.764	995.7	2.25	0.798	0.800	0.0712	4.24
40	9.730	992.2	2.28	0.653	0.658	0.0696	7.38
50	9.698	988.0	2.29	0.547	0.553	0.0679	12.33
60	9.642	983.2	2.28	0.466	0.474	0.0662	19.92
70	9.589	977.8	2.25	0.404	0.413	0.0644	31.16
80	9.530	971.8	2.20	0.354	0.364	0.0626	47.34
90	9.466	965.3	2.14	0.315	0.326	0.0608	70.10
100	9.399	958.4	2.07	0.282	0.294	0.0589	101.33

*At atmospheric pressure.   [†]In contact with air.

Source: Adapted from Ref. [C.1].

Surface Tension

Surface tension of a fluid, σ, is the physical property that enables a drop of water to be held in suspension at a tap, a glass to be filled with liquid slightly above the brim and yet not spill, and a needle to float on the surface of a liquid. The surface-tension force across any imaginary line at a free surface is proportional to the length of the line and acts in a direction perpendicular to it [6.2]. In SI units, surface tension is expressed as newtons per meter (N/m). There is a slight decrease in surface tension with increasing temperature.

Vapor Pressure

Liquid molecules that possess sufficient kinetic energy are projected out of the main body of a liquid at its free surface and pass into the vapor. The pressure exerted by this vapor is known as the vapor pressure p_v. In SI units the vapor pressure is expressed as Pascals (Pa) or kilonewtons per square meter (kN/m^2).

REFERENCES

C.1 Vennard, J. K., and R. L. Street, (1975), *Elementary Fluid Mechanics*, 5th ed., John Wiley and Sons, New York.

C.2 Webber, N. B., (1971), *Fluid Mechanics for Civil Engineers*, SI ed., Chapman and Hall, London.

D
PRIMARY DRINKING WATER STANDARDS

TYPE OF CONTAMINANT	NAME OF CONTAMINANT	HEALTH EFFECTS OF CONTAMINANT	MAXIMUM CONTAMINANT LEVEL (MCL), mg/L
Inorganic Chemicals	Arsenic	Long-term—Sometimes can cause fatigue and loss of energy; dermatitis	0.05
	Barium	Long-term—Increased blood pressure and nerve block	1.0
	Cadmium	Long-term—Concentrates in the liver, kidneys, pancreas, and thyroid; hypertension suspected effect	0.010
	Chromium	Long-term—Skin sensitization and kidney damage	0.05
	Lead	Long-term—Brain and kidney damage; birth defects	0.05
	Mercury	Long-term—Toxic to central nervous system; may cause birth defects	0.002
	Selenium	Long-term—Red staining of fingers, teeth, and hair; general weakness; depression; irritation of the nose and mouth	0.01
	Silver	Long-term—Permanent grey discoloration of skin, eyes, and mucus membranes	0.05

TYPE OF CONTAMINANT	NAME OF CONTAMINANT	HEALTH EFFECTS OF CONTAMINANT	MAXIMUM CONTAMINANT LEVEL (MCL), mg/L			
	Fluoride		Fluoride Conc.			
	Annual Avg. of Max. Daily Air Temp.—°C	Long-term—Ingestion of excessive amounts may cause stained spots on teeth (mottling). Ingestion of optimum has beneficial effect of reducing the occurrence of tooth decay.	Lower	Optimum	Upper	MCL
	12.0–Below		0.9	1.2	1.7	2.4
	12.1–14.6		0.8	1.1	1.5	2.2
	14.7–17.6		0.8	1.0	1.3	2.0
	17.7–21.4		0.7	0.9	1.2	1.8
	21.5–26.2		0.7	0.8	1.0	1.6
	26.3–32.5		0.6	0.7	0.8	1.4
	Nitrates (as NO_3)	Short-term—Temporary blood disorder in infants		45		
Organic Chemicals	Chlorinated Hydrocarbons—Pesticides					
	Endrin	Long-term—Exposure may cause convulsions and liver damage. Lindane suspected of being carcinogenic.		0.0002		
	Lindane			0.004		
	Methoxychlor			0.1		
	Toxaphene			0.005		
	Chlorophenoxys—Herbicides					
	2,4-D	Long-term—Liver damage; gastrointestinal irritation		0.1		
	2,4,5-TP Silvex			0.01		
				0.01		
	Total Trihalomethanes					
	TTHMs	Long-term—Considered potential carcinogens		0.10		
Turbidity	Turbidity	Short-term—Interferes with the disinfection process by shielding organisms, thereby	Surface water exposed to significant sewage hazard or recreational use shall receive, as a			

		possibly exposing consumer to disease-causing organisms	minimum pretreatment, filtration and disinfection and meet a 0.5 TU monthly average Surface water not exposed to significant sewage hazard or recreational use shall meet a 1 TU monthly average. A 5 TU monthly average may apply at Department option, provided it does not: 1. Interfere with disinfection 2. Prevent maintenance of Cl_2 residual in system 3. Interfere with bacteriological tests
Microbiological Contaminants	Coliform bacteria	Short-term—Presence indicates that other disease-causing organisms may be in the water.	Multiple-tube technique for 10-mL portion: Coliforms shall not be present in: 1. More than 10% of portions per month 2. 3 or more portions in more than one sample when less than 20 samples are examined per month 3. 3 or more portions in more than 5% of the samples when 20 or more samples are examined per month

TYPE OF CONTAMINANT	NAME OF CONTAMINANT	HEALTH EFFECTS OF CONTAMINANT	MAXIMUM CONTAMINANT LEVEL (MCL), mg/L
			Membrane-filter technique: Coliforms shall not exceed:
			1. 1 per 100 mL as the arithmetic mean of all monthly samples
			2. 4 per 100 mL in more than one sample when less than 20 samples are examined per month
			3. 4 per 100 mL in more than 5% of the samples when 20 or more samples are examined per month
Radiological Contaminants	Natural: Gross alpha Combined Ra-226 and Ra-228	Long-term—Bone cancer	Natural: Gross alpha = 15 pCi/L Ra-226 + Ra-228 = 5 pCi/L Screening. If gross alpha exceeds 5 pCi/L, test for Ra-226. If Ra-226 exceeds 3 pCi/L, test for Ra-228.
	Synthetic: Gross beta Tritium Strontium 90	Long-term—Bone cancer	Synthetic: Gross beta = 50 pCi/L Tritium = 20,000 pCi/L Strontium-90 = 8 pCi/L

Source: Adapted from a publication issued by the State of California, Department of Health Services, Sacramento, California, 1984.

E

CONSTRUCTION AND USE OF LOG-CONCENTRATION-VERSUS-pH DIAGRAMS

Logarithmic concentration-versus-pH diagrams are useful in visualizing equilibrium relationships and in solving chemical equilibrium problems. They are constructed by plotting the log concentration of each species present against a master variable (usually pH). In the following discussion, the construction and use of log concentration diagrams will be illustrated for a monoprotic and diprotic acid using pH as the master variable.

Construction of a Log-Concentration-Versus-pH Diagram for a Monoprotic Acid

To illustrate the construction of a log-concentration-versus-pH diagram, a 0.010 M solution of a monoprotic acid HA will be used. Assume $K_a = 1.0 \times 10^{-5}$ and $K_w = 1.0 \times 10^{-14}$.

1. List the chemical species present in solution.

   $[H^+]$, $[OH^-]$, $[HA]$, and $[A^-]$

2. List applicable dissociation, equilibrium, and mass balance expressions.
   a. Dissociation expressions:

   $$[HA] \rightleftarrows [H^+] + [A^-] \qquad\qquad (E.1)$$
   $$[H_2O] \rightleftarrows [H^+] + [OH^-] \qquad\qquad (E.2)$$

   b. The equilibrium expressions for the above reactions are

   $$\frac{[H^+][A^-]}{[HA]} = K_a \qquad\qquad (E.3)$$

   $$[H^+][OH^-] = K_w \qquad\qquad (E.4)$$

   c. The mass balance for the weak acid is

   $$C_T = [HA] + [A^-] = 0.010 \text{ mol/L} \qquad\qquad (E.5)$$

   where C_T represents the total molar concentration of the acid species in solution.

3. Develop the relationships needed to plot the log concentration of $[H^+]$, $[OH^-]$, $[HA]$, and $[A^-]$ as a function of pH.
   a. For $[H^+]$ by definition

   $$\log[H^+] = -pH \qquad\qquad (E.6)$$

   b. For $[OH^-]$ by definition [see Eq. (E.4)]

   $$\log[OH^-] = pH - pK_w \qquad\qquad (E.7)$$

c. For [HA] the required relationship is obtained by substituting the value of [A$^-$] from Eq. (E.5) in Eq. (E.3):

$$[HA] = \frac{C_T[H^+]}{K_a + [H^+]} \tag{E.8}$$

or

$$\log[HA] = \log C_T - pH - \log(K_a + [H^+]) \tag{E.9}$$

d. For [A$^-$] the required relationship is obtained by substituting the value of [HA] from Eq. (E.5) in Eq. (E.3):

$$[A^-] = \frac{C_T K_a}{K_a + [H^+]} \tag{E.10}$$

or

$$\log[A^-] = \log C_T - pK_a - \log(K_a + [H^+]) \tag{E.11}$$

4. Draw and label the coordinate axes of the logarithmic-concentration diagram ($\log C_T$ from 0 to -14) versus pH (from 0 to 14) as shown in Fig. E.1(a).

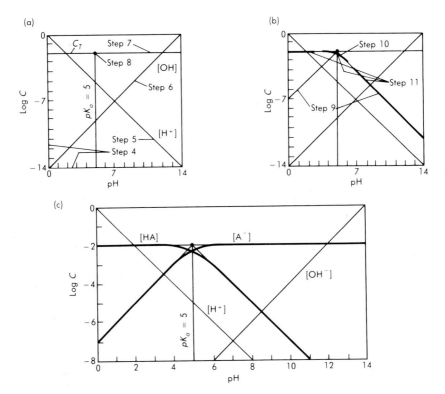

FIGURE E.1

Construction of a log-concentration-versus-pH diagram for a monoprotic acid.

5. Draw the $[H^+]$ line originating at $0,0$ with a slope of -1 and passing through the point $-7,7$ (Fig. E.1). If Eq. (E.6) is differentiated with respect to pH it is found that slope of the $[H^+]$ line is -1.

$$\frac{d \log [H^+]}{d\,pH} = -1$$

6. Draw the $[OH^-]$ line originating at $-14,0$ or $-7,7$ with a slope of $+1$ and terminating at the point $0,14$ (Fig. E.1a). If Eq. (E.7) is differentiated with respect to pH, it is found that the slope of the $[OH^-]$ line is $+1$.

$$\frac{d \log [OH^-]}{d\,pH} = +1$$

7. Draw a light horizontal line across the diagram at the concentration $\log C_T = -2.0$ $(\log 0.010 = \log 10^{-2})$.

8. Draw a light vertical line on the pH scale representing the pK_a value $(pK_a = 5)$. The intersection of the vertical pK_a line and the horizontal concentration line represents the system point (Fig. E.1a).

9. Draw light $45°$ lines passing through the system point to the left and right below the horizontal concentration line (C_T).

10. Locate a point 0.301 log-concentration units below the system point on the vertical line representing the pK_a value. Note that in either Eq. (E.8) or (E.10) when the value of the $[H^+]$ is equal to the K_a value then

$$[HA] = \frac{C_T}{2} \quad \text{and} \quad [A^-] = \frac{C_T}{2}$$

Because the log of 0.5 is -0.301, the curves for $[HA]$ and $[A^-]$ must intersect at a point -0.301 log-concentration units below the system point.

11. The curve for $[HA]$ is drawn by noting that
  a. When $[H^+] \gg K_a$, $[HA]$ from Eq. (E.9) is

  $$\log[HA] \simeq \log C_T$$

  If the above expression is differentiated with respect to pH then

  $$\frac{d \log [HA]}{d\,pH} = 0$$

  Thus the $[HA]$ line must be horizontal to the left of the system point.
  b. When $[H^+] \ll K_a$, $[HA]$ from Eq. (E.9) is

  $$\log [HA] \simeq \log C_T - pH - \log K_a$$

  If the above expression is differentiated with respect to pH then

  $$\frac{d \log [HA]}{d\,pH} = -1$$

Thus the [HA] line must fall off at a slope of -1 to the right of the system point.

c. The [HA] curve is completed by connecting the horizontal concentration line to the left of the system point to the $45°$ line located on the right of the system. The completed curve should pass through the point located -0.301 log-concentration units below the system point (see Fig. E.1). Note the curve passing through the point located -0.301 log-concentration units below the system point should connect to the horizontal concentration line approximately 1 pH unit to the left of the pK_a value and should connect to the $45°$ line approximately 1 log-concentration unit below the horizontal concentration line.

12. The curve for $[A^-]$ is completed in a manner similar to that for $[HA]$ by noting that

a. When $[H^+] \ll K_a$, $[A^-]$ from Eq. (E.11) is

$$\log[A^-] \simeq \log C_T$$

and

$$\frac{d \log[A^-]}{d\,pH} = 0$$

b. When $[H^+] \gg K_a$, $[A^-]$ from Eq. (E.11) is

$$\log[A^-] = \log C_T - pK_a + pH$$

and

$$\frac{d \log[A^-]}{d\,pH} = +1$$

c. The completed curve for $[A^-]$ is shown in Fig. E.1(c).

Construction of a Log-Concentration-Versus-pH Diagram for a Diprotic Acid

The construction of a log-concentration-versus-pH diagram for a diprotic acid will be illustrated using a $0.001\ M$ solution of carbonic acid $[H_2CO_3^*]$. Assume $K_1 = 4.47 \times 10^{-7}$, $K_2 = 4.68 \times 10^{-11}$, and $K_w = 1.00 \times 10^{-14}$.

1. List the chemical species in solution.

$$[H^+], [OH^-], [H_2CO_3^*], [HCO_3^-], \text{ and } [CO_3^{-2}]$$

2. List applicable dissociation, equilibrium, and mass balance expressions.

a. Dissociation expressions:

$$[H_2CO_3^*] \rightleftharpoons [HCO_3^-] + [H^+] \qquad (E.12)$$

$$[HCO_3^-] \rightleftharpoons [CO_3^{-2}] + [H^+] \qquad (E.13)$$

$$[H_2O] \rightleftharpoons [H^+] + [OH^-] \qquad (E.14)$$

b. The corresponding equilibrium expressions:

$$\frac{[HCO_3^-][H^+]}{[H_2CO_3^*]} = K_1 \qquad (E.15)$$

$$\frac{[CO_3^{-2}][H^+]}{[HCO_3^-]} = K_2 \qquad (E.16)$$

$$[H^+][OH^-] = K_w \qquad (E.17)$$

c. The mass balance for the carbonate ion is

$$C_T = [H_2CO_3^*] + [HCO_3^-] + [CO_3^{-2}] \qquad (E.18)$$

3. Develop the relationships needed to plot the log concentration of $[H^+]$, $[OH^-]$, $[H_2CO_3^*]$, $[HCO_3^-]$, and $[CO_3^{-2}]$ as a function of pH.
  a. For $[H^+]$ by definition

$$\log[H^+] = -pH \qquad (E.6)$$

  b. For $[OH^-]$ by definition [see Eq. (E.4)]

$$\log[OH^-] = pH - pK_w \qquad (E.7)$$

  c. For $[H_2CO_3^*]$ the required relationship is obtained by substituting for $[HCO_3^-]$ and $[CO_3^{-2}]$ from Eqs. (E.16) and (E.18) in Eq. (E.15).

$$[H_2CO_3^*] = \frac{C_T[H^+]^2}{[H^+]([H^+]+K_1)+K_1K_2} \qquad (E.19)$$

  d. For $[HCO_3^-]$ the required relationship is obtained by substituting for $[H_2CO_3^*]$ and $[CO_3^{-2}]$ from Eqs. (E.15) and (E.18) in Eq. (E.16).

$$[HCO_3^-] = \frac{K_1C_T[H^+]}{[H^+]([H^+]+K_1)+K_1K_2} \qquad (E.20)$$

  e. For $[CO_3^{-2}]$ the required relationship is obtained by substituting for $[H_2CO_3^*]$ and $[HCO_3^-]$ from Eqs. (E.15) and (E.18) in Eq. (E.16).

$$[CO_3^{-2}] = \frac{K_1K_2C_T}{[H^+]([H^+]+K_1)+K_1K_2} \qquad (E.21)$$

4. Draw and label the coordinate axes of the logarithmic-concentration diagram (Fig. E.2).
5. Draw the $[H^+]$ line as described previously in the development of the log-concentration-versus-pH diagram for a monoprotic acid.

6. Draw the $[OH^-]$ line as described previously in the development of the log-concentration-versus-pH diagram for a monoprotic acid.

7. Draw a light horizontal line across the diagram at the concentration $\log C_T = -3.0$ $(\log 0.0010 = \log 10^{-3})$.

8. Draw two light vertical lines on the pH scale corresponding to the two pK values:

$$pK_1 = 6.35$$
$$pK_2 = 10.33$$

9. Draw 45° lines passing through the system point on either side of the pK values.

10. Locate a point 0.301 log-concentration units below each system point on the pK lines. As noted previously, these points represent the condition where the concentration of $[H_2CO_3^*]$ and $[HCO_3^-]$ and $[HCO_3^-]$ and $[CO_3^{-2}]$ are equal.

   a. When $[H^+] = K_1$, the term K_2 can be neglected because $K_1 \gg K_2$ and the values for $[H_2CO_3^*]$ and $[HCO_3^{-2}]$ from Eqs. (E.19) and (E.20) are as given below:

   $$[H_2CO_3^*] \simeq \frac{C_T}{2}$$

   $$[HCO_3^-] \simeq \frac{C_T}{2}$$

   b. Also, when $[H^+] = K_2$, the term K_2^2 can be neglected and the values for $[HCO_3^-]$ and $[CO_3^{-2}]$ from Eqs. (E.20) and (E.21) are as

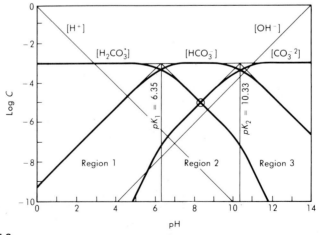

FIGURE E.2

Construction of a log-concentration-versus-pH diagram for a diprotic acid.

given below:

$$[HCO_3^-] \simeq \frac{C_T}{2}$$

$$[CO_3^{-2}] \simeq \frac{C_T}{2}$$

11. Sketch the curve for $[H_2CO_3^*]$ as described previously by referring to Eq. (E.19) and by noting that
   a. When $[H^+] \gg K_1$ (region 1, Fig. E.2)

   $$\log[H_2CO_3^*] \simeq \log C_T$$

   $$\frac{d \log[H_2CO_3^*]}{d\,pH} = 0$$

   Thus to the left of the vertical pK line the $[H_2CO_3^*]$ line has a slope of 0.
   b. When $K_1 > [H^+] > K_2$ (region 2, Fig. E.2)

   $$\log[H_2CO_3^*] \simeq \log C_T - pH - \log K_1$$

   $$\frac{d \log[H_2CO_3^*]}{d\,pH} = -1$$

   Thus to the right of the vertical pK_1 line the $[H_2CO_3^*]$ line drops off at slope of -1.
   c. When $[H^+] \ll K_2$ (region 3, Fig. E.3)

   $$\log[H_2CO_3^*] \simeq \log C_T - 2pH - \log K_1K_2$$

   $$\frac{d \log[H_2CO_3^*]}{d\,pH} = -2$$

   Thus to the right of the vertical pK_2 line the $[H_2CO_3^*]$ line drops off at a slope of -2.
12. The concentration curves for $[HCO_3^-]$ and $[CO_3^{-2}]$ are drawn in a manner similar to that for $[H_2CO_3^*]$.

Solution of Equilibrium Problems Using Log-Concentration-Versus-pH Diagrams

The procedure followed in using log-concentration-versus-pH diagrams is illustrated below.

Determine the pH of a Solution Containing NaA

To illustrate the use of the logarithmic-concentration diagrams, it will be instructive to determine the pH of a 0.01 M solution of NaA using Fig. E.1(c) developed for a monoprotic acid HA. The salt NaA will dissociate completely in water to form Na^+ and A^-. In turn, the A^- will react with the H^+ in solution to form HA.

1. List all of the chemical species in solution.

   $[H^+]$, $[OH^-]$, $[HA]$, $[A^-]$, and $[Na^+]$

2. Develop species mass balances.

   $C_T = [HA] + [A^-] = 0.01$ mol/L

   $C_T = Na^+ = 0.01$ mol/L

3. Prepare a change balance.

   $[H^+] + [Na^+] = [A^-] + [OH^-]$

   Because Na^+ is not included in Fig. E.1(c), some other relationship is needed. The needed relationship is known as the *proton condition*.

4. Develop the proton condition. The proton condition is obtained by substituting in the change balance, corresponding values from the species mass balance. Substituting for $[Na^+]$ and $[A^-]$ yields the proton condition:

   $[H^+] + [HA] = [OH^-]$

   The proton condition represents a balance on the protons that were gained or lost relative to the zero proton level. The zero proton level corresponds to the proton condition before the salt is added to the water (e.g., H_2O, NaA). The species with an excess of protons relative to the zero level are $[H^+]$ and $[HA]$; the species with a deficiency of protons is $[OH^-]$.

5. Using the proton condition developed in step 4, find the pH at which this relationship is satisfied. By inspection, when the value of $[OH^-]$ is large the corresponding value of $[H^+]$ is negligible. Thus

   $[HA] \simeq [OH^-]$

   This relationship is satisfied at a pH value of 8.6. At pH 8.6, the concentration of $[H^+]$ is three orders of magnitude smaller than the corresponding values for $[HA]$ and $[OH^-]$. Thus, the approximate relationship used to find the pH is acceptable.

Determine the pH of a Solution Containing NaHCO₃

To illustrate the solution procedure for a diprotic acid, the pH of a 0.001 M solution of sodium bicarbonate ($NaHCO_3$) will be determined using Fig. E.2.

1. List all of the chemical species in solution.

   $[H^+]$, $[OH^-]$, $[H_2CO_3^*]$, $[HCO_3^-]$, $[CO_3^{-2}]$, and $[Na^+]$

   Note because there are six unknowns, six equations are needed to solve the problem analytically.

2. Develop species mass balances.

$$C_T = [H_2CO_3^*] + [HCO_3^-] + [CO_3^{-2}] = 0.001 \text{ mol/L}$$
$$= [Na^+] = 0.001 \text{ mol/L}$$

3. Prepare a change balance.

$$[H^+] + [Na^+] = [HCO_3^-] + 2[CO_3^{-2}] + [OH^-]$$

Because Na^+ is not included in Fig. E.2, the proton condition must be used.

4. Develop the proton condition. Substitute for $[Na^+]$ and $[HCO_3^-]$ from the mass balances developed in step 2 and the charge balance developed in step 3.

$$[H^+] + C_T = C_T - [H_2CO_3^*] - [CO_3^{-2}] + 2[CO^{-2}] + [OH^-]$$

Solving yields the proton condition:

$$[H^+] + [H_2CO_3^*] = [CO_3^{-2}] + [OH^-]$$

5. Using the proton condition developed in step 4, find the pH for which this relationship is satisfied. Following either the $[H_2CO_3^*]$ line from left to right or the $[CO_3^{-2}]$ line from right to left, the proton condition is satisfied at a pH value of about 8.3. At a pH value of 8.3 the $[OH^-]$ concentration is approximately one order of magnitude smaller. If a more exact solution is needed it must be obtained by trial and error by summing the concentration values for $[H_2CO_3^*]$, $[CO_3^{-2}]$, and $[OH^-]$ to the left of the point of intersection.

REFERENCES

For additional details on the development and application of log-concentration-versus-pH diagrams the following references are recommended:

E.1 Bulter, J. N., (1982), *Carbon Dioxide Equilibria and Their Applications*, Addison-Wesley Publishing Company, Reading, Mass.

E.2 Loewenthal, R. E., and G. v. R. Marias, (1976), *Carbonate Chemistry of Aquatic Systems: Theory and Applications*, Ann Arbor Science Publishers, Ann Arbor, Mich.

E.3 Snoeyink, V. L., and D. Jenkins, (1980), *Water Chemistry*, John Wiley & Sons, New York.

E.4 Stumm, W., and J. J. Morgan, (1980), *Aquatic Chemistry*, 2d ed., Wiley Interscience, New York.

F

DISSOLVED OXYGEN CONCENTRATION VALUES AS A FUNCTION OF SALINITY AND BAROMETRIC PRESSURE*

SALINITY

TEMP, °C	DISSOLVED OXYGEN CONCENTRATION, mg/L SALINITY, PARTS PER THOUSAND									
	0	5	10	15	20	25	30	35	40	45
0	14.60	14.11	13.64	13.18	12.74	12.31	11.90	11.50	11.11	10.74
1	14.20	13.73	13.27	12.83	12.40	11.98	11.58	11.20	10.83	10.46
2	13.81	13.36	12.91	12.49	12.07	11.67	11.29	10.91	10.55	10.20
3	13.45	13.00	12.58	12.16	11.76	11.38	11.00	10.64	10.29	9.95
4	13.09	12.67	12.25	11.85	11.47	11.09	10.73	10.38	10.04	9.71
5	12.76	12.34	11.94	11.56	11.18	10.82	10.47	10.13	9.80	9.48
6	12.44	12.04	11.65	11.27	10.91	10.56	10.22	9.89	9.57	9.27
7	12.13	11.74	11.37	11.00	10.65	10.31	9.98	9.66	9.35	9.06
8	11.83	11.46	11.09	10.74	10.40	10.07	9.75	9.44	9.14	8.85
9	11.55	11.19	10.83	10.49	10.16	9.84	9.53	9.23	8.94	8.66
10	11.28	10.92	10.58	10.25	9.93	9.62	9.32	9.03	8.75	8.47
11	11.02	10.67	10.34	10.02	9.71	9.41	9.12	8.83	8.56	8.30
12	10.77	10.43	10.11	9.80	9.50	9.21	8.92	8.65	8.38	8.12
13	10.53	10.20	9.89	9.59	9.30	9.01	8.74	8.47	8.21	7.96
14	10.29	9.98	9.68	9.38	9.10	8.82	8.55	8.30	8.04	7.80
15	10.07	9.77	9.47	9.19	8.91	8.64	8.38	8.13	7.88	7.65
16	9.86	9.56	9.28	9.00	8.73	8.47	8.21	7.97	7.73	7.50
17	9.65	9.36	9.09	8.82	8.55	8.30	8.05	7.81	7.58	7.36
18	9.45	9.17	8.90	8.64	8.39	8.14	7.90	7.66	7.44	7.22
19	9.26	8.99	8.73	8.47	8.22	7.98	7.75	7.52	7.30	7.09
20	9.08	8.81	8.56	8.31	8.07	7.83	7.60	7.38	7.17	6.96
21	8.90	8.64	8.39	8.15	7.91	7.69	7.46	7.25	7.04	6.84
22	8.73	8.48	8.23	8.00	7.77	7.54	7.33	7.12	6.91	6.72
23	8.56	8.32	8.08	7.85	7.63	7.41	7.20	6.99	6.79	6.60
24	8.40	8.16	7.93	7.71	7.49	7.28	7.07	6.87	6.68	6.49
25	8.24	8.01	7.79	7.57	7.36	7.15	6.95	6.75	6.56	6.38
26	8.09	7.87	7.65	7.44	7.23	7.03	6.83	6.64	6.46	6.28
27	7.95	7.73	7.51	7.31	7.10	6.91	6.72	6.53	6.35	6.17
28	7.81	7.59	7.38	7.18	6.98	6.79	6.61	6.42	6.25	6.08
29	7.67	7.46	7.26	7.06	6.87	6.68	6.50	6.32	6.15	5.98
30	7.54	7.33	7.14	6.94	6.75	6.57	6.39	6.22	6.05	5.89
31	7.41	7.21	7.02	6.83	6.65	6.47	6.29	6.12	5.96	5.80
32	7.29	7.09	6.90	6.72	6.54	6.36	6.19	6.03	5.87	5.71
33	7.17	6.98	6.79	6.61	6.44	6.26	6.10	5.94	5.78	5.63
34	7.05	6.86	6.68	6.51	6.33	6.17	6.01	5.85	5.69	5.54
35	6.93	6.75	6.58	6.40	6.24	6.07	5.92	5.76	5.61	5.46
36	6.82	6.65	6.47	6.31	6.14	5.98	5.83	5.68	5.53	5.39
37	6.72	6.54	6.37	6.21	6.05	5.89	5.74	5.59	5.45	5.31
38	6.61	6.44	6.28	6.12	5.96	5.81	5.66	5.51	5.37	5.24
39	6.51	6.34	6.18	6.03	5.87	5.72	5.58	5.44	5.30	5.16
40	6.41	6.25	6.09	5.94	5.79	5.64	5.50	5.36	5.22	5.09

BAROMETRIC PRESSURE

TEMP, °C	DISSOLVED OXYGEN CONCENTRATION, mg/L BAROMETRIC PRESSURE, MILLIMETERS OF MERCURY									
	735	740	745	750	755	760	765	770	775	780
0	14.12	14.22	14.31	14.41	14.51	14.60	14.70	14.80	14.89	14.99
1	13.73	13.82	13.92	14.01	14.10	14.20	14.29	14.39	14.48	14.57
2	13.36	13.45	13.54	13.63	13.72	13.81	13.90	14.00	14.09	14.18
3	13.00	13.09	13.18	13.27	13.36	13.45	13.53	13.62	13.71	13.80
4	12.66	12.75	12.83	12.92	13.01	13.09	13.18	13.27	13.35	13.44
5	12.33	12.42	12.50	12.59	12.67	12.76	12.84	12.93	13.01	13.10
6	12.02	12.11	12.19	12.27	12.35	12.44	12.52	12.60	12.68	12.77
7	11.72	11.80	11.89	11.97	12.05	12.13	12.21	12.29	12.37	12.45
8	11.44	11.52	11.60	11.67	11.75	11.83	11.91	11.99	12.07	12.15
9	11.16	11.24	11.32	11.40	11.47	11.55	11.63	11.70	11.78	11.86
10	10.90	10.98	11.05	11.13	11.20	11.28	11.35	11.43	11.50	11.58
11	10.65	10.72	10.80	10.87	10.94	11.02	11.09	11.16	11.24	11.31
12	10.41	10.48	10.55	10.62	10.69	10.77	10.84	10.91	10.98	11.05
13	10.17	10.24	10.31	10.38	10.46	10.53	10.60	10.67	10.74	10.81
14	9.95	10.02	10.09	10.16	10.23	10.29	10.36	10.43	10.50	10.57
15	9.73	9.80	9.87	9.94	10.00	10.07	10.14	10.21	10.27	10.34
16	9.53	9.59	9.66	9.73	9.79	9.86	9.92	9.99	10.06	10.12
17	9.33	9.39	9.46	9.52	9.59	9.65	9.72	9.78	9.85	9.91
18	9.14	9.20	9.26	9.33	9.39	9.45	9.52	9.58	9.64	9.71
19	8.95	9.01	9.07	9.14	9.20	9.26	9.32	9.39	9.45	9.51
20	8.77	8.83	8.89	8.95	9.02	9.08	9.14	9.20	9.26	9.32
21	8.60	8.66	8.72	8.78	8.84	8.90	8.96	9.02	9.08	9.14
22	8.43	8.49	8.55	8.61	8.67	8.73	8.79	8.84	8.90	8.96
23	8.27	8.33	8.39	8.44	8.50	8.56	8.62	8.68	8.73	8.79
24	8.11	8.17	8.23	8.29	8.34	8.40	8.46	8.51	8.57	8.63
25	7.96	8.02	8.08	8.13	8.19	8.24	8.30	8.36	8.41	8.47
26	7.82	7.87	7.93	7.98	8.04	8.09	8.15	8.20	8.26	8.31
27	7.68	7.73	7.79	7.84	7.89	7.95	8.00	8.06	8.11	8.17
28	7.54	7.59	7.65	7.70	7.75	7.81	7.86	7.91	7.97	8.02
29	7.41	7.46	7.51	7.57	7.62	7.67	7.72	7.78	7.83	7.88
30	7.28	7.33	7.38	7.44	7.49	7.54	7.59	7.64	7.69	7.75
31	7.16	7.21	7.26	7.31	7.36	7.41	7.46	7.51	7.46	7.62
32	7.04	7.09	7.14	7.19	7.24	7.29	7.34	7.39	7.44	7.49
33	6.92	6.97	7.02	7.07	7.12	7.17	7.22	7.27	7.31	7.36
34	6.80	6.85	6.90	6.95	7.00	7.05	7.10	7.15	7.20	7.24
35	6.69	6.74	6.79	6.84	6.89	6.93	6.98	7.03	7.08	7.13
36	6.59	6.63	6.68	6.73	6.78	6.82	6.87	6.92	6.97	7.01
37	6.48	6.53	6.57	6.62	6.67	6.72	6.76	6.81	6.86	6.90
38	6.38	6.43	6.47	6.52	6.56	6.61	6.66	6.70	6.75	6.80
39	6.28	6.33	6.37	6.42	6.46	6.51	6.56	6.60	6.65	6.69
40	6.18	6.23	6.27	6.32	6.36	6.41	6.46	6.50	6.55	6.59

*Source: J. Colt, Department of Civil Engineering, University of California, Davis.

G

MPN TABLES AND THEIR USE

When three serial sample volumes (e.g., dilutions) are used in the bacteriological testing of water, the resulting MPN values per 100 mL can be determined using Table G.1. The MPN values given there are based on serial sample volumes of 10, 1, and 0.1 mL. If lower or higher serial sample volumes are used, the MPN values given in Table G.1 must be adjusted accordingly. For example, if the sample volumes used are 100, 10, and 1 mL, the MPN values from the table are multiplied by 0.1. Similarly, if the sample volumes are 1, 0.1, and 0.01 mL the MPN values from the table are multiplied by 10.

In situations where more than three test dilutions have been run, the following rule is applied to select the three dilutions to be used in determining the MPN value [G.1]: Choose the highest dilution that gives positive results in all five portions tested (no lower dilution giving any negative results) and the two next higher dilutions. Use the results at these three volumes in computing the MPN value. In the examples given in the accompanying table, the significant dilution results are shown in boldface. The number in the numerator represents positive tubes, that in the denominator, the total tubes planted.

Example	1 mL	0.1 mL	0.01 mL	0.001 mL	Combination of positives	MPN/ 100 mL
(a)	**5/5**	**5/5**	**2/5**	**0/5**	**5-2-0**	4900
(b)	**5/5**	**4/5**	**2/5**	0/5	**5-4-2**	2200
(c)	**0/5**	**1/5**	**0/5**	0/5	**0-1-0**	18
(d)	5/5	3/5	1/5	1/5		
(e)	**5/5**	**3/5**	**2/5**	0/5	**5-3-2**	1400

In example (c), the first three dilutions are used so as to throw the positive result in the middle dilution. Where positive results occur in dilutions higher than the three chosen according to the above rule, they are incorporated into the result of the highest chosen dilution up to a total of five. The result of applying this procedure to the data of example (d) is shown in example (e).

REFERENCE

G.1 *Standard Methods for the Examination of Water and Wastewater,* (1980), 15th ed., American Public Health Association, New York.

TABLE G.1
Most Probable Number of Coliforms per 100 mL of Sample

10 mL	1 mL	0.1 mL	MPN	10 mL	1 mL	0.1 mL	MPN	10 mL	1 mL	0.1 mL	MPN	10 mL	1 mL	0.1 mL	MPN	10 mL	1 mL	0.1 mL	MPN	10 mL	1 mL	0.1 mL	MPN
0	0	0		1	0	0	2.0	2	0	0	4.5	3	0	0	7.8	4	0	0	13	5	0	0	23
0	0	1	1.8	1	0	1	4.0	2	0	1	6.8	3	0	1	11	4	0	1	17	5	0	1	31
0	0	2	3.6	1	0	2	6.0	2	0	2	9.1	3	0	2	13	4	0	2	21	5	0	2	43
0	0	3	5.4	1	0	3	8.0	2	0	3	12	3	0	3	16	4	0	3	25	5	0	3	58
0	0	4	7.2	1	0	4	10	2	0	4	14	3	0	4	20	4	0	4	30	5	0	4	76
0	0	5	9.0	1	0	5	12	2	0	5	16	3	0	5	23	4	0	5	36	5	0	5	95
0	1	0	1.8	1	1	0	4.0	2	1	0	6.8	3	1	0	11	4	1	0	17	5	1	0	33
0	1	1	3.6	1	1	1	6.1	2	1	1	9.2	3	1	1	14	4	1	1	21	5	1	1	46
0	1	2	5.5	1	1	2	8.1	2	1	2	12	3	1	2	17	4	1	2	26	5	1	2	64
0	1	3	7.3	1	1	3	10	2	1	3	14	3	1	3	20	4	1	3	31	5	1	3	84
0	1	4	9.1	1	1	4	12	2	1	4	17	3	1	4	23	4	1	4	36	5	1	4	110
0	1	5	11	1	1	5	14	2	1	5	19	3	1	5	27	4	1	5	42	5	1	5	130
0	2	0	3.7	1	2	0	6.1	2	2	0	9.3	3	2	0	14	4	2	0	22	5	2	0	49
0	2	1	5.5	1	2	1	8.2	2	2	1	12	3	2	1	17	4	2	1	26	5	2	1	70
0	2	2	7.4	1	2	2	10	2	2	2	14	3	2	2	20	4	2	2	32	5	2	2	95
0	2	3	9.2	1	2	3	12	2	2	3	17	3	2	3	24	4	2	3	38	5	2	3	120
0	2	4	11	1	2	4	15	2	2	4	19	3	2	4	27	4	2	4	44	5	2	4	150
0	2	5	13	1	2	5	17	2	2	5	22	3	2	5	31	4	2	5	50	5	2	5	180

10 ML	1 ML	0.1 ML	MPN	10 ML	1 ML	0.1 ML	MPN	10 ML	1 ML	0.1 ML	MPN	10 ML	1 ML	0.1 ML	MPN	10 ML	1 ML	0.1 ML	MPN	10 ML	1 ML	0.1 ML	MPN
0	3	0	5.6	1	3	0	8.3	2	3	0	12	3	3	0	17	4	3	0	27	5	3	0	79
0	3	1	7.4	1	3	1	10	2	3	1	14	3	3	1	21	4	3	1	33	5	3	1	110
0	3	2	9.3	1	3	2	13	2	3	2	17	3	3	2	24	4	3	2	39	5	3	2	140
0	3	3	11	1	3	3	15	2	3	3	20	3	3	3	28	4	3	3	45	5	3	3	180
0	3	4	13	1	3	4	17	2	3	4	22	3	3	4	31	4	3	4	52	5	3	4	210
0	3	5	15	1	3	5	19	2	3	5	25	3	3	5	35	4	3	5	59	5	3	5	250
0	4	0	7.5	1	4	0	11	2	4	0	15	3	4	0	21	4	4	0	34	5	4	0	130
0	4	1	9.4	1	4	1	13	2	4	1	17	3	4	1	24	4	4	1	40	5	4	1	170
0	4	2	11	1	4	2	15	2	4	2	20	3	4	2	28	4	4	2	47	5	4	2	220
0	4	3	13	1	4	3	17	2	4	3	23	3	4	3	32	4	4	3	54	5	4	3	280
0	4	4	15	1	4	4	19	2	4	4	25	3	4	4	36	4	4	4	62	5	4	4	350
0	4	5	17	1	4	5	22	2	4	5	28	3	4	5	40	4	4	5	69	5	4	5	430
0	5	0	9.4	1	5	0	13	2	5	0	17	3	5	0	25	4	5	0	41	5	5	0	240
0	5	1	11	1	5	1	15	2	5	1	20	3	5	1	29	4	5	1	48	5	5	1	350
0	5	2	13	1	5	2	17	2	5	2	23	3	5	2	32	4	5	2	56	5	5	2	540
0	5	3	15	1	5	3	19	2	5	3	26	3	5	3	37	4	5	3	64	5	5	3	920
0	5	4	17	1	5	4	22	2	5	4	29	3	5	4	41	4	5	4	72	5	5	4	1600
0	5	5	19	1	5	5	24	2	5	5	32	3	5	5	45	4	5	5	81				

H
DERIVATION OF MATERIALS BALANCE FOR A PLUG-FLOW REACTOR

The derivation of the materials balance equation for a plug-flow reactor can be illustrated by considering the reactor shown in Fig. H.1. For the differential volume element ΔV, the materials balance for reactant A is

$$\frac{\partial C_A}{\partial t}\Delta V = QC_A|_x - QC_A|_{x+\Delta x} + r_A\Delta V \qquad (H.1)$$

$$\text{Accum} = \text{Inflow} - \text{Outflow} + \text{Generation}$$

where

C_A = concentration of material A, g/m³

ΔV = differential volume, m³

Q = volumetric flow rate, m³/s

r_A = reaction rate, g/m³ · s

Referring to Fig. H.2, the term $QC_A|_{x+\Delta x}$ is equal to

$$Q\left(C_A + \frac{\Delta C_A}{\Delta x}\Delta x\right)$$

Substituting the above expression for $QC_A|_{x+\Delta x}$ in Eq. (H.1) results in

$$\frac{\partial C_A}{\partial t}\Delta V = QC_A - Q\left(C_A + \frac{\Delta C_A}{\Delta x}\Delta x\right) + r_A\Delta V \qquad (H.2)$$

Substituting $A\Delta x$ for ΔV yields

$$\frac{\partial C_A}{\partial t}A\Delta x = -Q\frac{\Delta C_A}{\Delta x}\Delta x + r_A A\Delta x \qquad (H.3)$$

Dividing by A and Δx yields

$$\frac{\partial C_A}{\partial t} = -\frac{Q}{A}\frac{\Delta C_A}{\Delta x} + r_A \qquad (H.4)$$

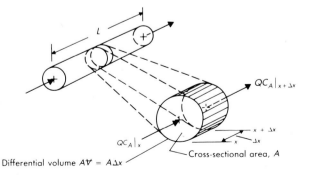

Differential volume $\Delta V = A\Delta x$

FIGURE H.1

Definition sketch for analysis of a plug-flow reactor.

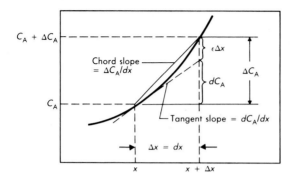

FIGURE H.2

Schematic representation of an infinitesimal as defined in calculus.

Taking the limit as Δx approaches zero yields

$$\frac{\partial C_A}{\partial t} = -\frac{Q}{A}\frac{\partial C_A}{\partial x} + r_A \qquad (H.5)$$

If steady-state conditions are assumed ($\partial C_A/\partial t = 0$) and the rate of reaction is defined as $r_A = -kC_A^n$, then integrating between the limits $C_A = C_{A_0}$ and $C_A = C_A$ and $x = 0$ and $x = L$ yields

$$\int_{C_{A_0}}^{C_A}\frac{dC}{kC^n} = -\frac{A}{Q}\int_0^L dx = -\frac{AL}{Q} = -\frac{V}{Q} = -\theta_H \qquad (H.6)$$

where θ_H is the hydraulic detention time.

Equation (H.6) is the basic materials balance equation for the analysis of a plug-flow reactor without dispersion. The same approach is used for a plug-flow reactor with dispersion.

I

SOLUTION OF A FIRST-ORDER ORDINARY DIFFERENTIAL EQUATION

The general form of the first-order ordinary differential equation is

$$\frac{dy}{dt} + P(t)y = Q(t) \tag{I.1}$$

Although a variety of methods can be used to solve Eq. (I.1), the method involving the use of an integrating factor of the form $\exp(\int P\,dt)$ is most commonly used. To illustrate the application of the integrating factor method of solution, two examples will be considered: (1) the non-steady-state solution for a reaction occurring in a CFSTR reactor and (2) the solution of the oxygen-sag differential equation.

CFSTR Reactor Analysis

Consider the CFSTR reactor shown in Fig. I.1. A materials balance on a reactive constituent C is

$$\frac{dC}{dt}V = QC_0 - QC + r_C V \tag{I.2}$$

Accum = Inflow − Outflow + Generation

Assuming $r_C = -kC$, Eq. (I.2) can be rearranged as

$$C' + \beta C = \frac{Q}{V}C_0 \tag{I.3}$$

where

$$C' = dC/dt$$
$$\beta = k + Q/V$$

Comparing Eq. (I.3) to Eq. (I.1), the appropriate integrating factor for

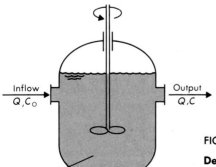

Inflow
Q, C_0

Output
Q, C

FIGURE I.1

Definition sketch for continuous-flow stirred-tank reactor.

the solution of Eq. (I.3) is

$$e^{\int P dt} = e^{\int \beta dt} = e^{\beta t} \tag{I.4}$$

Multiplying both sides of Eq. (I.3) by the above integrating factor yields

$$e^{\beta t}(C' + \beta C) = \left(\frac{Q}{V}\right) C_0 e^{\beta t} \tag{I.5}$$

The left-hand side of the above equation can be written as $(e^{\beta t}C)'$ by noting that

$$(e^{\beta t}C)' = e^{\beta t}C' + \beta C e^{\beta t} \tag{I.6}$$

Thus Eq. (I.5) can be written as

$$(e^{\beta t}C)' = \frac{Q}{V} C_0 e^{\beta t} \tag{I.7}$$

The differential sign can be removed by integrating Eq. (I.7) as follows:

$$e^{\beta t}C = \frac{Q}{V} C_0 \int e^{\beta t}\, dt \tag{I.8}$$

Integration of Eq. (I.8) yields

$$e^{\beta t}C = \frac{Q}{V} \frac{C_0}{\beta} e^{\beta t} + K \tag{I.9}$$

where K is the constant of integration. Dividing by $e^{\beta t}$ results in

$$C = \frac{Q}{V} \frac{C_0}{\beta} + K e^{-\beta t} \tag{I.10}$$

But when $t = 0$, $C = C_0$ and thus

$$K = C_0 - \frac{Q}{V} \frac{C_0}{\beta} \tag{I.11}$$

Substituting for K in Eq. (I.10) results in the following equation, which is the non-steady-state solution of Eq. (I.3):

$$C = \frac{Q}{V} \frac{C_0}{\beta} (1 - e^{-\beta t}) + C_0 e^{-\beta t} \tag{I.12}$$

Oxygen Sag Analysis

The basic oxygen-sag differential equation as derived in Chapter 8 in terms of the oxygen deficit D_{O_2} is [Eq. (8.19)]

$$D'_{O_2} + k_2 D_{O_2} = kL_i e^{-k\theta_H} \tag{I.13}$$

where

$$D'_{O_2} = \frac{dD_{O_2}}{d\theta_H}$$

Comparing Eq. (I.13) to Eq. (I.1), the appropriate integrating factor is

$$e^{\int P\,d\theta_H} = e^{\int k_2\,d\theta_H} = e^{k_2\theta_H} \tag{I.14}$$

Multiplying both sides of Eq. (I.13) by $e^{k_2\theta_H}$ and simplifying yields

$$\left(e^{k_2\theta_H} D_{O_2}\right)' = kL_i e^{(k_2-k)\theta_H} \tag{I.15}$$

Integrating Eq. (I.15) to remove the differential sign yields

$$e^{k_2\theta_H} D_{O_2} = \frac{kL_i}{k_2 - k} e^{(k_2-k)\theta_H} + K \tag{I.16}$$

Dividing by $e^{k_2\theta_H}$ yields

$$D_{O_2} = \frac{kL_i}{k_2 - k} e^{-k\theta_H} + Ke^{-k_2\theta_H} \tag{I.17}$$

Noting that when $\theta_H = 0$, $D_{O_2} = D_i$ (the initial deficit) and

$$K = D_i - \frac{kL_i}{k_2 - k} \tag{I.18}$$

Substituting for K in Eq. (I.17) and simplifying yields

$$D_{O_2} = \frac{kL_i}{k_2 - k} \left(e^{-k\theta_H} - e^{-k_2\theta_H}\right) + D_i e^{-k_2\theta_H} \tag{I.19}$$

Equation (I.19), known as the Streeter-Phelps oxygen-sag equation, is the same as Eq. (8.20) given in the text.

J

VALUES OF THE ERROR FUNCTION erf(ψ) AND THE COMPLEMENTARY ERROR FUNCTION erfc(ψ) FOR VALUES OF ψ VARYING FROM 0 TO 3.0

$$\text{erf}(\psi) = \frac{2}{\pi} \int_0^\beta e^{-\epsilon^2} d\epsilon$$

$$\text{erf}(-\psi) = -\text{erf}\,\psi$$

$$\text{erfc}(\psi) = 1 - \text{erf}(\psi)$$

ψ	erf(ψ)	erfc(ψ)
0	0	1.0
0.05	0.056372	0.943628
0.1	0.112463	0.887537
0.15	0.167996	0.832004
0.2	0.222703	0.777297
0.25	0.276326	0.723674
0.3	0.328627	0.671373
0.35	0.379382	0.620618
0.4	0.428392	0.571608
0.45	0.475482	0.524518
0.5	0.520500	0.479500
0.55	0.563323	0.436677
0.6	0.603856	0.396144
0.65	0.642029	0.357971
0.7	0.677801	0.322199
0.75	0.711156	0.288844
0.8	0.742101	0.257899
0.85	0.770668	0.229332
0.9	0.796908	0.203092
0.95	0.820891	0.179109
1.0	0.842701	0.157299
1.1	0.880205	0.119795
1.2	0.910314	0.089686
1.3	0.934008	0.065992
1.4	0.952285	0.047715
1.5	0.966105	0.033895
1.6	0.976348	0.023652
1.7	0.983790	0.016210
1.8	0.989091	0.010909
1.9	0.992790	0.007210
2.0	0.995322	0.004678
2.1	0.997021	0.002979
2.2	0.998137	0.001863
2.3	0.998857	0.001143
2.4	0.999311	0.000689
2.5	0.999593	0.000407
2.6	0.999764	0.000236
2.7	0.999866	0.000134
2.8	0.999925	0.000075
2.9	0.999959	0.000041
3.0	0.999978	0.000022

Name Index

Abraham, G., 380
Abufayed, A., 620, 670
Adelberg, E. A., 161
American Public Works Association, 558
American Society of Civil Engineers, 554, 703
American Water Works Association, 703
Amirtharajah, A., 554, 593
Anderson, J. L., 671
Anderson, R. B., 226
Applebaum, S. B., 159, 593
Ardern, E., 670
Arnett, R. C., 403
Arozarena, M. M., 707
Asano, T., 210, 673
Atkinson, B., 674
Ayers, R. S., 210

Baca, R. G., 403
Bachmat, Y., 438
Barnard, J. L., 620, 671
Barnes, D., 555
Barth, E. F., 671
Baumann, E. R., 555
Bayer, M. E., 139
Bell, B. A., 672
Benefield, L. D., 159, 593, 670
Benjamin, J. R., 706
Berg, G., 593
Bernado, L. D., 555
Bertheoux, P. M., 706
Bird, R. B., 302, 334
Blaney, H. F., 42
Bliss, P. J., 555
Bock, E., 674
Bohan, J. P., 403
Borchardt, J. A., 463, 555
Bouwer, H., 334, 438
Bowden, K. F., 381
Box, G. E. P., 706
Bredehoeft, J., 438
Brooks, N. H., 370, 381–382
Browman, M. G., 334
Brown, D. S., 438
Bryan, E. H., 671
Bryant, J. O., 161, 302
Bryers, J. D., 671
Buchan, L., 671
Buckingham, R. A., 381
Burchett, M. E., 381

Burdoin, A. J., 381
Burns, R. G., 334
Busch, A. W., 157, 160–161, 302, 608, 672
Buswell, A. M., 161, 671
Butler, J. N., 160, 733
Butterfield, C. T., 593

Camp, T. R., 160, 555
Canale, R. P., 403
Carlson, C. A., 382
Carman, P. C., 515, 555
Cartwright Aerial Surveys, Inc., 32, 283
Chambers, B., 671
Chambers, C. W., 593
Chapman, F. S., 556
Characklis, W. G., 674
Cheadle, R. F., 210
Cherry, J. A., 334, 438
Chesters, G., 334
Chiesa, S. C., 671
Chudoba, J., 671
Churchill, M. A., 381
Churchill, S. W., 266
Clark, R. M., 706
Clarkson, W. W., 672
Cleary, E. J., 226
Cleasby, J. L., 555
Cochrane Environmental Systems, 632
Collins, W. D., 168
Colt, J., 674, 707, 735
Contra Costa Water District, 701
Cook, T. M., 160
Cornell, A., 706
County Sanitation Districts of Los Angeles County, 699, 702
Cox, C. R., 209, 555
Craig, P. P., 227
Crane Co., 632
Crank, J., 334, 438
Craun, G. F., 160
Criddle, W. D., 42
Crites, R. W., 673, 707
Cruver, J. E., 555
CUES West, 25
Culp, G. L., 555
Culp, R. L., 555, 706

Danckwerts, P. V., 334
Davies, J. T., 302, 334

745

Subject Index

Index citations to individual geographic locations, lakes, and rivers have been grouped for convenience under the headings "Geographic locations," "Lakes," and "Rivers."

Absorbate, properties of, 323
Absorbent, properties of, 322
Absorption, gas, 318–322
  definition of, 318
  in a quiescent liquid, 319
  in an agitated liquid, 321
  of selected organics, 377
  two-film model, 321
Absorption system, soil, 660
Accumulation rate, 249. *See also* Materials balances
Accuracy of water analysis, 71
Acidity, definition of, 89
Acid rain, 44, 179–180, 218
Acids, 77–82
  definitions of, 77
  logarithmic concentration-versus-pH diagrams, 78, 87–89, 725–733
  monoprotic, 77–81
  pH of acid solutions, 85, 725–733
  polyprotic, 81–86
Activated carbon, 530–533. *See also* Adsorption
  characteristics of, 529
  granular, 530
  powdered, 530
  regeneration of, 532
  use in
    wastewater treatment, 458, 533
    water treatment, 447, 456–457, 533
Activated sludge, 599–621
  aeration systems for, 470. *See also* Aerators
  air, application rates, 601–602
  applications, 448
  cell concentration, 599, 604
  configurations
    batch, 596
    contact stabilization, 596, 614–615
    continuous flow stirred tank, 596, 607, 610–614
    extended aeration, 596
    minimal aeration, 596
    modified aeration, 617
    oxidation ditch, 596, 615
    plug flow (conventional), 600
    pure oxygen, 596, 615–616
    step aeration, 615–616
  description of process, 599–600
  design of, 610–614

energy requirements, 682
filamentous bulking, 618–619
flow sheets
  batch, 601
  contact stabilization, 615
  continuous flow stirred tank, 601
  denitrification, 620
  oxidation ditch, 615
  plug flow, 600–601
  pure oxygen, 615
  step aeration, 615
kinetic models
  cell growth rate, 803
  organic removal rate, 602
  oxygen uptake rate, 603–604
loading rates, 604
mean cell residence time, 605
mixed liquor suspended solids (MLSS), 599
mixed liquor volatile suspended solids (MLVSS), 599
nitrification, 595, 618
organic removal rate, 602
oxygen requirements, 603–604
oxygen transfer in, 473–475
process coefficient and parameter values, 604
process design and operating parameters, 608
secondary sedimentation, 497, 609–610
sludge age, 605
solids retention time (SRT), 605
stoichiometry, 602–603
Activation energy, 118, 240, 567
Activity coefficient, 52
Adenovirus disinfection, 564
Adsorption
  applications of, 447–448
  definition of, 322–327, 529–530
  factors affecting, 322–323
  indexes, 530
  models, 323–326
    determination of isotherm coefficients, 326–327
    Freundlich adsorption isotherm, 323, 533
    Langmuir adsorption isotherm, 324, 530–533
  of contaminants on soils, 424–425
Advanced wastewater treatment, 453
Advection, 307
Aerated lagoons, 448

749

ATOMIC NUMBERS AND ATOMIC MASSES*

Actinium	Ac	89	227.0278	Mercury	Hg	80	200.59
Aluminum	Al	13	26.98154	Molybdenum	Mo	42	95.94
Americium	Am	95	(243)	Neodymium	Nd	60	144.24
Antimony	Sb	51	121.75	Neon	Ne	10	20.179
Argon	Ar	18	39.948	Neptunium	Np	93	237.0482
Arsenic	As	33	74.9216	Nickel	Ni	28	58.70
Astatine	At	85	(210)	Niobium	Nb	41	92.9064
Barium	Ba	56	137.33	Nitrogen	N	7	14.0067
Berkelium	Bk	97	(247)	Nobelium	No	102	(259)
Beryllium	Be	4	9.01218	Osmium	Os	76	190.2
Bismuth	Bi	83	208.9804	Oxygen	O	8	15.9994
Boron	B	5	10.81	Palladium	Pd	46	106.4
Bromine	Br	35	79.904	Phosphorous	P	15	30.97376
Cadmium	Cd	48	112.41	Platinum	Pt	78	195.09
Calcium	Ca	20	40.08	Plutonium	Pu	94	(244)
Californium	Cf	98	(251)	Polonium	Po	84	(209)
Carbon	C	6	12.011	Potassium	K	19	39.0983
Cerium	Ce	58	140.12	Praseodymium	Pr	59	140.9077
Cesium	Cs	55	132.9054	Promethium	Pm	61	(145)
Chlorine	Cl	17	35.453	Protactinium	Pa	91	231.0359
Chromium	Cr	24	51.996	Radium	Ra	88	226.0254
Cobalt	Co	27	58.9332	Radon	Rn	86	(222)
Copper	Cu	29	63.546	Rhenium	Re	75	186.207
Curium	Cm	96	(247)	Rhodium	Rh	45	102.9055
Dysprosium	Dy	66	162.50	Rubidium	Rb	37	85.4678
Einsteinium	Es	99	(254)	Ruthenium	Ru	44	101.07
Erbium	Er	68	167.26	Samarium	Sm	62	150.4
Europium	Eu	63	151.96	Scandium	Sc	21	44.9559
Fermium	Fm	100	(257)	Selenium	Se	34	78.96
Fluorine	F	9	18.99840	Silicon	Si	14	28.0855
Francium	Fr	87	(223)	Silver	Ag	47	107.868
Gadolinium	Gd	64	157.25	Sodium	Na	11	22.98977
Gallium	Ga	31	69.72	Strontium	Sr	38	87.62
Germanium	Ge	32	72.59	Sulfur	S	16	32.06
Gold	Au	79	196.9665	Tantalum	Ta	73	180.9479
Hafnium	Hf	72	178.49	Technetium	Tc	43	(97)
Helium	He	2	4.00260	Tellurium	Te	52	127.60
Holmium	Ho	67	164.9304	Terbium	Tb	65	158.9254
Hydrogen	H	1	1.0079	Thallium	Tl	81	204.37
Indium	In	49	114.82	Thorium	Th	90	232.0381
Iodine	I	53	126.9045	Thulium	Tm	69	168.9342
Iridium	Ir	77	192.22	Tin	Sn	50	118.69
Iron	Fe	26	55.847	Titanium	Ti	22	47.90
Krypton	Kr	36	83.80	Tungsten	W	74	183.85
Lanthanum	La	57	138.9055	Uranium	U	92	238.029
Lawrencium	Lr	103	(260)	Vanadium	V	23	50.9414
Lead	Pb	82	207.2	Xenon	Xe	54	131.30
Lithium	Li	3	6.941	Ytterbium	Yb	70	173.04
Lutetium	Lu	71	174.97	Yttrium	Y	39	88.9059
Magnesium	Mg	12	24.305	Zinc	Zn	30	65.38
Manganese	Mn	25	54.9380	Zirconium	Zr	40	91.22
Mendelevium	Md	101	(258)				

*From *Pure Appl. Chem.*, **47**, 75 (1976). A value in parentheses is the mass number of the longest-lived isotope of the element.